Materials in Marine Technology

Robert Reuben

Materials in Marine Technology

With 261 Figures

Springer-Verlag
London Berlin Heidelberg New York
Paris Tokyo Hong Kong
Barcelona Budapest

Robert L. Reuben, BSc, PhD, CEng, MIM

Department of Mechanical Engineering, Heriot-Watt University, Riccarton,
Edinburgh EH14 4AS, UK

ISBN-13:978-1-4471-2013-1   e-ISBN-13:978-1-4471-2011-7
DOI: 10.1007/978-1-4471-2011-7

British Library Cataloguing in Publication Data
Reuben, Robert L.
   Materials in Marine Technology
   I. Title
   620.1
   ISBN-13:978-1-4471-2013-1

Library of Congress Cataloging-in-Publication Data
Reuben, Robert (Robert L.), 1953–
   Materials in marine technology/Robert Reuben
      p.   cm.
   Includes bibliographical references and index.
   ISBN-13:978-1-4471-2013-1
   1. Ocean engineering.   2. Materials.   I. Title.
TC1650.R48   1993
620'.4162–dc20   93-15462

Typeset by EXPO Holdings, Malaysia

69/3830–543210   Printed on acid-free paper

Preface

Any reader of this preface will probably first want to know the meaning of the title since, although the intention is obvious, the scope and method of the coverage could mean anything from a set of data tables to a detailed metallurgical treatise. I hope to have met some sensible balance between these extremes and to have written something which will be *used* by engineers rather than serving to emphasise the barriers which clearly exist between those who use materials and those who research them.

The book has been written in response to a real need that I have observed among engineers working in design and manufacturing activities related to the offshore and marine industries. I have often been asked by those practising engineers to whom I act as consultant whether there is a text which covers the applications of materials in the marine environment. Leaving aside my pecuniary interest in making the negative reply, it has to be said that, although many good texts on aspects of this subject exist, few seem to cover all of this ground, preferring rather to stay with one aspect, such as corrosion, or with one material or group of materials. Having tried, I hope with some success, to distil the wealth of available information into a single text, I think I can now see why this was so.

In my activities as a researcher and consultant to the marine and offshore industries and as a lecturer in mechanical and offshore engineering, I have observed that engineers need only know sufficient metallurgy and chemistry to converse sensibly with materials experts and to select materials, provided they take into account the possible pitfalls that lie in wait for the unwary. Therefore, the approach taken in this book has been to develop the chemical and metallurgical aspects only to such a degree as is necessary and to concentrate mainly on providing data and, more importantly, on communicating the accumulated experience of designers and manufacturers that can be found in the literature. It is inevitable in a book of this nature to have to rely on the published work of others. I have tried to ensure accuracy by referring to as many authorities as possible but accept that opinions and standards may change with time. I would welcome comments from readers which would help me to correct or update the book for any future edition.

The data are collected mostly in Chapter 4 and, although the experience aspect is spread throughout the book, the Case Studies in Chapter 7 provide a more detailed study of particular examples. I have referred to Codes and Standards on a number of occasions but the reader should recognise that these references do not constitute specifications. Designers should always check for details and updates in the current versions of the Code or Standard itself.

The structure of the book has been designed so that entry can be made directly to Chapter 4 where most of the data reside, Chapters 2 and 3 providing a detailed description of the properties tabulated and also indicating how these properties might be used in design. Chapters 5 and 6 refer mainly to aspects of manufacture and maintenance which are of particular importance in marine technology, with the Case Studies giving some indications of how the data might be used in specific situations. I have tried to make mention of 'new' materials such as composites throughout the book and have probably overcompensated so that the relative treatment of these materials exceeds their proportional use in marine environments. I make no apology for this, as I have tried to look forward to the time when the 'structural materials' such as concrete and steel will only be used where necessary.

Finally, it is hoped that the liberal use of schematic illustrations will make the book communicate more quickly, as well as making the text briefer and allowing the reader to concentrate on methodology.

Edinburgh, 1993 R. L. R.

Contents

1 The Marine Environment, Marine Structures and the Role of Materials Technology

The purpose of this chapter is to illustrate how the development of marine technology and materials technology have paralleled each other over the centuries and to highlight some milestones with particular relevance to the latter. In order that this relationship can be appreciated the chapter will also consider those aspects of the marine environment which are of most importance in design and materials selection.

For as long as people have been interested in venturing into and onto the sea and oceans there has been a need to support such activities with structures and hardware of various types. For any but the simplest and most temporary of such hardware, the properties and durability of materials in marine environments are of fundamental importance. Indeed, there are cases where engineering developments have been directly stimulated by developments in materials technology, as well as instances where the opposite has applied and the demands of offshore and marine engineers have stimulated developments in materials technology. The early story of marine technology is mainly concerned with transport, although exploitation (mostly of flora and fauna) is also recorded as far back as records go.

Until relatively recently the development of marine technology has been mainly concentrated on military, transport and trade, fishing and diving applications and has mostly involved ships or submersibles, though a number of other areas have been important from time to time, particularly on the coastal fringes.

Modern marine technology includes such motivations as oil and gas exploration and recovery, seabed mining, ocean power stations (wind, wave, tidal, thermal energy conversion and others), recreation, fish farming, and harbour and estuary development; thus the methodology of design has become somewhat different from that applied by naval architects, although this ancient branch of engineering has also seen considerable recent development.

This chapter concentrates on an overview of the design–environment–materials triangle suggested by Birchon [1] and shown in Fig. 1.1. This illustrates how the marine environment and the capabilities of materials are inextricably bound up with the process of engineering design, and how an alteration of any limb of the triangle affects the remaining two. The discussion in this chapter commences with a very brief description of the marine environment and its effect on engineering design. This is followed by a description of the development of marine engineering, highlighting the materials technology present at each stage and culminating with

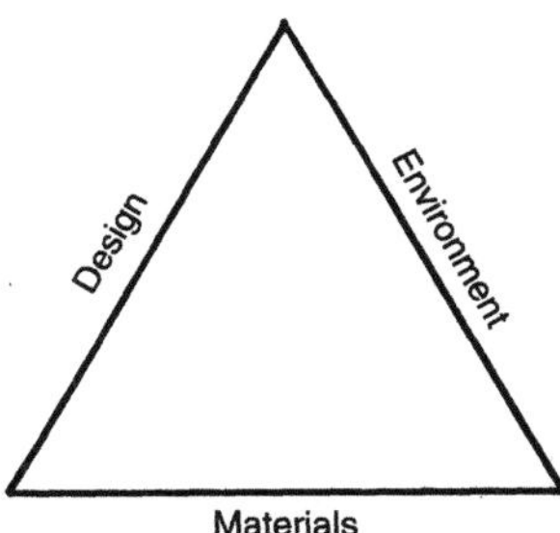

Fig. 1.1. Schematic interrelationship of environment, design and materials selection (from Birchon [1]).

the most modern types of marine construction and materials uses. The final section deals very broadly with the types of materials available and the selection process.

Throughout this book the broadest possible view is taken of marine construction, covering ships, platforms, subsea vehicles and habitats, and even some coastal and estuarine developments where the materials and techniques are similar to those used in the marine environment.

1.1 The Ocean Environment

As far as the designer is concerned, the ocean environment differs chemically, biologically and physically from most land-based ones. The chemical and biological differences are, of course, related to the presence of seawater and associated lifeforms. The physical differences are largely associated with waves and currents, but other effects may also need to be considered, such as those of wind loading in exposed locations and the movement of ice, sand or sediment. This section treats each of these physical and chemical aspects separately but very briefly, concentrating on those which affect the design of offshore and marine constructions.

1.1.1 The Chemistry of Seawater

The chemical composition of seawater can vary substantially, both in the short and long geographical ranges and also seasonally, because of a dynamic process which includes

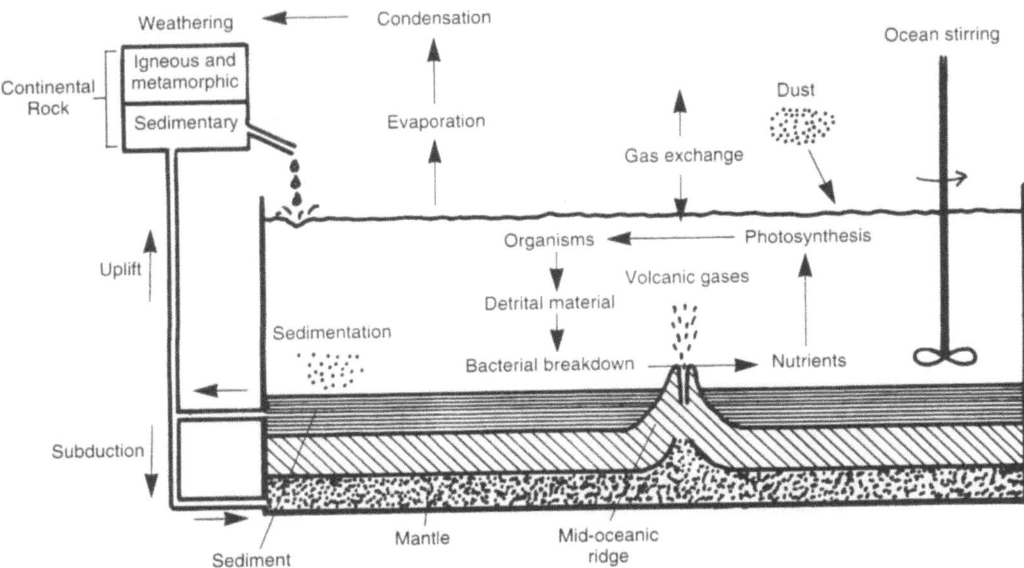

Fig. 1.2. Schematic description of the chemical mass balance of seawater (after Whitfield [2]).

the weathering of rocks, the flow of rivers, evaporation, rain, photosynthesis and volcanic reactions (Fig. 1.2).

A typical analysis of seawater such as that given in Table 1.1 shows that it contains predominantly sodium and chloride ions, but that magnesium and sulphate ions are also present in appreciable quantities. From the point of view of engineering, the most important chemical properties of seawater (and hence those which are most often measured) are salinity, pH, dissolved oxygen content and temperature, although these in turn also affect to a certain degree the biological distribution of species.

Salinity is defined as the total weight per thousand (parts of water) of dissolved inorganic salts, and is usually around 35 per thousand for open-ocean seawater [4]. The global and seasonal variation of surface salinity is quite small (typically 31 to 36 parts per thousand), especially when depth is taken into account (Fig. 1.3). Apart from its effect on corrosion, salinity affects the speed of propagation of sound (with consequent implications for refraction in

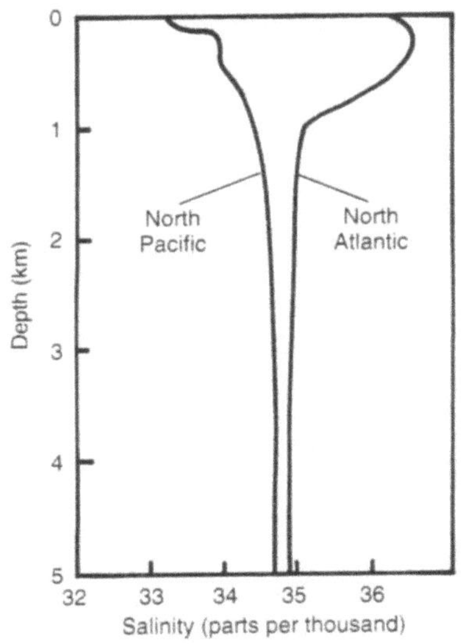

Fig. 1.3. Salinity–depth profiles for North Atlantic (right) and North Pacific (left) open-water sites (from Dexter [4]).

Table 1.1. Typical chemical constitution of seawater

Constituent	Concentration (parts per thousand by weight)
Sodium	10.8
Magnesium	1.3
Calcium	0.41
Potassium	0.40
Strontium	0.008
Chloride	19.3
Sulphate	2.7
Bromide	0.07
Carbon (as bicarbonate, carbonate and dissolved carbon dioxide)	0.02–0.03

Source: Morgan [3].

underwater acoustics) and, to a lesser extent, seawater density and hence buoyancy forces.

Variations in temperature have a much larger effect on marine design than do those in salinity. Temperature can affect, among other things, rates of marine growth, material properties, corrosion rates and the feasibility of diver deployment. Figures 1.4 and 1.5 summarise seasonal variations in seawater temperature according to global position and depth.

Dissolved oxygen content can have a considerable effect on corrosion (see Chapter 3), and is subject to rather greater variations than are the inorganic ions. Both salinity and

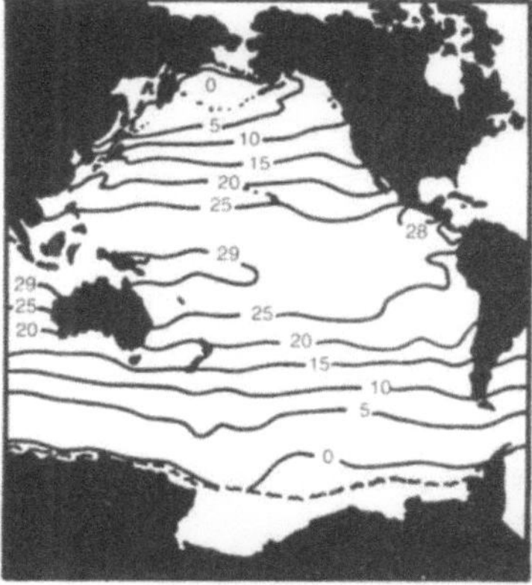

Fig. 1.4. Sea surface temperatures in the Pacific Ocean for February (left) and July (right). Figures on contours are in degrees Celsius and broken lines represent the edge of floating ice (from Dexter [4]).

temperature affect the solubility of oxygen in seawater according to:

$$\ln[O_2] = A_1 + A_2\frac{100}{T} + A_3\ln\frac{T}{100} + A_4\frac{T}{100}$$
$$+ S\left[B_1 + B_2\frac{T}{100} + B_3\frac{T^2}{100}\right]$$

where the salinity is in parts per thousand, temperature is in Kelvins, giving oxygen concentration in millilitres per litre. For atmospheric air at 100 per cent relative humidity, the constants are given by (e.g. Dexter [4]):

A_1	−173.429 2
A_2	249.633 9
A_3	143.384 3
A_4	−21.849 2
B_1	−0.033 096
B_2	0.014 259
B_3	−0.001 700

As an example, at 31 parts per thousand salinity the oxygen solubility can vary between 8.3 and 4.4 ml/l between zero and 30 °C. Global variations in surface dissolved oxygen are usually close to or above saturation at the prevailing temperature and salinity. In well-stirred oceans, such as the North Sea, there is little effect of depth on dissolved oxygen, most variation being due to variations in temperature and salinity. Otherwise, the tendency is for dissolved oxygen to drop to a minimum under mixing effects and then to increase steadily under the temperature effect on solubility. In the Atlantic Ocean surface oxygen levels are generally lower than in the Pacific and the minimum is not as intense. However, deep-Atlantic oxygen concentrations are generally higher than those in the deep Pacific.

The pH of seawater is controlled largely by carbonaceous equilibria and their effect on carbon dioxide. Carbon dioxide (and hence its acidic ionic dissociation product) is incorporated by air–sea interchange, as is oxygen, but also (more importantly) through photosynthesis in the water column. This latter effect is controlled largely by the photosynthesis–oxidation cycle:

Photosynthesis

$$CH_2O + O_2 \rightleftharpoons CO_2 + H_2O$$

Biochemical
oxidation

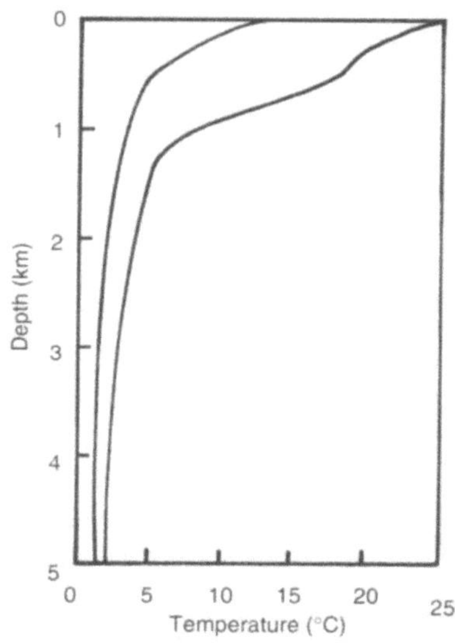

Fig. 1.5. Temperature–depth profiles for North Atlantic (right) and North Pacific (left) open-water sites (from Dexter [4]).

where CH_2O represents a typical carbohydrate molecule. Biochemical oxidation proceeds either through respiration or by decomposition of organic matter.

The pH of seawater in the oceanic regions is around 8, with a range from 7.5 to 8.3. This range has very little implication for marine design.

1.1.2 Water Depth and Seabed

Water depth is perhaps the most important design parameter in marine technology, as it radically affects the feasibility of whole categories of conceptual design. It is most important to designers who wish to operate at the seabed and so have to decide at what point direct bottom contact becomes feasible and also at what point compliance becomes necessary in the design. Water depth also has an effect on ambient pressure, and an easily remembered equation for the total (including atmospheric) pressure p_H (in bar) at a depth D (in metres) is:

$$p_H = 1 + 0.1D$$

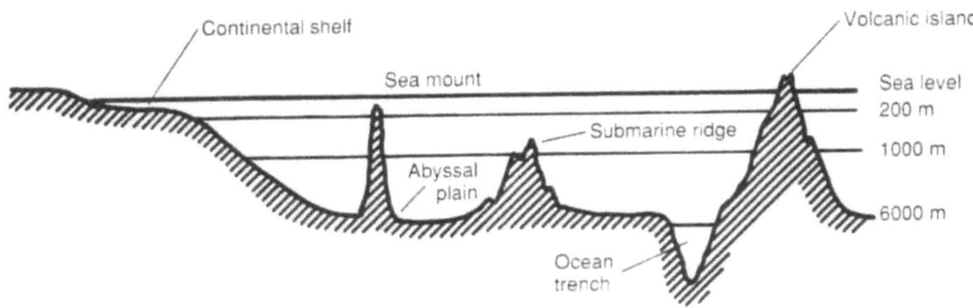

Fig. 1.6.   Schematic diagram of different types of seabed features (after Morgan [3]).

Fig. 1.7.   Schematic representation of the development of a subsea pipeline freespan.

Apart from the well-known effect of hydrostatic pressure on divers and submarine hulls, it has some implications for the corrosion of aluminium alloys (see Chapter 4) and for the structural integrity of foams for insulation and buoyancy.

Broadly, ocean depths are in the region of 2.5 to 6 km. with exceptional areas (trenches) from 7.5 to 11 km [3]. Asperities in the form of submarine ridges and sea mounts are also present in the oceans where the water depth may locally become much less than the average. The continental margins are made up of relatively flat regions close to shore (usually around 70 km), with an increase in height close to the shore and a gradual slope to the deep ocean (Fig. 1.6).

Continental shelf depths vary over their width from zero to around 200 m. This is only a rough value, since some shelves are deeper than this and, indeed, not all continental margins have appreciable shelves.

The seabed itself can be covered by a variety of different sediments of mineral or biological origin [3]. The mechanical (and, on occasion, the biochemical) behaviour of marine sediments can be an important part of marine design, particularly if the sediments are mobile under the influence of bottom currents, resulting for example in scour around structural legs or the development of freespans on subsea pipelines (Fig. 1.7). The underlying geological structure of the seabed can also be of some importance in foundation design for the normal soil- and rock-mechanics reasons. Of course, geological conditions are often the reason why a structure is to be located at the site for reasons of exploitation.

1.1.3 Biological Considerations

The biological marine environment can be of importance in marine exploitation on a number of counts, which range from the obvious effects on fishing to the rate and distribution of fouling. The latter, which as far as marine design is concerned is perhaps the most important biological consideration, can be defined as the settlement of flora and fauna on exposed surfaces. Fouling can occur at the macro and micro levels, and the implications range from enhanced corrosion to increased structural loading. Clogging of filters (at the micro level) and intakes (at the macro level) can also present difficulties, and this is usually coped with by the judicious use of screens and back-washing. Fouling can also produce unacceptable drag on ships' hulls, and can be a problem when inspection needs to be carried out *in situ*, as on offshore platforms. In this latter case it is often necessary to remove attached organisms, and, depending upon the species involved, this can vary from being a relatively easy task to one requiring the use of a high-speed water jet with entrained abrasive particles.

The behaviour of micro-organisms can also affect marine design. The most important such organisms are the sulphate-reducing bacteria (discussed in Chapter 3), which under certain conditions can produce a very corrosive environment, particularly for steels.

In some areas, the marine fauna can present a distinct danger to human beings, and so must be considered before divers are deployed. These animals vary from the obvious (usually biting) animals to those which may simply damage diving equipment either actively or passively [5].

Finally, designers for the marine environment need also to be aware of the inverse effect, namely the effect that their activities may have on marine life and on the marine environment generally. This is a particularly crucial point when the designer's motive is an exploitative one, and apparently harmless activities such as gravel recovery can result in serious disruption of the seabed and of the water column, even a long way from the site of intervention.

1.1.4 Marine Atmospheres

The constitution of the marine atmosphere has implications mainly for corrosion, but wind is also discussed under this heading. From the point of view of corrosivity, the principal factors are the temperature and the amounts of moisture in the form of condensation, rain and wind-borne spray. Clearly some of these factors are dependent upon height above sea level, but others are features of the weather and are hence seasonal and geographical. The level of air-borne pollution, particularly carbon dioxide and sulphur dioxide, can also be an important factor in marine atmospheric corrosion. All of these are discussed further in Chapter 3.

The exposed nature of all but the most marginal marine constructions means that winds usually impinge on structures in an unattenuated condition. According to Patel [6], wind forces account for about 15 per cent of the total fluid loading on structures, and the overturning moment resulting from the wind load is, of course, proportionately greater as the water depth increases for rigid offshore structures.

Design for wind loading in the marine environment is ostensibly the same as on land, important considerations being drag coefficients and the possibility of fatigue loading from the non-steady element of the wind velocity, or even the excitation of natural frequencies of components due, for example, to vortex shedding. As with wave and current data, it is common to see wind data expressed as a power spectrum. This reflects the random nature of wind (and other fluid) loading, and can be used directly to obtain a spectral density of response in a way similar to that described for wave loading in section 1.1.6.

1.1.5 Marine Currents

Like wind, submarine and surface currents can produce steady and non-steady loading on marine structures. The steady loading is simply due to hydrodynamic drag, but vortex shedding may further give rise to dynamic loads, with consequent fatigue implications. Seabed currents are also important in producing sediment movements, which may bury seabed structures or leave them (or cause then to become) exposed to other influences.

Currents can arise from tides, wind or circulation effects, and, depending on the water depth and the source of the current, there can be a substantial variation with depth. In general, current velocities are rather low (rarely greater than 2.5 m s^{-1}), but their effect on structural loading is greater than wind loading in proportion to the greater density of water compared with that of air. In the relatively shallow regions of continental shelves the dominant component of bottom current is usually tidal. For a given location, a very rough estimate of the tidal current can be made by assuming it to behave as a simple, progressive wave so that the current velocity, c, is quite simply related to the tidal amplitude, a, by:

$$c = \sqrt{(g/D)} \times a$$

where g is the acceleration due to gravity and D is the water depth. This equation, though barely adequate for design purposes, does illustrate the effect of water depth on tidal current.

1.1.6 Waves

Waves arise from the interaction of wind with the free surface of water, producing disturbances which, when they encounter a structure, can result in very high loads or displacements, depending upon the compliance of the struc-

ture. Waves are the single most important factor considered by offshore and marine designers, from the points of view of both static and dynamic (fatigue) loading.

Although small-amplitude waves in deep water can often be described as sinusoidal, their form in shallow water is more nearly trochoidal [7]. In fact, water waves are three-dimensional, with finite crest lengths, and are even more irregular in the area in which they are being generated, owing to local wind effects. Nevertheless, it is usual for designers to characterise and forecast waves in terms of a set of wave heights (H) and wave periods (T), the wave length (L) being related to the period through

$$L = T^2 g/2\pi$$

The wave steepness, defined as H/L, is an important factor in governing wave stability; the wave breaks when the steepness exceeds about 1/7.

Water waves propagate at a speed (celerity) which can be approximated by

$$c = \sqrt{gL/2\pi}$$

provided that bottom effects can be neglected (depth greater than $L/2$), but the actual movement of the water particles, being orbital in nature, is more tortuous than simple wave propagation, and the orbits become flatter as depth decreases. It is these movements which give rise to structural loading, and the overall energy associated with a unit length of crest per wave length is given approximately by

$$E = 2000H^2T^2$$

for deep-water waves.

Spectral wave data are used in much the same way as indicated for wind data above, where a spectral density of water elevation can be used along with a structure transfer function to obtain a spectral density of structural response, which represents the fatigue loading (Fig. 1.8).

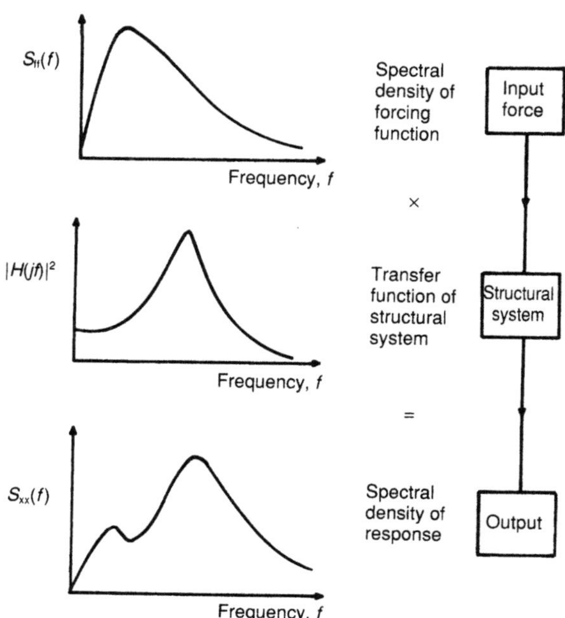

Fig. 1.8. Illustration of the spectral method of obtaining structural response to random loading such as wind and waves (from Hallam et al. [7]).

1.1.7 Ice

In polar regions, the ocean surface may become frozen for a good part of the year and this must be taken into consideration in designing for these latitudes.

The major danger from ice on structures is the loading resulting from the approach of moving ice on the structure. The magnitude of this effect depends upon the thickness and condition of the ice, the speed of ice movement, the geometry of the encounter and the shape of the structure. Many Arctic structures are of the artificial-island type, but there are some narrower kinds such as lighthouses and bridge piers (which are subject to river ice). Some of the principal types of defence against ice loading are illustrated in Fig. 1.9. As can be seen, one of the basic strategies is to try to cause the ice to break in either upward or downward flexure, but it may fail if broken ice does not clear the structure quickly, so that the design load must be that for a vertical sided structure. The other main strategy is to encourage the formation of a grounded ice rubble pile, which carries and dissipates the loading from following ice. Taking this to its extreme produces the spray ice islands which are deliberately manufactured from grounded ice.

Design against ice loading needs also to be considered for ice-breaking vessels (Fig. 1.10). These have sloping bows made from very heavy steel plate. The breaker can be operated in a continuous mode for thinner, softer ice, where it can break ice as it proceeds, or in ramming mode, where the vessel is ridden up onto the ice which then breaks under its weight [9].

Under cold conditions, ice may accrete from the atmosphere onto the exposed surfaces of structures and ships' superstructures, resulting in increased static and aerodynamic loading.

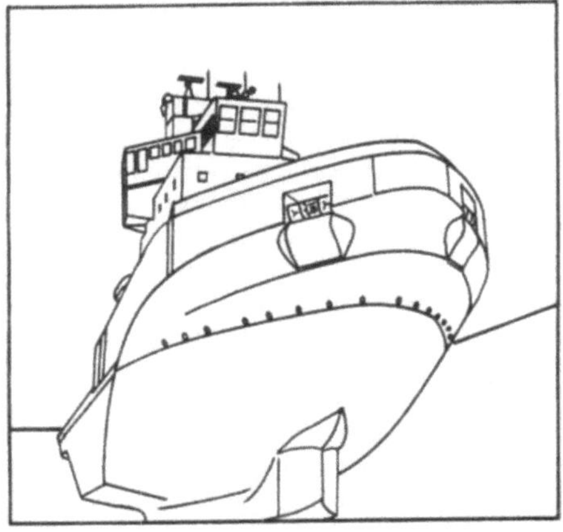

Fig. 1.10. Bow design of a modern icebreaker (from Harbron [9]).

1.2 The Development of Marine Structures and Materials

Apart from the first equipment used for diving, the earliest structures to be used in the marine environment were undoubtedly ships. Though early ships were constructed mainly from wood, they could still be extremely sophisticated (Fig. 1.11). Furthermore, materials selection factors were working even then, since British and French ships used oak whereas Spanish ones used mahogany for the hulls. Mahogany was selected not simply because of its superior resistance to dry rot: availability was a key factor, the

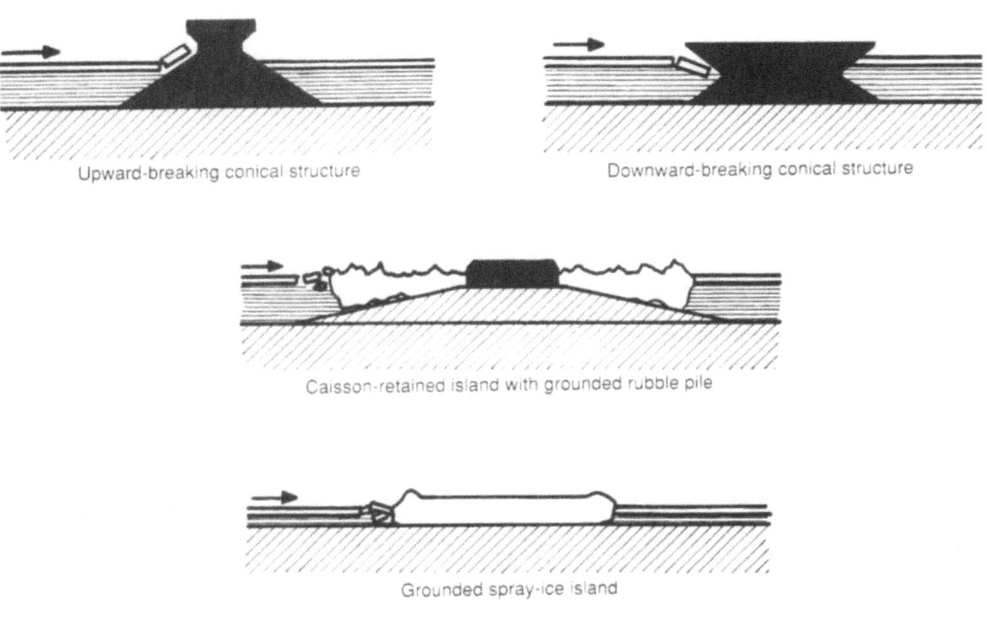

Fig. 1.9. Some strategies for coping with ice loading (after Sanderson [8]).

Spanish having access to Cuban and Honduran timber [9, 10]. All ships of this type used pine for the masts and yards, presumably for the same reasons as pine is used today.

Conde [11] further reports that fasteners for the hulls of such ships, from nails to nuts and bolts, could be fabricated from a range of materials including hardwoods, wrought iron, copper and bronze. Cannon were also, of course, made of metal, usually cast iron or copper alloys. The dangers of teredo attack (see Chapter 3) were also well known, as were the disadvantages of a fouled hull on manoeuvrability, and copper sheathing of hulls appeared as early as the early seventeenth century in eastern waters and about a hundred years later in Europe [11]. It is interesting to note that, around this period, in 1756, Smeaton developed a new marine jointing material for use in his Eddystone Lighthouse, namely Portland cement [11].

Apart from these modest uses of metals in marine engineering, the next great leap forward came with the beginning of the use of iron as a hull material. This was marked by the launch in 1819 of the fast passage barge *Vulcan*, which according to Walker [12] opened the 'Iron Age' of British shipbuilding and spurred considerably the evolution of modern naval architecture and shipbuilding practice. Figure 1.12 highlights the radical changes in design and construction practice which took place with the change from timber to iron. In 1860, the first iron-hulled armoured fighting ship, HMS *Warrior*, emerged. A comparison of its structure (Fig. 1.13) with that shown in Fig. 1.11 further illustrates the radical changes in design capability brought about by what was essentially a change in Young's modulus of the structural material. It is worth noting that a number of the iron ships constructed in this period (including HMS *Warrior*) are still intact today.

Quality of plate production was a problem with iron, and the increased strength of steel plates (and hence lighter hulls) led to their almost total takeover from iron by about 1890 [12]. An interesting artefact from this period was pro-

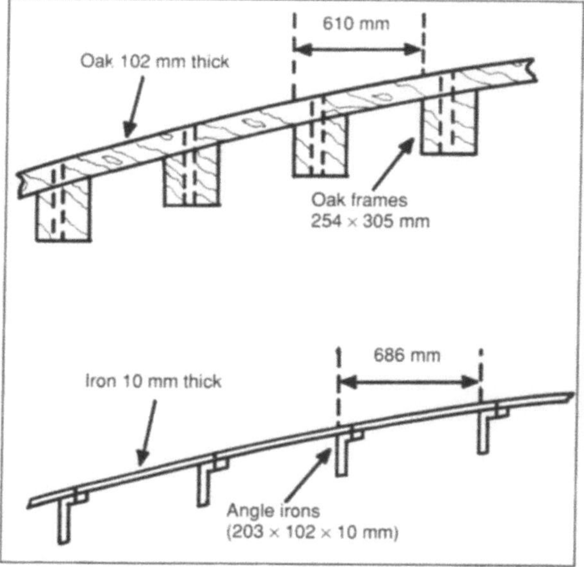

Fig. 1.12. Comparison of structural elements for similar ships fabricated from wood (top) and iron (bottom) (from Walker [12]).

vided by the recovery in 1982 of the Royal Navy's first submarine boat, *Holland I*, which was launched in 1902 and sank on the way to the breaker's yard in 1913. Apart from their surprise at the relatively low rate of corrosion affecting the hull, the investigators [14] made a number of other interesting observations regarding the materials used for this vessel. Table 1.2 shows the chemical composition of some of the metallic materials found in *Holland I*. Of particular note are the wrought-iron hull plates, and the steel deck plates.

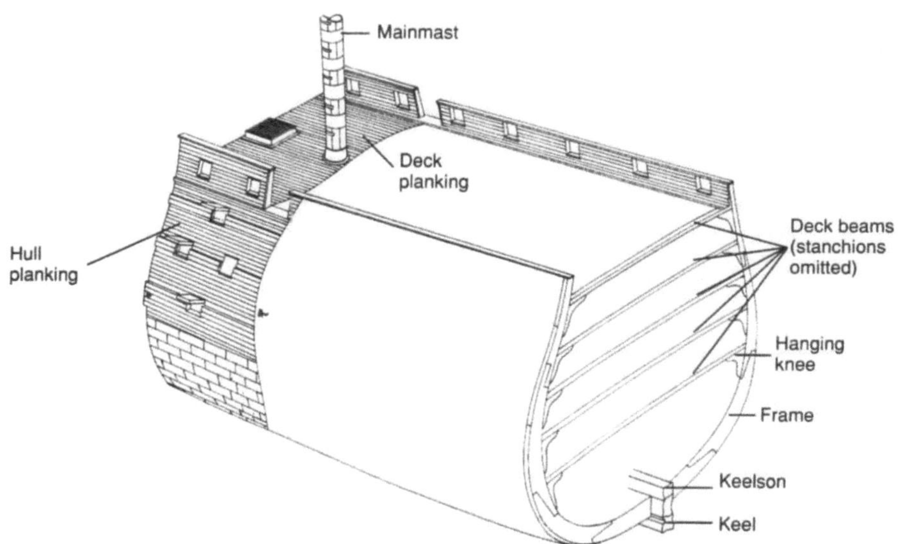

Fig. 1.11. Some structural aspects of an early wooden Spanish military ship (simplified from Harbron [10]).

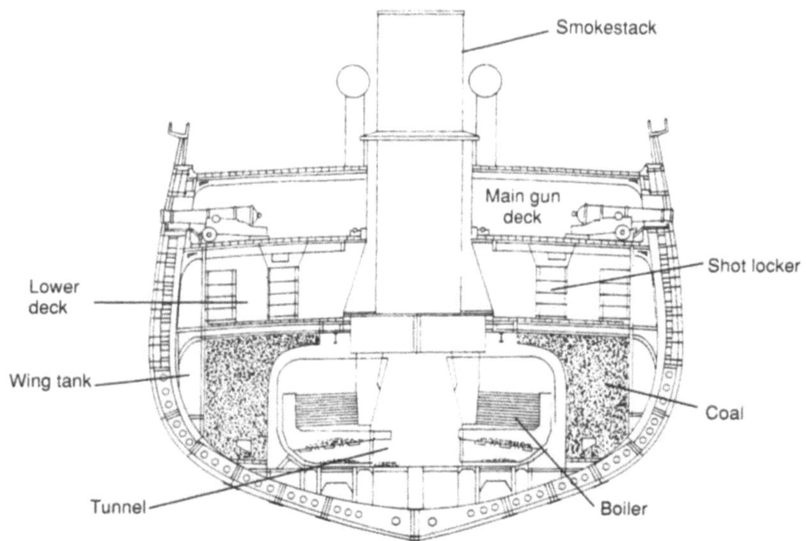

Fig. 1.13. Midships cross-section of the first iron-hulled armoured fighting ship, HMS *Warrior* (after Brownlee [13]).

Once the two main marine constructional materials (concrete and steel) had arrived, the next major advance in marine technology was a manufacturing process, namely welding. Although arc welding developed between 1880 and 1910, the first all-welded ocean-going ship was not built until 1921. However, it was not until the Second World War that welding became widespread in ship construction and although the technology of welding was not lacking, there was insufficient design experience of large fabricated structures [15]. This experience was provided, however, by the failure by fracture of more than two hundred ships between 1942 and 1952 [11], perhaps the most spectacular of these being the breaking in half of the T2 Tanker *Schenectady* while it lay in calm conditions at a fitting-out quay in 1943. The investigations of these failures led to the development of new structural steels for which toughness became a controlled property, and to this day improvements in the strength of such materials are not a sufficient reason to adopt them for constructional use: it is also necessary to demonstrate that toughness is not compromised in use, particularly when they are welded.

Despite its marine beginnings, developments in concrete technology have been driven far more by the requirement of large, land-based structures. The reason for this may be the undoubted durability of concrete and its quite acceptable performance in coastal applications such as sea defences and inshore structures. One of the best known of these latter structures is the Tongue Sands Tower, a fort

Table 1.2. Chemical composition of some of the metallic materials found on the recovered submarine *Holland I*

Brasses	Cu(%)	Zn(%)	Fe(%)	Sn(%)	Mn(%)	Pb(%)	
Gearwheel	bal.	39.2	3.8	0.8	0.5	0.7	
Torpedo tube	bal.	35.5	3.4	1.2	0.4	0.7	
Bronzes	**Cu(%)**	**Sn(%)**	**Fe(%)**	**Zn(%)**	**Pb(%)**		
Periscope turret	bal.	9.3	2.8	1.8	2.1		
Conning-tower	bal.	11.3	1.2	1.0	0.6		
Nickel silver	**Cu(%)**	**Zn(%)**	**Ni(%)**	**Fe(%)**	**Sn(%)**	**Al(%)**	**Pb(%)**
Resistor	55.8	26.2	17.9	0.35	0.06	0.05	0.04
Ferrous alloys	**C(%)**	**Mn(%)**	**P(%)**	**S(%)**	**Fe(%)**		
Hull plate	0.12	0.34	0.06	0.11	bal.		
Deck plate	0.21	0.45	0.09	0.09	bal.		

Source: Waite and McKendrick [14].

built in the Thames Estuary and recently surveyed as part of a UK research programme on concrete use in the oceans. Browne and Domone [16] quote one of the earliest examples of an offshore structure as being the Nab Defence Tower in the Solent, placed (as a gravity structure) in about 1920 and still in sound condition.

Development of these two main structural materials and of the associated design and construction methods continues for their use as offshore oil and gas platforms. The need for design and construction methods that take account of fatigue and fracture considerations is highlighted in the enquiries into two serious rig accidents [17, 18], and such methods continue to be developed. The relatively massive and rigid constructions used from the 1960s to the 1980s are giving way to those of cleaner, lighter and more compliant design for use in marginal fields and deep water. The drive is now towards higher-strength steels and towards other materials modifications which can result in weight saving, for example the use of aluminium alloys and structural composites [19]. There is also considerable interest in minimising the amount of offshore equipment which needs to be held above the water line, opening the way for new methods and materials for subsea construction. The use of concrete (in particular, prestressed concrete) to construct buoyant structures continues to develop with such diverse applications as hulls for tension leg platforms and LNG carriers.

A major series of advances in design confidence with composite materials for marine use has come with the Royal Navy's development of GRP hulls for mine countermeasure vessels (outlined in Case Study 7.3). It should be said, however, that these developments were made for entirely different reasons than for structural efficiency or weight saving.

The need for coastal barrages continues to attract novel design solutions. For example, the Dutch have recently built an enormous surge barrier to protect the Eastern Scheld estuary. The concrete piers were laid onto plastic sand- and gravel-filled mattresses [20]. The Thames Barrier, consisting of a series of 'rising sector' gates, of principally steel and concrete construction, includes some interesting materials choices, not least of which is the use of stainless steel for the distinctive coverings to the pier buildings. The hinging mechanism, borne on 8168 studbolts, was required to be extremely corrosion-resistant but also to be of the highest possible strength consistent with this, a combination which was met with a precipitation-hardened nickel–copper alloy [21].

As developments continue in the oceans, the requirements for new levels of property attainment will become apparent, and it is no more possible to speculate what these will be than it would have been to predict the brittle fracture problem in welded steel ships. However, some of the challenges can be foreseen, and indeed materials engineers are currently working to solve some of these. For example, ocean thermal energy conversion (OTEC) requires the development of materials resistant to corrosion in both cold, deep seawater and surface seawater; however, the implications of this for materials selection are not simple, because experience of the former environment is limited [22]. The carrier pipes for OTEC will need to be very long and probably of large diameter, making for further likely challenges in materials selection. The interest in recovery of manganese nodules and other minerals from the seabed also presents some design challenges, not least because of

the very large depths at which such deposits are to be found [23]. The continued development of high-performance surface craft calls for innovation in materials use for extreme strength-to-weight ratios, a recent example being the novel design of a composite flap for a hydrofoil craft [24].

It has to be hoped, however, that progress in marine technology will switch from the exploitative to the sustainable, and that sufficient effort will be made to develop offshore (as opposed to coastal) and subsea farming activities for both energy and food.

1.3 The Range of Material Properties and the Selection Process

Materials selection is quite often not a conscious process in the minds of designers, and heavy reliance is usually put on previous direct experience of similar designs and applications. Formal selection procedures, even when computerised, can therefore sometimes seem turgid and unnecessary, since, for example, no one would consider the full range of available materials (from ceramics to wood) when designing a pipeline.

Nevertheless, it is sometimes refreshing to remind oneself of the full range of available materials and the constraints on their use, especially when carrying out particularly innovative design work, but also for some detailed aspects for which there may be a number of satisfactory solutions whose practicability or feasibility changes with small changes in economic or technical conditions.

The purpose of this section, therefore, is to review, in an extremely broad fashion, the range of available engineering materials and their capabilities in preparation for the more detailed examination in later chapters. This section should also provide a guide a Chapter 4 and is arranged in broadly the same way.

The coarsest classification of materials is usually carried out by differentiating between metals and non-metals, and, whereas division within the metals can clearly be carried out in terms of the basis metals, usually with extra groups in the ferrous metals, there is much less agreement as to how to divide the non-metals. The approach in this book is to divide the non-metals into inorganic materials, polymers and composites and to have separate classes for cement and concrete and timber (Table 1.3). Within each of these classes, the divisions are relatively simple and conventional and are shown in Table 1.4.

When selecting a material from the vast array available, an engineer will usually ask (consciously or unconsciously) whether the application is an established or a novel one, and this immediately sets up barriers to selection (which may or may not be a good thing). Examples of established marine use might be ship hulls, pipelines, offshore platforms, seawater piping or coating systems and it must be accepted that, unless something is radically different about the design requirements (as, for example, in the MCMVs of Case Study 7.3), most of the feasible materials choices have been tried, so that selection becomes a matter of being aware of how these choices have been made in the past and avoiding known pitfalls. Often, of course, this job is helped considerably by a code of practice.

Even when a new or radical design is contemplated, the selection process will start with a search for similarities with established uses, though this should also be tempered

Table 1.3. Classification of main groups of materials of importance in marine technology

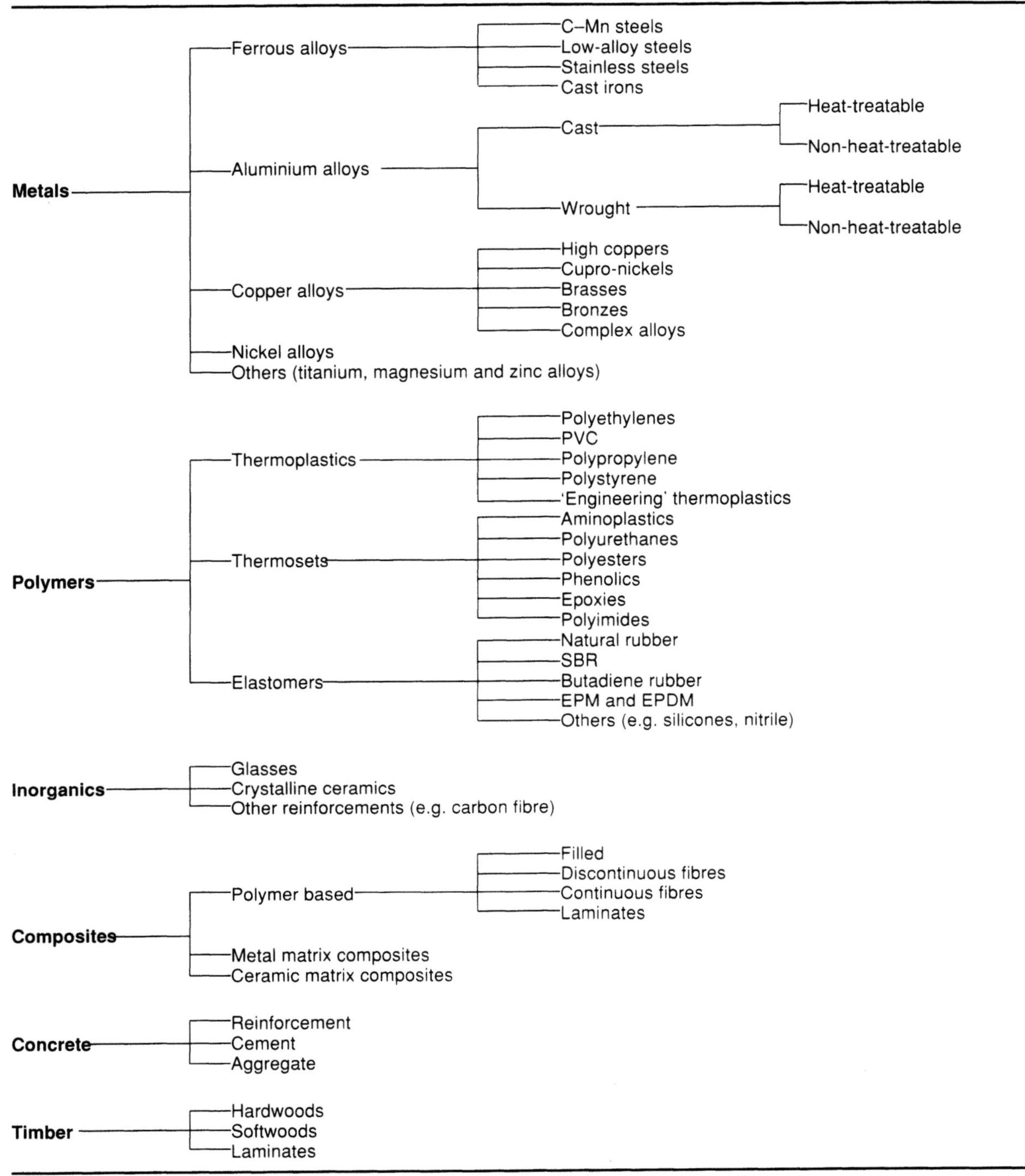

with an awareness of the differences which may affect the choice. Such 'coarse' materials selection is illustrated in Fig. 1.14.

As well as the obvious considerations of materials properties, effective design must also take cognisance of the 'fabricability' of any proposed materials. This interrelationship of materials properties, fabricability and design can be summarised in the three questions:

How will it work?

How will it be made?

What will it be made of?

Clearly, these questions must be asked in parallel, along perhaps with the economic question, 'Is it worth making?'

As will be seen, fabricability is a very important issue in marine technology, but for the moment some general

Table 1.4. Subclasses in the materials hierarchy shown in Table 1.3

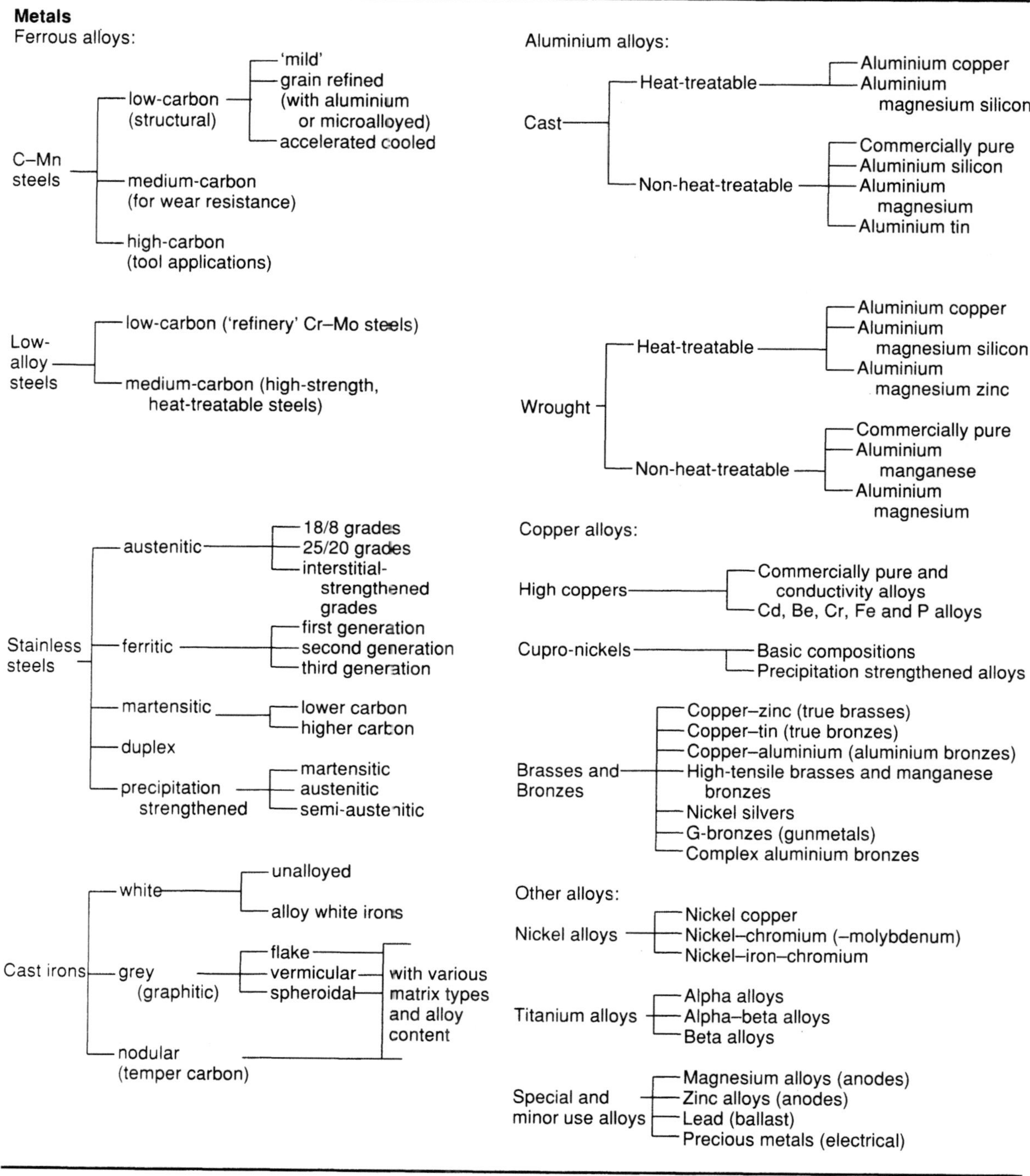

continued on next page

aspects are summarised in Table 1.5. Of course, this table is very general and to say, for example, that steels can be fabricated by welding must be qualified by the details given in Chapter 5. As can be seen, however, it is of overriding importance that constructional materials (i.e. those used to make structures as opposed to components) should be fabricable with ease, although the methods of fabrication may be radically different. The materials must also be cheap, though this cheapness quite often comes with the widespread use brought about by the application. It is worth reflecting upon the fabricability and cheapness issues both with regard to the two main marine constructional materials (steel and concrete) and with regard to the possible uses of composite materials.

The cost of a material is partly controlled by the amount of energy required to make it from its raw components, but

Table 1.4. (*continued*)

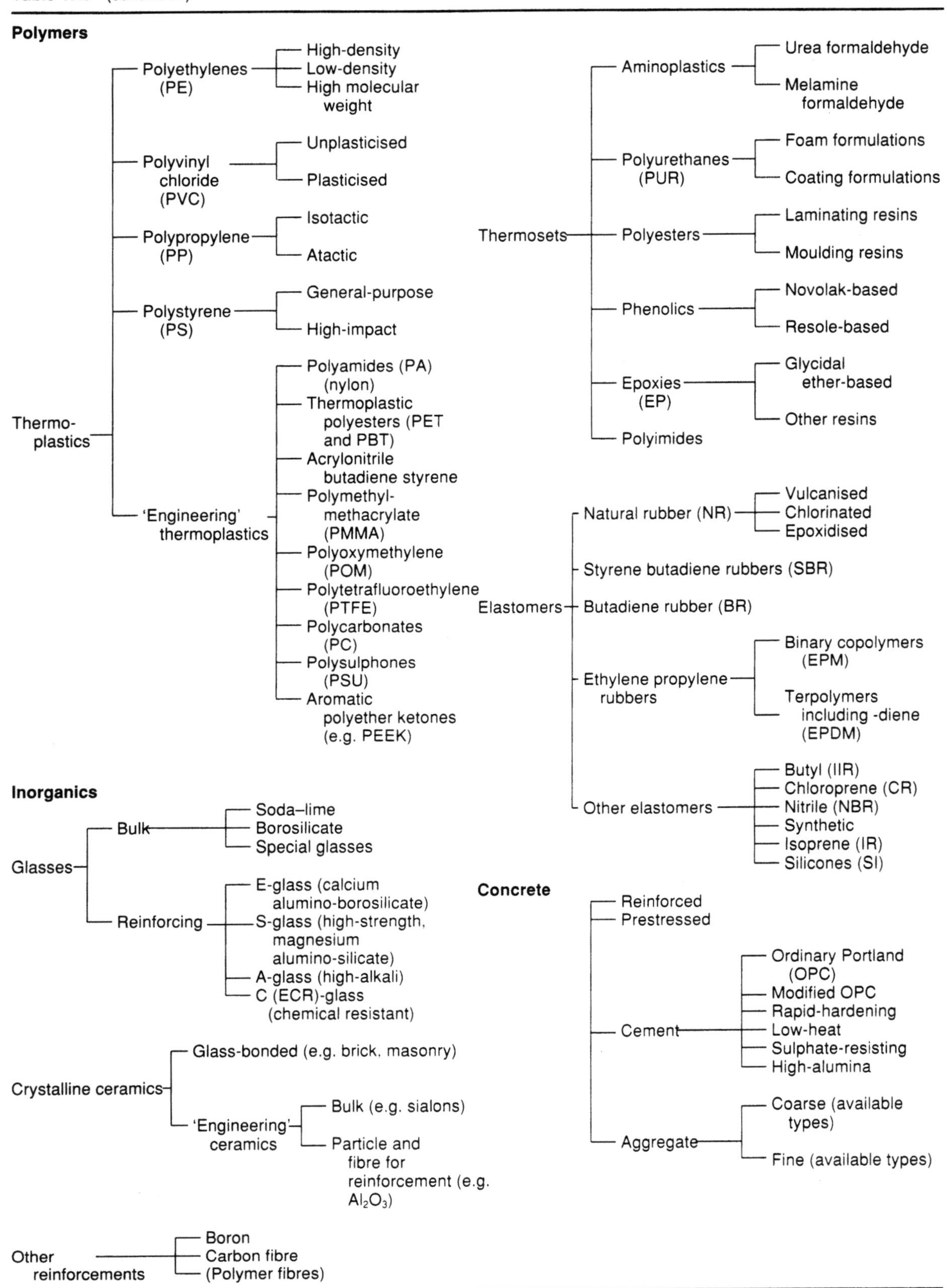

continued on next page

Table 1.4. (*continued*)

Composites

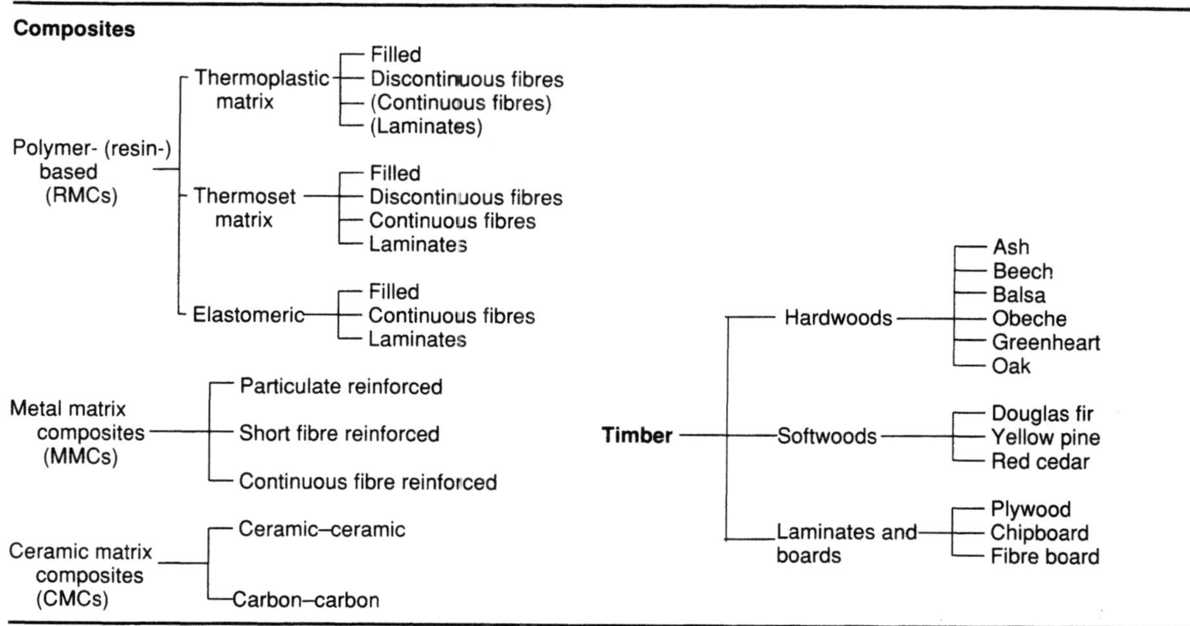

even if this was not so it is still of interest to know how energy-intensive materials production actually is. This might also be considered in the light of recycling opportunities where metals, thermoplastics and glasses feature well but other materials such as concrete, elastomers and thermosets do not, although this might be tempered somewhat by re-use (for another application) of these latter materials.

For general mechanical use, coarse selection can usually be carried out on a relatively small number of properties. For most marine applications, coarse materials screening can, in fact, be made on the basis of strength, stiffness and density. The absence of corrosion resistance from this list is not an indication of its lack of importance but rather to indicate that lack of corrosion resistance can be overcome by other means. Toughness is also absent, mainly because it is such an elusive property and is much less tightly tied to generic material type than is, for example, density. Figures 1.15, 1.16 and 1.17 summarise the available range in density, strength and Young's modulus, while Figs 1.18 and 1.19 show the available range in marine corrosion resist-

ance and toughness (though these latter two sets of data are rather more qualitative, for the reasons outlined above).

Once the properties required have been isolated and quantified (albeit in a relatively coarse way), the question arises as to how they should be combined to choose the best material for the intended application. For this it may be prudent to include fabricability as a property, although the quantification of this would be subjective at best. Cost, not only of the material but of the manufacture into the product, must also be considered at this stage.

Property combinations may be simple, such as Young's-modulus-to-density ratio, or more complex combinations may be considered, according to function and depending upon whether the design is stress limited or deflection limited. A relevant although uncomplicated example of this process is given by Crane and Charles [26] for the comparison of various candidate materials for ships' hulls. The analysis was carried out first of all on the basis of the bending deflection of those materials as compared with that of ordinary mild steel for ships of identical dimensions and

Table 1.5. Fabricability of various classes of materials (a, a process commonly used for the material class; b, one which may be used although some precautions or modifications may be necessary; c, a process which either cannot or is rarely used for the material). All processes are secondary, i.e. they are those used to make products rather than bulk materials

Material type	Casting	Moulding	Cutting	Forging	Sheet forming	Welding	Brazing/ soldering	Adhesive bonding	Application as coating
C–Mn steels	b	c	a	a	a	a	a	c	c
Low-alloy steels	b	c	b	a	b	b	b	c	c
Stainless steels	b	c	a	a	a	a	a	b	a
Cast irons	a	c	a	c	c	b	c	c	c
Aluminium alloys	b	c	a	a	b	b	b	a	b
Copper alloys	a	c	a	a	a	a	c	a	a
Thermoplastics	c	a	b	c	a	a	c	b	a
Thermosets	c	a	b	c	b	c	c	a	a
Elastomers	c	b	b	c	c	b	c	a	a

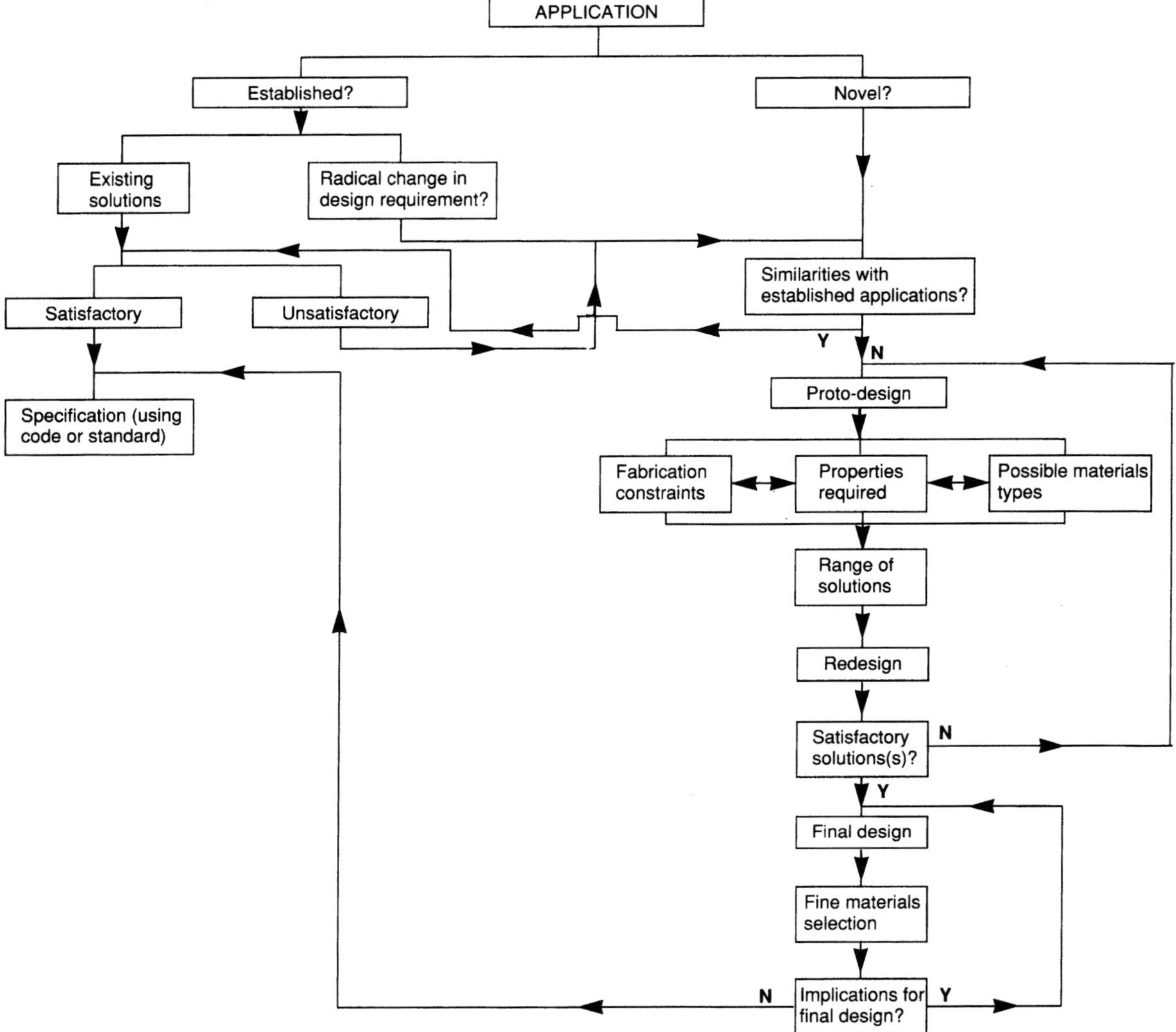

Fig. 1.14.   Schematic representation of materials selection considerations in the design of a new structure or component.

with the saved weights transferred to the cargoes. The results are given in Table 1.6, and show clearly that the adoption of reinforced plastics for larger hulls requires a radical change in design to improve the stiffness, especially if the goal is weight reduction. Another way of looking at the selection is to consider the weight saving which might be made for ships of the same outer dimensions and longitudinal bending moment, trimming the material effective thickness to compensate for the change in material strength. Again using mild steel as the baseline, Table 1.6 shows the potential weight savings on this basis. Of course, this analysis does not take into account the increased deflections in materials of lower stiffness; in fact, Charles and Crane indicate that it is likely that the lightest GRP ship satisfying the deflection rules of the classification societies would probably be heavier than the equivalent steel ship.

A number of semi-quantitative methods also exist for optimising a material selection, and these sometimes form the basis of computerised selection packages. One of these is illustrated below to indicate the way in which the process works, but it is perhaps best to regard this as an example of how to proceed with optimising selection rather than a rigid recipe for all situations. As was seen above for the case of ships' hulls, the selection process owes much to the subjective decision of the designer as to what properties and combinations are important, and what relative weightings to give them.

The process [26] is summarised in Tables 1.7 to 1.9, and consists initially of a coarse screening procedure where a set of requirements (ranked roughly according to importance – primary, secondary, etc.) are considered in relation to whether these are acceptable, over-provided or under-provided, under-provision in a primary property normally being considered as a cause for rejection of the material. Cost should also be brought in at this stage, at least to act as a rejector of materials of excessive cost. This process will act to reject a great number of materials (in this case materials M1, M3 and M5); the next stage usually requires some quantification of the properties, and here cost may be considered as a property. These properties can then be nor-

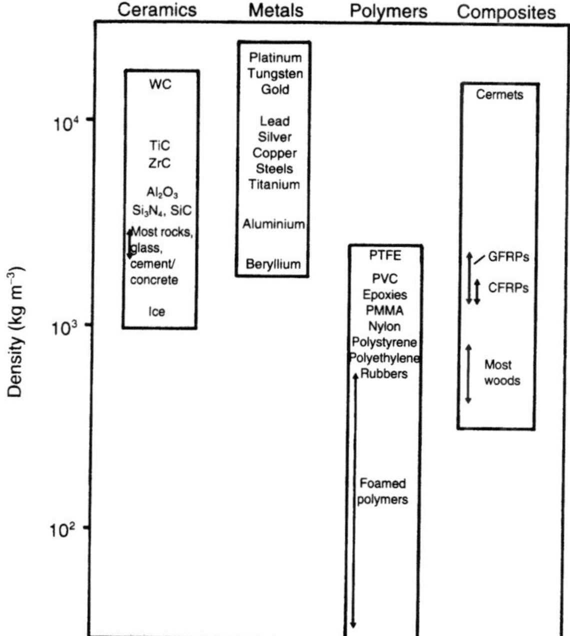

Fig. 1.15. Overview of the range of densities available in each of the broad classes of materials (after Easterling [25]).

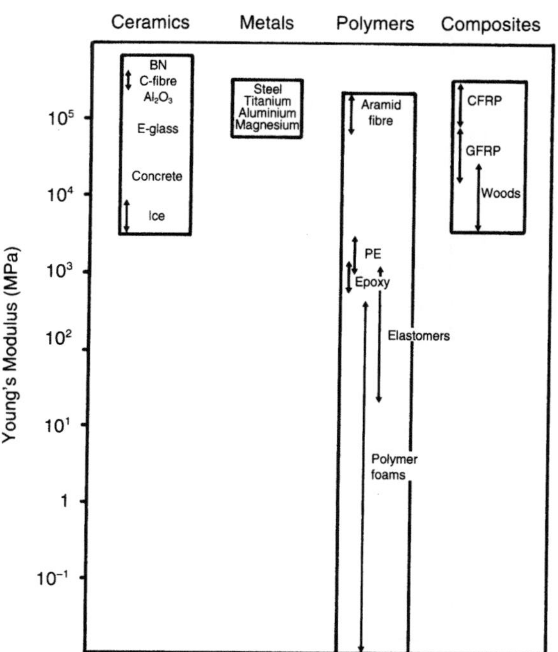

Fig. 1.17. Overview of the range of Young's modulus available in each of the broad classes of materials (after Easterling [25]).

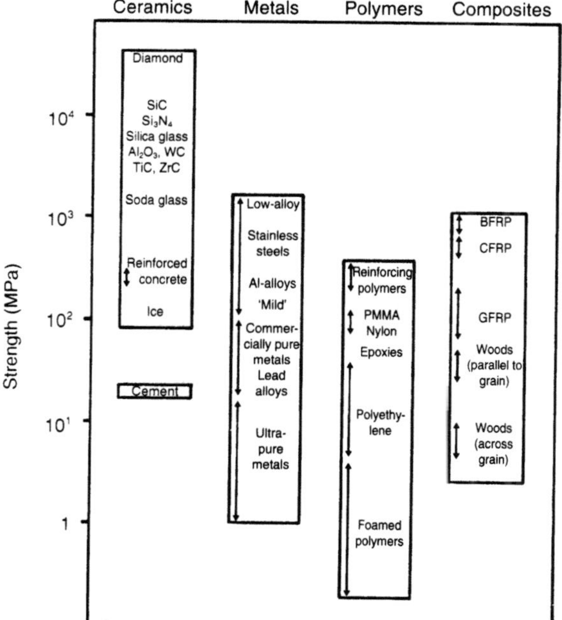

Fig. 1.16. Overview of the range of strengths available in each of the broad classes of materials (after Easterling [25]).

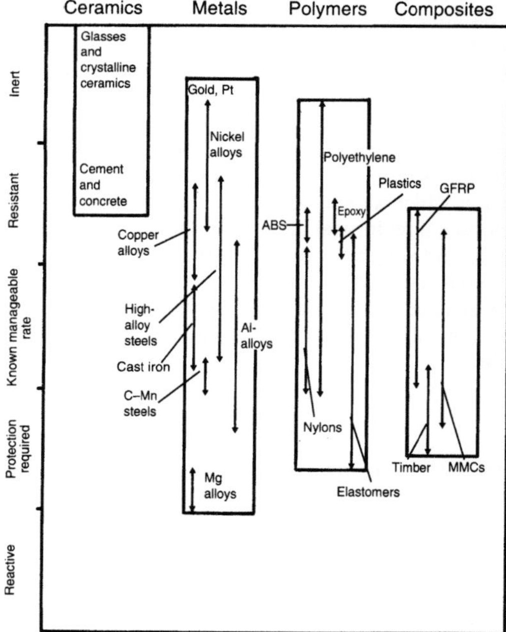

Fig. 1.18. Overview of the range of corrosion resistance available in each of the broad classes of materials. Resistance is classified in terms of marine exposure including such factors as atmospheric resistance, resistance to immersion, biological and ultra-violet exposure.

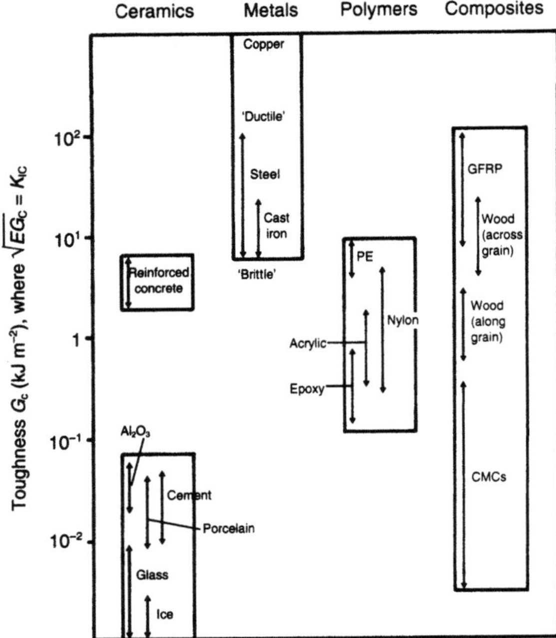

Fig. 1.19. Overview of the range of toughness available in each of the broad classes of materials.

malised (since they will all be in different units) and this can be done by scaling them in proportion to the most desirable value of the particular property displayed in the group. It is sometimes helpful at the normalisation stage also to make the most desirable property the largest, so that cost, for example, may be expressed as a reciprocal or as the normalised cost subtracted from unity. After this the properties should be weighted (Table 1.9), a subjective decision which should account for such factors as having taken, say, a reciprocal of some of the properties. The normalised and weighted values can then be combined to give an overall rating which, if the prior process has been correctly arranged, should consist simply of adding. although some logical operators such as ANDs and ORs may be helpful.

Table 1.6. Comparison of candidate ship hull materials on a deflection and on a weight-saving basis

Material	E_1/E_2	σ_2/σ_1	δ_2/δ_1	d_2/d_1	W_2/W_1
Higher-tensile steel	1	1.2	1.2	1.0	0.8
Aluminium alloy (N8)	3	0.6	1.8	0.35	0.6
GFRP – 30% chopped mat	20	0.5	10	0.19	0.4
GFRP – 50% woven cloth	15	1.0	15	0.2	0.2
CFRP (approx.)	2	2.7	5.4	0.2	0.07
				Deflection limited	Weight limited

Source: Crane and Charles [26].

Table 1.7. Coarse materials screening procedure (A, acceptable; o, over-provision; e, excessive; u, under-provision)

Material	Primary			Secondary			Cost
	P1	P2	P3	P4	P5	P6	
M1	A	o	A	A	A	A	e
M2	A	A	A	o	A	A	A
M3	u	A	A	A	A	A	A
M4	A	o	A	A	o	o	A
M5	A	A	A	A	A	A	e
M6	A	A	A	A	u	A	A

Source: Crane and Charles [26].

Although such a procedure can readily be computerised for handling large numbers of materials choices, it is still recommended that designers selecting from scratch should carry out at least a simple version of this process manually, because this can provide much information on the subjective elements of selection which, of course, cannot be simulated without recourse to a full-blown expert system.

Table 1.8. Fine materials selection data. Hypothetical example based on a subset of material M2 in Table 1.7 (weighting factors are given in square brackets at the top of the property column)

Material	Primary properties			Secondary properties		
	Strength (MPa) [20]	Toughness (MPa√m) [20]	Cost (£/kg) [10]	Corrosion rate (mm/yr) [4]	Weldability [3]	Density (kg/m³) [1]
M2a	468	75	5.63	0.1	Excellent	7.891
M2b	591	28	5.91	0.1	Poor	7.885
M2c	635	95	7.05	0.05	Good	7.885

Table 1.9. Normalised and weighted data from Table 1.8 (cost, density and corrosion rate normalised as reciprocal and overall rating calculated by adding all normalised and weighted values and dividing by the maximum attainable rating of [20] + [20] + [10] + [4] + [3] + [1] = 58)

Material	Strength	Toughness	Cost	Corrosion resistance	Weldability	Density	Overall rating
M2a	14.7	15.8	10	2	3	0.999	0.80
M2b	18.6	5.89	9.52	2	1	1	0.65
M2c	20	20	7.98	4	2	1	0.95

References

1. Birchon D. The use and abuse of materials in ocean engineering. Proceedings of the Institution of Marine Engineers 1971; 185(22): 241–271

2. Whitfield M. The salt sea – accident or design? New Scientist 1982; 94(1299): 14–17

3. Morgan N. Ocean environments. In: Morgan N (ed.) Marine technology reference book. Butterworths, London, 1990, pp 1/1–1/31

4. Dexter SC. Marine corrosion – seawater. In Metals handbook, vol. 13. ASM, Metals Park, Ohio, 1987, pp 893–926

5. Miller JW (ed.). NOAA diving manual. US Department of Commerce/National Oceanic and Atmospheric Administration, US Government Printing Office, Washington, 1979

6. Patel MH. Offshore structures. In Morgan N (ed.) Marine technology reference book. Butterworths, London, 1990

7. Hallam MG, Heaf NJ, Wootton LR. Dynamics of marine structures. CIRIA Underwater Engineering Group, London, 1978 (report no. UR8)

8. Sanderson TJO. Ice mechanics – risks to offshore structures. Graham & Trotman, London, 1988

9. Harbron JD. Modern icebreakers. Scientific American 1983; 249(6): 53–59

10. Harbron JD. The Spanish ship of the line. Scientific American 1984; 251(6): 122–131

11. Conde JFG. New materials for the marine and offshore industry. Transactions of the Institution of Marine Engineers 1985; 97, paper 24

12. Walker FM. Iron's contribution to modern shipbuilding. Metals and Materials 1990; 6(12): 778–782

13. Brownlee W. HMS Warrior. Scientific American 1987; 257(6): 86–92

14. Waite DF, McKendrick RD. Holland I – an underwater miracle. Metallurgist and Materials Technologist 1984; 16(2): 70

15. Masubuchi K. Analysis of welded structures. Pergamon, Oxford, 1980

16. Browne RD, Domone PLJ. The long-term performance of concrete in the marine environment. Proceedings, Conference on Offshore Structures, Institution of Civil Engineers, London, 1975, pp 49–59

17. Cmnd. 3409. Report of the enquiry into the causes of the accident to the drilling rig 'Sea Gem'. HMSO, London, 1967

18. Almar-Naess A, Haagensen PJ, Lian B, Moan T, Simonsen T. Investigation of the Alexander L Kielland failure – metallurgical and fracture analysis. Proceedings, 14th Offshore Technology Conference, Houston, Texas, OTC 4236, 1982, pp 79–83

19. Offshore Supplies Office, Department of Energy. Study on lightweight materials for offshore structures, Wimpey Offshore Ltd, 1987 (report no. WOL 161/87)

20. Anon. An odd navy that fights the tides. New Scientist 1982; 94(1299): 24

21. Brookes KJA. Thames barrier success hinges on nickel alloys. Metallurgist and Materials Technologist 1983; 15(1): 33–34

22. Larsen-Basse J, Park Y. Corrosion in slowly flowing ocean thermal energy conversion seawater. Materials Performance 1989; 28(2): 46–51

23. Amann H, Oebius HV, Gehbauer F, Schwarz W, Weber R. Soft ocean mining. Proceedings, 23rd Offshore Technology Conference, Houston, Texas, OTC 6553, 1991, pp 469–480

24. Oken S, Deppa RW, Taylor DW. Development of an advanced composite hydrofoil control flap. Proceedings, 4th Conference on Fibrous Composites in Structural Design, San Diego, 1978, pp 659–674

25. Easterling KE. Tomorrow's materials. Institute of Metals, London, 1988

26. Crane FAA, Charles JA. Selection and use of engineering materials. Butterworths, London, 1984

2 Mechanical Properties and Design for Marine Use

This chapter reviews the mechanical properties of materials which are of most importance in marine design. Of course, such properties are often important outside the marine area, but the emphasis here will be on those aspects which are most relevant and on describing the ways in which property data can be incorporated into design for this particular environment. Collected data on mechanical properties are included in Chapter 4 and properties related to corrosion and environmental resistance are covered in Chapter 3, again with data being collected in Chapter 4.

Mechanical properties can be conveniently classified as those related to deformation behaviour, those related to fracture and those related to surface integrity (excluding corrosion mechanisms). The first two of these are substantially more important in marine environments and are therefore considered first, with relatively less attention being devoted to surface properties. Under each of the above classifications will be considered the appropriate group of properties, the ways in which these can be measured and, finally, the ways in which they are applied in mechanical design.

A number of the Case Studies in Chapter 7, notably 7.1, 7.2, 7.3 and 7.6, are relevant to the material covered in this chapter and it should also be helpful in interpreting some of the property data in Chapter 4.

able to yield before fracture, so that stress–strain curves are normally linear elastic in nature right up to failure. An important feature of such materials is that their tensile properties are defect-dominated, so that a number of tests on nominally similar test pieces will usually show a wide scatter in results (e.g. Holloway [1]). Furthermore, the geometry of uniaxial tension is such that gripping stresses are likely to cause premature failure, and it is far more common to use three-point bend or compression specimens when dealing with brittle materials. Timber properties are rather more directional and also more variable than manufactured materials, so that special test methods and precautions need to be taken in dealing with these. Finally, the most modern materials are those whose properties can be directly engineered by mixing materials of different component properties. Such composite materials are finding wide use in all branches of engineering, marine applications being no exception. The major factor which needs to be taken into account when dealing with these materials is their anisotropy, although some of the more sophisticated composite structures have additional local reinforcement, so that mechanical properties may vary from place to place within such a structure.

In the following, a brief summary of the key elements of mechanical testing and the quantification of deformation behaviour for each of these classes of materials is given.

2.1 Properties Related to Deformation Behaviour

Most of the properties related to deformation behaviour can be derived from the familiar load–extension or stress–strain curve. Such a curve illustrates how a material will behave in simple uniaxial tension, and this is fundamental to design against excessive deformation, yield, or plastic collapse. In some cases it is not appropriate to test in uniaxial tension (e.g. very brittle materials), and tests based on bending provide a useful alternative.

The essential types of stress–strain behaviour can be described as elastic, plastic and viscoelastic–plastic. Most metals behave in an elastic–plastic manner and some show the well-known discontinuous yielding phenomenon characteristic of the 'mild' steels. Plastics generally behave in a viscoelastic–plastic manner so that time is an important variable in mechanical testing (usually characterised as strain rate). Ceramics and other brittle materials are rarely

2.1.1 Metallic Deformation Properties

Because of the elastic–plastic behaviour of metals, engineers require a reproducible method of describing both the elastic and the plastic behaviour of metals, and also the point at which the transition from elastic to plastic deformation takes place. In general, designers prefer to keep metallic components and structures within the elastic range, although plastic behaviour is also of interest where cold forming is to be carried out, as well as for plastic-collapse analysis. Local yielding may also be permissible and, indeed, desirable in some structures.

As is well known, the nominal uniaxial tensile stress–strain curve is used to obtain information regarding deformation behaviour of metals. The data normally recorded are: Young's modulus, the yield stress (or proof stress at a given plastic strain, typically 0.2 per cent), the ultimate tensile strength (UTS) and some measure of ductility, usually either percentage plastic elongation or reduction in cross-sectional area at failure. In addition, a quantity known as the 'flow stress' (usually the mean of yield stress and UTS) is often used for plastic collapse analysis.

In polycrystalline metals, directionality of mechanical properties is not normally a matter of concern. One notable exception to this – steel plate is expected to experience through-thickness stresses. It is a feature of modern steel-making practice that unwanted dissolved sulphur is precipitated as manganese sulphides, and these spherical inclusions are rolled into a pancake shape during the manufacture of plate [2]. For thick plates in critical offshore structural nodes, special tensile testing (Fig. 2.1) can be applied to find the through-thickness ductility as a measure of resistance to lamellar tearing (see Chapter 5).

The use of Young's modulus as a design quantity is confined to the region of elastic deformation, although this is sufficient for most engineering purposes, because stresses are normally kept substantially below yield. Being the slope of the stress–strain curve, Young's modulus is a measure of the rigidity of a material, in other words it is a measure of the extent to which materials of identical section geometry will deflect under a given load. Young's modulus, along with the other elastic properties of a solid, is related to the strength of bonding between individual atoms in the solid, so that metals have in general quite high moduli compared with, for example, plastics, whereas the moduli of metals are somewhat lower than those of the covalent crystalline solids. However, the relatively high moduli of metals and other desirable properties such as toughness combine to make them the most suitable structural materials for most applications. Apart from the obvious control of dimensional stability, deflections are important in dynamic behaviour and in elastic buckling analysis.

The yield stress (or proof stress, usually measured at 0.2 per cent strain) is used as an absolute design maximum for structural purposes and various safety factors are normally applied to ensure that the stress remains well below yield, even accounting for dimensional discontinuities and production variations. Of course, in stress states more complex than simple uniaxial tension a yield criterion is required to establish whether yielding is likely to take place. For example, a rectangular section in simple bending will show first yield at a bending moment of

$$M_y = \sigma_y t^2 b/6$$

where σ_y is the yield stress and the dimensions t and b are the thickness and breadth of the section respectively. However, it is not until the bending moment reaches a value of $1.5M_y$ that plastic collapse, or general yield, can be said to have taken place, this latter condition being defined as that at which it is no longer possible to trace a path across the load-bearing section from load point to load point without encountering plastically deformed material [3].

Many design guides now include plastic collapse analysis for structures with and without defects. The buckling of pipelines during the laying operation provides a typical marine example of such an analysis. Such buckling can occur as a result of excessive bending during the laying operation and may render the pipeline subsequently unfit for service. Jensen and Pedersen [4] have discussed typical design criteria for pipe buckling during laying in terms of hydrostatic pressure (at the seabed) and bending moment as a function of pipe dimensions and mechanical properties.

Apart from stress–strain data, material deformation processes have the additional dimension of time and the increase of strain with time at constant stress is normally referred to as 'creep', the complementary effect of reduction of stress with time at constant strain being called 'stress relaxation'. Metals can deform by creep, although the process is a thermally activated one based on vacancy diffusion within the metal lattice. This results in viscoplastic behaviour which does not recover with time after the removal of the load. However, the temperatures at which creep is important in metals are usually about one-half to three-quarters of the melting temperature (in Kelvins) so that only metals with low melting points, such as lead, creep appreciably at the temperatures commonly encountered in marine environments. Creep of metals is therefore of interest to marine designers only in such areas as turbines, boilers, flare boom tips and internal parts of engines, although stresses in the latter two are not normally high.

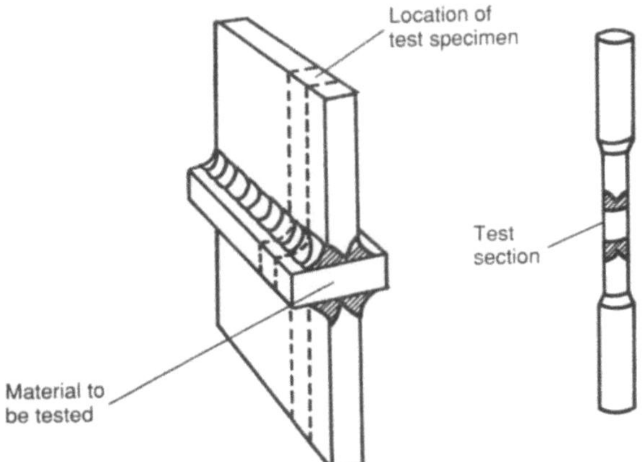

Fig. 2.1. Orientation of specimens for through-thickness ductility tension test. (After Easterling [2].)

2.1.2 Polymer Deformation Processes

Polymer deformation processes depend critically on the type of polymer involved, and in particular whether it is an elastomer, a thermoplastic or a thermoset (see Chapter 4).

Elastomers are well known for their very high extensibility. As can be seen from Fig. 2.2, the elastic behaviour is far from being linear, with typically two regions of different curvature. Because of the uses to which elastomers are put, the emphasis in mechanical testing is on strain rather than on stress. In fact, elastomer strengths are among the lowest of all engineering materials. Hysteretic effects, as illustrated in Fig. 2.2, are of interest in that they indicate the degree of energy absorption provided by the material. Furthermore, it is usual to carry out separate tests for different deformation modes such as shear and compression, again from the point of view of the major application of the material (Fig. 2.3). Shape is also used to alter energy storage capabilities in elastomeric structures (Fig. 2.4). Like

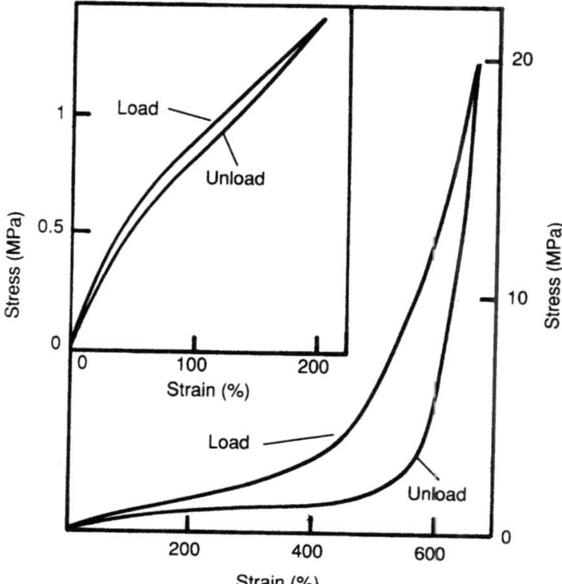

Fig. 2.2. Extension–retraction curves from 200 and 600 per cent strain in unfilled natural rubber vulcanisate. (After Lake [5].)

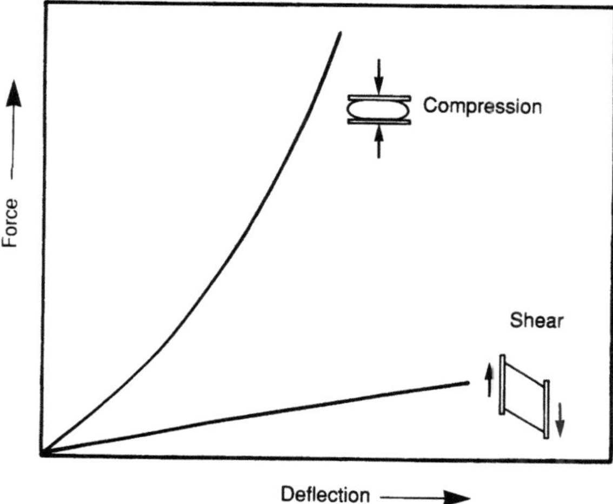

Fig. 2.3. Force–deflection curves for an elastomer in shear and compression. (After Lake [5].)

other polymers, the definition of modulus for elastomers is difficult, mainly owing to the lack of a well-defined linear portion to the stress–strain curve and, in fact, the complex modulus (see p. 22) is much more useful in energy-absorbing design. Some of the aspects of marine design using elastomers are explored in Case Study 7.6.

The most important feature of thermoplastic stress–strain behaviour is viscoelasticity. This can give real design difficulties, particularly when manufacturers' data sheets are the only available source of strength and modulus data. The lengths of time used for mechanical testing are rarely those in which the designer is interested, so that some information on the effect of strain rate on modulus and on yield behaviour is necessary. Furthermore, the temperature

stability of thermoplastics is such that creep properties are highly temperature-dependent, and this adds an additional dimension of difficulty to mechanical design using these materials. Finally, the engineer also needs to be aware of the possible crystallinity and glass transition effects which may affect thermoplastics over the design temperature range.

A number of different models for the viscoelastic behaviour of thermoplastics exist [6], the most widely used being the Maxwell, Voigt–Kelvin and four-parameter models. The principal features of these are illustrated in Table 2.1, and Fig. 2.5 shows the strain–time behaviour predicted by the most complete of these models, namely the four-parameter model which has been suggested as 'a crude qualitative description' of the viscoelastic response of a linear amorphous polymer [6], but which nevertheless is far too sophisticated for any but the most exacting engineering design, especially when it is considered that the viscoelastic response represents only part of the stress, strain, time and temperature variations of the deformation behaviour of a thermoplastic material (Fig. 2.6).

The simplest way of dealing with this problem is to employ the method of pseudo-elastic design, which centres around the choice of an elastic modulus so as to give safe design for the lifetime of the component. Of course, even once the time dimension has been fixed by the expected life of the component, it is still necessary to arrive at an acceptable limiting strain and hence find the allowable stress. This limiting strain may be dictated by the design itself, or may be reached by some other criterion, for example using a secant modulus on an isochronous curve based on the design life (Fig. 2.7). The secant modulus is 0.85 of the initial tangent modulus and a limiting stress and strain can be obtained by the intersection of the secant modulus line with the stress–strain curve. However, this may be excessively conservative for some thermoplastics, and Crawford [7] gives a number of examples where limiting strains are set rather higher than the secant limit.

Thermosetting plastics behave rather differently from thermoplastics owing principally to the three-dimensional nature of the bonding (see Chapter 4). They are relatively

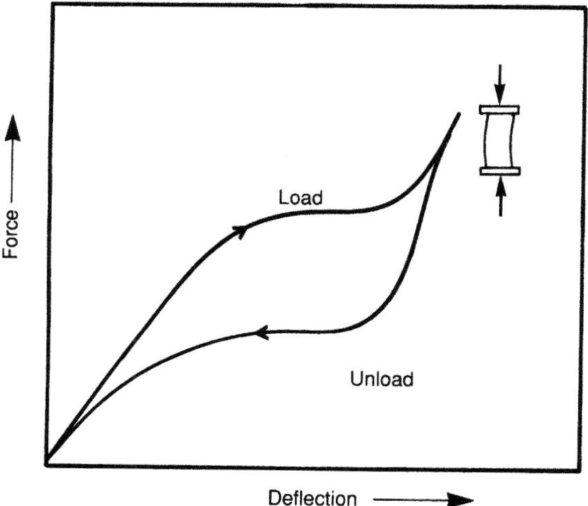

Fig. 2.4. Force–deflection–recovery curve for an elastomer in compression and buckling. (After Lake [5].)

Table 2.1. Summary of stress–strain models for polymers

Model	Stress–strain relationship
Maxwell	$\dot{\sigma} + (E/R)\sigma = E\dot{\epsilon}$

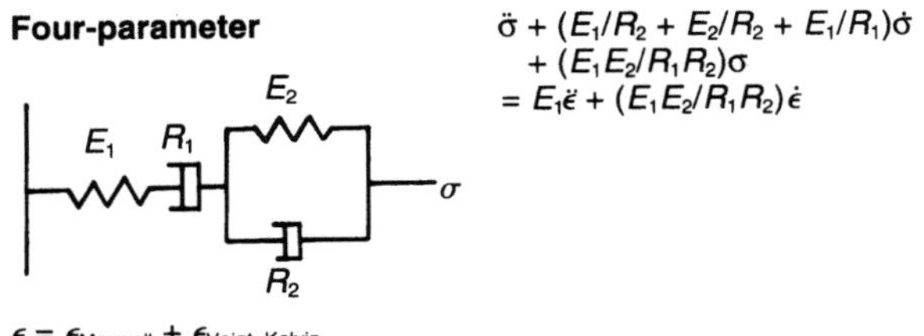

$\sigma = \sigma_{spring} = \sigma_{dashpot}$
$\epsilon = \epsilon_{spring} + \epsilon_{dashpot}$

| **Voigt–Kelvin** | $\dot{\epsilon} + (E/R)\epsilon = \sigma/R$ |

$\epsilon = \epsilon_{spring} = \epsilon_{dashpot}$
$\sigma = \sigma_{spring} + \sigma_{dashpot}$

| **Four-parameter** | $\ddot{\sigma} + (E_1/R_2 + E_2/R_2 + E_1/R_1)\dot{\sigma}$ $+ (E_1 E_2/R_1 R_2)\sigma$ $= E_1\ddot{\epsilon} + (E_1 E_2/R_1 R_2)\dot{\epsilon}$ |

$\epsilon = \epsilon_{Maxwell} + \epsilon_{Voigt-Kelvin}$

Source: Adapted from Throne and Progelhof [6].

stiff, brittle materials and are rarely used in engineering applications without reinforcement, even if this is only wood flour. Nevertheless, matrix resin properties are of interest, notably in composites design, and although tensile tests may be used (especially for the foregoing purpose), it is sometimes more appropriate to use a quantity such as modulus of rupture (see p. 23) to quantify strength.

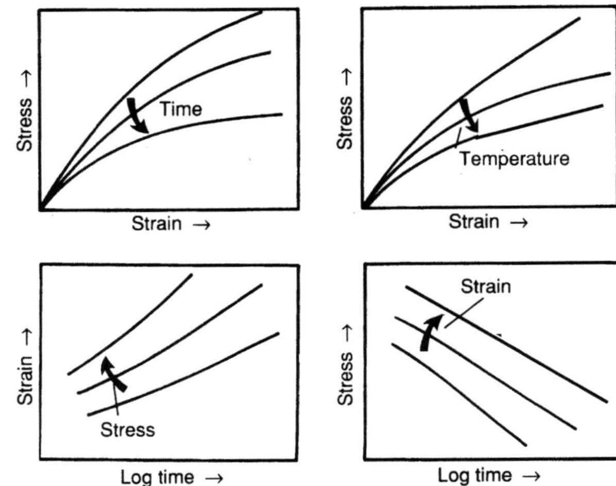

Fig. 2.6. Summary of stress–strain–time–temperature behaviour of a viscoelastic material.

In some plastics applications, it is important to know the dynamic mechanical response of the material [8, 9]. Testing can be carried out in any geometry and a complex modulus is obtained, reflecting the magnitude and phase relationship between the stress and the strain. The complex modulus comprises a real part (M'), sometimes called the 'storage modulus', and an imaginary part (M''), sometimes called the 'loss modulus', for the particular geometry of deformation (e.g. E' and E'' for tension). The complex modulus was originally used as an analysis tool for plastics materials where a study of its behaviour over a range of temperature yields useful structural information. However, in the form of the loss factor,

$$\tan \delta = M''/M'$$

presented as a function of frequency at a temperature of interest, the complex modulus can provide information on areas where a particular material might be a useful damping medium (indicated by peaks in the loss-factor–frequency curve).

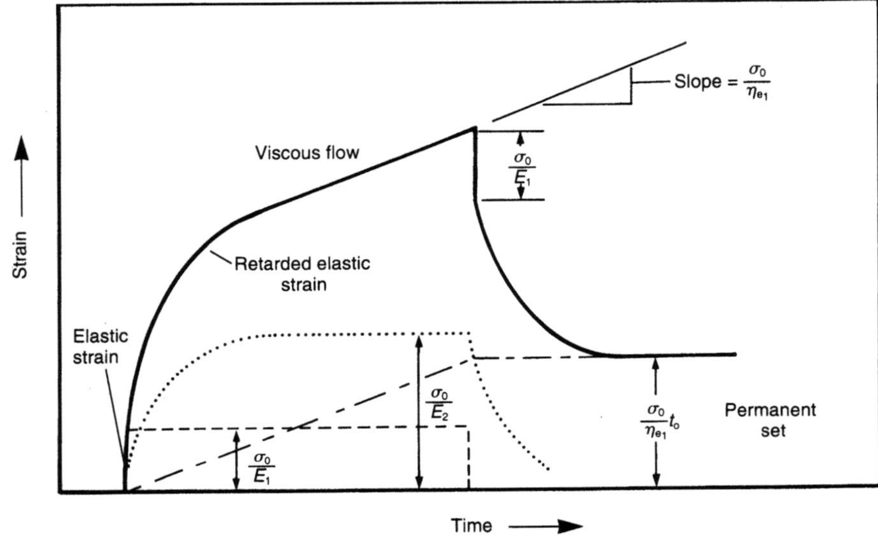

Fig. 2.5. Strain–time behaviour for a viscoelastic material according to the four-parameter model showing the component responses given by the Voigt–Kelvin (dotted), the Maxwell spring (dashed) and the Maxwell dashpot (dash–dotted) elements. (After Throne and Progelhof [6].)

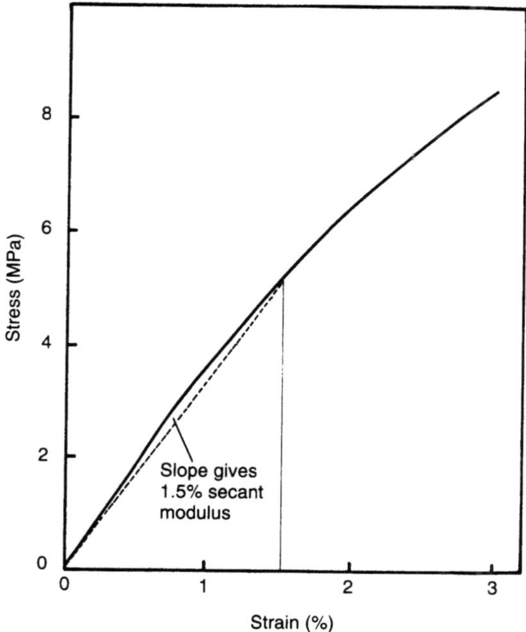

Fig. 2.7. One-year isochronous curve for polyethylene, showing 1.5 per cent secant modulus. (After Crawford [7].)

2.1.3 The deformation of concrete

Concretes are brittle materials and are hence rarely used in tension. Although the stress–strain behaviour of cement paste and aggregate are sensibly linear, that of concrete involves creep (Fig. 2.8), while the long-term deformation of concrete is also affected by microcracking and shrinkage effects. It is normal to use a secant modulus for quantifying the elastic behaviour of concrete, but this is a somewhat ill-defined quantity [10] and the stress, σ_s, at which the modulus is taken requires to be stated. Of course, since concrete cures over a relatively long period, it can be expected that its properties will vary with time. Ultimately, however, because of the way in which concrete is used in construc-

tion, engineers are most interested in the compressive strength of the fully hardened material. Such compression tests involve crushing of cubes or cylinders between steel platens and recording the stress at which crushing takes place. For measurement of the tensile strength of concrete, a flexure test is normally used. This consists of loading a plain beam in bending and recording the nominal extreme fibre stress at which failure takes place (modulus of rupture).

Most structural concrete is either reinforced or pre-stressed, and is therefore a composite material in an even wider sense than the paste–aggregate mixture which goes to make up the unreinforced material.

The addition of reinforcing bars introduces a capacity of the concrete to carry tensile stress. This is classically illustrated by the load deflection behaviour of a reinforced beam (Fig. 2.9). As the beam is loaded, it moves from a region where both steel and concrete are intact to one where there is some tensile cracking of the concrete on the reinforcement side of the beam, lowering its bending stiffness. Once the steel yields, there is a further reduction in bending stiffness and finally the beam will collapse as compressive failure of the concrete on the upper surface occurs. The beam should be designed such that yield of the steel occurs before compressive failure of the concrete.

The principle of prestressing in concrete is rather simpler, in that the aim is to keep the concrete in compression under all envisaged loads by the use of steel tendons. The tendons can be prestressed with the concrete being cast directly onto them. Once the concrete has hardened, the prestressing jacks can be removed. Alternatively, the tendons can be run in ducts prior to concrete casting and then stressed after hardening, a combination of wedging and grout injection (Fig. 2.10) being used to transfer the tension to the concrete. The two methods are known as pre- and post-tensioning respectively, the latter being preferred for larger structures.

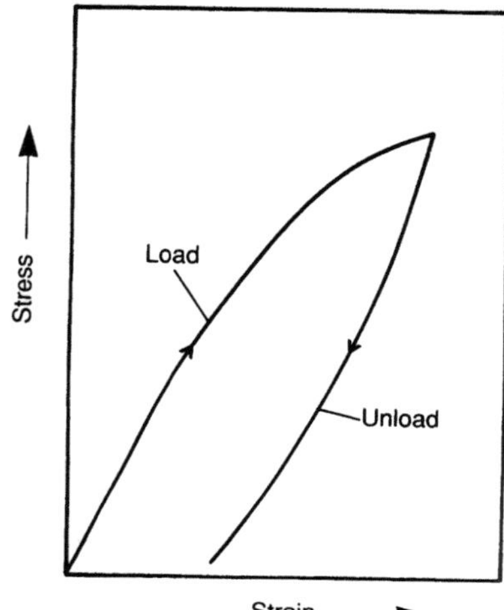

Fig. 2.8. Typical loading and unloading stress–strain curve for concrete. (From Neville [10].)

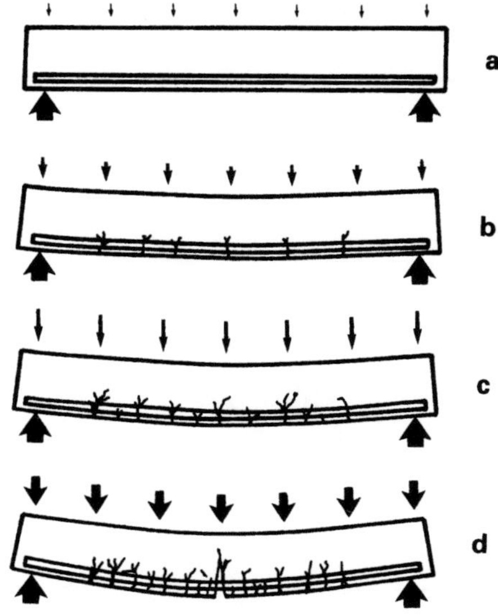

Fig. 2.9. Behaviour of a concrete beam as loading is increased: **a** light loading with steel and concrete intact; **b** increased loading with cracking of concrete on tension side; **c** heavy loading where reinforcing steel yields; **d** severe loading where concrete fails on compression side. (From Weidmann et al. [11].)

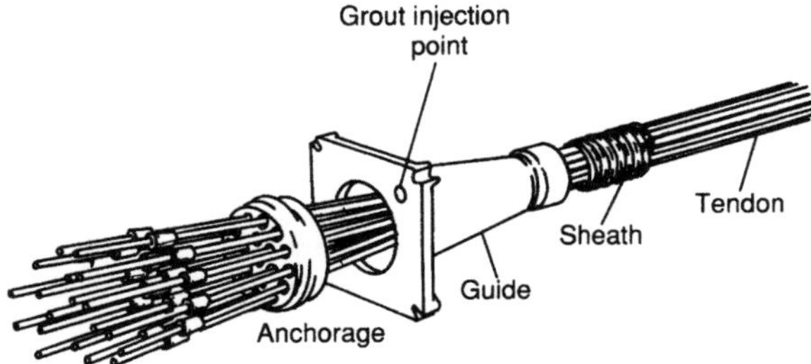

Fig. 2.10. Schematic view of the end of a post-tensioning tendon. (From Weidmann et al. [11].)

2.1.4 The Deformation of Timber

Timber is one of the few natural materials which still retains some engineering application. The reason for the decline in use of natural materials is that their properties contain a degree of unpredictability which is not normally considered to be acceptable in modern design. Despite this, the cheapness and versatility of timber has retained its use and it will be the reluctance to use bio-resources which will result in its decline rather than any engineering consideration.

Like concrete, timber is rather complex structurally, and is normally regarded as an organic composite material. Again as with concrete, the movement of moisture within the structure can have a profound effect on properties. Normal types of mechanical testing applied to timber are bending, compression, shear parallel to the grain, and cleavage [12].

The composite analogy is apt because timber is highly anisotropic in its structure (like some fibre-reinforced composites) and this leads to an anisotropy in deformation properties. This, along with natural variations, makes testing a more statistical exercise than in, say metals. Sample size has also been found to have an effect on measured deformation properties of otherwise identical samples of material. A selection of typical data for compressive strength of commercial timbers is shown in Fig. 2.11.

2.1.5 The Deformation of Ceramics and Glasses

Most ceramics and glasses are compounds of metallic and non-metallic elements and are predominantly ionically or covalently bonded. This results in relatively high moduli of elasticity, but also a marked reluctance to undergo plastic deformation. This means that the materials are brittle, and tensile tests are not usually helpful in estimating mechanical properties. Modulus of rupture, as described in Section 2.1.3, is the most commonly used mechanical test for ceramics, although compression is also used if the application renders this appropriate. Figure 2.12 shows typical stress–strain behaviour for two ceramic materials, and it is clear that the simplicity of these curves means that properties can be suitably summarised with the modulus of elasticity and the modulus of rupture.

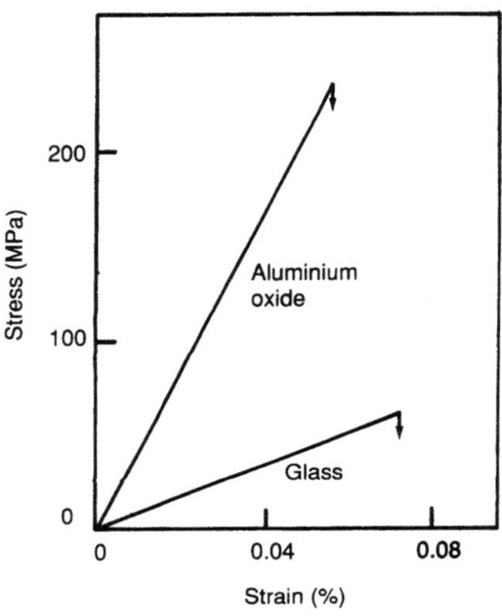

Fig. 2.12. Typical stress–strain curves for oxides. (From Callister [13].)

2.1.6 The Deformation of Composite Materials

Composites represent a major advance in the science of materials, because there is far more scope for adjusting structure and hence properties than in traditional materials. The technology of composites is now such that properties can be engineered precisely and locally within components and structures and this ties the design, manufacture and materials selection function more than ever before. This is not to say that bulk composites are not made and used in engineering structures, but rather that the approach of having a single set of bulk mechanical properties is only one of the options. For this reason, this section looks not only at ways of testing the properties of made-up composite materials but also at ways of predicting the properties of composites from those of their component parts.

Composites are classified in terms of the matrix material, the dispersant normally being referred to as reinforcement. The morphology of the dispersant is also of some fundamental importance, giving rise to such diverse composites as plywood (a laminate), filled polymers (with particulate

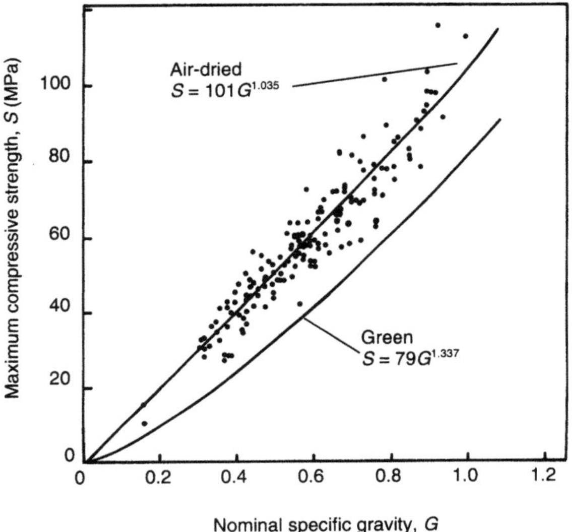

Fig. 2.11. The relationship between maximum compressive strength and specific gravity in air-dried and green timber of all species. (Data points for green timber omitted for clarity [12].)

dispersant) and glass-fibre-mat-reinforced resins. The major composites of importance in modern engineering design are based on resin, metal and ceramic matrices. By far the most widely investigated of these are the resin matrix composites, but there is also considerable interest in the potential of metal matrix composites for marine applications.

Reinforcement materials are normally chosen from those of high modulus or strength but, as is often the case with such materials, they may be brittle. The major reinforcement forms are fibre and granule, although the former can range from woven glass matting to silicon carbide whiskers and the latter from powdered clay fillers to boron nitride particulates. Figure 2.13 shows the range of sizes, shapes and forms in which reinforcements can be applied to matrix materials. In order to understand, and more importantly quantify, the effect of reinforcing a matrix, it is necessary to know the mechanical properties of these reinforcements. It is quite feasible to test fibre tows (untwisted bundles of continuous fibres), braids and weaves, but it is not generally possible to load particulate reinforcement in a way which can generate any useful data. Furthermore, it is not normal to test braids and weaves, because the presence of a matrix would introduce constraints to movement of the individual fibres, so that the fibre and tow data are much more useful. Because of the nature of the materials involved, most fibre tests produce linear stress–strain curves up to fracture but (again because of the type of materials involved) the data usually show a wide degree of scatter, and a large number of tests are usually required to obtain a meaningful result, as can be seen from Fig. 2.14. Fortunately, the number of individual fibres involved in the made-up composite means that the inverse effect on material reliability is not a problem.

The testing of matrix materials is, by contrast, rather simpler than that of reinforcements. The reason for this is that matrices based on metals, polymers, and ceramics or glasses are tractable using the data for the unreinforced materials. One possible exception to this is the carbon–graphite matrix used for carbon–carbon composites. However, at present these materials are of little importance in the marine environment, being confined largely to high-temperature applications.

If the analysis of composite properties is sufficiently accurate, it should not be necessary to carry out mechanical tests on final made-up composites, especially when the properties are to vary throughout the component. Nevertheless, for quality control or for bulk materials with constant properties, mechanical testing may be applied. The types of tests which are used are largely those which might be expected from the foregoing discussion on other types of materials, but cognisance must be taken of the fact that composites are neither homogeneous nor isotropic.

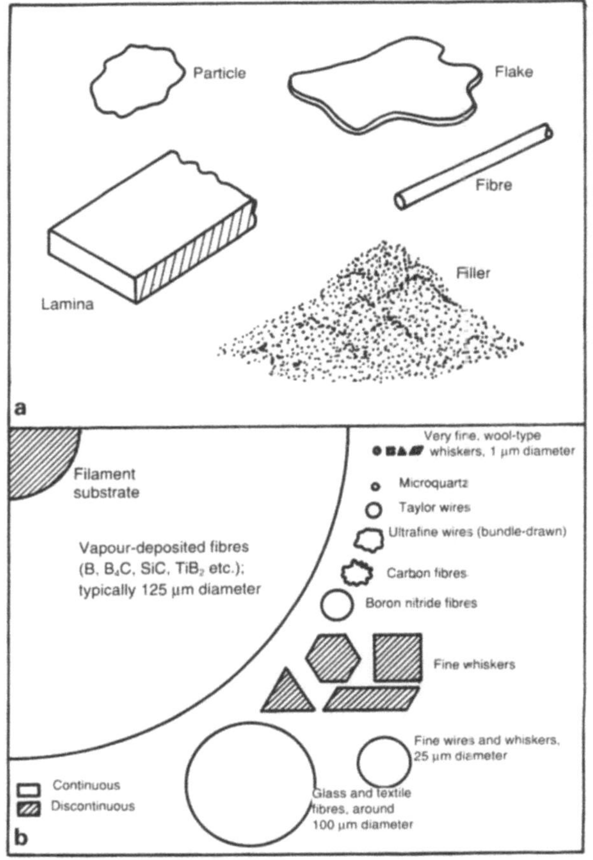

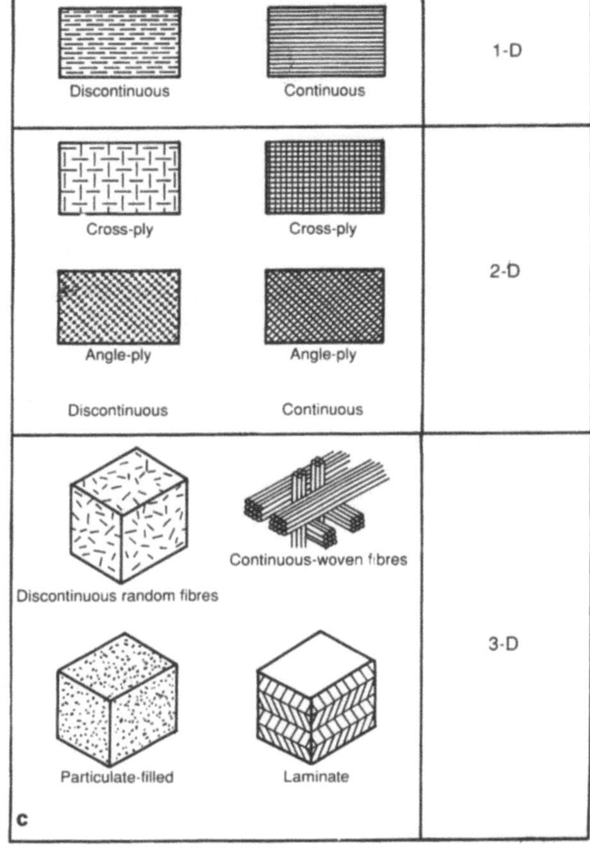

Fig. 2.13. Summary of composite morphology. **a** Reinforcement shapes. **b** Fibre types and sizes. **c** One-, two- and three-dimensional arrangements. (From Weeton et al. [14].)

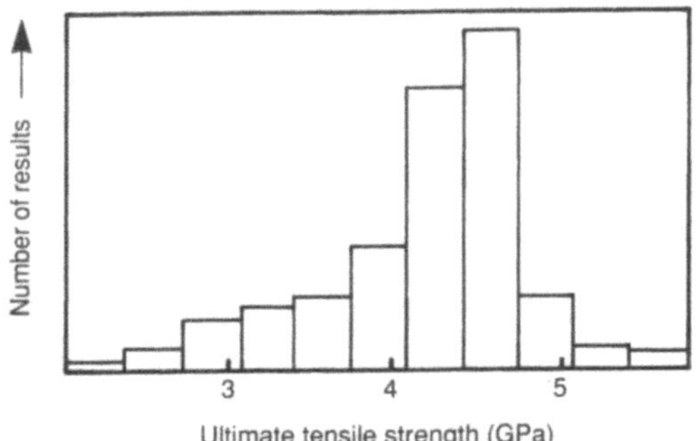

Fig. 2.14. Histogram of tensile strengths of boron fibres. (From Schoenberg [15].)

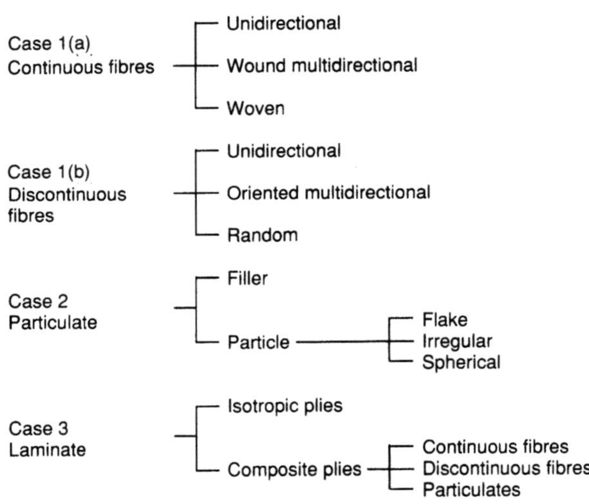

Fig. 2.16. Summary of the micromechanical categories of composites.

Furthermore, most fibres and quite a number of matrix materials are brittle and this can give rise to variability in measured mechanical properties, especially if hand lay-up is used. For this reason, a statistical approach is often applied to composites testing. Tension, compression, flexural and shear testing are all used for composites, and their interpretation is broadly similar to that used for the homogeneous materials already described. However, for laminates, it must be remembered that appropriate beam theory has to be applied for tests involving bending (see Fig. 2.15).

One of the great values of composite materials is the fact that mechanical properties can be altered by varying the size, amount, orientation and distribution of fibres within a matrix. In order to exploit this versatility, it is important to have an idea of how the properties change with these variables. Over the past years there have arisen a number of simplified approaches to composites development which allow the prediction of composite properties from the properties of the component materials. It would seem highly likely that the addition of a second material to a matrix produces a composite with mechanical properties between those of the two components and, whereas this is perhaps not always true for properties such as toughness, the deformation properties of the composite can often be described simply in terms of those of the components. Such an approach consists of predicting a set of 'effective' homogeneous properties for the composite.

From the point of view of their micromechanical description, most composites can be regarded as belonging to one of the four categories shown in Fig. 2.16, defined by the morphology and distribution of the reinforcement, namely discontinuous fibres, continuous fibres, laminates and par-

ticulates. The discontinuous fibres may be oriented in a particular direction, or can be randomly oriented in two or three dimensions.

To build up a three-dimensional structure a number of plies may be used in opposing directions to produce the final desired properties, in which case the composite material is called a 'laminate'. A shorthand designation system has evolved to describe make-up of laminates to avoid the tedious description of repeating groups of plies and pairs of plies. Numerical subscripts are used for repeating plies and the subscript 's' to denote repetition of a pattern in reverse order (symmetric laminate) with a multiplier if appropriate. Two examples should suffice to clarify this notation:

$$[0°/0°/45°/-45°/90°/90°/-45°/45°/0°/0°]$$
becomes $[0°_2/\pm 45°/90°]_s$

and

$$[45°/-45°/0°/0°/45°/-45°/0°/0°$$
$$/0°/0°/-45°/45°/0°/0°/-45°/45°]$$
becomes $[\pm 45°/0°_2]_{2s}$.

The prediction of mechanical properties within the ply (or indeed in the bulk material if lamination is not used) is simplest for cases 1(b) and 2 in Fig. 2.16 and, indeed, for case 2 has been applied for many years to two-phase metallic materials (e.g. Gurland [17]). These, and simple models for fibre reinforcement (e.g. McDaniels et al. [18]), are based on the original 'law of mixtures' proposed by Jech et al. [19].

For the case of unidirectional fibres, the elastic modulus in tension can be very simply written [20]:

$$E_c = E_f V_f + E_m V_m$$

where the subscripts c, f and m refer to the composite, the fibres and the matrix respectively. A similar relationship can be written for Poisson's ratio for the composite in terms of those for the matrix and the reinforcement [21]. For polymeric matrices, it is often sufficient to use the first term of the above equation, which reflects the relatively small contribution made by the matrix to the elastic modulus.

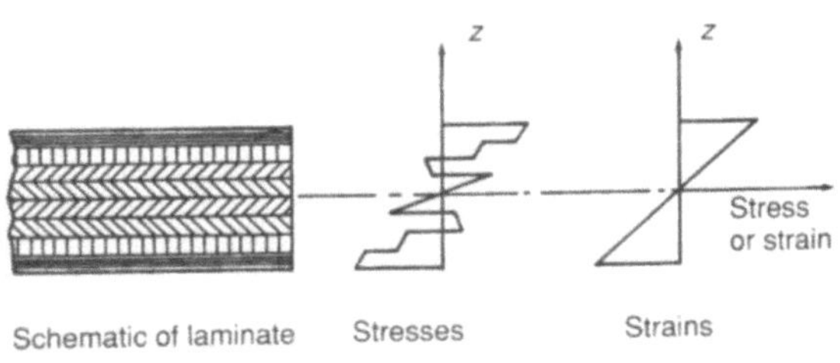

Fig. 2.15. Stresses and strains in a laminate under flexural loading. (From Benham and Crawford [16].)

If the matrix has deformed to the extent that it is behaving in a plastic manner, the tensile modulus can be written

$$E_c = E_f V_f + (\sigma_m^* /e)\, V_m$$

where σ_m^*/e is the slope of the stress–strain curve for the matrix at the strain in question.

The failure strength of an aligned fibre composite depends upon the relative properties of fibres and matrix and also upon the continuity of the fibres (e.g. McCrum et al. [22]).

In the case of continuous fibres, two approaches can be identified, depending upon which is the more brittle, the fibre or the matrix. Where the fibre is more brittle, the failure stress is given by

$$\sigma^* = (1 - V_f)\, \sigma_m^*$$

at lower volume fractions of fibres and by

$$\sigma^* = V_f \sigma_f^* + (1 - V_f)\sigma_m'$$

at higher volume fractions where the meanings of the parameters in the above two equations are illustrated in Fig. 2.17.

As can be seen from Fig. 2.17, the tensile strength of a series of composites using brittle fibres in a ductile matrix goes through a minimum at a critical volume fraction [23] below which the composite is weaker than the unreinforced matrix. This value varies with composite and matrix properties and can be calculated from Fig. 2.17 by considering the volume fraction of reinforcement at which the composite strength reaches the strength of the unreinforced matrix:

$$V_{f,\,crit} = (\sigma_m^* - \sigma_m')/(\sigma_f^* - \sigma_m')$$

although it should be noted that this is usually less than 0.1 for practical composite materials.

If the matrix is more brittle than the reinforcement, then the failure stress can be written

$$\sigma^* = V_f \sigma_f' + (1 - V_f)\, \sigma_m^*$$

at low-volume fractions, and as

$$\sigma^* = V_f \sigma_f^*$$

at higher-volume fractions. The parameters involved in the above equations, component stress–strain curves and the composite series strength are shown in Fig. 2.18.

With discontinuous fibres, the situation is a little more complex, since the stress varies along the fibres, from zero at the ends to a maximum at the middle, owing to the method of load transfer from the matrix to the fibres (Fig. 2.19). A parameter, δ, can be introduced to correct for the shortness of the fibres [22]:

$$\delta = 1 - [\tanh\,(na)\,/na]$$

where $n = \sqrt{[2G_m/(E_f \ln\,(2R/d)]}$ and $a = l/d$, the fibre aspect ratio,

and l and d are the fibre length and diameter, $2R$ is the distance from the fibre to its nearest neighbour and G_m is the shear modulus of the matrix material.

Hence, the parameter δ depends upon the tensile modulus of the fibre and the shear modulus of the matrix. As can be seen from the above, as na becomes large, δ tends towards unity and the expression for the tensile modulus of the composite

$$E_c = \delta V_f E_f + (1 - V_f)\, E_m$$

reduces to that for continuous fibres.

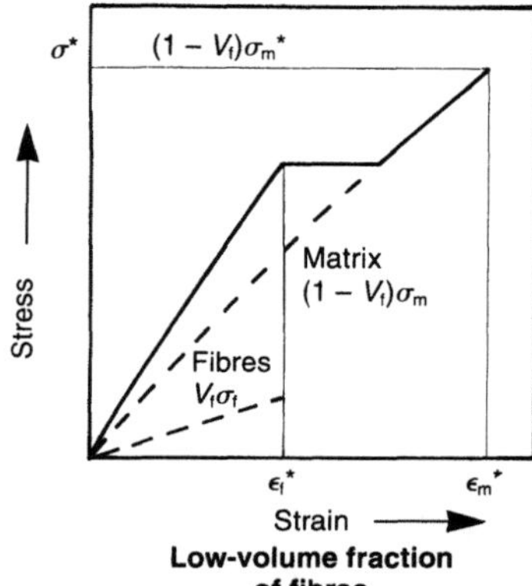

Low-volume fraction of fibres

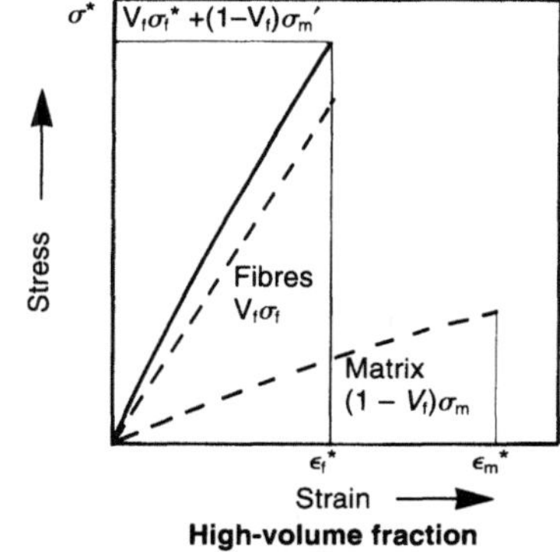

High-volume fraction of fibres

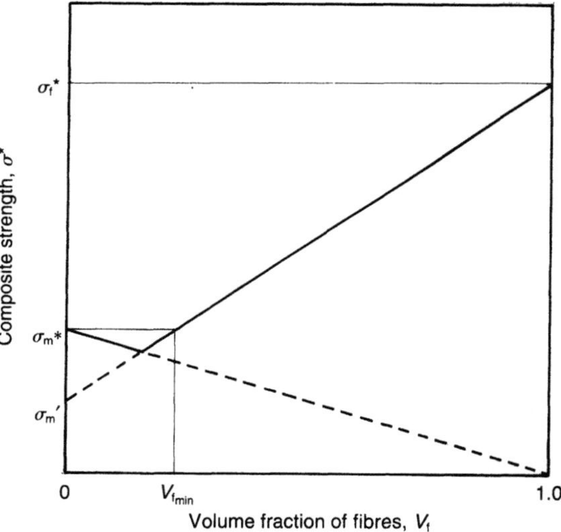

Fig. 2.17. Tensile strengths of a series of composites employing continuous, aligned fibres where the fibres are more brittle than the matrix. (Adapted from McCrum et al. [22].)

The tensile strength of the short-fibre composite in which the fibres are more brittle then the matrix can be given by

$$\sigma^* = (l_{crit}/2l)V_f \sigma_f^* + (1 - V_f)\, \sigma_m^*$$

at low-volume fractions, and

$$\sigma^* = (1 - l_{crit}/2l)\, V_f \sigma_f^* + (1 - V_f)\, \sigma_m'$$

for larger-volume fractions, where l_{crit} is a critical fibre length below which the fibres can pull out of the matrix by shear without reaching their fracture strength. Clearly, the use of fibres shorter than the critical length would be inefficient, and it is helpful to know that this value can be calculated from

$$l_{crit} = \sigma_f{}^* / 2T_i d$$

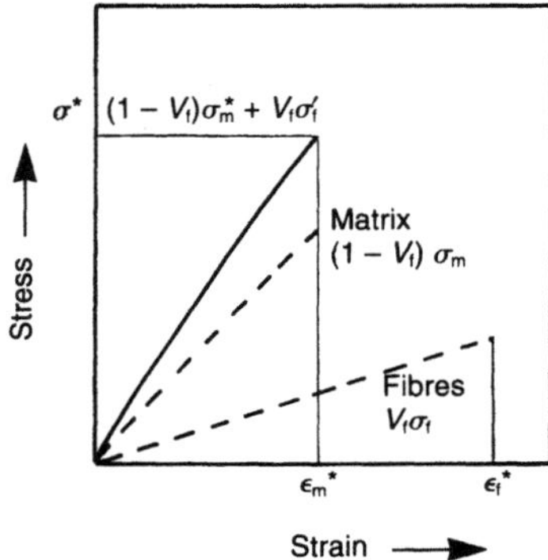

Low-volume fraction of fibres

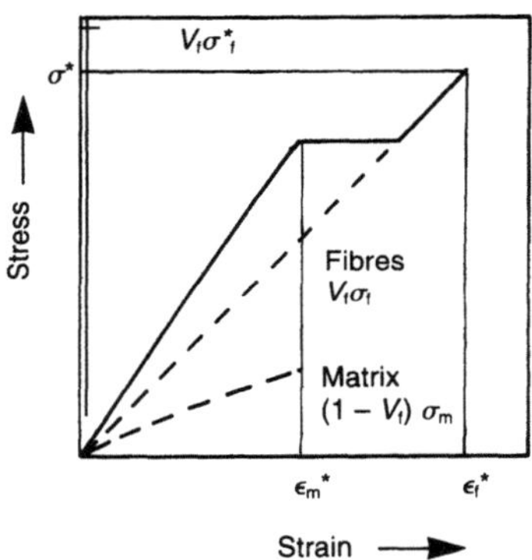

High-volume fraction of fibres

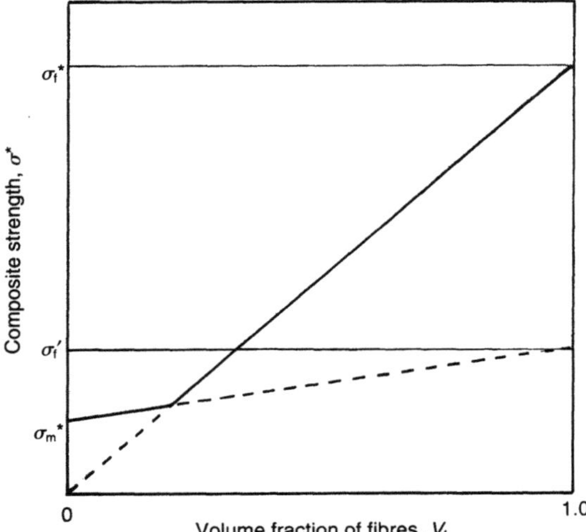

Fig. 2.18. Tensile strengths of a series of composites employing continuous, aligned fibres where the matrix is more brittle than the fibres. (Adapted from McCrum et al. [22].)

where T_i is the interfacial shear strength of the bond between fibre and matrix (or the matrix shear strength if this is less). The strength of the composite series can be seen in Fig. 2.19, and comparison with Fig. 2.17 shows that shorter fibres result in reduced strength for a given amount of reinforcement and that the matrix dominates the failure strength over a wider range of fibre volume fraction. Because of this, the case of short fibre reinforcement for brittle matrix materials (relative to the reinforcement) is not a common one for structural use and is not pursued here.

The above equations for unidirectional fibre reinforcement along the line of the applied stress are summarized in Table 2.2.

When fibres are disposed at an angle to the tensile axis the strength of the composite is reduced due to the fact that failure shifts from fibre tension to interfacial or matrix failure, and it is normal to produce composites with a series of plies which show different fibre orientations, thus reproducing (at least partially) the isotropic macroscopic behaviour of other solids. There are two ways to proceed from this point, depending upon whether the material is a laminate or whether the fibres are arranged in a three-dimensional weave.

It is not within the scope of this text to become involved in the details of composites analysis, and the reader is referred to reviews by Chamis [24] and Rosen [25]. A simplified set of equations can, however, be derived for laminates based on linear elastic behaviour of plies and assuming complete interfacial bonding with limitations as to failure modes in each direction and mode of resistance to deformation within each ply [24]. Uniaxial strengths in the plane of fibres and transverse to the fibres along with the various elastic moduli can be derived, and a summary of these is given in Table 2.3.

Table 2.2. Summary of strength relationships in unidirectional composites

Continuous fibres where fibres are more brittle than matrix	$\sigma^* = (1 - V_f)\sigma_m^*$ **(low V_f)** $\sigma^* = V_f\sigma_f^* + (1 - V_f)\sigma_m'$ **(higher V_f)**
Continuous fibres where matrix is more brittle than fibres	$\sigma^* = V_f\sigma_f' + (1 - V_f)\sigma_m^*$ **(low V_f)** $\sigma^* = V_f\sigma_f^*$ **(higher V_f)**
Discontinuous fibres where fibres are more brittle than matrix	$\sigma^* = (l_{crit}/2l)\,V_f\sigma_f^* + (1 - V_f)\sigma_m^*$ **(low V_f)** $\sigma^* = (1 - l_{crit}/2l)\,V_f\sigma_f^* + (1 - V_f)\,\sigma_m'$ **(higher V_f)**

σ^*	= failure stress of composite
σ_f^*	= failure stress of fibre
σ_m^*	= failure stress of matrix
σ_f'	= stress in fibre at a strain corresponding to the failure stress of the more brittle matrix
σ_m'	= stress in matrix at a strain corresponding to the failure stress of the more brittle fibre
V_f	= volume fraction of fibres
l	= fibre length (discontinuous fibres)
l_{crit}	= critical fibre length = $\sigma_f^*/2T_i d$
d	= fibre diameter
T_i	= fibre–matrix interfacial shear strength

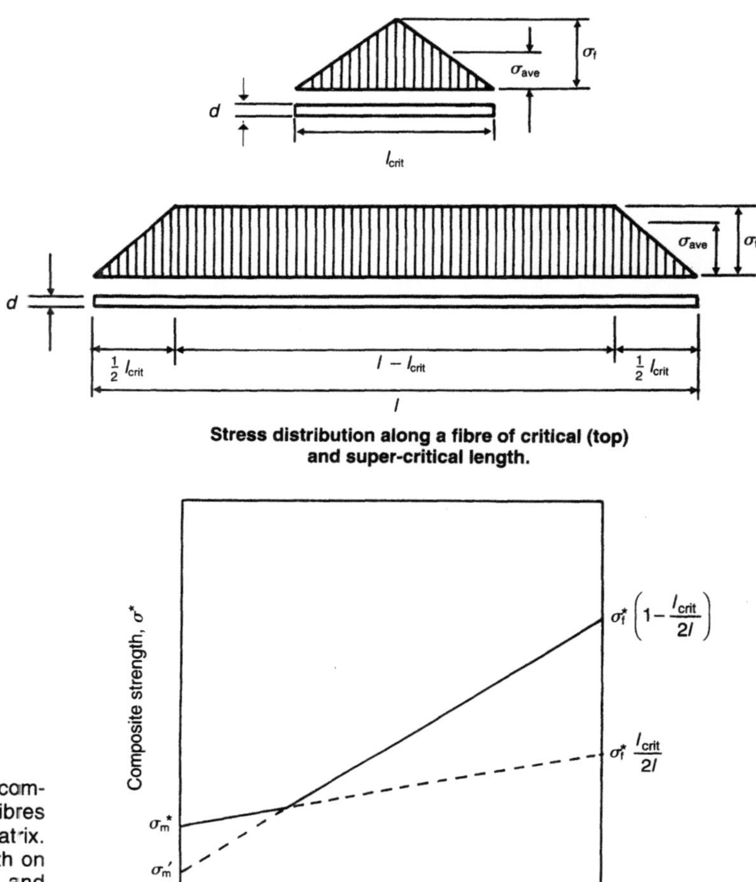

Stress distribution along a fibre of critical (top) and super-critical length.

Fig. 2.19. Tensile strengths of a series of composites employing discontinuous, aligned fibres where the fibres are more brittle than the matrix. Top diagrams show the effect of fibre length on average fibre stress for fibres of critical and super-critical length. (Adapted from McCrum et al. [22] and Weeton [20].)

2.1.7 The Use of Deformation Property Data in Design

The data presented in literature and standards for the deformation properties of materials are usually related to very simple loading conditions (most commonly uniaxial tension). This section is concerned with illustrating how such data can be used in design, particularly in cases where the stress state is not simple. It should be appreciated that such analyses are only suitable for static loading and for relatively ductile materials. The situation for other cases is covered in section 2.2.7.

For metals, it is normal to use one of the yield criteria in which yielding can be identified as having occurred once a certain combination of the principal stresses σ_1, σ_2 and σ_3, reaches the yield stress. The Tresca and Von Mises criteria are commonly used:

Tresca criterion:

$$\sigma_y = (\sigma_1 - \sigma_3)$$

Von Mises criterion:

$$2\sigma_y^2 = (\sigma_1 - \sigma_2)^2 + (\sigma_2 - \sigma_3)^2 + (\sigma_3 - \sigma_1)^2$$

These can be represented in principal stress space as the well-known hexagonal or elliptical yield envelopes (or loci in two dimensions).

The calculation of principal stresses in any but the simplest of geometries usually requires the use of computer-aided methods such as finite-element analysis. Such methods are an integral part of stress analysis, and most design offices have packages capable of performing these for at least some of the cases where analytic methods are unwieldy.

Of course, even when the principal stresses are known, most design codes use the concept of allowable stresses which specify acceptable factors of safety applied to the yield stress. A simple example is given by pressurised pipelines where the yield condition is, of course, dominated by the hoop stress. The ANSI/ASME Code [26], for example, calls for a number of factors to be applied to the specified minimum yield stress of the pipe material for gas transmission applications to account for such effects as wall thickness tolerance, temperature and longitudinal joint efficiency.

When polymers are under consideration, it is usual to have to take account of the possibility of creep, as well as any other strength limitations. For example, McCrum et al. [22] indicate that a modified form of the von Mises criterion can be used:

$$6k^2 = (\sigma_1 - \sigma_2)^2 + (\sigma_2 - \sigma_3)^2 + (\sigma_3 - \sigma_1)^2$$

where the constant (for metals), k, has been employed to avoid its identification with yield stress. In polymers, k will

Table 2.3. Selection of composite micromechanics equations

Longitudinal tension

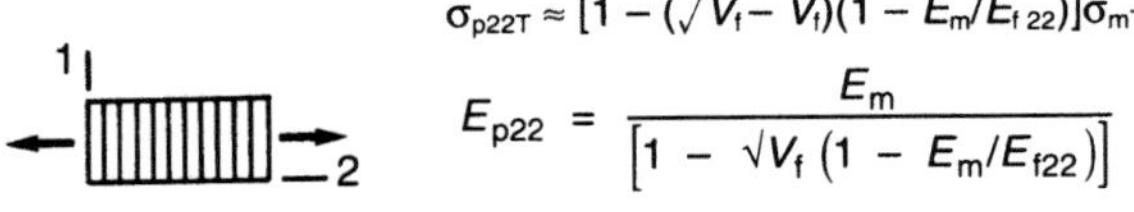

$$\sigma_{p11T} \approx V_f \sigma_f$$

$$E_{p11} = V_f E_{f11} + (1 - V_f) E_m$$

Transverse tension

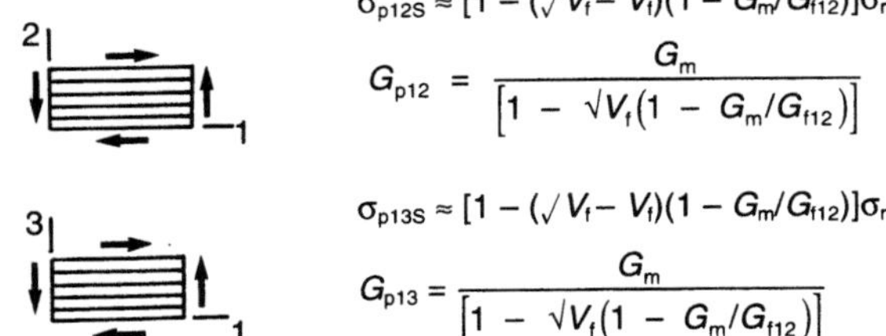

$$\sigma_{p22T} \approx [1 - (\sqrt{V_f} - V_f)(1 - E_m/E_{f22})]\sigma_{mT}$$

$$E_{p22} = \frac{E_m}{\left[1 - \sqrt{V_f}\left(1 - E_m/E_{f22}\right)\right]}$$

Shear

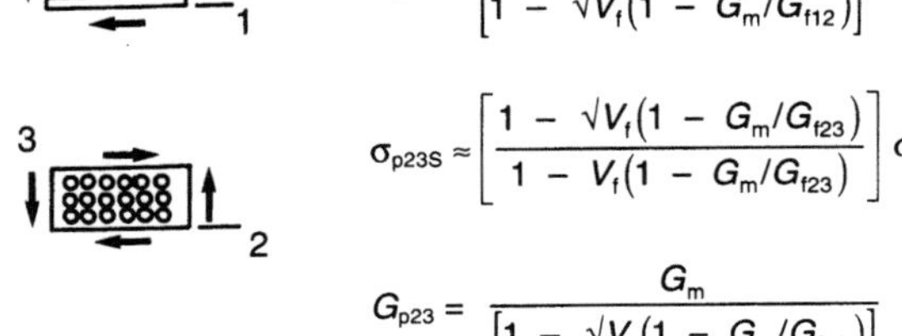

$$\sigma_{p12S} \approx [1 - (\sqrt{V_f} - V_f)(1 - G_m/G_{f12})]\sigma_{mS}$$

$$G_{p12} = \frac{G_m}{\left[1 - \sqrt{V_f}(1 - G_m/G_{f12})\right]}$$

$$\sigma_{p13S} \approx [1 - (\sqrt{V_f} - V_f)(1 - G_m/G_{f12})]\sigma_{mS}$$

$$G_{p13} = \frac{G_m}{\left[1 - \sqrt{V_f}(1 - G_m/G_{f12})\right]}$$

$$\sigma_{p23S} \approx \left[\frac{1 - \sqrt{V_f}(1 - G_m/G_{f23})}{1 - V_f(1 - G_m/G_{f23})}\right]\sigma_{mS}$$

$$G_{p23} = \frac{G_m}{\left[1 - \sqrt{V_f}(1 - G_m/G_{f12})\right]}$$

Source: After Chamis [24].

Subscript notation

Direction 1
Direction 2
Direction 3

Matrix (m)
Ply (p)
Fibre (f)

G = shear modulus
E = tensile modulus
σ_p = ply strength
V_f = volume fraction of fibres (zero voidage assumed)

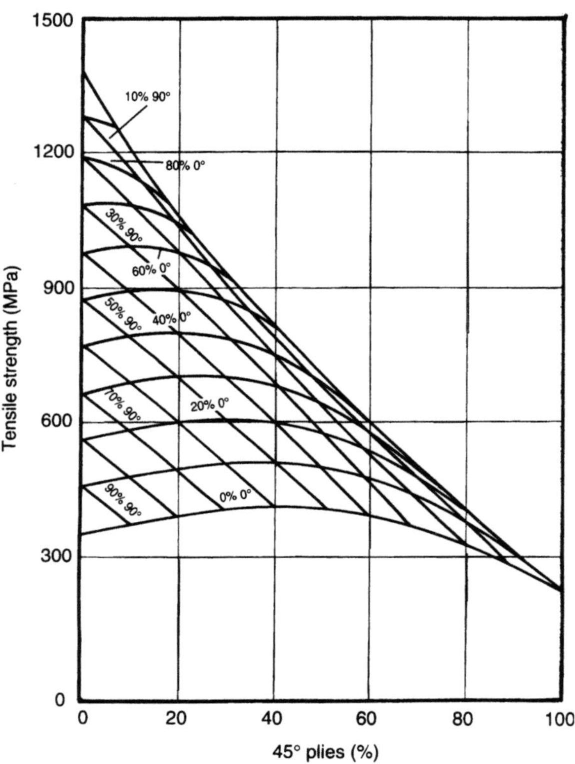

Fig. 2.20. Typical carpet plot for the strength of laminar composites containing plies at 0°, 45° and 90°. (From Tohlen [27].)

vary with hydrostatic pressure and also with temperature and strain rate.

In composite design the yield criterion must be modified to account for the anisotropy of the material. For laminates, the Tsai–Hill criterion, expressed as [16]

$$(\sigma_{pa}/T_{pa})^2 - (\sigma_{pa}\sigma_{pe}/T_{pa}^2) + (\sigma_{pe}/T_{pe})^2 + (t/S)^2 = 1$$

can be used where the direct stresses and strengths (σ and T respectively) are expressed parallel (pa) and perpendicular (pe) to the fibres. S and t are the shear strength and shear stress respectively across the matrix fibre interfaces.

For specific angles of reinforcement of the laminae, carpet plots (an example of which is shown in Fig. 2.20) can be used to simplify the design process and choose suitable levels of reinforcement.

Failure criteria for concrete are difficult to formulate, most probably because fracture processes are involved and a stress analysis approach such as the von Mises criterion is not applicable to this type of failure except in a very conservative sense. Nevertheless, Berg [28] has formulated a criterion to predict failure of concrete in terms of three parameters derived from uniaxial compression and uniaxial tension data. These parameters are σ_{cr} and σ_{pr}, the stresses at the onset of crack propagation and the maximum nominal stress in a prism compression test, and σ_{cl}, the cleavage strength of the concrete in the direction normal to the applied compressive stress. Neville [10] has pointed out that this last parameter is approximately the same as the uniaxial tensile stress. The criterion can be stated as

$$\sigma_1/\sigma_{pr} = \sqrt{(1 + 2Kn_3 + K^2n_3^2 - n_2^2/c^2 - n_3^2/c^2)}$$

where $\quad n_2 = \sigma_2/\sigma_{cl}; \quad n_3 = \sigma_3/\sigma_{cl}$

and $\quad c = \sigma_{pr}/\sigma_{cl}; \quad K = (\sigma_{pr} - \sigma_{cr})/\sigma_{pr}.$

2.2 Fracture and Fatigue

In modern design, particularly in the marine environment, it is necessary not only to consider the possibility of lack of strength in a structure, but also to consider the possibility of crack propagation by fracture or by fatigue. Such a possibility is enhanced by the presence of defects and, in the case of fatigue, by non-steady loading. The genesis and control of defects is a separate subject which relates to manufacture and inspection, and a treatment of this aspect is postponed until Chapters 5 and 6.

Before making any use of fracture (and to some extent fatigue) data in design, it is necessary to have an appreciation of the science of fracture mechanics. Furthermore, some of the consequences of fracture mechanics are of importance in corrosion and welding considerations to be dealt with in Chapters 3 and 5 respectively. Fracture mechanics is also very useful in some aspects of structural reliability analysis, and this is dealt with briefly in Chapter 6.

The basis of fracture mechanics centres around the behaviour of cracks under opening stresses. For this purpose, cracks can be defined as discontinuities of zero radius of curvature, which brings them out of the regime where their influence might be tractable by the methods of stress concentration factors. For example, the problem of a round bar containing a surface notch of root radius of curvature r can be treated by a designer using a stress concentration factor diagram such as that in Fig. 2.21. However, it is clear from these data that when the radius of curvature reaches zero the stress concentration factor becomes infinite, suggesting that the stress at the root of the notch is infinite. This problem is only partly due to the fact that the SCF solutions are obtained from linear elastic analysis, and the difficulty is primarily one of approach. An alternative method is required, based on finding the criterion for crack opening rather than trying to find local values of stress which are only informative below yield.

Fracture mechanics can be approached conveniently from an appreciation of the Griffith criterion first proposed in 1920 [30] to describe the behaviour of glass. This criterion considers the energy balance for the propagation of a crack in a perfectly elastic medium where the crack can be considered to be unstable (i.e. will propagate) if the amount of strain energy which is released in an incremental step of growth exceeds the amount of energy consumed. In the case of a perfectly elastic medium, the only energy which is consumed during propagation is that required to produce the two new surfaces (no permanent, plastic deformation takes place), so that the Griffith criterion can be written

$$\sigma_f = \sqrt{(2E\gamma/\pi a)}$$

where γ is the surface energy of a unit area of free surface of the fracturing solid, E is Young's modulus for the material and a is the dimension of the crack in the propagating direction, which, in the case of the original Griffith formulation, was the depth of a single-edged notch. The terminology for crack length in fracture mechanics normally uses a for surface-breaking defect dimension and $2a$ for embedded defect dimension.

If the crack propagates with some plastic distortion, the Griffith criterion can be modified to account for the total amount of plastic work involved in the production of the two new surfaces of unit area γ_p:

$$\sigma_f = \sqrt{[E(2\gamma + \gamma_p)/\pi a]}$$

which can be expressed in the more general form

$$\sigma_f = \sqrt{K/\pi a}$$

where K is some constant, characteristic of the material.

These criteria can also be expressed in a form which indicates the maximum size of defect which can be tolerated at a given applied stress, σ:

$$a_{crit} = K/\pi\sigma^2$$

Such an approach represents a substantial improvement on the stress concentration factor approach, because it now provides a method of calculating fracture stresses in terms of defect geometry and some property of the material which can, in principle, be obtained from a fracture test on a specimen with known defect geometry.

Despite the disadvantage of the SCF method, fracture mechanics in fact advanced considerably with the analyses of stress distributions around cracks, of which Westergaard's solution gives a notable example. This distribution applies to a crack embedded in a infinite medium in a uniform biaxial stress field (Fig. 2.22), and it can be seen (e.g. Knott [3]) that the solution has direct components in the directions x_1 and x_2 as well as shear components. However, some interesting observations can be made simply from the direct stress, expressed in terms of distance from the crack tip, r:

$$\sigma_{11} = \sigma \sqrt{(a/2r)} \cos(\theta/2) [1 + \sin(\theta/2) \sin(3\theta/2)] + \dots$$

where higher terms in the series are not given for clarity. The type of opening provided by this stress is defined as Mode I, the others being shear opening, Mode II, and anti-plane shear opening, Mode III (Fig. 2.23).

Taken together, the Griffith and Westergaard solutions to the fracture problem make it clear, in general if not in detail, how the phenomenon of the stress intensity factor has arisen. Defined in Mode I opening, the stress intensity factor is given by:

$$K_I = \sigma Y \sqrt{(\pi a)}$$

where Y is a factor accounting for the geometry of the defect and applied loading.

In linear elastic fracture mechanics, design consists simply of ensuring that K_I is below a critical value K_{Ic} at which fracture occurs, this being measured by a standard fracture test [31]. Values of Y can be calculated from the methods of stress analysis, but, as is the case for stress concentration

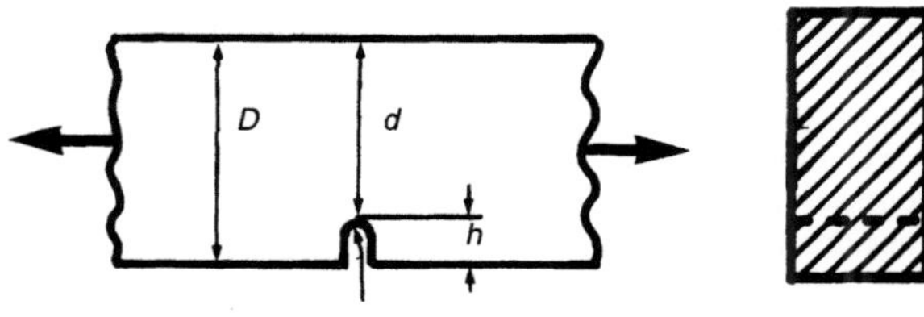

Fig. 2.21. Stress concentration factor for a bar in tension containing a U-shaped notch. (From Young [29].)

Stress concentration factor, K_t:

$$K_t = K_1 + K_2\left(\frac{h}{D}\right) + K_3\left(\frac{h}{D}\right)^2 + K_4\left(\frac{h}{D}\right)^3$$

where $K_1 = 0.721 + 2.394 \sqrt{(h/r)} - 0.127\ h/r$
$K_2 = 1.978 - 11.489 \sqrt{(h/r)} + 2.211\ h/r$
$K_3 = -4.413 + 18.751 \sqrt{(h/r)} - 4.596\ h/r$
$K_4 = 2.714 - 9.655 \sqrt{(h/r)} + 2.512\ h/r$

for $0.5 \le h/r \le 4.0$

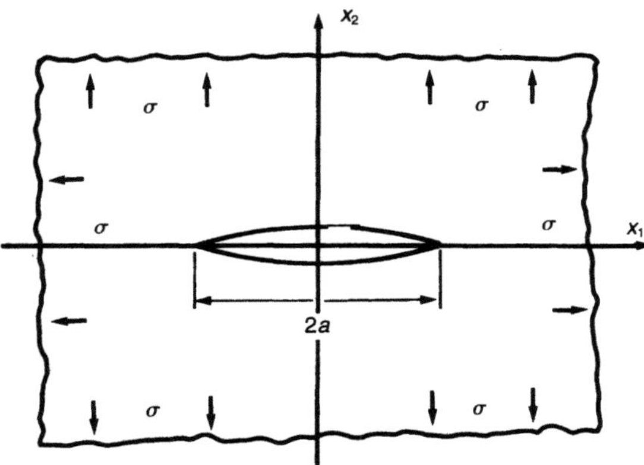

Fig. 2.22. Geometry for Westergaard's solution for the stress distribution around a crack subjected to a uniform biaxial stress field in an infinite sheet. (From Knott [3].)

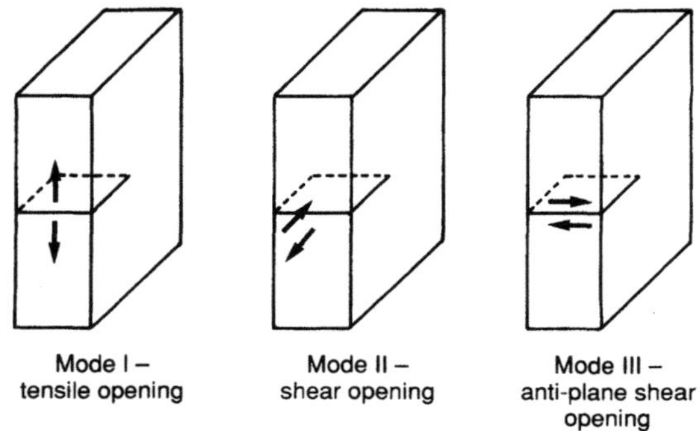

Fig. 2.23. Mode classification for crack opening.

factors, collected values exist for many geometries [32] and many others can be obtained by superposition.

Of course no material is perfectly elastic, and some degree of yielding is always possible at the root of a sharp defect. Methods exist to cope with small amounts of yielding by modifying the linear elastic approach, so that the 'plastic' stress intensity factor K_p can be expressed in terms of the 'elastic' one, K_e (e.g. [33]), as:

$$K_p = K_e \left[1 - \tfrac{1}{2}(\sigma/\sigma_y)^2\right]^{-1/2}$$

At some stage, however, a different approach is required and this transition can be defined in terms of the stress state at the crack tip.

Considering a through crack embedded in an infinite plate, this triaxiality transition can be identified in terms of the dimensionless group of parameters:

$$K_{Ic}^2/\sigma_y^2 B$$

where the dimension B is measured along the crack front as shown in Fig. 2.24 for Mode I opening. The plane strain condition can be regarded as one where the plastic zone ahead of the crack is confined, and this is typified by a triaxial stress state. The plane strain condition is indicated by low values (less than about 1) of the dimensionless parameter indicated above, i.e. by low toughness and high yield stress, and by relatively thick sections of material to provide constraint against yielding in the B direction. As

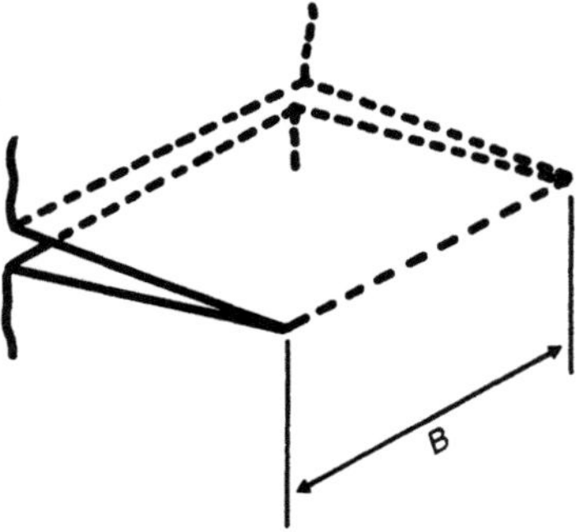

Fig. 2.24. Crack front dimension, B, for triaxiality assessment.

$K_{Ic}^2/\sigma_y^2 B$ becomes larger, a transition (often called the 'triaxiality' or 'plane-stress–plane-strain transition') is gradually traversed until plane stress conditions are reached at values of $K_{Ic}^2/\sigma_y^2 B$ greater than about 3. In the plane stress condition linear elastic fracture mechanics are no longer applicable and the required approach uses the strain-based fracture parameter δ_c, or critical crack opening displacement. This is measured in a similar way to K_{Ic}, except that the degree of crack opening at failure is recorded rather than the load (from which the stress intensity factor is obtained).

The transition from plane stress to plane strain conditions can be illustrated usefully by a failure assessment diagram such as that shown in Fig. 2.25, although it should be noted that this diagram is only applicable to through thickness cracks in infinite plate. Nevertheless, some useful observations can be made, and an example of the transition in structural steel is given in Case Study 7.2.

General yielding fracture mechanics is somewhat more complicated than plane strain, but some useful semi-empirical methods have been devised. An example of this approach is given in Case Study 7.1 for the case of a pipeline girth weld.

Fatigue is a process whereby cracks can propagate under cyclic loading. The mechanism by which this occurs can be understood qualitatively by considering the strain field around a crack tip as the applied stress goes through a single cycle, as shown in Fig. 2.26.

For fatigue cracks to propagate, stress reversals are not necessary, nor is the existence of a pre-crack (although this considerably aids incubation).

It has been found that the most important variable affecting fatigue life is the stress range (the difference between the maximum and the minimum stress) and this has led to the conventional S–N representation of fatigue data where the logarithm of stress range, S, is plotted against the logarithm of the number of cycles to failure at this stress range, N.

Under many circumstances, fatigue data of this type can be represented as a straight line in log–log space:

$$\lg (S) = m \lg (N) + \lg (S_o)$$

However, many other factors are important in determining fatigue life. In particular, surface condition is extremely important both at the 'macroscopic' level (for example the stress concentrations at the welds in tubular joints for off-

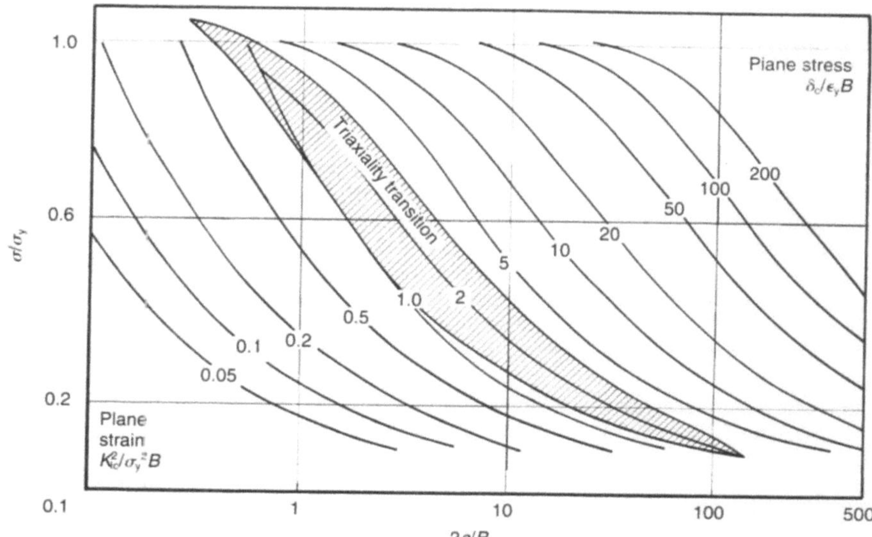

Fig. 2.25. Failure assessment diagram with respect to triaxiality in an infinite plate with a through-thickness crack. (After Gray and Spence [33].)

shore structures) and at the 'microscopic' level (for example small discontinuities due to weld defects). The main effect of surface condition is to reduce incubation time and hence to shift the entire S–N curve towards the S-axis. Other factors affecting fatigue life include stress ratio, corrosion, geometry, and residual stress distribution.

The phenomenon of fatigue can also be examined from the point of view of fracture mechanics using the Paris law, which states that the rate of crack growth per cycle under non-steady loading can be described by a relationship of the form (e.g. Osgood [35]):

$$da/dn = C\,\Delta K^m$$

where K is the range in stress intensity factor resulting from the stress range and any increment of crack growth, and C

and m are material constants which should be independent of test geometry. The advantage of this formulation is that the incorporation of the stress intensity factor should, in principle, account for the effect of geometry. Quite a body of the literature fatigue data (particularly for metals) is now in the form of a–N data, although the S–N curve remains a very useful design tool.

In fact the two formulations are equivalent, and this can be demonstrated by substituting

$$\Delta K = \Delta \sigma Y \sqrt{(\pi a)} = SY\sqrt{(\pi a)}$$

into the Paris equation. Integrating then gives an equation for the number of cycles, N', required to propagate a fatigue crack from a size a_i to a size a_f under a stress range S:

$$N' = \frac{S^{-m}}{C} \int_{a_i}^{a_f} (\pi a)^{-m/2} Y^{-m}\, da$$

If a_i is taken as an initial crack size characteristic of the as-new condition of the structure and a_f as the largest tolerable defect at peak stress in the cycle, then N' is the fatigue life N for a stress range S and a plot of lg S vs. lg N would be linear with an intercept dependent upon the value of the integral.

Clearly this formulation contains a term which is dependent upon geometry as well as the embodiment of the S–N approach, and hence provides a more fundamental analysis for fatigue problems, provided data for the material parameters C and m are available. It is interesting to note that the integrated form of the Paris law indicates that the slope of S–N curves should be dependent only upon the material and not upon geometry, and this is broadly observed in empirical S–N fatigue data. Care should, however, be taken when using the integrated form of the Paris law that a reasonable initial crack size is chosen, because the integral is highly sensitive to the value of this parameter, becoming infinite at crack sizes approaching zero, and for this reason the fracture mechanics approach to fatigue is more often used for the evaluation of known defects rather than for design, where predicted fatigue lives would be critically dependent upon the envisaged initial defect size.

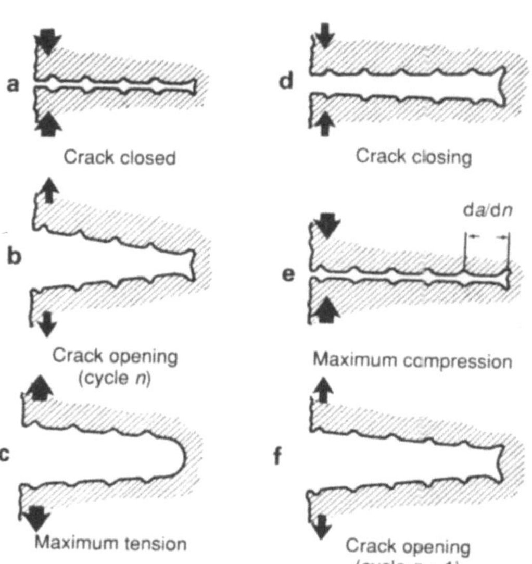

Fig. 2.26. Schematic representation of fatigue crack propagation under cyclic stress. (From Engel and Klingele [34].)

2.2.1 Fracture and Fatigue in Metals

Fatigue and fracture have been studied in metals more extensively than in any other material type. The reason for this is obviously that metals have traditionally been the materials subjected to large static and dynamic loads. In fact, owing to the magnitude and cyclic nature of wave loading, much of the development of the science of fatigue and fracture has been promoted by problems encountered in marine technology.

The difficulty of brittleness in metals was first recognised in ferritic steels, particularly in early welded ships (see Chapters 1 and 5), and a number of methods were developed essentially as quality controllers to give an indication as to whether a steel was brittle or 'ductile', although the latter term is an inexact one which could refer to performance in a tensile test. Perhaps the most enduring of these quality control toughness tests is the Charpy impact test, which utilises a swinging pendulum to break a notched bar of the material under test (Fig. 2.27). The pendulum is given a specific amount of potential energy, and the toughness is measured in terms of the amount of energy absorbed as the pendulum swings past the specimen breakage point.

In mild steels it has long been known that toughness (measured by impact energy) is dependent upon temperature, there being a point (referred to as the 'ductile–brittle transition temperature') below which there is a fairly abrupt change from high to low impact energy, and Fig. 2.28 shows this transition: first in a schematic way and then in a more realistic one as it might be encountered in a quality control situation.

Apart from the quotation of a minimum impact energy at a specific temperature, the Charpy (and related tests) are of no direct quantitative help in designing against fracture. It must be said, however, that many design guides call for such tests as fracture control measures and, provided that the specified energies and temperatures are conservatively chosen, there is no danger in this practice. It must be said that toughness is a rather elusive parameter in ferritic steels,

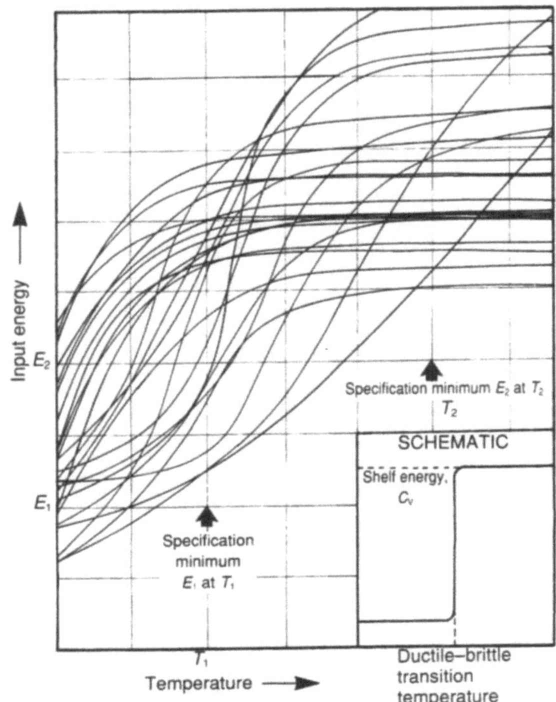

Fig. 2.28. The 'ductile–brittle' transition with temperature in a ferritic steel shown schematically (inset) and for production batches of a nominally similar material (adapted from steel producers' literature).

and it is much more difficult to predict toughness from a knowledge of the metallurgy of a steel than it is, for instance, to predict strength.

The measurement of quantitative fracture parameters (such as K_{Ic} or δ_c) is usually carried out in tension or in

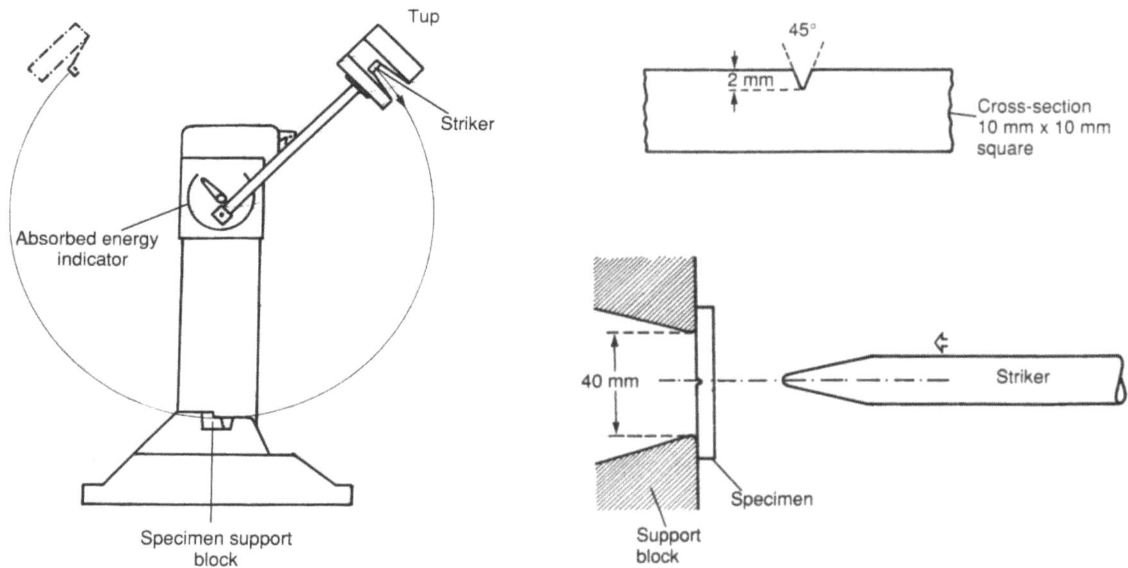

Fig. 2.27. Schematic view of the Charpy impact test showing overall arrangement (left), detail of specimen notch (top right) and striking geometry in plan view (bottom right). (Adapted from Davies [36].)

three-point bending with Mode I crack opening, and two types of standard test pieces are shown in Fig. 2.29.

In order to obtain reproducible measurements of toughness it is necessary to ensure that cracks are as sharp as can be produced in the material, and this is normally done by first manufacturing a stress concentrator in the form of a notch and growing a fatigue crack from the root of this notch. The toughness test then consists of gradually loading the specimen containing the fatigue crack (usually in Mode I opening), as one might do in a tensile test, and measuring the crack opening as a function of load using a transducer such as a 'clip gauge'. The type of load crack-opening curve depends somewhat upon the toughness of the specimen, and can vary from a linear elastic form corresponding to plane strain 'brittle' fracture to an elastic–plastic curve corresponding to plane stress 'ductile' fracture. In metals, the respective microscopic appearances of such fractures are quite different, with brittle fracture corresponding to cleavage across grains on a single set of atomic planes and ductile fracture to a void coalescence type of mechanism associated with a high degree of plastic strain (see, for example, Engel and Klingele [34]).

The interpretation of a plane strain curve is based on linear elastic fracture mechanics and consists of calculating the value of K_I (from a known geometric parameter Y, tabulated in, for example BS 5447 [32]) and using $K_I = \sigma Y \sqrt{(\pi a)}$ at which unstable fracture took place and designating this as K_{Ic}. Under ductile conditions, it is necessary to use the crack opening, δ_c, at which fracture commences as the toughness parameter. Thus the plane strain fracture toughness is a stress-based measurement whereas the plane stress fracture toughness is a strain-based measurement.

Fatigue testing of metals has in the past followed the S–N approach, where specimens were subjected to alternating stresses of a single range and the number of cycles to failure recorded at that stress range. The test is then repeated for a variety of values of S and, because of the statistical nature of fatigue crack initiation, often a number of times at each stress range. Even with very high loading frequencies this is a rather time-consuming task, and fatigue data are therefore very expensive to produce. Furthermore, changes of geometry (and often scale), environment, surface condition, stress ratio and other factors need also to be tested over the range of S-values of interest, and hence complete fatigue data sets are very rare.

For high-frequency loading, such as in machinery applications where large numbers of cycles can be expected in a fairly short time, it is more useful to consider a fatigue limit approach where the design is kept in a regime of stress range and mean stress, under which failure by fatigue will not occur however many cycles are encountered. There are a number of empirical ways of doing this, perhaps one of the commonest being the Goodman diagram. A plot of the fatigue-safe region in stress-range–mean-stress space is shown in Fig. 2.30. Note that one of the effects of corrosion in association with fatigue is to remove the fatigue limit (see Chapter 3), so this approach is of limited use where corrosion fatigue is taking place.

The requirement to carry out fatigue tests over a range of geometry can be overcome by using the fracture mechanics approach but, as pointed out earlier, it is always necessary to bear in mind that this approach does not account for initiation phenomena and hence is not always easily applicable to design. Probabilistic approaches are available which use a distribution of initial defect sizes, and these are discussed further in Chapter 6.

One of the most comprehensive data sets for fatigue is that which has been accumulated for structural steel in marine environments. Because of the relatively low frequencies of wave loading, a typical structure will only encounter around 10^7 cycles at its highest stress range, but loading is of a random nature, involving a mixture of stress ranges and frequencies. One way of coping with this is to use a cumulative damage criterion, the simplest and most widely used of these being Miner's rule, which is stated as follows:

$$D = \Sigma n/N$$

where D is the fatigue damage (a value of unity indicating failure), and n and N are the number of cycles encountered and the number of cycles to failure respectively at a given

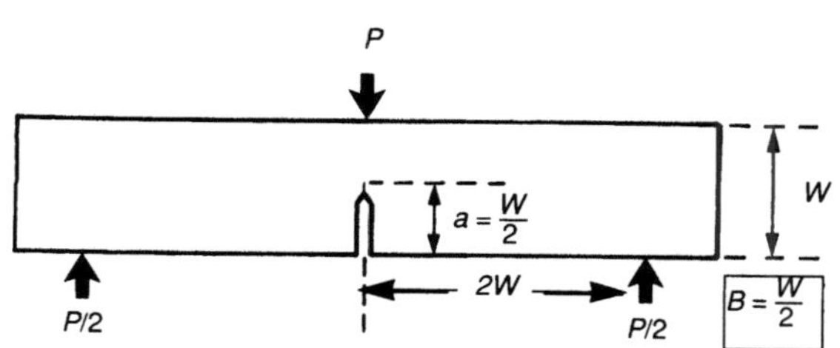

Single-edge notch (three-point bend) test piece

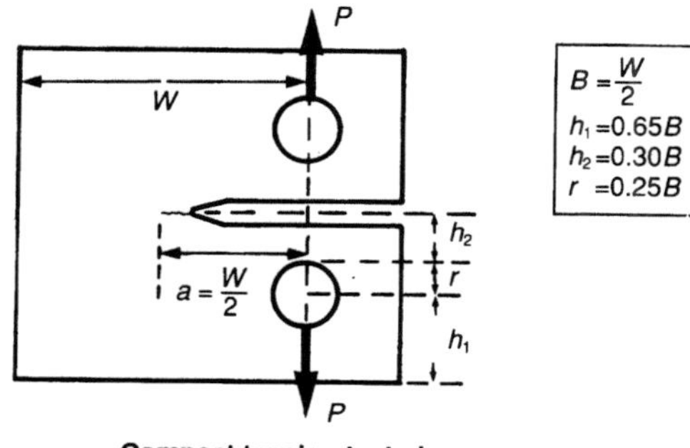

Compact tension test piece

Fig. 2.29. Three-point bend and compact tension fracture mechanics test pieces. (From Knott [3].)

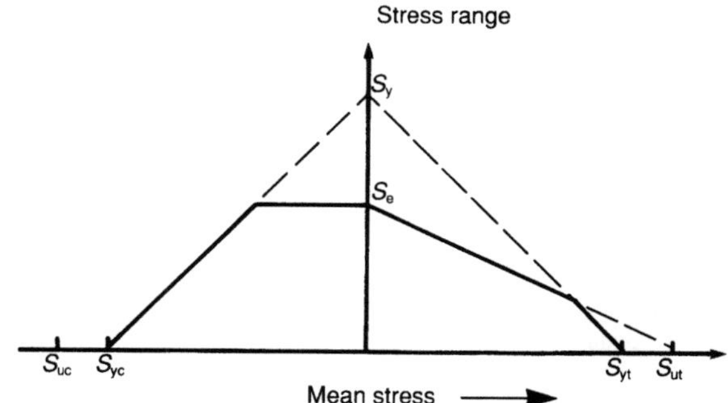

Fig. 2.30. Construction of fatigue diagram for assessing effect of mean stress on fatigue limit. (After Shigley [37].) S_x is the stress range corresponding to: y, yield; yt, yield in tension; yc, yield in compression; e, fatigue limit at zero mean stress; ut, ultimate tension; uc, ultimate compression.

stress range, the summation being over all stress ranges encountered. Obviously, the use of this rule requires that the wave load time history is discretised into blocks of stress range and the n- and N-values for each stress range evaluated before summation. This process is illustrated in Fig. 2.31, and an extension of Miner's rule which allows direct use of the continuous loading history is described in Case Study 7.2.

2.2.2 Fracture and Fatigue in Polymers

The phenomenology of fatigue and fracture in polymers is more complicated than that in metals, and this is due first to the temperature sensitivity (to both high and low temperatures) of polymers and further to their viscoelastic behaviour.

As far as fracture is concerned, the first of these complications can be examined by knowing the glass (and other) transition temperatures of the polymer along with more extensive testing over the temperature range of interest. Most care over toughness is, of course, exercised for thermoplastics in the glassy or crystalline state and for thermosets. Both temperature and strain rate effects can be taken into account quantifying the plastic behaviour in a modulus of toughness, which is defined as the integral of the stress–strain curve up to fracture (e.g. Petrie [38]). This provides a good comparison between materials, and is perhaps more useful than impact testing since it allows a variation in loading rate.

Because of the substantially smaller strength/toughness ratio in plastics, it is quite possible that failure in a plastic containing a defect can be caused simply by the reduction in section rather than by stress intensification and unstable propagation. This can be illustrated by plotting failure stress for an infinite plate with a through thickness crack against crack size, as shown in Fig. 2.32. For 'ductile fracture', the failure stress is proportional to crack size, whereas for brittle fracture the failure stress is proportional to the inverse square root of crack size. Clearly, the material will fail by the mode which has the lower failure stress. Thus, two materials which have the same value of K_{Ic} (e.g. polystyrene and LDPE) can show very different resistances to crack propagation. The reason for this can be seen clearly in Fig. 2.32, where the LDPE ductile failure curve is below the brittle fracture curve for a much wider range of crack sizes.

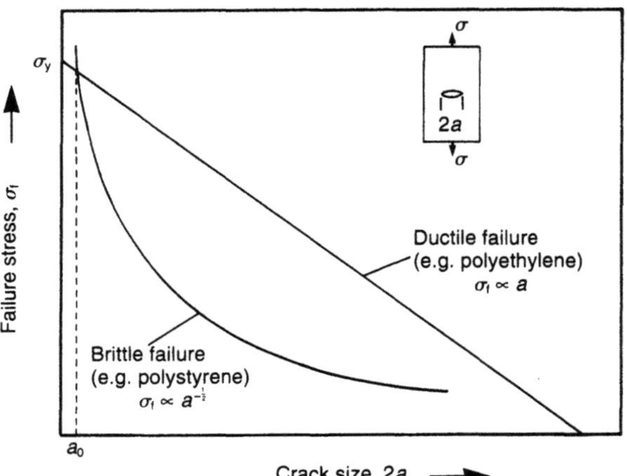

Fig. 2.32. Failure stress for through cracked plastic plates for ductile and brittle materials. (From Crawford [7].)

Plastics are often ranked by the use of impact tests in a similarly qualitative way to impact tests for metals. As with metals, such data are not directly applicable to the calculation of fracture stress. The parameter normally used for this purpose is the strain energy release rate, G_c (identifiable from the Griffith analysis described earlier), given by

$$\sigma_f = \sqrt{(G_c E/\pi a)}$$

and comparison with K_{Ic} shows these two fracture parameters to be related by

$$K_{Ic} = \sqrt{(EG_c)}$$

It is easily seen (e.g. Crawford [7]) that the measured energy U_c in a Charpy-type test is related to G_c by

$$G_c = U_c/BD\phi,$$

where BD is the cross-sectional area of the specimen and ϕ is a geometrical function related to the way in which the compliance, C, (reciprocal of stiffness) of the specimen varies with crack depth:

$$\phi = C(\delta C/\delta a)^{-1}$$

so that Charpy data can be used to give a quantitative measure of impact toughness.

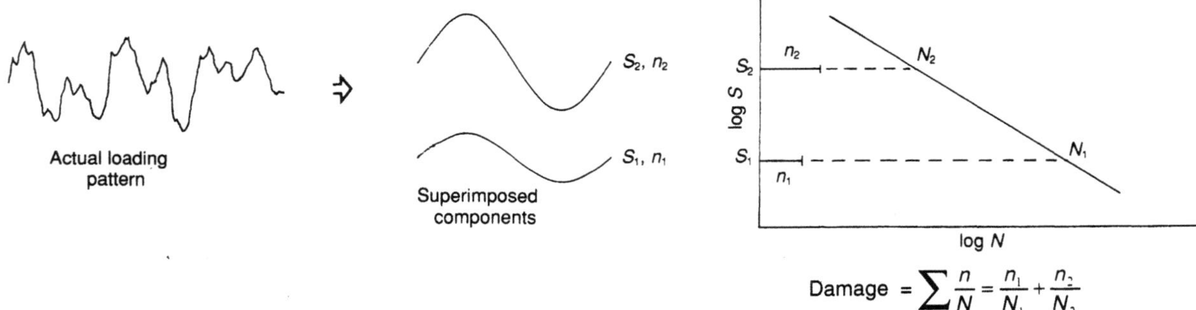

Fig. 2.31. Schematic illustration of the Palmgren–Miner rule for cumulative fatigue damage.

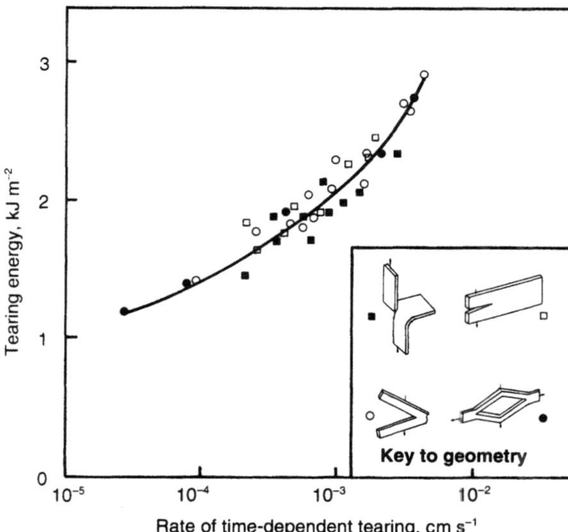

Fig. 2.33. Tearing energies for a variety of geometries of an SBR vulcanisate. (Adapted from Lake [5].)

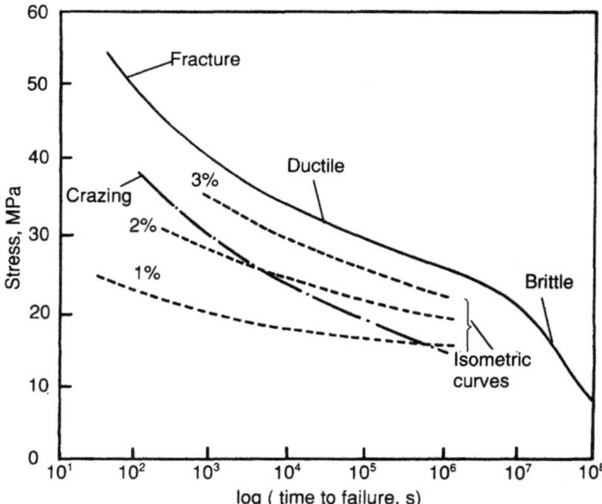

Fig. 2.34. Typical creep rupture curves for plastic material. (From Crawford [7].)

Such an approach has been found very useful for rubber (Lake [5]), where the quantity, G_c, is, for historical reasons, referred to as 'tearing energy', T. T is used to quantify fracture properties for rubber over a range of geometries, as illustrated in Fig. 2.33.

It has also been observed that plastics are very susceptible at room temperature to plane stress–plane strain transition and that this is in turn strongly affected by loading rate. For example, McCrum et al. [22] indicate that G_c for polycarbonate can change from about 10 kJ m^{-2} to about 1.5 kJ m^{-2} when going from thin (plane stress) to thick (plane strain) specimens. Furthermore, they note that increasing strain rate results in an increase in fracture toughness, but that the resulting increase in fracture stress is usually less than the increase in yield stress with strain rate. On applying this observation to Fig. 2.32, it can be seen that increasing strain rate (i.e. impact loading) is more likely to cause fracture in such materials than is relatively slow loading.

Polymers will creep under static loading and, if the loading is maintained for long enough, will fail by creep rupture. A typical creep rupture curve for a plastic is shown in Fig. 2.34, although, in some plastics, whitening or crazing may take place in advance of actual rupture. Fracture mechanics have been used to treat creep in plastics using a law very like the Paris law for fatigue in metals:

$$da/dt = AK^n$$

where A and n are material constants.

This equation can be integrated [7] to find the time to failure from an initial defect size, a_i, to a critical defect size, a_f:

$$t_f = \frac{2\left(a_i^{1-m/2} - a_f^{1-m/2}\right)}{A(Y\sigma)^m \pi^{m/2}(m-2)}$$

A phenomenon, mentioned above, and shown by plastics and not metals, is crazing. This occurs in glassy plastics and takes the form of a series of small proto-cracks which are, however, still held together with some molecular material. Subsequent failure occurs by extension and linking of the proto-cracks. The failure locus for biaxial loading of a material which exhibits crazing (Fig. 2.35) illustrates the importance of a hydrostatic component to the stress state in inhibiting crazing, in that below the pure shear line ($\sigma_1 = -\sigma_2$) crazing does not take place.

Fatigue in polymers is a more complex phenomenon than in metals because of the effect of frequency of loading on the temperature of the plastic and, further, because of the viscoelastic nature of plastic deformation processes.

Phenomenologically, fatigue in plastics is similar to metals, with crack propagation proceeding with each loading cycle, so that a striation- and beach-marked surface is produced, which, as in metals, is immediately recognisable as a fatigue fracture surface. This physical similarity means that fatigue data for plastics can still be presented by means

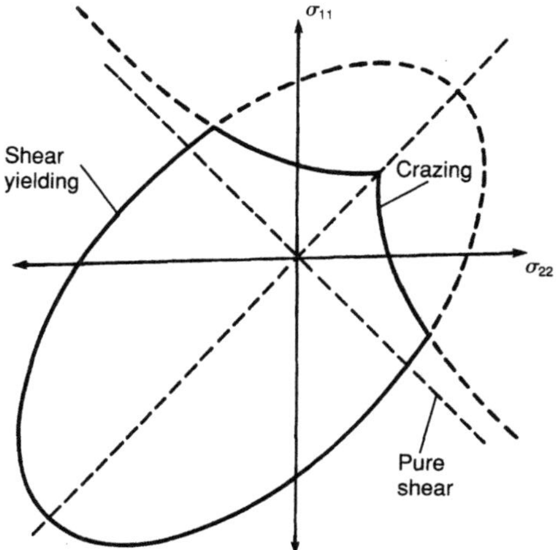

Fig. 2.35. Failure locus for polymethylmethacrylate (biaxial loading). (After McCrum et al. [22].)

of an *S–N* curve, but some factors, such as frequency of loading, may assume greater importance than in metals.

The hysteretic nature of the deformation process in plastics and rubbers means that cyclic loading can produce internal heating, and it can be shown (e.g. Hertzberg and Manson [39]) that the specimen temperature rise in a polymer under cyclic loading is given by

$$T = KS^2fd^2D/E_dA$$

where K = a constant of proportionality
     S = stress range
     f = test frequency
     d = specimen diameter
     D = damping capacity
     E_d = dynamic modulus of elasticity
     A = a parameter related to the heat transfer
          characteristics of the specimen

This means that some care needs to be taken in interpreting polymer fatigue data, since the thermal effect will result in an apparent decrease in fatigue life with increasing test frequency at a given stress range, as already indicated. This also gives problems with cumulative damage criteria, because the damage will be sensitive to order of loading as well as frequency.

A useful way of treating fatigue in polymers is shown in Fig. 2.36, where a single mechanical fatigue curve is superimposed on a series of thermal fatigue curves. Such data are of course acquired by carrying out tests at constant frequency, and these curves give thermal fatigue failures at the higher stress ranges but eventually settle to a common (for all frequencies) mechanical fatigue curve which can then be used in design, provided that frequencies and stress ranges are low enough not to cause thermal fatigue failures in service.

The fracture mechanics approach has also been used for fatigue in polymers and the Paris law and modifications have been applied to a wide range of plastics materials [40]. The da/dN–K curves in polymers are generally of similar shape to those in metals, although the additional dimension of temperature has to be considered, and, as for *S–N* data, care must be taken over hysteretic heating.

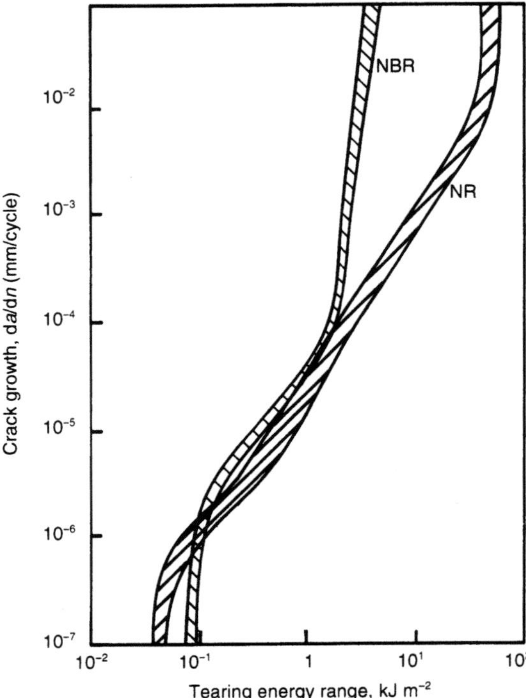

Fig. 2.37. Crack growth rate as a function of tearing energy for two engineering elastomers. (From Stevenson [41].)

More recently, a model known as the 'crack layer model' [40] has been proposed for fatigue crack propagation in polymers. This model is based on the thermodynamics of irreversible processes and expresses the crack propagation rate in terms of the energy release rate. It has been successfully applied to a number of polymeric materials.

Because rubbers require such extensive deformation before the onset of plastic deformation, plastic zones are absent at crack tips in rubber, and this in some ways simplifies the fracture mechanics of such materials. As pointed out above, it is traditional to use the tearing energy, T, to describe the fracture mechanics of rubbers. Graphs of da/dn vs. range of tearing energy (Fig. 2.37) show a characteristic shape not unlike that in other materials, and this shape is independent of loading geometry [41]. Rubbers can also show time-dependent crack growth under static load, although some of them appear to be immune to this.

2.2.3 Fracture and Fatigue in Concrete

The inhomogeneous and only partially controlled nature of the concrete microstructure makes stress analysis and hence fracture mechanics a difficult prospect (Roelfstra and Sadouki [42]).

Fracture toughness measurement in concrete is not generally as well standardised as it is in metals. In particular, size effect in testing is quite significant in concrete [43], and also there is generally an interest in modes other than Mode I because of the nature of the loading encountered in designs employing concrete.

At present, the commonest fracture mechanics specimen is the double cantilever [44], where the load and the stress intensity factor are related through semi-empirical

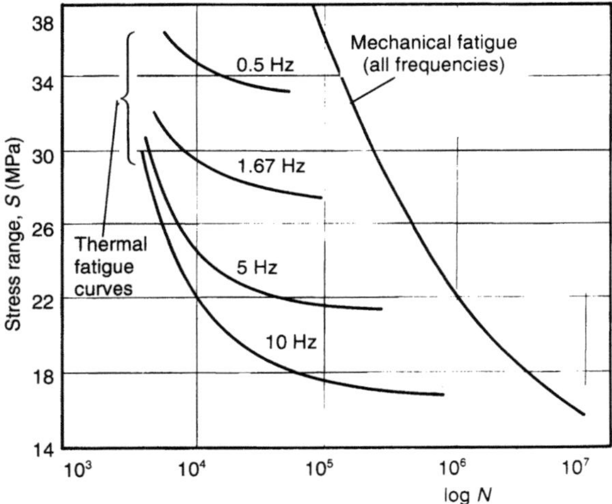

Fig. 2.36. Effect of cyclic frequency on the interaction of thermal and mechanical components of fatigue failure in acetal. (From Moet and Aglan [40].)

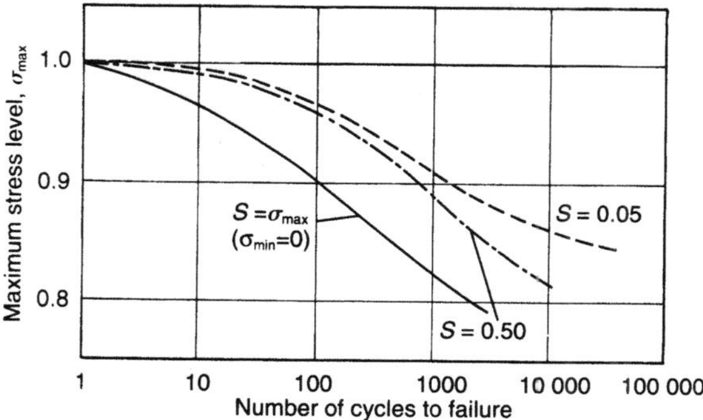

Fig. 2.38. Typical flexural fatigue curve for plain concrete under sawtooth loading. Vertical axis is maximum stress and curves are for various values of stress range, S. (After Hilsdorf and Kesler [46].)

equations much as they are in metal fracture tests. Because concrete is normally to be found in compression, the relevance of such a test to design is to some extent called into question. However, if one considers the simple 'lattice' model of concrete originally outlined by Baker [45], it can be seen that at least Modes I and II are operative in concrete in compression owing to the inhomogeneity of the material.

Fatigue in concrete is, again, mechanistically unlike that in steel, but is often treated by the same methods. For plain concrete, it is possible to carry out fatigue tests under varying flexural stresses, and generally in such cases concrete 'fatigue' damage can be separated into a time-dependent and a cycle-dependent part [46, 47], and a typical curve for flexural fatigue is shown in Fig. 2.38.

In structural applications, the fatigue of steel and of interfacial bonds between steel and concrete is also of concern, and the cracking of concrete around reinforcing bars can cause stress concentrations and hence initiate fatigue. For reinforced concrete, testing is usually carried out on the

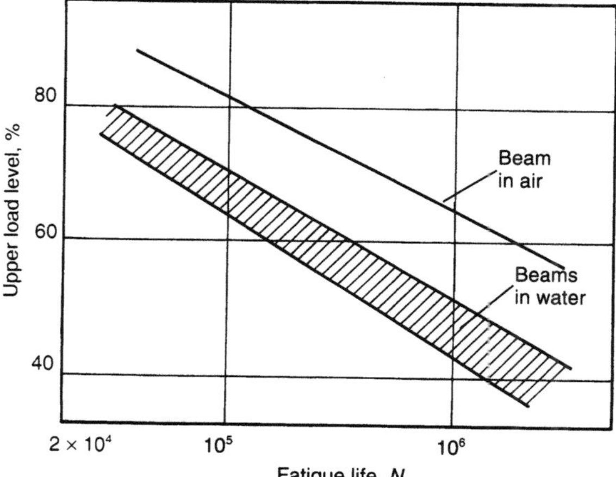

Fig. 2.39. Fatigue data for reinforced concrete beams in air and in water where the upper load level is expressed as a percentage of the ultimate for the beam and the lower load level is fixed at 10 per cent of ultimate. (From Nishibayashi et al. [48].)

entire assembly, although tests on reinforcing bars are carried out to assess the importance of various parameters on their fatigue life. Figure 2.39 shows a typical fatigue curve for a reinforced concrete beam.

2.2.4 Fracture in Timber

The orthotropic nature of timber makes the application of fracture mechanics rather more involved than in isotropic materials. For example, the relationship between strain energy release rate and stress intensity factor must be modified to use the appropriate elastic constant, and also there is little or no interaction between the basic modes of crack opening.

As is well known, timber will fracture along the grain irrespective of the opening force and original notch geometry, and some applications of fracture mechanics in Mode I opening have been successful, subject to the inherent variability of timber as a natural material [49].

Fatigue, in the classical sense, is not a phenomenon which affects timber.

2.2.5 Fracture in Ceramics

The very brittle nature of ceramics makes these probably the simplest materials to treat using linear elastic fracture mechanics. In fact Griffith's original approach was intended for application to glasses, for which it is quite adequate, although the linear elastic fracture toughness, K_{Ic}, is much more commonly used than the more fundamental modulus and surface energy. In the more modern engineering ceramics fracture toughness is generally higher than in glasses, but linear elastic theory is still usually applicable.

Fatigue is not normally considered as a design criterion for ceramic materials, owing to their very low tolerance of defect size.

2.2.6 Fracture and Fatigue in Composites

As with strength and elastic modulus, it is possible to predict (although to a lesser degree of precision) the fracture toughness of composite materials from the properties of the constituents.

For example, Chamis [24] gives longitudinal and transverse in-plane fracture toughness parameters for unidirectional composites with full thickness penetration defects in Mode I opening, and these are shown in Table 2.4. The same author gives simplified equations for in-plane and through-thickness uniaxial impact resistance.

LEFM and its modifications for the formation of a damage zone (the damage zone in composites can be treated in a similar way to the plastic zone in metals) can be used to predict fracture in composites which have suffered damage, for example by an impact (Whitney [50]), although the tabulations of geometric factors for isotropic materials (e.g. Rooke and Cartwright [33]) are not applicable. Whitney et al. [51, 52] have proposed an alternative method of dealing with fracture in composites by considering the stress distribution ahead of a sharp crack in much the same way as Westergaard's solution led to the definition of K_{Ic}. In calculations employing such 'stress fracture' criteria, failure is considered to occur when the stress ahead of the crack reaches some point or average value which exceeds the ultimate tensile strength of the composite.

The complexity of composite geometry and their directionality make the failure modes by fracture and by fatigue very complex and difficult to describe. In particular, the fatigue failure event is not simply characterised by the propagation of a single crack, but by the buildup of a very

large number of relatively short cracks which can be any of the following (Fig. 2.40): layer cracks, delamination, fibre breakage and fibre–matrix interfacial debonding. The same is true to a certain extent under static loading. Nevertheless, S–N curves do find some applications for composites, as do Goodman diagrams, and some examples are described by Kim [53].

As suggested above, there is still something of a problem in defining the point of failure in a composite where fatigue damage is normally accumulated as a density of cracks. One way of treating this is to carry out partial fatigue experiments and measure the residual strength of the material after a certain loading history. This approach is limited by the fact that it is impracticable to investigate a wide range of the experimental parameters. One or two methods are currently under development to deal with fatigue in composite as a series of local failures which eventually accumulate until the properties of the composite are no longer acceptable. Such 'damage accumulation models' [55] aim to express the fatigue damage (for example, as cracks per centimetre) in terms of the loading history. It has been found that residual stiffness is quite closely related to fatigue damage but, more significantly, some success is claimed [54] in simulating the cracking process in a simply loaded $(0°/90°)_s$ laminate, and Fig. 2.41 shows the relation between simulated and experimentally measured damage.

Table 2.4.   Fracture properties of composites

Longitudinal tensile toughness

$$S_{p11T} = \frac{\sigma_{p11T}}{1 + \sqrt{2\left[(E_{p11}/E_{p22}) - v_{p12}\right] + E_{p11}/G_{p12}}}$$

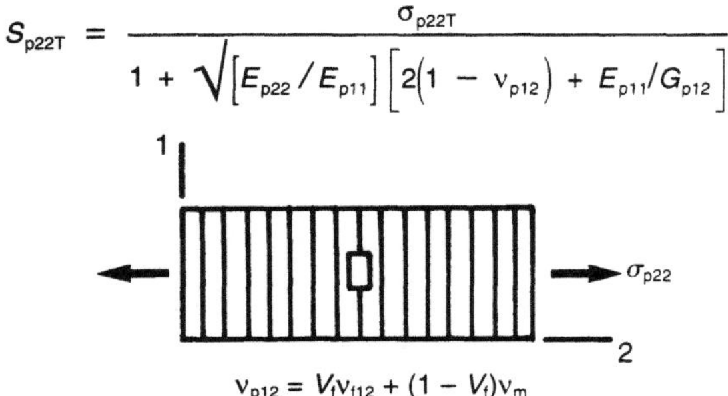

Transverse tensile toughness:

$$S_{p22T} = \frac{\sigma_{p22T}}{1 + \sqrt{\left[E_{p22}/E_{p11}\right]\left[2(1 - v_{p12}) + E_{p11}/G_{p12}\right]}}$$

$$v_{p12} = V_f v_{f12} + (1 - V_f)v_m$$

Source: After Chamis [24].
Subscripts and other definitions are as for Table 2.3, and
S_p = uniaxial 'fracture toughness' (defined as the far field stress required to produce additional damage in a composite which already contains a defect)
v_p = ply Poisson ratio

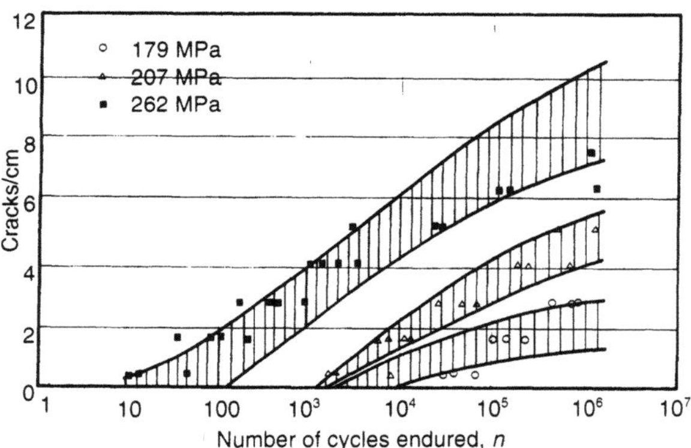

Fig. 2.41.   Experimental fatigue damage (shaded) in a simple $(0°_2/90°_3)_s$ laminate compared with simulated figures (points) using a damage accumulation model. (From Wang [54].)

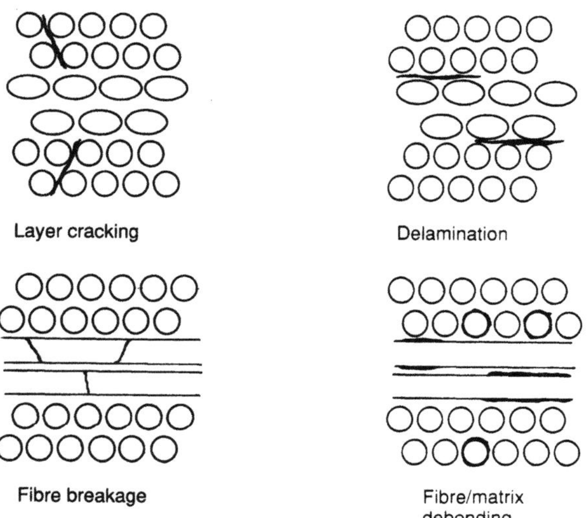

Fig. 2.40.   Schematic view of some of the types of cracking which can contribute to fatigue failure in composites.

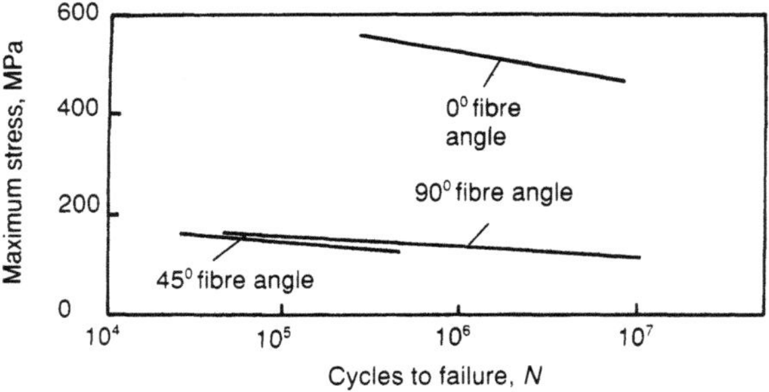

Fig. 2.42.   Rotating bending fatigue performance for 50 per cent volume alumina fibre–aluminium metal matrix composite. (From Romine [55].)

For metal matrix composites, the potential aerospace applications have meant that fatigue resistance has been an important development criterion, and this makes the application of such materials to marine technology rather easier. For example, Romine [55] reports that the fatigue resistance imparted to aluminium by aluminium oxide (alumina) fibre reinforcement has meant that they can replace steel in a number of areas subject to fatigue loading at half the weight. Some typical data for alumina-fibre-reinforced aluminium are shown in Fig. 2.42.

2.2.7 Use of Fracture and Fatigue Data in Design

The danger of fracture in engineering materials has traditionally been treated as a 'factor of safety' problem. In other words, provided that the stresses are kept sufficiently below yield, most engineering materials are unlikely to suffer static fracture. There is still an element of this in fracture-safe design, but there is also a growing awareness among design engineers that the conventional stress analysis approach needs to be complemented by a consideration of fracture mechanics.

In more critical applications such as offshore structures, pressure vessels and aerospace, fracture considerations are always important, mainly because of the use of thick sections but also because of the nature of the loading.

Many design codes quote the toughness requirement in terms of impact data of the Charpy or Izod type. As pointed out above (p. 34) these are of limited use in quantitative design, although their use normally ensures that a steel, for example, will be tough enough to make failure by fracture unlikely. Some correlations between impact data and quantitative fracture parameters do exist (for example in BS PD6493 [56]) although such correlations are of limited applicability and it is best to use fracture mechanics tests to allow calculations of fracture loads or stresses.

Fracture toughness data (as K_{Ic}, δ_c or G_c) are available for many modern materials, and these can be used to make quantitative assessments of defect tolerance under well-defined stress systems. In more complex geometries where more than one opening mode is possible analysis becomes rather more difficult, because neither the geometric factor solutions nor the Modes II and III opening factors are generally available. Nevertheless, much can be done with the stress intensity factor approach and much-used designs, such as tubular joints, have been the subject of detailed study, so that SIF data are readily available in a form which can be easily accessed by designers. An example of this application is given in Case Study 7.2.

Of course, defects need not necessarily lead to failure by fracture, and it is quite possible that plastic collapse can take place more easily than fracture in a structure containing defects. A number of methods exist for handling this situation, and the two most important were developed for pressure vessels. One such procedure (the 'two-criteria procedure') consists of identifying the degree to which the defective component has approached failure by brittle fracture and also by plastic collapse, and interpolates between the two by use of a failure envelope. The method can therefore be used to assess defect severity in relatively tough materials where a linear elastic approach would be needlessly conservative. Case Study 7.1 provides an example of the use of the two-criteria procedure.

Fatigue data are often still represented by S–N curves for a range of geometries and stress ratios. The use of these curves in design is obvious and is usually conservative, provided that an appropriate set of fatigue data is consulted. The fracture mechanics approach is also valuable, but has the drawbacks that it is only really appropriate to propagating cracks and assumes that initiation has taken place. Case Study 7.2 compares fracture mechanics and S–N approaches for fatigue in offshore structures subjected to wave loading, and much of the analysis there can be applied to other marine components.

2.3 Wear Properties

As mentioned in the introduction to this chapter, a detailed study of wear is outwith the scope of this text, but a brief description of the phenomena involved is included as a guide to the data and to provide some background to the conjoint corrosion and wear phenomena discussed in Chapter 3.

It is difficult to separate wear from corrosion and oxidation effects in environments where both are taking place, because inevitably the exposure of fresh material by wear will introduce the possibility of enhanced interaction with the environment. However, in cases of heavy surface loading the mechanical component may be predominant. Lubrication, and the behaviour of lubricant films, are an essential factor in the study of tribological phenomena, but again their discussion is outwith the scope of this text and the reader is referred to Arnell et al. [57] for an introduction to the subject and from which most of the material in this section has been obtained.

Mechanical wear phenomena can be broadly subdivided into adhesive and abrasive wear and surface (or contact) fatigue.

Although it is a somewhat simplistic view, the difference between adhesive and abrasive wear can be visualised in terms of the nature of the contact between the wearing surfaces. The basis of adhesive wear is that cold welds are formed between the wearing surfaces, and the subsequent shearing of these bonds will result in the removal of material from one or both of the surfaces. In abrasive wear, a hard material ploughs material from softer surfaces. The hard material may be a 'third body' which ploughs from one or both of the sliding surfaces or may be one of the sliding surfaces, ploughing material from the softer surface.

Contact fatigue and delamination wear involve the deformation of the underlying strata of the wearing surfaces. The main difference between these mechanisms and the two described above is that the former can operate through a lubricant film. Contact fatigue can take place under rolling or sliding action. In the rolling case, any subsurface element is subjected to cyclic shear stresses for each passage of the rolling element which, if it is above the fatigue limit, will lead to subsurface cracking which eventually spreads to the surface resulting in spalling. In sliding contact, the mechanism is similar, except that the regularity of loading is altered. The delamination theory of spalling involves a number of different stages. Initially the harder surface causes any asperities on the softer surface to be flattened until the softer surface is relatively plane and is loaded in a manner similar to that described for contact fatigue. Subsurface cracking is then initiated by the triaxial compressive stresses, and these cracks grow and coalesce until they reach a critical size at which they can break through to the surface.

From experiment, it is known that mechanical wear is proportional to the normal load, W, and also to the slid distance, and this is largely irrespective of the mechanism (within the mechanical mechanisms). It is normal to express wear as volume per unit sliding distance, Q, so that the general form of the wear law is

$$Q = KW$$

The constant of proportionality, K, contains information on the surface properties and also on the mechanism of the wear. One of the most important of these surface properties is hardness, H, so that this is sometimes separated from the constant of proportionality:

$$K = K'/H$$

The alleviation of wear is therefore to a large extent dependent upon the use of hard materials, or materials which will harden under deformation without fracture. However, the mechanisms which lead to contact fatigue or delamination are rather more difficult to resist, and this is often attacked by design to reduce contact loads and frequencies, rather than by selection of new materials. In abrasive wear, surface hardness is of paramount importance, but there is also some benefit in having a soft component to at least one of the surfaces which allows embedment of hard third-body particles. Plain-bearing design also benefits from good conformity of surfaces, thus providing for reduced contact loads [58].

References

1. Holloway DG. The physical properties of glass. Wykeham, London, 1973
2. Easterling K. Introduction to the physical metallurgy of welding. Butterworths, London, 1983
3. Knott JF. Fundamentals of fracture mechanics. Butterworths, London, 1981
4. Jensen JJ, Pedersen PT. The buckling of submarine pipelines. In: de la Mare (ed.) Advances in offshore oil and gas pipeline technology. Oyez, London, 1985, pp 41–60
5. Lake GJ. Aspects of design for high severity applications. In: Stevenson A (ed.) Rubber in offshore engineering. Adam Hilger, Bristol, 1984, pp 150–165
6. Throne JL, Progelhof RC. Creep and stress relaxation. In: Engineered materials handbook, vol. 2: Engineering plastics. ASM International, Metals Park, Ohio, 1988, pp 659–678
7. Crawford RJ. Plastics engineering. Pergamon, Oxford, 1987
8. Driscoll SB. Physical, chemical, and thermal analysis of thermoplastic resins. In: Engineered materials handbook, vol. 2: Engineering plastics. ASM International, Metals Park, Ohio, 1988, pp 533–543
9. Turner S. Mechanical testing. In: Engineered materials handbook, vol. 2: Engineering plastics. ASM International, Metals Park, Ohio, 1988, pp 544–558
10. Neville AM. Properties of concrete. Pitman, London, 1975
11. Weidmann G, Lewis P, Reid N (eds). Structural materials. Open University/Butterworths, Milton Keynes and London, 1990
12. Building Research Establishment, Princes Risborough Laboratory. The strength properties of timber. MTP Construction/HMSO, London, 1974
13. Callister WD. Materials science and engineering. Wiley, New York, 1991
14. Weeton JW, Peters DM, Thomas KL (eds). Engineers' guide to composite materials. ASM, Metals Park, Ohio, 1987
15. Schoenberg T. Boron and silicon carbide fibers. In: Engineered materials handbook, vol. 1: Composites. ASM International, Metals Park, Ohio, 1988, pp 58–59
16. Benham PP, Crawford RJ. Mechanics of Engineering Materials. Longmans, Harlow, 1987
17. Gurland J. A structural approach to the yielding of two-phase alloys with coarse microstructures. Materials Science and Engineering 1979; 40: 59–71
18. McDaniels DL, Jech RW, Weeton JW. Analysis of stress-strain behaviour of tungsten-fiber-reinforced copper composites. Transactions of the Metallurgical Society of the AIME 1965; 233: 636–642
19. Jech RW, McDaniels DL, Weeton JW. Fiber reinforced metallic composites. In: Proceedings, 6th Sagamore Ordinance Materials Research Conference, report MET 661–601. Syracuse University Research Institute, N.Y., 1959
20. Weeton JW. Introduction and comments on composite design equations. In: Weeton JW, Peters DM, Thomas KL (eds) Engineers' guide to composite materials. ASM, Metals Park, Ohio, 1987, pp 3-1–3-3
21. Weeton JW. Calculations of composites properties. In: Weeton JW, Peters DM, Thomas KL (eds) Engineers' guide to composite materials. ASM, Metals Park, Ohio, 1987, pp 3-3–3-7
22. McCrum NG, Buckley CP, Bucknall CB. Principles of polymer engineering. Oxford University Press, Oxford, 1988
23. Kelly A, Davies GJ. The principles of the fibre reinforcement of metals. Metals Reviews 1965; 10 (37): 1–77
24. Chamis CC. Simplified composite micromechanics equations for mechanical thermal and moisture-related properties. In: Weeton JW, Peters DM, Thomas KL (eds) Engineers' guide to composite materials. ASM, Metals Park, Ohio, 1987, pp 3-8–3-24
25. Rosen BW, Hashin Z. Analysis of material properties. In: Engineered materials handbook, vol. 1: Composites. ASM International, Metals Park, Ohio, 1988, pp 185–205
26. ANSI/ASME. Gas transmission and distribution piping systems, code for pressure piping. B31.8-1986. ASME. New York, 1986
27. Tohlen ME. Generation of design allowables. In: Engineered materials handbook, vol. 1: Composites. ASM International, Metals Park, Ohio, 1988, pp 308–312
28. Berg OY. Strength and plasticity of concrete. Doklady Akademii Nauk USSR 1950; 70 (4): 617–620
29. Young WC. Roarke's formulas for stress and strain. McGraw-Hill, New York, 1989
30. Griffith AA. Philosophical Transactions of the Royal Society 1920; A221: 163
31. BS 5447. Methods of test for plane strain fracture toughness (K_{1c}) of metallic materials. BSI, London, 1977
32. Rooke DP, Cartwright DJ. Compendium of stress intensity factors. HMSO, London, 1976
33. Gray TGF, Spence J. Rational welding design. Butterworths, London, 1982
34. Engel L, Klingele H. An atlas of metal damage. Wolfe/Hanser, Munich, 1981
35. Osgood CC. Fatigue design. Pergamon, Oxford, 1982
36. Davies AC. The science and practice of welding. Cambridge University Press, Cambridge, 1981
37. Shigley JE. Mechanical engineering design. McGraw-Hill, Tokyo, 1981
38. Petrie SP. Crazing and fracture. In: Engineered materials handbook, vol. 2: Engineering plastics. ASM International, Metals Park, Ohio, 1988, pp 734–740
39. Hertzberg RW, Manson JA. Fatigue of engineering plastics. Academic Press, New York, 1980
40. Moet A, and Aglan H. Fatigue failure. In: Engineered materials handbook, vol. 2: Engineering plastics. ASM International, Metals Park, Ohio, 1988, p 741–750
41. Stevenson A. Design principles and fatigue life determinations

for structural rubber bearings. In: Stevenson A (ed.) Rubber in offshore engineering. Adam Hilger, Bristol, 1984, pp 14–37

42. Roelfstra PE, Sadouki H. Fracture processes in numerical concrete. In: Wittmann FH (ed.) Fracture toughness and fracture energy of concrete. Elsevier, Amsterdam, 1986, pp 105–116

43. Tian ML, Huang SM, Liu EX, Wu LY, Long KQ, Yang ZS. Fracture toughness of concrete. In: Wittmann FH (ed.) Fracture toughness and fracture energy of concrete. Elsevier, Amsterdam, 1986, pp 299–306

44. Brown JH. Magazine of Concrete Research 1972; 24: 191–199

45. Baker ALL. Magazine of Concrete Research 1959; 11: 119–126

46. Hilsdorf HK, Kesler CE. Fatigue strength of concrete under varying flexural stresses. Journal ACI Proceedings 1966; 63 (10): 1059–1075

47. Award ME, Hilsdorf HK. Strength and deformation characteristics of plain concrete subjected to high repeated and sustained loads. In: Fatigue in concrete, American Concrete Institute publication SP-41. ACI, Detroit, 1974, pp 1–14

48. Nishibayashi S, Inoue S, Yamura K. Fatigue characteristics of reinforced concrete in water. Proceedings 2nd International Conference – Concrete in Marine Environment, St Andrews by-the-Sea, Canada. ACI, Detroit, 1988, pp 543–562

49. Williams JG, Birch MW. Mixed mode fracture in anisotropic media. In: The properties of wood in relation to its structure. NATO Science Committee, Washington, 1975

50. Whitney JM. Fracture analysis of laminates. In: Engineered materials handbook, vol. 1: Composites. ASM International, Metals Park, Ohio, 1988, pp 252–258

51. Whitney JM, Nuismer RJ. Stress fracture criteria for laminated composites. Journal of Composite Materials 1974; 8(3): pp 253–265

52. Nuismer RJ, Whitney JM. Uniaxial failure of composite laminates containing stress concentrations. In: Fracture mechanics of composites, ASTM STP 593. ASTM, Philadelphia, 1975, pp 117–142

53. Kim RY. Fatigue strength. In: Engineered materials handbook, vol. 1: Composites. ASM International, Metals Park, Ohio, 1988, pp 436–444

54. Wang ASD. Strength, failure and fatigue analysis of laminates. In: Engineered materials handbook, vol. 1: Composites. ASM International, Metals Park, Ohio, 1988, pp 236–251

55. Romine JC. Continuous aluminium oxide fiber MMC's. In: Engineered materials handbook, vol. 1: Composites. ASM International, Metals Park, Ohio, 1988, pp 874–877

56. BS PD6493. Guidance on some methods for the derivation of acceptance levels for defects in fusion welded joints. BSI, London, 1980

57. Arnell RD, Davies PB, Halling J, Whomes TL. Tribology – principles and design applications. Macmillan, London, 1991

58. Crane FAA, Charles JA. Selection and use of engineering materials. Butterworths, London, 1984

3 Marine Corrosion and Biodeterioration

Corrosion is probably the best known of the deterioration mechanisms which can affect materials in the marine environment and much attention is rightly paid to its avoidance. Strictly, corrosion is a phenomenon which only affects metals, but there are a variety of 'weathering' and other environmental deterioration phenomena which can affect plastics and other non-metallic materials, and these are also dealt with in this chapter. Deterioration due to the effects (direct or indirect) of living organisms are also discussed, although quantitative data in this area are somewhat limited. Despite the fact that it is not strictly a deterioration phenomenon, marine fouling is also included in this latter category.

This chapter commences with a brief review of the elements of corrosion science covering both thermodynamic and kinetic considerations, followed by a description of the various corrosion and biodeterioration mechanisms. The various types of exposure which might be expected in engineering for the marine environment are then described followed by a discussion of the various avoidance and monitoring strategies, along with the methods used to study corrosion and biodeterioration. The discussion of the susceptibility of both metallic and non-metallic materials to the various degradation mechanisms identified in this chapter is postponed until Chapter 4.

3.1 The Science of Corrosion

As far as marine applications are concerned, corrosion can be defined as 'any electrochemical process which leads to the formation of a dissolved metal ionic species in aqueous solution'. This dissolved ion may then go on (as part of either the same reaction, or a different one) to form an inorganic compound, but the fundamental process of wastage of a metal, M, of valency z, is the reaction

$$M = M^{z+} + ze^-$$

and this is normally referred to as an 'anodic' reaction (as it is the type which would take place at the anode of an electrochemical cell). The study of the electrochemistry of corrosion reactions is rather involved, and it is not within the scope of this book to deal with this in detail; however, the following two subsections briefly lay out some of these fundamentals, first in the equilibrium case and then in the non-equilibrium one.

3.1.1 Equilibrium Electrodics

The electrochemical metal wastage reaction shown above can be treated thermodynamically in a way exactly analogous to any other chemical equilibrium, with electrochemical potential (a concentration of electrons) being regarded in the same way as conventional concentration in the 'equilibrium constant'. It can be seen, therefore, that the standard electrochemical potential of a metal is a measure of its propensity to undergo the oxidation reaction, all other concentrations being held at standard values. Thus, the thermodynamic behaviour of an electrochemical half-cell,

$$M = M^{z+} + ze^-,$$

is summarised by its electrochemical potential,

$$E = E^0 + \frac{RT}{zF} \ln \frac{\left[M^{z+} \right]}{\left[M \right]}$$

where $E^0 = $ standard electrochemical potential of the reaction (i.e. that which would prevail if the pure metal was brought into equilibrium with a solution of its own ions at unit concentration)
$R = $ universal gas constant
$T = $ temperature
$F = $ Faraday constant

As indicated above, this is a simple extension of equilibrium thermodynamics to reactions which involve electrons, and the concentration of electrons is simply a potential. The derivation of the above equation (called the 'Nernst equation') can be found in most standard physical chemistry texts (e.g. Moore [1]) and is more commonly seen in the form which recognises that the concentration of the metal before it dissolves is usually unity:

$$E = E^0 + \frac{RT}{zF} \ln \left[M^{z+} \right]$$

Metals can therefore be classified in terms of their ability to undergo such a reaction, and the standard electrochemical potential is used for this purpose. Table 3.1 shows some selected values of electrochemical potentials from Bard et al. [2]. Metals which oxidise (i.e. give up electrons) readily are often referred to as 'base', whereas those which oxidise less readily are referred to as 'noble'. It is tempting to think of the noble metals as being corrosion-resistant and the base metals as being less resistant, but unfortunately, although there is some truth in this, inherent nobility is not

Table 3.1. Selected values of standard electrochemical potentials

Half-cell reaction	E_0 (V)
$Li^+ + e^- \rightarrow Li$	-3.045
$Na^+ + e^- \rightarrow Na$	-2.714
$Mg^{2+} + 2e^- \rightarrow Mg$	-2.356
$Al^{3+} + 3e^- \rightarrow Al$	-1.67
$Ti^{2+} + 2e^- \rightarrow Ti$	-1.63
$Ni^{2+} + 2e^- \rightarrow Ni$	-0.257
$2H^+ + 2e^- \rightarrow H_2$	0.000
$Cu^{2+} + 2e^- \rightarrow Cu$	0.159
$Fe^{3+} + e^- \rightarrow Fe^{2+}$	0.771
$O_2 + 4H^+ + 4e^- \rightarrow 2H_2O$	1.229
$Cl_2 + 2e^- \rightarrow Cl^-$	1.3583
$Au^{3+} + 3e^- \rightarrow Au$	1.52

Source: Bard et al. [2].

the only thing which can confer corrosion resistance. A glance at the electrochemical series (Table 3.1) shows this clearly when it is considered that metals like titanium and aluminium are, in fact, rather resistant to corrosion. Further, most metallic materials are alloys, and it is difficult to explain the resistance of brasses (alloys of copper and zinc) in simple electrochemical terms.

The analogy with conventional chemical thermodynamics can be taken still further when it is considered that an isolated anode immersed in an electrolyte in which it is able to corrode will do so only until the reactants and products

in the metal wastage reaction are in equilibrium, when net corrosion will cease. This rarely occurs in practice because the ionic species is usually swept, or diffuses, away from the reaction surface and, more importantly in view of their mobility, electrons can be consumed by a variety of possible cathodic reactions:

$$C^{x+} + xe^- = C$$

or

$$C + ye^- = C^{y-}$$

The two most important cathodic reactions in marine environments are the oxygen reduction reaction,

$$O_2 + 2H_2O + 4e^- = 4OH^-$$

and the hydrogen evolution reaction,

$$2H^+ + 2e^- = H_2$$

The standard electrochemical potentials of these two reactions are given in Table 3.1.

The net result of this is that the thermodynamic severity of aqueous corrosion of a metal is only defined once the anodic and cathodic reactions are known and quantified. Pourbaix has carried out a detailed thermodynamic study of aqueous corrosion, culminating in his *Atlas of Electrochemical Equilibria* [3] which contains the useful potential–pH diagrams of which Fig. 3.1 is an example. These diagrams are derived from purely thermodynamic considerations and are a very powerful aid to understanding the thermodynamics of corrosion, in particular metal–electrolyte systems. It is therefore worth describing in

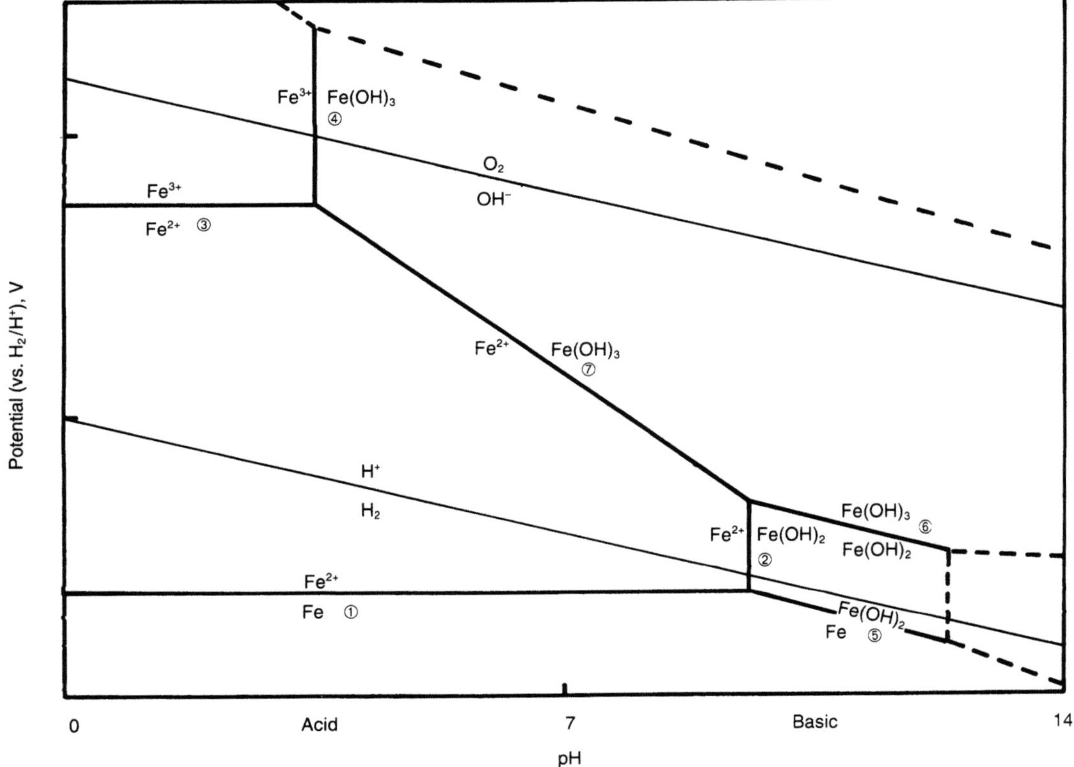

Fig. 3.1. Pourbaix diagram for the iron–water system at 25 °C. (After Pourbaix [3].)

detail how these diagrams can be derived from thermodynamic data. The example used here is corrosion of iron in water at 25 °C. The first stage in the corrosion of iron in water is the dissolution of ferrous ions, which can be represented by the following equilibrium:

$$Fe^{2+} + 2e^- = Fe$$

$$E = -0.44 + 0.0295 \lg [C_{Fe^{2+}}] \quad (1a)$$

Once the concentration of ferrous ions reaches a sufficiently high value, a hydroxide may be precipitated according to the following equilibrium

$$Fe(OH)_2 + 2H^+ = Fe^{2+} + 2H_2O \quad (2a)$$
$$K = 10^{13.29}$$

and this equilibrium can be rewritten as

$$\lg C_{Fe^{2+}} = 13.29 - 2\,pH$$

If the surface concentration of ferrous ion is not allowed to build up and is, say, kept to a value of 10^{-6} gram–equivalent per litre then the pH–potential boundary for the above two equilibria can be described as follows:

$$E = -0.62 \quad \text{for } Fe/Fe^{2+} \quad (1b)$$

$$pH = 9.6 \quad \text{for } Fe^{2+}/Fe(OH)_2 \quad (2b)$$

These two lines can be seen in Fig. 3.1 as two of the boundaries of the Fe^{2+} 'phase field'. It can be seen that those equilibria which involve electrons are potential-sensitive and those which involve protons are pH-sensitive.

Ferrous ions can also be further oxidised to ferric ions and this, and the associated hydroxide equilibrium can be represented by

$$Fe^{3+} + e^- = Fe^{2+} \quad (3a)$$
$$E = 0.771 + 0.509 \lg C_{Fe^{3+}}/C_{Fe^{2+}}$$

$$Fe(OH)_3 + 3H^+ = Fe^{3+} + 3H_2O \quad (4a)$$

$$K = 10^{4.62}$$

Again, if the concentration of ferric ions is considered to be maintained at 10^{-6} gram-equivalent per litre then two lines on the Fe^{3+} field can be drawn:

$$E = 0.771 \quad \text{for } Fe^{2+}/Fe^{3+} \quad (3b)$$

$$pH = 2 \quad \text{for } Fe^{3+}/Fe(OH)_3 \quad (4b)$$

To complete the diagram it is necessary to obtain expressions in pH–potential space for the $Fe^{2+}/Fe(OH)_3$, the $Fe/Fe(OH)_2$ and the $Fe(OH)_2/Fe(OH)_3$ boundaries. This can be done by combining some of the above equilibria. For example, the addition of reactions (1) and (2) using Hess's law gives

$$Fe(OH)_2 + 2H^+ + 2e^- = Fe + 2H_2O \quad (5a)$$

yielding

$$E = -0.051 - 0.059\,pH \quad \text{for } Fe(OH)_2/Fe \quad (5b)$$

and in similar fashion from (4) + (1) − (2):

$$E = 0.262 - 0.059\,pH \quad \text{for } Fe(OH)_3/Fe(OH)_2 \quad (6)$$

and, from (4) + (1):

$$E = 1.40 - 0.177\,pH \quad \text{for } Fe^{2+}/Fe(OH)_3 \quad (7)$$

and, as can be seen, the last three reactions are both potential- and pH-sensitive, since they involve both protons and electrons.

It is also helpful to include the likely cathodic reactions on the Pourbaix diagram, and these are given by the following:

for oxygen reduction,

$$O_2 + 2H_2O + 4e^- = 4OH^-$$

$$E = 1.227 - 0.059\,pH \quad \text{for } O_2/OH^-$$

and, for hydrogen evolution,

$$H^+ + e^- = \tfrac{1}{2}H_2$$

$$E = -0.59\,pH \quad \text{for } H_2/H^+$$

The resulting simplified version of the Pourbaix diagram for iron in water shown in Fig. 3.1 is very helpful in understanding a number of corrosion phenomena which occur in steels when immersed in an electrolyte. For example, depression of the potential of iron in water can put it in an area where corrosion will not occur (cathodic protection), but can also lead to the possibility of hydrogen evolution from water at the iron surface (over-protection). Increasing the pH to more alkaline values leads to a region where hydroxides are more stable with the possibility of passivation of the surface. A similar effect might be expected by raising the potential, provided that the pH is not too low. It should be emphasised that these are qualitative predictions only, and the diagram in Fig. 3.1 is limited by the following:

1. The diagram only covers the Fe/H_2O system. 'Impurities' such as alloying elements in steel and anions other than OH^- in the water can have a marked effect. This is especially so for the effect of chloride ions on regions of passivity.

2. The diagram is limited by assumptions regarding the concentration of dissolved iron near the surface, and also by the fact that local pH may be rather different than that in the bulk electrolyte.

3. The oxide and hydroxide equilibria for iron are rather more complex than suggested above, and account should be taken of the ageing of hydroxides to oxides.

4. The diagram takes account of thermodynamic factors only, and it therefore merely serves to show whether or not corrosion has the potential to occur and gives no indication of how long this will take.

Pourbaix shows several versions of his diagram, taking some of these considerations into account, and modifications accounting for such factors as marine fouling have also recently been published [4].

3.1.2 Electrode Kinetics

In order to assess the need for corrosion protection it is first necessary to have an idea of the likely severity of the problem. This is only partly addressed by considering the thermodynamics of the corrosion reaction, since reactions which are thermodynamically feasible may proceed only at a very slow rate and hence may not be of concern.

The description of electrode kinetics centres around finding out what is the corrosion current (as opposed to the corrosion potential in the case of thermodynamics), since each electron flowing in a corrosion cell is associated with oxidation (and hence dissolution) of a metallic species.

Clearly, for corrosion to occur, a cell is required, in other words current must flow between a cathodic and an anodic site as shown in Fig. 3.2. These sites may be on the same metallic surface or not, provided that there is electric contact between the cathode and the anode. From Faraday's laws it is known that the mass, m, of metal of atomic mass, M, which dissolves when a corrosion current, I, passes for a time, t, is given by:

$$m = \frac{MIt}{zF}$$

where z is the valency of the dissolved ions and F is the Faraday constant. It can therefore be seen that the rate of metal loss (as mass per unit time) will be proportional to corrosion current. However, if corrosion is taking place more or less uniformly over a surface, it is the rate of loss of metal thickness which is often of more concern. Hence, the important corrosion parameter is current density, i, which is the corrosion current per unit corroding area.

When current is drawn from a half-cell reaction, the potential changes in much the same way as the free energy of a chemical reaction changes when the reaction is driven. In electrochemistry this phenomenon is known as 'polarisation' and is dominated by the slowest step in the electrochemical reaction. Polarisation can be quantified as a relationship between electrode potential and corrosion current (or current density) and, from a simple application of the principle of Le Chatelier, it can be appreciated that a depression of potential will stifle anodic reactions and will enhance cathodic reactions. When an electrode reaction is at equilibrium its forward and reverse rates are equal, although not necessarily zero. This exchange rate is normally referred to as the 'exchange current density', and is dependent upon the nature of the reaction and also upon the solution and the electrode material involved. For example, the exchange current density for the hydrogen evolution reaction from molar HCl is about 10^{-3} A cm^{-2} on platinum and about 10^{-8} A cm^{-2} on tin [5].

Polarisation is related to one of the steps of reaction in going from the oxidised to reduced species (or vice versa for anodic reactions). Broadly such reactions consist of transport of reactants and adsorption onto the corroding surface, various steps of the reaction (e.g. electronation and de-electronation, recombination, etc.) and transport of product species away from the reaction surface. For example, the apparently simple iron dissolution reaction

$$Fe = Fe^{2+} + 2e^-$$

is not likely to simply involve the simultaneous tunnelling of two electrons between an ion and the metal, because the probability of such simultaneous events is very low (Bockris [6]). The mechanism which has been elucidated through years of detailed experimentation is that the reaction occurs through the following steps:

$$Fe + H_2O = FeOH + H^+ + e^-$$

$$FeOH \rightarrow FeOH^+ + e^-$$

$$FeOH^+ + H^+ = Fe^{2+} + H_2O$$

with the middle step being the rate-determining one. The story of how this mechanism was discovered gives an interesting insight into the vast field of corrosion science, and the reader will find a very interesting account in Bockris [6].

In order to see how reaction mechanisms are important to engineers, the example of another important reaction, the de-electronation of water (more commonly known as the 'oxygen reduction reaction') is as follows (again from Bockris [6]).

Fundamentally, the oxygen reduction reaction is written

$$O_2 + 4H^+ + 4e^- = 2H_2O$$

although, by using Hess's law and the well-studied and well-known water solubility product, it is often useful to write the reaction in a modified form for neutral and alkaline solutions (with an appropriate change in electrochemical potential but no change in rate-determining step) as:

$$O_2 + 2H_2O + 4e^- = 4OH^-$$

Bockris reports that there is far less information regarding oxygen reduction (partly due to the experimental difficulties involved), but proposes three possible mechanisms which fit the observed polarisation data.

Two important types of electrochemical polarisation can be identified with regard to corrosion phenomena – activation and concentration polarisation.

3.1.2.1 Activation Polarisation

Activation polarisation arises through reaction rate control at the electrode surface, and can be described by a modification of the fundamental principle of reaction kinetics, that the forward and reverse rates in a reaction can be described in terms of the probability of attaining the necessary energy to jump the barrier between reactant and product energetic states. For an electrochemical reaction which can occur in either the forward (electronation) or reverse (de-electronation) direction, the overall electrode kinetic behaviour can be expressed as the reaction current density:

$$i = i_0(e^{(1-\mu)nF/RT} - e^{-\mu nF/RT})$$

(this is known as the 'Butler–Volmer equation'), where the first term results from the de-electronation component and the second from the electronation component [6]. The meaning of the remaining terms can be inferred from the following developments.

Recasting of the Butler–Volmer equation leads to the Tafel equation, which gives more specifically the deviation

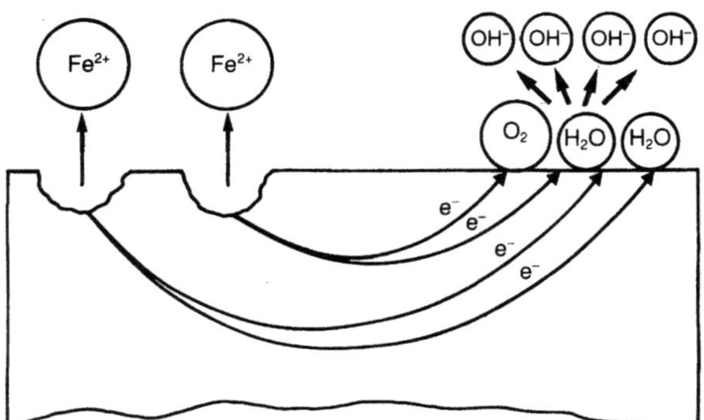

Fig. 3.2. Schematic of a balanced anode–cathode corrosion cell using the example of corrosion of steel in an electrolyte involving the transfer of four electrons: $2Fe + O_2 + 2H_2O = 2Fe^{2+} + 4OH^-$.

from equilibrium potential, n_a (the overpotential), under conditions of activation control (e.g. Fontana and Greene [5]):

$$n_a = \pm \, \beta \, \lg \, (i/i_0)$$

where $\beta = \dfrac{2.303 \, RT}{\alpha z F}$

and α is a factor accounting for the geometry of the energy barrier, giving a value of β which usually lies between 50 and 15 mV. This means that a half-cell can have its potential substantially altered by drawing current in either the forward or the backward direction. This can be represented by a diagram of potential–current space as shown in Fig. 3.3, where it should be noted that the scale in current density is logarithmic.

The way in which activation polarisation affects corrosion can be illustrated by the example of the immersion of zinc in a strong acid such as HCl. In this case, both the zinc dissolution and the hydrogen evolution are under activation control (at least at the currents and potentials of interest), and if the two half-reactions,

$$\text{Zn} = \text{Zn}^{2+} + 2e^- \quad \text{and} \quad \text{H}^+ + e^- = \tfrac{1}{2}\text{H}_2$$

are drawn in the same current density–potential space as shown in Fig. 3.4 then some conclusions as to the cell behaviour can be drawn.

It is clear that there must be some interaction between the two systems (hydrogen reaction and zinc reaction), because they will both take place on the same surface. It is probably safe therefore to assume that both reactions will be at the same potential, because the metal is a good conductor and will produce only very small potential drops between anodic and cathodic sites as electrons move between these. Thus, the corrosion current density, i_{corr} (and hence also the hydrogen evolution current density), will be given by the intersection between the two curves and the corrosion reaction will take place at a potential E_{corr}. The corrosion current density for acid corrosion of a metal, M, will thus depend upon the exchange current density for hydrogen on the metal surface and the polarisation behaviour of that metal's dissolution reaction (E_0, i_0 and β).

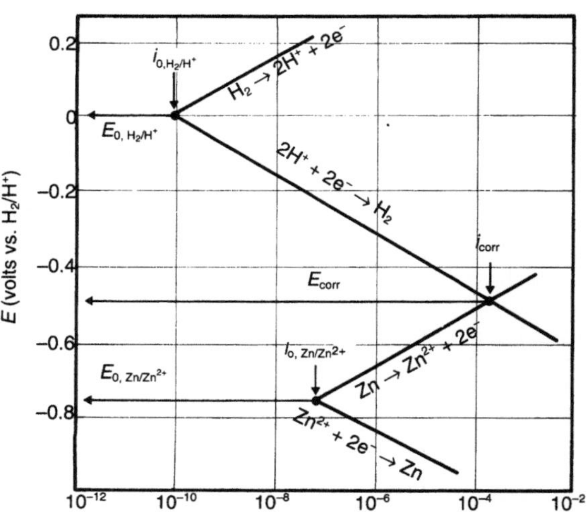

Fig. 3.4. Interaction of two activation polarisation curves giving corrosion current and potential. (After Fontana and Greene [5].)

3.1.2.2 Concentration Polarisation

Activation control only occurs when no impediment to corrosion other than the rate of the critical step of the reaction is present, since the Butler–Volmer equation is derived from considerations of reaction kinetics. However, there are other ways in which reactions can be rate-limited, and the most important of these in corrosion science are the supply of reactants, the removal of products and the introduction of a physical barrier on the surface.

The first of these two effects has already been seen in Section 3.1.1, where the Pourbaix diagram was derived, assuming that the dissolved ion concentration at the electrode had a specific value. Clearly, if the mechanism by which the ions were removed from the surface was rate-determining, the potential would be dependent upon the diffusion steady-state. Polarisation dominated by diffusion, i.e. concentration polarisation, can be understood quite simply in terms of the supply, for example, of a cathodic species from a bulk solution to a surface at which the reduction reaction is proceeding (as is the case with the oxygen reduction reaction).

As the rate at which the diffusing species is used at the surface increases, there is more likelihood of the reaction being dominated by the rate of its supply from the bulk of the solution. By linearising the concentration–distance curve, as is common in heat transfer for the temperature profile close to surfaces, the diffusion flux (J_c), which can be related to the corrosion current (i_c), is given by

$$J_c = i_c / zF = \frac{D[C_s - C_0]}{\delta}$$

where δ is the diffusion boundary layer thickness analogous to the heat transfer boundary layer thickness. As the reaction rate increases, the surface concentration, C_s, decreases until, in the limit, it reaches zero, at which point no further increase in reduction rate is possible. The result-

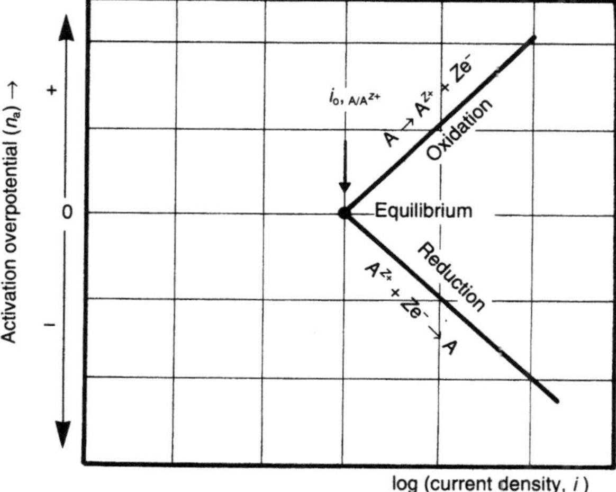

Fig. 3.3. Typical activation polarisation curve.

ing diffusion-limiting current density is therefore given (in magnitude) by

$$i_L = \frac{DzFC_0}{\delta}$$

Although δ cannot easily be determined, the concept of limiting current density is an extremely useful one in understanding many corrosion phenomena (e.g. Bockris [6]).

For instance, the effect of concentration polarisation on the current–potential relationship can be seen by applying the foregoing considerations to the Nernst equation, realising that when the reduction rate is zero the surface concentration (i.e. that which determines the potential in the Nernst equation) is equal to the bulk concentration, C_0, and thus the concentration overpotential, n_c, at a current density, i, where the surface concentration is C_i, can be expressed in terms of the reference state ($i = 0$):

$$n_c = \frac{RT}{zF} \ln\left(C_i / C_o\right)$$

which can be written in terms of current (Fig. 3.5) as

$$n_c = \frac{RT}{zF} \ln\left(1 - i/i_L\right)$$

Of course, this presupposes that there is no kinetic impediment to corrosion at low reduction rates, but in practice it is far more likely that activation and concentration polarisation will both occur, the former being more dominant at low reduction rates and the overall equation being given by

$$n_{tot} = -\beta \log\left(i/i_o\right) + \frac{2.303\, RT \log\left(1 - i/i_L\right)}{zF}$$

giving an overall current–potential curve as shown in Fig. 3.6, which also illustrates the effect of increasing i_L. From this it can be seen that the effect of concentration polarisation only becomes important at current densities relatively close to i_L and that it rapidly becomes the dominant effect thereafter. Furthermore, the above equation for the diffusion-limited current density indicates that i_L is increased by increasing diffusion coefficient (i.e. increasing temperature), increasing bulk concentration of diffusing species and decrease of the boundary layer thickness (e.g. convection, turbulence). The importance of i_L on corrosion can be appreciated, at least qualitatively, by the example of

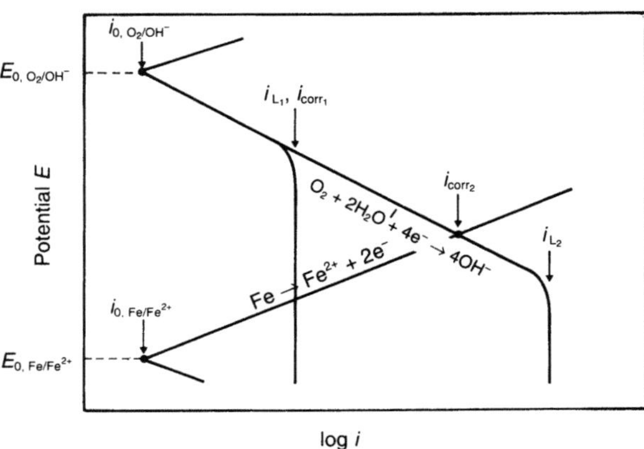

Fig. 3.6. Interaction of two polarisation curves where the cathodic reaction shows concentration and activation polarisation and the anodic reaction shows activation polarisation only. Example of steel in aerated seawater.

steel corroding in an oxygenated electrolyte such as seawater. The iron dissolution reaction can be regarded as showing activation polarisation only, whereas the oxygen reduction reaction shows both activation and concentration polarisation and the two reactions are shown in Fig. 3.6. Clearly, the rate of corrosion of steel in seawater will depend upon the concentration of dissolved oxygen, and the degree of mixing at the steel–seawater interface. The effect of temperature is a little more subtle, because although the polarisation slopes will change and the diffusion coefficient will be affected, the solubility of oxygen in brine decreases with increasing temperature [7].

3.1.2.3 Other Kinetic Effects

Surface films also have an influence on the kinetics of corrosion. The degree to which corrosion is impeded is very dependent upon the integrity of the film, its thermodynamic stability and its permeability to oxygen and water. Apart from applied films, the only naturally forming ones of sufficient integrity to be relied upon to limit corrosion are those known as 'passive' films.

Certain metals (e.g. iron, chromium, nickel and titanium), while polarising normally at low potentials, exhibit a sudden transition at which the current density drops dramatically at a potential normally referred to as the 'primary passivation' potential (E_{pp}) (Fig. 3.7).

As the potential is raised above E_{pp} there is a passive region in which there is little or no increase in corrosion current, until a potential is reached at which the current again begins to increase with potential and the metal exhibits transpassive behaviour.

The mechanism of passivation is known to be associated with the formation of a thin film on the metal surface which is continuous, electronically conducting and mechanically stable [6]. The electronic conductivity is of fundamental importance, in that it ensures that there is no potential drop between the erstwhile corroding surface and the solution and hence prevents the passage of dissolving ions. Because passive films have very particular properties, their effectiveness and range of stability are strongly influenced by the chemical environments on both sides of the film. In particu-

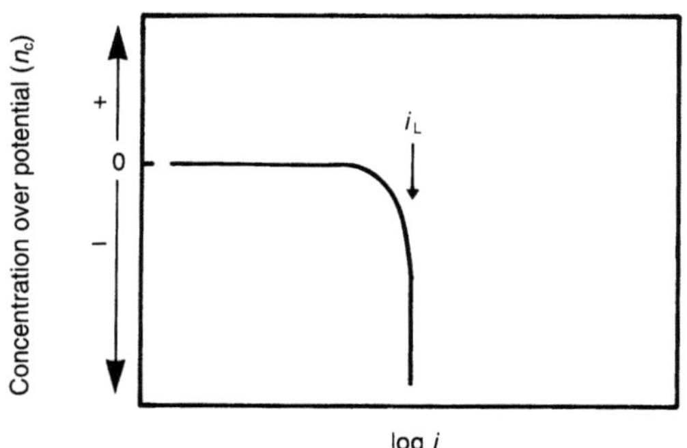

Fig. 3.5. Typical concentration polarisation curve. (From Fontana and Greene [5].)

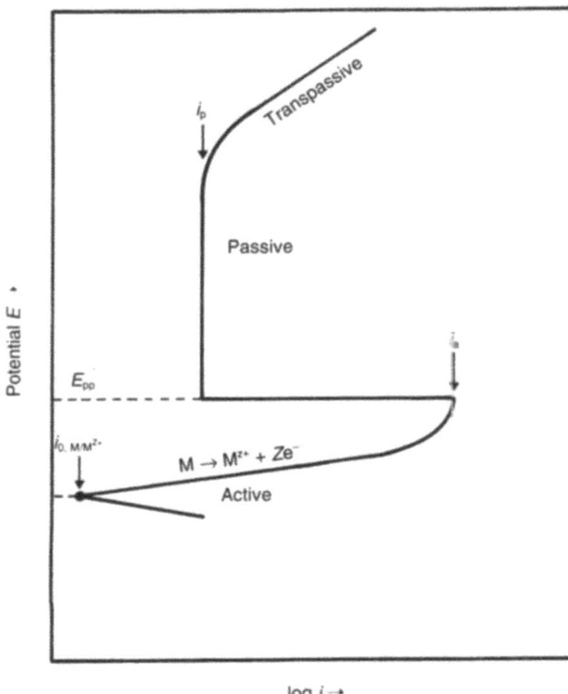

Fig. 3.7. Schematic of a potentiostatic polarisation curve for a metal which shows passivation behaviour.

lar, chloride ions are known to severely compromise the passivating properties of metals, as for example is the case with reinforcing steel within concrete, which can be depassivated if contacted by seawater.

3.2　The Morphology of Corrosion

The foregoing considerations provide the basis of understanding, in phenomenological terms, most of the some-

times subtle effects which render specific metals and alloys suitable for marine service and others less so.

The designer's problem is not, of course, one of elucidating the mechanisms of corrosion, but rather one of avoidance and the main interest is in the macroscopic effects of corrosion and the degree to which it affects function. To this end, it is very important to know how the metal loss resulting from the anodic dissolution is distributed over the surface under attack, as this will affect the degree to which integrity is compromised. This section therefore builds on the scientific principles established in Section 3.1 and examines the factors which influence the morphological distribution of anodic and cathodic sites.

3.2.1　General Corrosion

This is the term normally used (e.g. Fontana and Greene [5]) to describe the morphologically simple situation where anodes and cathodes are relatively evenly distributed over the corroding surface (Fig. 3.8).

It is tempting to consider general corrosion as a broad-fronted phenomenon which affects all parts of the exposed surface equally, resulting in a predictable rate of loss of wall thickness which may or may not be constant with time (Fig. 3.8). In fact, such broad-fronted attack does occur, but there are usually variations in its rate in both the short and the long range, so that a high degree of conservatism is normally required in the interpretation of data even for this simple morphology. For example, Schumacher [8] shows some data for average pitting penetration and average metal loss for carbon steel in seawater where a rate of wall thickness loss of around 70 μm/per year has superimposed on it local areas in which the loss in thickness is considerably greater than this (for example, the maximum pit depth at the 16-year data point was 3.9 mm).

Corrosion data are normally expressed either as rate of weight loss per unit area (e.g. milligrams per square metre per day) or as rate of wall thickness loss (e.g. micrometres per year), and it is important to know under which conditions these have been measured to ensure that adequate provision is made for corrosion allowance (see Section 3.5).

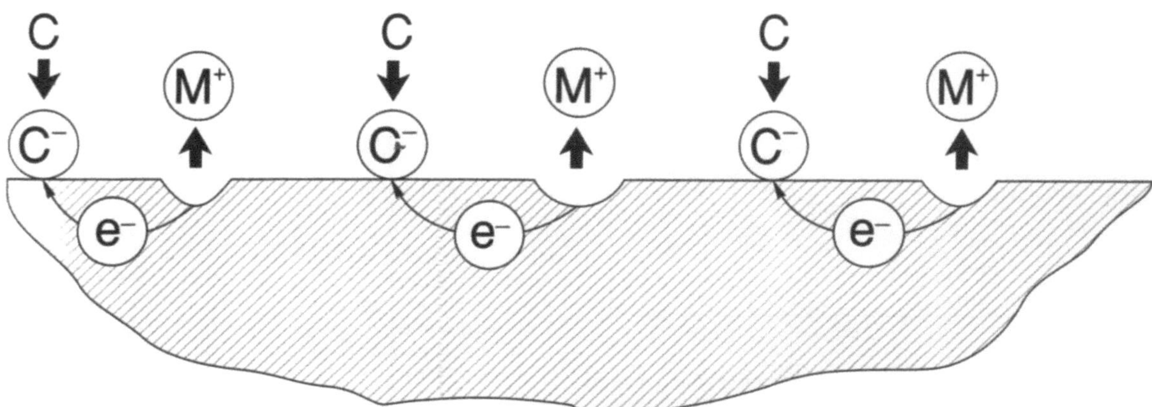

Fig. 3.8.　Schematic illustration of general corrosion where anodes and cathodes are evenly distributed over the metallic surface.

3.2.2 Effect of Material or Environmental Heterogeneity

Heterogeneity, in either the environment or the corroding material, can have a substantial effect on corrosion, because the short-circuiting of the two half-cells which form the heterogeneity can produce a corrosion reaction.

The simplest case of such a galvanic corrosion cell occurs when two dissimilar metals are brought together in an electrolyte. If both of these metals exhibit simple activation polarisation and an activation-polarised cathodic reaction is also present, then a potential–current density diagram can be used to illustrate the galvanic effect (Fig. 3.9).

In the absence of the nobler metal (B), the baser metal (A) can be expected to corrode at a current density of i_{cA} in conjunction with the cathodic reaction due to species C. It can be expected that species C will show a different exchange current density when reduced on A than when on B, so that the addition of the second anodic material will give rise to two additional polarisation curves (Fig. 3.9).

If the second metal is relatively inert, so that it simply provides an alternative site for cathodic reduction, there can still be enhancement of corrosion through an increase in the exchange current density. It should be noted that the reduction current will include a small contribution from the reduction which is taking place on the original anode (A) at the mixed potential. Thus, the coupling of A to a non-corroding species B results in a shift in corrosion potential from E_{cA} to E_{cA-B} and an increase in corrosion current density from i_{cA} to i_{cA-B} (Fig. 3.9b). It should be noted that inherent nobility (i.e. position in the electrochemical series) is only one part of the effect on galvanic compatibility. It is clear that exchange current density and other kinetic effects such as passivation are also of importance and 'engineering galvanic series' have been published which account for such effects and allow a realistic practical assessment of galvanic compatibility (Table 3.2). These are referred to again later in this section.

When the nobler metal is relatively active, the situation is slightly more complicated (Fig. 3.9c). Here, the corrosion rates of the two metals in the uncoupled state would be i_{cA} and i_{cB} respectively and the mixed corrosion potential occurs where the total reduction rate (sum of curves C(A) and C(B)) equals the total oxidation rate (sum of curves A and B). This results in a corrosion current density of i_{cA-B} for the baser metal in the presence of the nobler and a current density of i_{cB-A} for the nobler metal in the presence of the baser. In other words, the corrosion rate of the baser

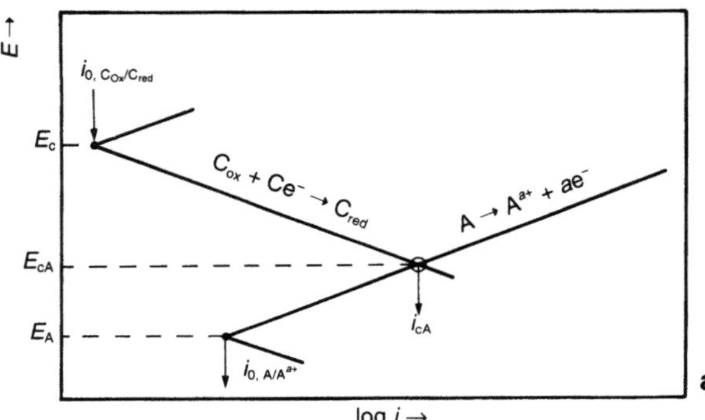

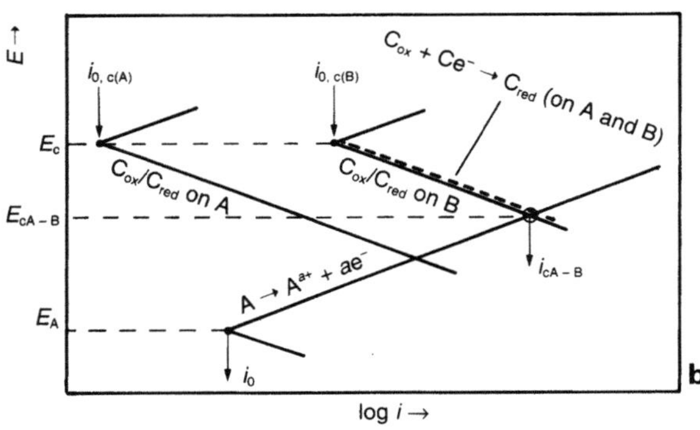

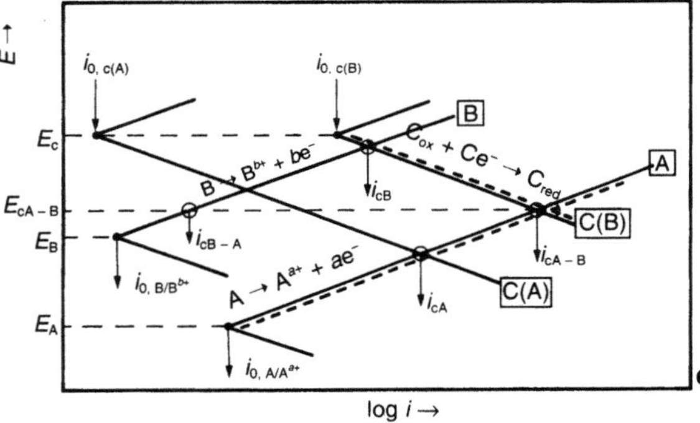

Fig. 3.9. Illustration of galvanic corrosion using mixed potential theory. (After Fontana and Greene [5].) **a** Mixed potential in absence of nobler metal (B). **b** Mixed potential when nobler metal is relatively inert. **c** Mixed potential when nobler metal is relatively active.

Table 3.2. Galvanic series for engineering alloys in seawater

Alloy series	Electrode potential (volts vs. SCE)	
Graphite	$0.2 \rightarrow 0.3$	
Platinum	$0.19 \rightarrow 0.24$	
Ni–Cr–Mo and Ni–Fe–Cr	$0.09 \rightarrow -0.03$	
Titanium	$0.06 \rightarrow -0.04$	
Austenitic stainless steels	$\begin{cases} 0.06 \rightarrow -0.1 \\ -0.36 \rightarrow -0.56 \end{cases}$	(passive) (active)
Nickel–copper alloys	$-0.03 \rightarrow -0.13$	
Silver and silver brazing alloys	$-0.1 \rightarrow -0.2$	
Nickel	$-0.1 \rightarrow -0.2$	
Ni–Cr alloys	$\begin{cases} -0.15 \\ -0.36 \rightarrow -0.46 \end{cases}$	(passive) (active)
Ni–Al bronze	$-0.15 \rightarrow -0.22$	
Cupronickels	$-0.18 \rightarrow -0.28$	
Lead	$-0.19 \rightarrow -0.24$	
Martensitic stainless steels	$\begin{cases} -0.2 \rightarrow -0.35 \\ -0.46 \rightarrow -0.56 \end{cases}$	(passive) (active)
Tin, silicon and manganese bronzes	$-0.24 \rightarrow -0.33$	
Admiralty and aluminium brasses	$-0.28 \rightarrow -0.35$	
Copper	$-0.3 \rightarrow -0.36$	
Tin	$-0.31 \rightarrow -0.33$	
Naval, yellow and red brasses	$-0.3 \rightarrow -0.4$	
Aluminium bronze	$-0.31 \rightarrow -0.42$	
Austenitic nickel cast iron	$-0.43 \rightarrow -0.54$	
Low–alloy steels	$-0.58 \rightarrow -0.62$	
C–Mn steels and cast irons	$-0.6 \rightarrow -0.71$	
Aluminium alloys	$-0.76 \rightarrow -1$	
Zinc	$-0.98 \rightarrow -1.02$	
Magnesium	$-1.6 \rightarrow -1.63$	

Source: LaQue [7].

metal is increased and that of the nobler metal is decreased.

The above considerations have implicitly assumed that the two metals are of approximately equal exposed area. The effect of surface area is more easily appreciated by considering plots of potential versus current (as opposed to current density) (Fig. 3.10).

For the simple case of the inactive noble metal illustrated schematically in Fig. 3.10, it can be seen that the effect of increasing cathodic area is to increase the exchange current and hence increase the corrosion current on the anodic material.

If one of the metals passivates then the galvanic effect can of course be beneficial, because the coupling of a noble metal to a base one may raise its potential into a region of passivity. This is the case, for example, for titanium in acid solution when coupled to platinum [6]. Galvanic corrosion is thus a complex interaction of electrode potentials, reaction kinetics, alloy composition, passivation characteristics, mass transport and solution properties. Geometric effects and the resistance of the coupling are also of importance [9].

If the geometric and electrochemical characteristics of the system are sufficiently well known, it is possible to calculate the potential distribution over the corroding surface using the Poisson equation (e.g. [10]),

$$\nabla . \sigma \nabla \phi + i_P = 0$$

with appropriate anodic and cathodic boundary conditions, and hence infer the corrosion rate over the entire surface.

This is only occasionally done for galvanic effects (Astley and Rowlands [11]), but is more common in cathodic protection analysis (see Section 3.4.1.2).

More often, a general guide to galvanic effects is adequate. This starts with an engineering galvanic series (Table 3.2) which represents the likely direction of any effect, thus identifying the nobler and the baser members of the couple.

The area effect is the next most important, and can be accounted for using tables which represent the results of cumulative experience of the scale of galvanic acceleration effects. BS PD6484 [12] contains detailed information on likely galvanic effects, but a useful summary can be found in Astley and Rowlands [11], reproduced as Table 3.3. The first column in this table shows the one-year 'uniform' corrosion rate in static seawater, and this and the acceleration factors do not include any effect such as crevice corrosion or pitting. As can be seen, cathode/anode size ratios of about 0.1 rarely result in enhanced corrosion due to galvanic effects, whereas ratios of 10 rarely do not. Of course, the anode area may become covered in an insulating deposit and thus the data should be interpreted with care. The table is included for illustrative purposes only, and the reader is referred to the original paper for details.

Heterogeneities in environment can also give rise to enhanced corrosion. For example, a metal which is immersed in two different environments will have each surface at a different potential. Even in the case of a change in

Table 3.3. General guide to corrosion rate acceleration factors due to galvanic coupling in static seawater

Material corroding (A)	Uncoupled corrosion rate* (mm/year)	Cathode member of couple (C)	Acceleration factor[†] C/A area ratio		
			10	1	0.1
Aluminium alloys	0.02	More noble materials	11	2	1
Low alloy and mild steels	0.04	More noble materials	12	2	1
Ni-Resist	0.03	More noble materials	9	2	1
Copper alloys	0.005	Carbon	(60)	(5)	(1)
		Monels Inconel 625 Stainless steels Titanium	15	2	1
		Copper alloys[‡]	5	2	1
Monel 400/K500 alloys	0.02	Carbon	(30)	(4)	(1)
		Inconel 625 Stainless steels Titanium	6	2	1
Stainless steels	0.001	Inconel 625	(6)	(2)	(1)
Inconel 625	<0.001	Titanium, carbon	...	1	...

Source: Astley and Rowlands [11].

*Uncoupled corrosion rates relate to uniform corrosion and are quoted for an exposure period of one year.

[†]Factors given relate to uniform corrosion and may therefore underestimate maximum depth of penetration arising from localised corrosion effects such as pitting and crevice corrosion. Such effects are often dependent upon minor environmental changes (such as sulphide pollution or deposit attachment) and the metallurgical condition of the material.

In flowing seawater (1 to 2 ms^{-1}), factors for aluminium alloys could be about ten times higher. The corrosion rate of uncoupled aluminium alloy is the same in flowing and static seawater. Available evidence indicates that factors for materials other than aluminium alloys will be the same in flowing and in static seawater, relative to the uncoupled corrosion rates in flowing and static seawater respectively. Uncoupled corrosion rates in flowing seawater are: low-alloy steels, Ni-Resist 0.10–0.15 mm/yr, copper alloys 0.01–0.05 mm/yr, Monels 0.002 mm/year.

Factors in parentheses have been derived on the basis of a limited amount of evidence.

[‡]The order of copper alloy nobility (i.e. most noble alloys likely to be the cathodes, least noble the anodes) is, in order of increasing nobility: group 1: LG4 gunmetal, phosphor bronzes, Si–Al bronzes; group 2: Ni–Al bronzes, 10% Al bronzes, copper; group 3: 10% cupronickel, Hidurons 191 and 501; group 4 (most noble): 30% cupronickel, 2% Al brass. Materials in the same group are usually compatible, so a unity rating applies. It should be noted that at a C/A area ratio of 10, there is a possibility of higher factors (i.e. 10–20) for group 1 anodes in combination with group 3 or 4 cathodes.

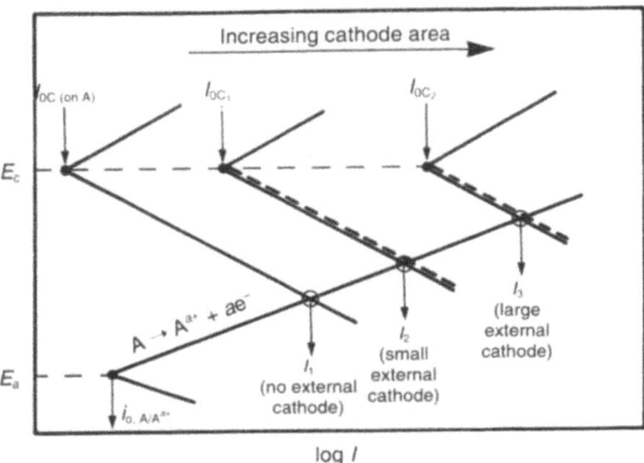

Fig. 3.10. Effect of variation in anode-to-cathode area on galvanic effect. (After Fontana and Greene [5].)

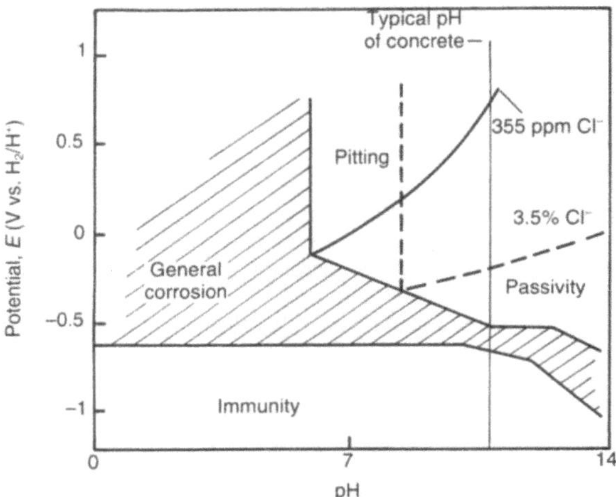

Fig. 3.12. Modification of the iron–water Pourbaix diagram to account for the presence of chloride ions. (From Wilkins and Lawrence [14].)

concentration of an active element in the environment, the potential of the corroding surface will change according to the Nernst equation. The effect of differential environment is amply illustrated by the differential aeration cell which can easily occur in marine environments. If access of oxygen is restricted to part of the surface, then the majority of the cathodic reaction will take place outside this area, and the sheltered area will develop a relatively high concentration of ferrous ions (Fig. 3.11).

Often, the cathodic area is much larger than the anodic area and this means that the effect of diffusion limited current density as described in Section 3.1.2 is somewhat reduced and the corrosion rate in the sheltered region can become very large. Fangteng [13] has studied differential aeration cells in flowing seawater, and has found that the effect of flow on the local Fe^{2+} concentration at the anode is the dominant factor and that increased flow over the cell causes an increase in the iron dissolution over-potential and hence increased corrosion in the anodic area.

Another example of a differential aeration cell which has some other interesting aspects is the corrosion of reinforcing steel inside concrete. Carbon–manganese steels are normally passive when embedded in concrete because of the high pH engendered by the existence of inter-pore water which is saturated with a basic oxide (lime). As can be seen from the Pourbaix diagram (Fig. 3.1), the potential range over which this passivity is maintained is quite large. When chloride ions are present, the diagram requires modification, and the work of Wilkins and Lawrence [14] (Fig. 3.12) addresses this problem. This means that, at a

typical pH inside concrete of about 12.5, normally passive steel may be subject to corrosion if chloride ions are present in sufficient quantity.

Both of the possible cathodic reactions (hydrogen evolution and oxygen reduction) are potential- and pH-dependent, as described in Section 3.1.1. As can be seen from the foregoing Pourbaix diagrams, the potential of steel in concrete (unless subjected to cathodic overprotection) is unlikely to be high enough for hydrogen evolution, so that the most likely cathodic reaction will be the oxygen reduction reaction. Of course, the concrete cover will severely limit the rate of oxygen supply to the steel surface, so that the cathodic reaction will be heavily concentration-polarised, although this will be offset to a large degree if the anode-to-cathode size ratio is small.

The basic model for the differential environment cell produced by chloride contamination of concrete (e.g. Browne and Domone [15]) is shown in Fig. 3.13.

The mechanism has been described as one where locally poor cover can lead to penetration of chloride ions to the steel surface and resulting depassivation and iron dissolution. The cathodic reaction can then be provided by the

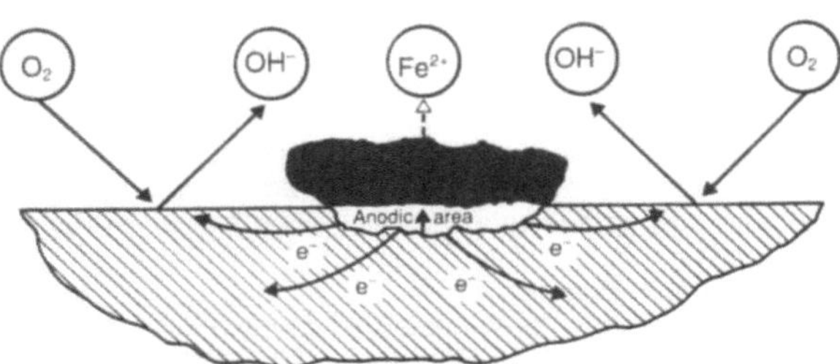

Fig. 3.11. Schematic of differential aeration cell set up at a sheltered site on a corroding steel surface.

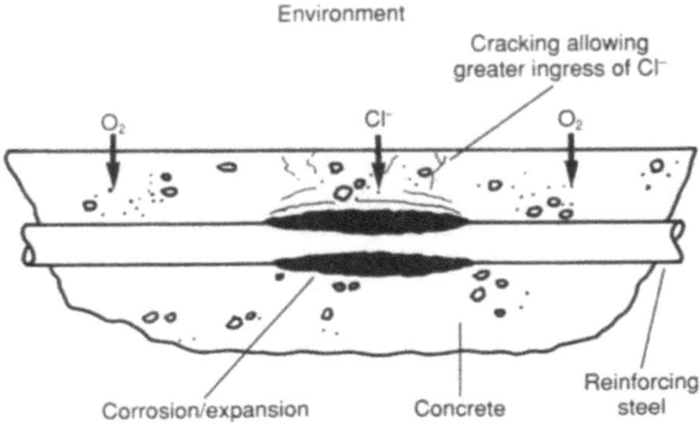

Fig. 3.13. Model for the differential environment cell set up as a result of chloride contamination of concrete. (After Browne and Domone [15].)

remaining area of passive steel, which, being large, can result in the formation of enough corrosion product at the anode to cause the concrete cover to spall. The spalling of the cover can then lead to enhanced chloride penetration, and acceleration of the deterioration mechanism. The quality, thickness and constitution of the cover are all important factors in determining the permeability to water (Powers et al. [16]) for the initial depassivation and also to oxygen (Gjorv et al. [17]) for the maintenance of the reaction.

Local metallic heterogeneities can also give rise to enhanced corrosion. Local heterogeneities are normally defined as those due to the phase structure of the metal as opposed to global heterogeneities caused by the coupling of two separate metals. The difference is only one of scale, and so the term 'micro-galvanic' has been used to describe such short-range corrosion effects.

Micro-galvanic effects are specific to the particular alloy and often lead to intergranular attack. For example, the $CuAl_2$ and Al_xCu_xMg precipitates in aluminium–copper alloys are cathodic to the regions adjacent to the grain boundaries on which they can form under certain conditions of heat treatment (Hatch [18]). These regions can show depletion of the solute (here copper or magnesium), which enhances the galvanic effect and can lead to intergranular corrosion penetration in extreme cases.

Another example of a micro-galvanic effect is in cast iron, where the relatively large cathodic graphite precipitates (see Chapter 4) enhance the rate of corrosion of the ferritic component of the iron. However, cast irons do not corrode particularly quickly compared with steels, and this is due to the formation of a layer of graphite on the surface due to selective removal of iron. In fact this graphitic layer can cause macro-galvanic problems if new cast iron is connected to old cast iron, for example as part of a repair (see, for example, BS PD6484 [12]).

Welding can produce a variety of galvanic corrosion problems. Heat-affected zone phase balance, the possible existence of cathodic defects and differences between weld metal and parent metal composition can all adversely affect corrosion from the galvanic point of view, but other mechanisms such as stress corrosion cracking can also be encouraged by welding (Rothwell and Turner [19]).

3.2.3 Effect of Surface Layers (Non-mechanical)

The interaction between surface layers and corrosion can lead to a series of localised corrosion problems which can render otherwise resistant metals liable to substantial damage. There are a variety of mechanisms by which such localised damage can occur, and these are mostly associated with the effect of local film breakage. Because of this, localised corrosion is often mechanically assisted, but this aspect is dealt with in Section 3.2.4.

The simplest form of such localised corrosion enhancement is due to galvanic action between the corroding metal and its corrosion products. For example, steel covered with mill scale is known to be more susceptible to pitting in quiet seawater than non-scaled steel, and the pitting cell can be envisaged to be formed at a break in the scale (Fig. 3.14).

Important aspects of this mechanism include the hydrolysis of ferrous ion inside the pit brought about by the migration of chloride ion into the pit. The chloride ions are attracted into the pit because of the preponderance of cath-

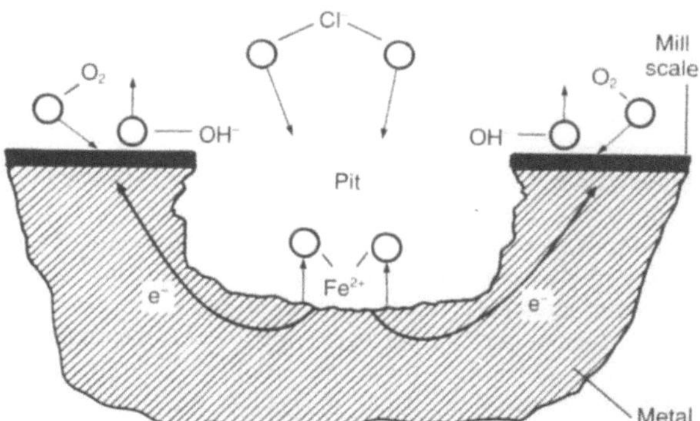

Fig. 3.14. Pit formation on a scaled surface in quiet seawater.

odes outside the pit and preponderance of anodes inside it. This produces a net positive charge inside the pit to which the chloride ions are attracted. The hydrolysis reaction

$$Fe^{2+}(Cl^-)_2 + H_2O = FeOH_2 + 2(H^+Cl^-)$$

then locally increases the pH leading to enhanced metal dissolution and intensification of the pitting effect. The galvanic cell eventually can be described as (Wilhelm [20])

$$Fe\,|\,1M\ HCl,\ pH\ 2\,|\,0.2M\ HCl,\ 30\ ppm\ O_2,\ pH\ 7\,|\,Fe_2O_3$$

Similar cells can be set up in many metal–oxide systems.

The stoichiometric and defect structures of the oxide are also of considerable importance in so far as they affect their dielectric properties and this, along with the potentials of the passivated and active surfaces, can be used to explain the pitting tendency of alloys in seawater (Wilhelm [19]). The pitting mechanism does not depend upon whether the oxide is a passive film, a thermally produced scale (as is mill scale) or an applied layer such as paint, except that the dielectric and electrodic properties of the layer will be different. Because of the diffusive nature of the mechanism, geometry, aeration and flow are also of importance in influencing localised corrosion of this type. For example, pits are more likely to be initiated on the upper surfaces of components because the concentrated solution in which hydrolysis occurs is denser than the environmental electrolyte. Oxides which can self-heal in aerated solutions (e.g. stainless steel) are more resistant to pitting in aerated conditions where there is good flow to prevent the oxygen concentration gradient which initiates the whole mechanism. The existence of crevices or other sheltered areas on the surface enhances the tendency to pit to such a great extent that pitting in sheltered areas is often called 'crevice' or 'gasket' corrosion.

In metals which form an adherent film the existence of surface heterogeneities in structure is very important in influencing the tendency to pit, because such heterogeneities can provide the local discontinuity in the surface film which acts as an initiation site. One example of this is given for inappropriately welded stainless steel in Chapter 5, but many others can be quoted.

3.2.4 Mechanically Assisted Corrosion

There are a number of corrosion processes in which mechanical action can enhance or localise the chemical effect and, as with localised corrosion, these are often much more damaging than the chemical effect alone.

One class of such mechanisms involves the removal by wear of the corrosion products which normally limit the corrosion rate after a period of time. For example, the corrosion rate of wrought iron and mild steel in seawater shows a decelerating behaviour (Fig. 3.15), and this gradual reduction is associated with the formation of a corrosion product layer on the surface. Although, for design purposes, this is normally modelled as a straight line with a non-zero intercept, the continuous removal of product layer would tend to alter the corrosion rate towards the tangent to the initial portion of the curve. Also, direct removal of metallic material by wear may play some part in the process. Such erosion-corrosion (or corrosion-erosion, usually depending upon whether the mechanical or electrochemical effect is the dominant) can be assisted by a number of different mechanical effects.

In flowing fluids a variety of different erosional effects can occur, depending upon the fluid-mechanical conditions. The removal of metal ions as they dissolve is one possible mechanism of enhancement of the corrosion rate, and this may also improve the supply of cathodic oxygen. Although not strictly corrosion-erosion, this stirring mechanism is responsible for part of the velocity effect observed in corrosion tests. As the fluid velocity increases there is the possibility of removing corrosion product through impingement and turbulence, and some corrosion products (e.g. siderite which is formed during carbonic acid corrosion of steels as described in Section 3.3.4) are particularly susceptible to this kind of enhancement. The removal mechanism may be by dissolution or by mechanical means or a combination of both. The entrainment of solid particles in the flow, especially if these are sharp, can result in substantial increases in corrosion-erosion rate. However, the most severe type of corrosion-erosion occurs during cavitation, when cavities formed in high velocity flows due to locally low pressure collapse when they encounter a region of higher pressure. Depending upon the differential pressure and the location of collapse, substantial damage to containment walls can arise. In this last case the mechanical component is usually dominant over the electrochemical.

Direct contact between two solid surfaces can also lead to enhanced corrosion by a product removal mechanism. Short-range reciprocating surface movement, such as might occur in vibration of a bolted joint or between the strands of wire rope, can give rise to an enhanced corrosion usually

called 'fretting' corrosion. The two surfaces need not be immersed in an electrolyte, since retention of liquid by capillary action is normally sufficient to maintain an aqueous environment even with intermittent immersion.

One of the most potentially catastrophic forms of corrosion occurs through the interaction of a surface tensile stress with a highly localised corrosion reaction. Stress corrosion cracking (SCC) can arise from a number of different detailed mechanisms (and some of these remain to be fully elucidated), but some features are common to all. Firstly, cracking is normally considered to initiate at local damage in passive or protective layers, and subsequent penetration can be either transgranular or intergranular, travelling roughly at right angles to the surface tensile stress (Fig. 3.16).

It is a feature of SCC that it affects specific alloy systems in specific environments, and just why particular combinations are affected is one of the puzzles which has led to an active research being maintained into the detailed mechanisms, on the grounds that this will eventually allow the development of resistant alloys. A large number of the systems which show SCC are now known, and some of these are collected in Table 3.4. Of particular importance in marine environments are the susceptibilities of stainless steels in warm chloride-containing solutions and of carbon–manganese and low-alloy steels in hydrogen sulphide-

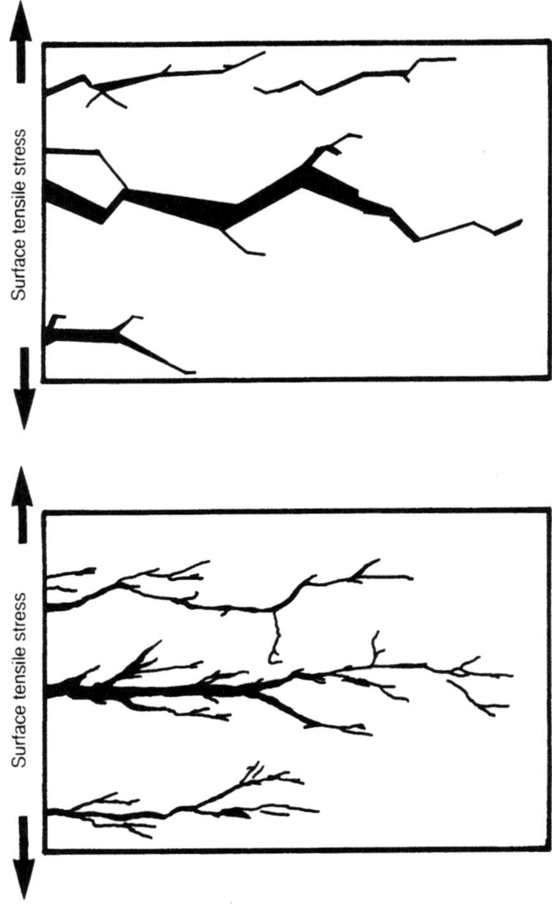

Fig. 3.16. Schematic morphology of intergranular (top) and transgranular (bottom) stress corrosion cracking.

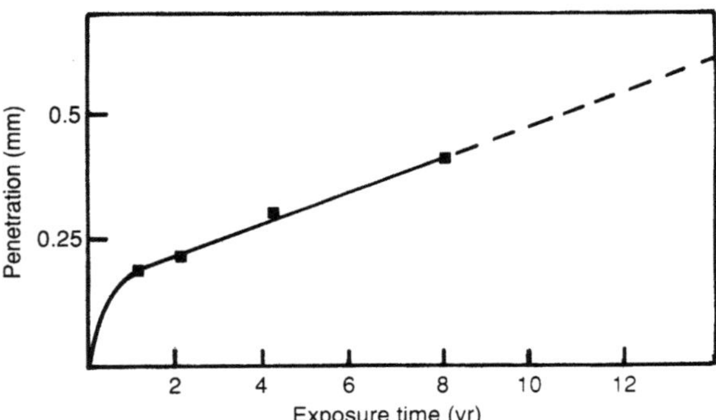

Fig. 3.15. Corrosion rate of wrought iron at mean tide in the Pacific Ocean. (From Schumacher [8].)

Table 3.4. Some alloy–electrolyte combinations in which stress corrosion cracking has been observed

Corrosive	Susceptible alloy types
Ammonia	Ferritic steels
	Brasses with 70–80% Cu and Zn, Sn or Pb
	Brasses with 59–93% Cu and Al, Zn or As
Chlorine	Austenitic stainless steels types 302, 304, 321, 347, 316, 317
Hydrogen sulphide	Ferritic steels
Polythionic acid	Stainless steels types 302, 304, 321, 347
Sodium chloride	Austenitic stainless steels types 302, 304, 321, 347, 316, 317

Source: Engel and Klingele [21].

containing solutions, although the latter is not strictly a stress-corrosion system, as is explained on p. 65. These statements are very general, and are only useful as a general warning. For instance, not all stainless steels are equally susceptible to SCC in chloride solutions, and many other factors such as the temperature, chloride concentration, applied stress, surface condition, and other aspects of the solution chemistry are all relevant. These are discussed in more details in Chapter 4 and the issue of steels in hydrogen sulphide is discussed further in Section 3.3.6.

Detailed mechanisms for SCC are still the subject of much debate (e.g. Staehle [22]), but the general tenor of the various models which have been proposed is that film formation prevents general dissolution and that localised attack takes place at the crack tips by some interaction between the applied stress (and the resulting intensification due to the presence of the crack) and the film, for example the local breakage of the film in the simplest case.

The corrosive component of SCC is not always an anodic one (i.e. dominated by dissolution processes at the crack tip), but may be cathodic, involving particularly the hydrogen evolution reaction. This reaction, as explained in Section 3.3.5, produces dissolved hydrogen at the crack tip, which can embrittle the plastic zone and allow crack propagation in steps, controlled by the stress intensity factor and the degree of hydrogen penetration. A distinction is sometimes drawn in the literature between this phenomenon and the anodic dissolution mechanism, and it is sometimes referred to as hydrogen embrittlement to amplify this distinction.

SCC (in its anodic and cathodic forms) is often treated as a mechanical phenomenon, and much of the current work is concentrated on measuring the chemical and mechanical conditions under which SCC takes place. Thus, instead of finding data on the susceptibility or non-susceptibility of an alloy to SCC in a given environment, it is more likely that a value of stress intensity factor, K_{ISCC}, will be found below which SCC will not occur, and it is important to be able to interpret this in terms of design. Since stress corrosion, unlike fracture, is a time-dependent process, some modification to the standard fracture toughness test outlined in Chapter 2 is necessary to establish criteria for SCC. Early fracture mechanics tests for SCC (e.g. Brown [23] and Knott [24]) consisted of pre-cracking standard fracture toughness test pieces and measuring the time to failure

under dead-loading conditions for a variety of values of initial stress intensity factor, K_i (since clearly the stress intensity factor will increase with time as the cracks propagate). The value of K_i at which no failure was observed was taken as K_{ISCC}.

Now, one of the most widely accepted ways of measuring stress corrosion susceptibility is by carrying out a rising load K_{ISCC} test consisting of three-point bending a notched and pre-cracked specimen in the relevant environment (e.g. Opoku and Clark [25]). Such testing is carried out with a very slow rate of loading (corresponding typically to a rate of 1 MPa m$^{1/2}$/min) and, as with fracture toughness testing, the load–displacement records can be used to determine the load and hence the value of K_I at which cracking commences (Fig. 3.17).

Fatigue crack propagation can also be aggravated by corrosion at the crack tip. Mechanisms which have been proposed to explain this include enhanced dissolution of heavily strained material, local film breakage at stress concentrations, crevice effects and pumping of electrolyte into and out of the crack tip under opening and closing displacements. Hydrogen can also influence fatigue crack growth rates if it is able to diffuse into the crack tip region, giving

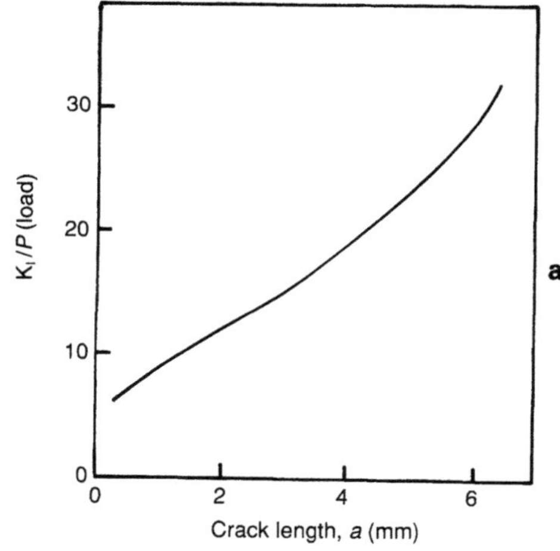

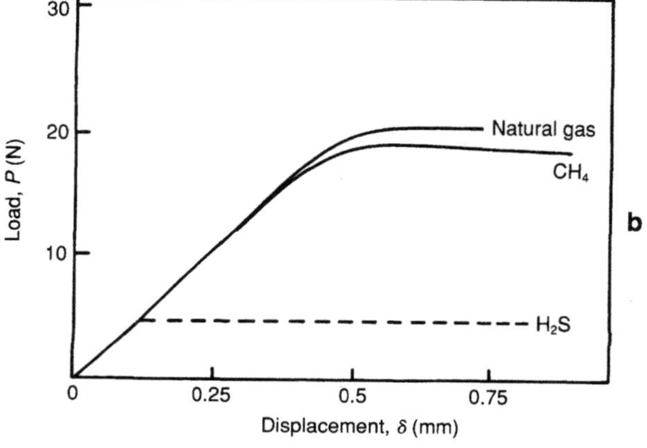

Fig. 3.17. Typical rising load SCC test: **a** calibration of three-point bend specimen of height 12.5 mm, thickness 10 mm and span 40 mm; **b** test results for AISI 4340 steel in natural gas, methane and hydrogen sulphide. (From Opoku and Clark [25].)

rise to a combined fatigue and hydrogen embrittlement effect. In fact, more generally, all of the environmentally assisted cracking mechanisms can take place under alternating stresses as well as static ones.

Early investigations of the corrosion fatigue problem centred mainly around the comparison between S–N data for fatigue tests carried out in air and in the environment in question, and such data are still used by designers. Since corrosion is a time-dependent phenomenon and fatigue crack growth is dependent on numbers of stress cycles experienced, it might be expected that the superposition of the two phenomena would render corrosion fatigue sensitive to frequency, with the corrosive component being more important at lower frequencies. This is certainly observed (Gilbert [26]) and, further, corrosion has also been observed to remove the fatigue limit. The overall effect of corrosion is generally to lower fatigue strength, but in some circumstances initiation is more difficult than purely mechanical fatigue at higher stresses (Fig. 3.18).

For most engineering alloys in seawater, the slope of the S–N curve becomes more steeply negative and the fatigue limit disappears. For steels, a substantial drop in fatigue strength with frequency has been observed [28], and this is as might be expected, with the smallest corrosive effect in the high-frequency and high-stress-range regime.

For structural steels there is also considerable interest in the effect that cathodic protection has on corrosion fatigue. Certainly, if the potential is too low, one might expect a reduction in fatigue life from hydrogen evolution, but, for properly protected structures, the fatigue life should lie somewhere between that seen in air and that seen in seawater under freely corroding conditions. Such data as are available suggest that cathodic protection at –0.85 V results in as good a fatigue performance as in air [29], and possibly a better one. Hooper and Hartt [30] have found a detrimental effect on fatigue life by overprotecting at – 1.25 V (Fig. 3.19). Congleton et al. [31] have investigated the effect of cathodic protection on corrosion fatigue of rope wires and have found a similar effect with potential.

Fatigue crack propagation and initiation are both affected by corrosion. The effect on propagation is seen more clearly by fracture mechanics tests for fatigue, as outlined in Chapter 2. Plots of da/dN against ΔK isolate the propagation component and, as can be seen from Fig. 3.20, cor-

rosion and cathodic protection have similar effects on propagation as were seen on the S–N curves.

When corrodants which can produce stress corrosion or hydrogen embrittlement are present, the phenomenon of corrosion fatigue begins to resemble incremental stress corrosion and the behaviour becomes dominated by this under

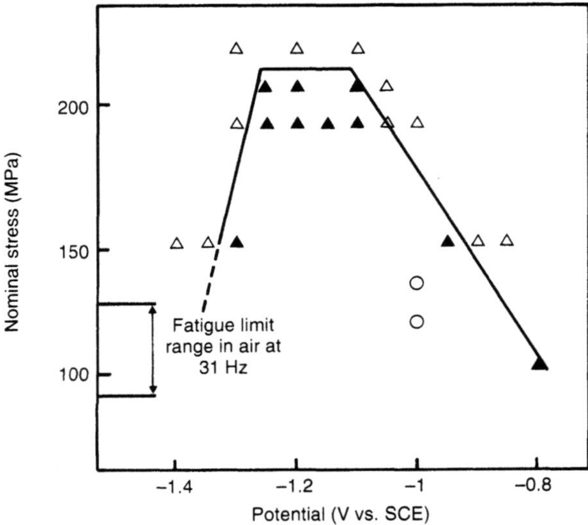

Fig. 3.19. Effect of CP potential on the corrosion fatigue strength of AISI 1018 steel – open points denote failure and solid points no failure, circles denote 3% NaCl solution and remainder are seawater. (From Hooper and Hartt [30].)

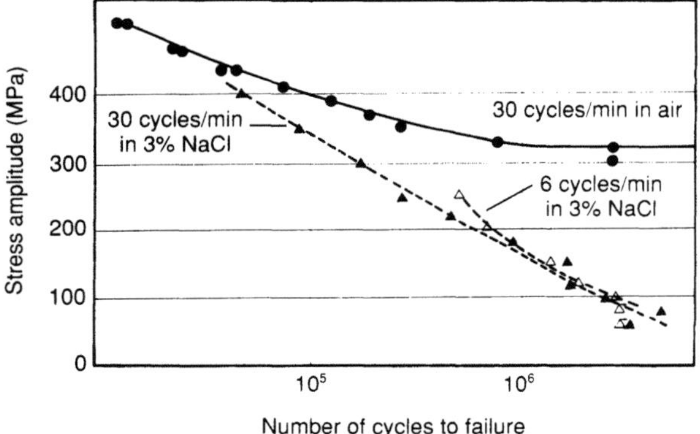

Fig. 3.18. Effect of corrosion and frequency of loading on the fatigue performance of a 0.17C–1.35 Mn–0.35Si steel. (From Jaske et al. [27].)

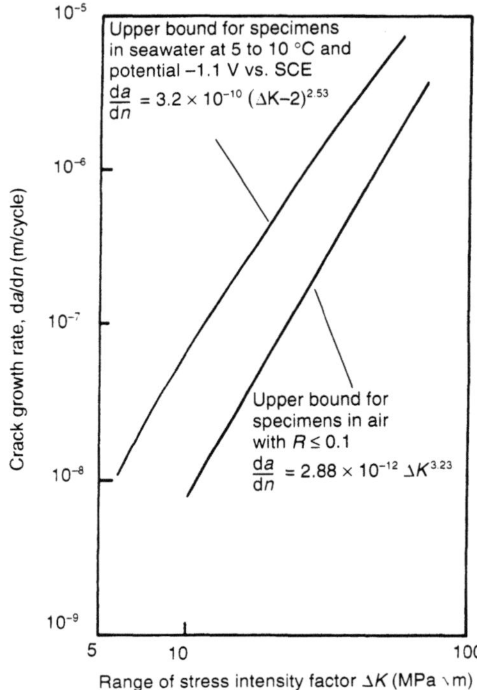

Fig. 3.20. Corrosion fatigue crack growth rates for BS 4360 Grade 50D structural steel at a frequency of 0.1 Hz. (From Jaske et al. [27].)

more severe conditions of SCC. As well as propagation, initiation of corrosion fatigue is aided by the presence of SCC corrodants, and overall S–N behaviour shows a substantial reduction in fatigue life.

The interaction between environmentally assisted cracking and corrosion fatigue is most easily understood from a fracture mechanics point of view. The effect depends upon whether the stress intensity factor during the cycle exceeds K_{ISCC} or not (Congleton [32]). If it does, da/dN is seen to increase with decreasing frequency, because cycles which have significant hold times where $K_{\mathrm{I}} > K_{\mathrm{ISCC}}$ will have a larger time-dependent (rather than cycle-dependent) component of crack propagation. In such circumstances, a substantial effect of stress ratio (R) on fatigue life would also be expected. If K_{I} does not exceed K_{ISCC} during the cycle, it seems [32] that a decrease in crack propagation rate with increasing frequency is again likely, but that the effect of stress ratio on the environmental contribution is weak or absent.

3.2.5 Deterioration of Polymers

Polymers cannot be said to suffer from corrosion in the sense of dissolution of the material into an electrolyte by electrochemical reactions. However, they can suffer from environmental deterioration, and it is pertinent in this chapter to consider how this can occur.

The main processes by which polymers can deteriorate are physical or chemical in nature, rather than electrochemical as is the case in metallic corrosion. Thus, polymers can undergo swelling or dissolution in the presence of certain environments, and the molecular structure of the polymer may be attacked by chemical reactions such as oxidation, or by heat which can, for example, cause partial reversal of polymerisation reactions. In addition, radiation (even in the form of sunlight) can accelerate degradation processes such as oxidation. Pollutants (especially the oxides of nitrogen and sulphur) and some solvents can cause degradation by dissolution. The absorption of water and other substances can also lead to reduction of function. Because of the nature of the weathering process, it is difficult to measure directly the degree of attack, but it is clear from the foregoing general discussion that the standard method used for metallic corrosion (i.e. loss of thickness) will not be useful.

The principal degradation mechanisms for polymers in the marine environment are the absorption of ultra-violet radiation, water absorption, biodeterioration and environmental stress cracking.

Photo-decomposition commences with the production of free radicals as photons are absorbed by polymer molecules. For example, hyperoxide groups can be split by the following reaction:

$$\mathrm{ROOH} \rightarrow \mathrm{RO^{\cdot}} + {^{\cdot}\mathrm{OH}}$$

and the free radical produced can subsequently react with oxygen, the overall process being known as 'photo-oxidation' (Mills [33]). For example, the rate of oxygen uptake due to purely thermal effects and under photo-assisted conditions can be compared in Fig. 3.21. Photo-assisted degradation is normally overcome by the use of absorbing additives in the polymer.

The absorption of water is another example of the fact that polymers are generally more permeable than metals to

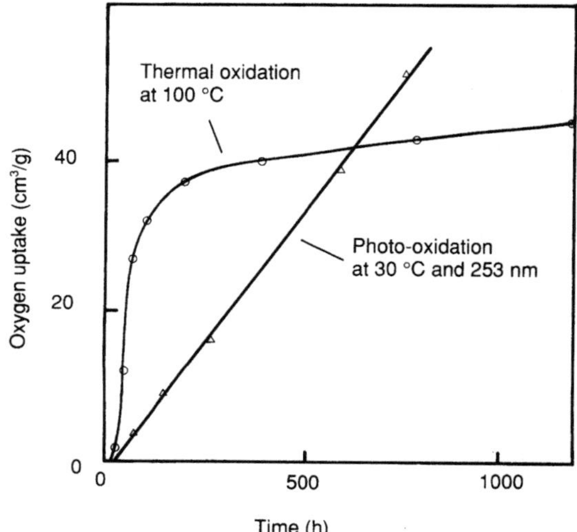

Fig. 3.21. Comparison of the rates of uptake of oxygen by polyethylene film due to purely thermal effects and under photo-assisted conditions. (From Mills [33].)

many common molecules. This means that water can dissolve and diffuse in many plastics and this can result in a loss of property. In particular, the glass transition temperature may be reduced, and also the strength. Irreversible mechanical damage (such as microcracking, especially under conditions of alternating wetting and drying) and chemical damage to the polymer structure may also result (Allen and Bauer [34]).

Failure of plastics after absorption of moisture can occur by an enhancement of creep, by loss of strength or by enhanced crack propagation during fatigue. For fatigue, there is evidence of enhancement of crack propagation rates in some polymeric materials such as nylon 6/6, whereas others such as polycarbonates and other polyesters seem not to be affected, or affected in a way which depends on ΔK [34]. For epoxy resins, Comyn [35] reports cases of both toughening and embrittlement in the presence of water, highlighting the need for very specific data for these materials. Resin–matrix composites can suffer from additional damage mechanisms besides those above for the resin itself. These additional mechanisms can occur at the interface between fibres and matrix, for example delamination, or in the fibre itself, for example stress cracking (Allen and Bauer [34]).

Environmental stress cracking is often presented as the polymeric equivalent to stress corrosion cracking in metals, and indeed the overall phenomenological effect is similar. Although this phenomenon has been known to occur in aqueous solutions (e.g. polythene in detergent–water solutions, Crawford [36]), it is not of sufficient importance in marine environments to warrant detailed description.

The biodeterioration of polymers is described in Section 3.3.5.

3.2.6 Deterioration of Other Materials

Timber, being a biological material, is particularly susceptible to biological and water absorption phenomena. In fact, it can be argued that timber is 'designed' to be biodegradable, and so major steps have to be taken to halt or retard this process. The hydrolytic, oxidative and photochemical degradative processes are usually of secondary importance to the biological ones in timber, but these last processes are discussed separately in Section 3.3.5 and discussion of the others here does not imply that they are of more concern in the marine environment.

The physical and chemical degradation mechanisms of timber mostly involve water transport (either into or out of the timber) and this requires some discussion. To begin with, timber, unlike engineered materials, contains moisture in its natural state and is normally dried to use. Water is held in the cell walls and also in the cell cavities as free moisture, and this latter is usually lost first on drying but has relatively little effect on strength. Drying beyond the fibre saturation point, however, causes shrinkage of the wood and an increase in strength [37]. Thus, moisture content of the wood cell wall has a substantial effect on its strength (for a given type of timber), and it is usual to express this variation [37] as

$$\log S = \log S_p + k\,(M_p - M)$$

where S is any strength property corresponding to a moisture content M and the corresponding subscripted values refer to the fibre saturation point. The constant k is characteristic of the species, and results must be interpreted in the light of the statistical nature of timber properties.

The drying behaviour of timber can be reversed in humid or wet conditions, so that loss of strength in marine conditions will depend not only upon the moisture–strength rela-tionship but also upon the permeation properties of the timber in question, as well as any protective measures which have been taken.

The effects of ultra-violet light and weather (e.g. rain) are generally slight, usually only amounting to about 1 mm of surface loss every 20 years (Illston et al. [38]). Resistance of timber to acids is also generally good, although alkalis dissolve both the lignin and hemicellu-loses. Iron salts can cause hydrolytic degradation, and in particular the corrosion of metallic fasteners can cause local decay and failure of the fastening [38].

Ceramic materials are generally regarded as being immune to corrosion or weathering processes, but concrete, owing to its particular nature, should be considered separately from other inorganic materials. The question of corrosion of embedded steel has already been described in Section 3.2.2, and so the following discussion is confined to degradative processes which affect the concrete itself. It should be pointed out that these are secondary processes, the corrosion of reinforcing steel normally being regarded as the main mechanism of structural degradation.

Fookes et al. [39] have identified the main deterioration mechanisms and these are illustrated in Fig. 3.22. The rela-tive importance of these is dependent primarily upon the climate, but also upon the locally available materials and workmanship. Apart from reinforcement corrosion, Fookes et al. suggest that chemical decomposition of the cement paste is an important degradation mechanism. These authors also emphasise the importance of climate on deteri-oration, and suggest that degradation is worst in hot, dry climates.

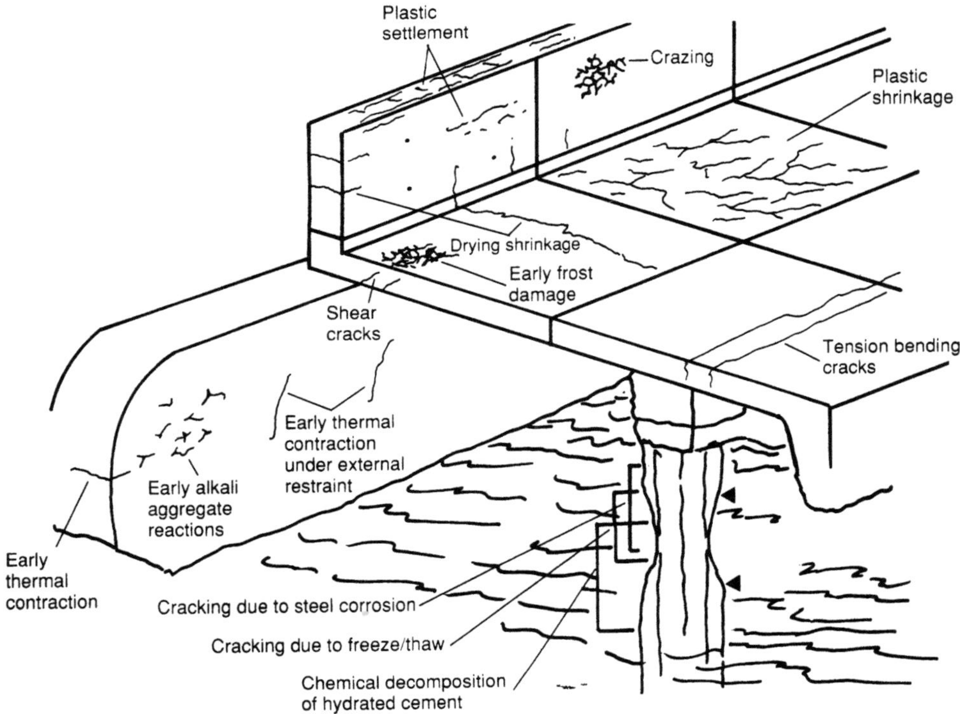

Fig. 3.22. Some deterioration mechanisms in marine concrete. (From Fookes et al. [30].)

3.3 Corrosivity and Aggressiveness of Specific Environments

In the previous section some of the mechanistic principles of corrosion and deterioration of materials in marine environments were discussed. The purpose of this section is to outline what it is about certain specific environments which needs to be known in order to assess their aggressiveness towards materials. The environments mentioned all have relevance to marine technology, although they may not be directly associated with seawater.

3.3.1 The Marine Atmosphere

From the point of view of corrosion, atmospheres are generally classified as marine, industrial, rural or desert (Griffin [40]). Of these, the marine atmosphere is generally considered to be the most aggressive, certainly for mild steels (Brown and Masters [41]).

Apart from the qualitative aspects indicated above, a number of other parameters affect the corrosivity of the marine atmosphere. These include moisture, temperature, weather factors (such as wind and rain), location relative to sea, attitude of corroding surface, pollutants and biological organisms.

Moisture is one of the most important of these features, both as humidity and rainfall. In atmospheric corrosion studies it is normal to use a quantity such as time of wetness to quantify the moisture aspect of the corrosivity. The time of wetness is normally measured by the use of moisture sensors on the exposed surface (Sereda et al. [42]), but takes no account of the chemical or electrochemical content of the spray or dew.

The principal chemical conditions which affect marine atmospheres are the airborne levels of salt and contaminants such as sulphur and carbon dioxides. Solids, such as sand, dust and ice will also have an effect, although this will in turn be linked to wind speed. The airborne salt level is, of course, related to height above sea level and distance from the sea, and at close distances even a few tens of metres can make a substantial difference to corrosion rates. Sulphur dioxide contamination has been found to increase the corrosion rates of zinc and of steel in marine atmospheres [41]. As well as the effects of distance and height already referred to, the orientation of the corroding surface is also important, as this affects exposure to spray, runoff, solar and wind drying and washing by rain, as well as any abrasive effects. It is particularly important that differences in these factors be taken into account when interpreting test data for a particular application.

For ultra-violet degradation of plastics, the solar irradiance is of some importance. Generally plastics are most affected by radiation in the wavelength range from 290 to 400 nm, with a peak sensitivity which depends upon the particular plastic. Natural daylight has a fairly flat spectrum in the 400 to 800 nm range, with a gradual drop towards 250 nm (Delre and Miller [43]).

Temperature has an effect on the brittleness and strength of plastics which is due simply to the physical transitions which occur in the material. Many can become brittle on exposure to low temperatures, and elevated temperature can lead to loss of strength (see Chapter 4 for more detail).

In ferritic steels also low temperatures can lead to embrittlement (Chapters 2 and 4).

3.3.2 Seawater (Including Spray, Splash, Immersion and Handling)

Some of the aspects of seawater chemistry have been described in Chapter 1, but here those parameters will be dwelt upon which particularly affect corrosion. Perhaps the most important factor affecting corrosivity of seawater is its oxygen content. For carbon steels the effect of oxygen is to increase substantially corrosion rates, but other metals such as stainless steels generally perform less well in poorly aerated conditions. Other dissolved gases such as carbon dioxide, ammonia and hydrogen sulphide have effects which are usually specific to the type of alloy in question [44].

The salinity of seawater does not vary much in the open sea, but can be locally reduced at river water inflows or increased in relatively enclosed seas with high temperatures (for example the Red Sea). The major effect of salinity is that it substantially increases water conductivity, with the result that large surfaces can be brought into electrolytic contact. This also has consequences for cathodic protection, as is described in Section 3.4.1.2. The presence of chloride ions also has some particular implications for the corrosion of alloys which are protected by passive films, as described in Section 3.1.2. Finally, calcium, magnesium and strontium carbonates can produce 'chalking' (precipitation of solid) on cathodic surfaces (owing to the change in pH brought about by the cathodic reaction), which can lead to a reduction in cathodic protection load or even a reduction in free corrosion rate.

The interdependence of salinity, oxygen solubility and temperature is described in detail by Dexter [52] and is covered in Chapter 1.

Corrosivity of seawater is also controlled by some physical parameters. The most obvious of these is the corrosion effect exhibited at the water surface, where intermittent wetting and drying can occur and the seawater is generally well aerated, normally referred to as the 'splash zone'. For metals susceptible to corrosion acceleration by oxygen, the splash zone is the area in the marine environment with the highest free corrosion rate. It is also difficult to protect, because it may be subject to abrasion and also cannot be cathodically protected.

Water depth can also have some effect on corrosivity, through its effects on oxygen content, temperature and hydrostatic pressure and the variations in these quantities have been discussed in Chapter 1. Generally, increasing temperature results in an increase in corrosion rate, but this must be balanced against the fact that the solubility of oxygen in seawater decreases with increasing temperature. Hydrostatic pressure can also have an effect upon chemical potentials of aggressive agents. For example, the highest partial pressure of hydrogen sulphide produced in an anaerobic bacterial environment under normal conditions is one atmosphere. If, however, the bacterial reaction takes place on the seabed then the chemical potential of dissolved hydrogen sulphide will be that in equilibrium with the local hydrostatic pressure of hydrogen sulphide. Finally, hydrostatic pressure is known to affect pitting phenomena. For example, Beccaria and Poggi [45] have found that the susceptibility of aluminium alloys to pitting in seawater is increased by increasing hydrostatic pressure. This effect was attributed to increased thermodynamic activity of the chloride ions and to increased solubility of the passivating oxides.

In some circumstances it may be necessary to handle seawater, and this can involve avoidance measures that differ from those employed in the case of environmental exposure to seawater. A few examples of this are cooling, desalination units, fire systems and hydrocarbon reservoir injection. For many of these applications water treatment is required, and indeed some of these are specifically for such treatment. In general, seawater is taken into such systems in its raw form, when it is often at its most corrosive. Subsequent treatments remove aggressive agents such as dissolved gases, and also extract biological contaminants. Often inhibitors and/or biocides are added. In some cases, such as desalination or cooling applications, the water may subsequently suffer an increase in temperature.

The velocity of seawater, whether in a handling system or in the environment, can, at low velocities, lead to reduced dissolved metal ion concentration and enhanced oxygen concentration at the corroding surface. The first of these will always enhance the corrosion rate (see Section 3.1.1), whereas the second will depend on whether the alloy passivates or not. Higher velocities can also result in corrosion-erosion, depending upon the velocity and the susceptibility of the alloy.

Oldfield et al. [46] have reviewed data and proposed a model for corrosion in deaerated seawater for handling systems. This model covers such variables as alloy type, temperature, oxygen content and flow conditions. For example, for carbon steels where the corrosion rate is limited by the diffusion of oxygen to the surface, Oldfield et al. suggest that a pipe carrying seawater with oxygen concentration C_0 (ppb) travelling at a speed U_0 (cm/s) will show a corrosion rate given by:

$$\text{corrosion rate (mm/yr)} = \frac{0.0565\, C_0 U_0}{Re^{0.125} Pr^{0.75}}$$

where the flow effect is contained in Reynolds' number (Re) and the Prandtl number (Pr) contains part of the temperature effect (related to diffusion of oxygen) and the remaining temperature effect is contained in the dissolved oxygen content. They suggest that their model is applicable to relatively scale-free metal, or where turbulence or other mechanical or thermal action results in removal of any surface scale.

The addition of biocides (particularly chlorine) during water treatment to prevent fouling may have its own implications as regards corrosion. Goodman [47] has reviewed the effect of chlorination on seawater chemistry and on corrosivity.

3.3.3 Mineral, Mud and Hydrocarbon

Many marine activities are associated with the winning and transportation of seabed and sub-seabed resources, and it is pertinent to consider the possible corrosion processes associated with these activities.

Any process which involves the piping of a solid-laden fluid runs a considerable risk of corrosion-erosion, and the particle velocity in the fluid will of course influence the erosion rate. One example of this might be the entrainment of sand and other mineral matter during hydrocarbon production. The drilling of oil and gas wells involves the use of mud, which is usually a slurry of heavy oxides in a fluid carrier. The production of any mineral material from the seabed may also involve the pumping of suspended solids [48].

The corrosivity of seabed sediments is usually expressed in terms of the chemical and physical characteristics of the overlying seawater. The factors which influence the corrosivity of sediments may be positive or negative, in other words the sediment may be more or less corrosive than the seawater. These factors are often biological in nature and these aspects are dealt with in Section 3.3.5. As with metals buried in soils, the main non-biological factors influencing corrosion in marine sediments are oxygen content and access (often related to the particle size distribution), pH, temperature, salt and contaminant content, and resistivity [49]. As with seawater, it is normal to protect carbon–manganese steels cathodically when they are buried in sediments and the resistivity is an important factor in the design of such CP systems. King [50] has considered the physical and biological factors in the corrosivity of UK North Sea marine sediments and has mapped this corrosivity for the area in question.

Hydrocarbons themselves (i.e. oil and gas) are not corrosive to metals, and any corrosivity is associated with water content, oxygen, hydrogen sulphide and carbon dioxide, all of which are dealt with in other sections of this chapter. The high pressures at which oil and gas are produced and transported can, however, mean that very low concentrations of contaminants can assume very large partial pressures, and any chemical potential is enhanced accordingly.

Plastics can be susceptible to damage by organic chemicals. The effect may be either physical (e.g. absorption, permeation) or chemical (e.g. chain scission). This degradation can often be assisted by stress, producing cracking, crazing or brittle failure (Tonyali [51]), although swelling may also be a problem for some polymers. A specific example of this is included in part of Case Study 7.5.

3.3.4 Carbon Dioxide

When carbon dioxide dissolves in water it produces the weak acid H_2CO_3:

$$CO_2 + H_2O = H^+ + HCO_3^-$$

and

$$HCO_3^- = H^+ + CO_3^{2-}$$

In seawater, carbon dioxide is produced by air–sea exchange and by photosynthesis:

$$CH_2O + O_2 \rightarrow CO_2 + H_2O$$

The pH of seawater (controlled principally by inorganic carbon content) ranges between about 7.5 and 8.3, and, apart from aluminium alloys, this has little direct effect on corrosivity (Dexter [52]). The high levels of magnesium, calcium and carbonate ions in seawater mean that a small increase in hydroxyl ion concentration such as occurs when the oxygen reduction process takes place (particularly when accelerated by cathodic protection) can result in the precipitation of scales of aragonite (calcium carbonate) or, more exceptionally, bruceite (magnesium hydroxide) onto the cathodic surface. This can result in a stifling of the oxygen reduction reaction, which can either reduce the corrosion rate in freely corroding conditions or reduce the cathodic protection load. The thermodynamics of this effect are dependent upon local chemistry and temperature, and it is not normal for designers to rely upon its taking place.

In conditions of elevated pressure, for example in natural gas production or drilling, dissolved carbon dioxide can considerably increase corrosion rates in carbon–manganese steels, most probably through its effect on cathodic reaction rates (Eriksrud and Sondvedt [53]). DeWaard and Milliams [54] have quantified this effect as a function of temperature and carbon dioxide partial pressure for natural gas pipelines, and have produced an empirical equation for the corrosion rate, v, in millimetres per year:

$$\log(v) = 7.96 - \frac{2.32 \times 10^{3}}{t + 273} - 5.55 \times 10^{-3} t + 0.67 \lg(P_{CO_2})$$

where t is the temperature ($^\circ$ C) and the partial pressure of carbon dioxide (P_{CO_2}) is in bar. This relationship is still widely used in the prediction of carbonic corrosion rates in oil and gas transportation and production systems and in assessing the corrosivity of formation waters from oil and gas reservoirs. The formation of iron carbonate scales during this type of corrosion is recognised as being an important factor in the typical 'mesa' type morphology [54] which is commonly taken as being characteristic of 'sweet' or carbonic corrosion. It is further recognised that these scales can reduce substantially the corrosion rate from that predicted by the deWaard and Milliams equation at higher temperatures, but also that the effectiveness of such protection is compromised by erosive effects including rapid flow (Eriksrud and Sondtvedt [53]}).

3.3.5 Biological and Microbiological Environments

Biological and microbiological factors can be of considerable importance in determining corrosion and degradation susceptibility of materials in the marine environment. These factors affect all materials to a greater or lesser extent, and can be either physical or chemical in nature. The phenomenon of fouling, although not strictly a degradation process, is also dealt with in this section.

When a solid is immersed in seawater, it will almost immediately (within about 2 hours) form a film of non-living organic material. Shortly afterwards, a bacterial film begins to colonize this layer, producing a layer of slime (mostly protein-rich polymeric by-products of bacterial activity) which alters the biochemistry of the surface and forms a receptive surface for macrofouling organisms (Dexter [52]).

Although microbes may produce metabolic products which influence the corrosivity of the environment either locally or generally, they may also play a role in increasing corrosion by affecting the corrosion half-cell reactions themselves. As far as metallic corrosion is concerned, the most important micro-organisms are the sulphate-reducing bacteria of the species *Desulfovibrio desulfuricans*, although a number of others of the fungi, algae, aerobic and anaerobic bacteria may play some part (Tiller [55]).

The sulphate-reducing bacteria, as their name suggests, carry out the dissimilation of organic matter by reducing sulphate ions. They also, incidentally, can take up molecular hydrogen, so that the effect of their metabolism upon the environment can be represented by the reaction:

$$SO_4^{2-} + 4H_2 \rightarrow S^{2-} + 4H_2O$$

Sulphate-reducing bacteria (SRBs) thrive under anaerobic (oxygen deficient) conditions where there is not normally a cathodic reaction (oxygen reduction or hydrogen evolution) which can proceed at a high enough current to produce substantial corrosion. It has traditionally been supposed that SRBs contribute to the mechanism in such environments to enhance the corrosion rate. For example, the following mechanism has been proposed by Kuhr and Flugt (e.g. see Tiller [55]):

$$4Fe \rightarrow 4Fe^{2+} + 8e^-$$
$$8H_2O \rightarrow 8H^+ + 8OH^-$$
$$8H^+ + 8e^- \rightarrow 8H$$
$$SO_4^{2-} + 8H \rightarrow S^{2-} + 4H_2O$$
$$Fe^{2+} + S^{2-} \rightarrow FeS$$
$$3Fe^{2+} + 6OH^- \rightarrow 3Fe(OH)_2$$

This produces an overall reaction:

$$Fe + 4H_2O + SO_4^{2-} \rightarrow 3Fe(OH)_2 + FeS + 2OH^-$$

However, it would appear that the corrosion rate is independent of enzyme activity, which is to be expected if the bacterial action to control the corrosion rate through its metabolic behaviour is as indicated above. Alternative mechanisms have been proposed which recognise the fact that iron sulphide is cathodic to mild steel and can act as a cathodic depolariser, e.g. King and Miller [56].

Apart from the effect upon corrosion kinetics outlined above, SRBs can metabolise the sulphate ions present in seawater to produce hydrogen sulphide solutions. This is particularly likely in stagnant seawater and it is, in principle, possible for all of the sulphate ion present in the seawater (about 2500 mg/l) to be converted to dissolved hydrogen sulphide (about 1000 mg/l), which is, in fact, close to saturation (the solubility of H_2S in water is between 2 and 4 g/l, depending upon temperature and salt content [57]). Field measurements of hydrogen sulphide production are rare, but Wilkinson [58] and Gooch [59] have reported possible levels of hydrogen sulphide from putrefying organic matter in seawater between 50 and 170 mg/l, the former author suggesting that a doubling time of about 10 h is possible for H_2S concentration. The metabolism of SRBs requires some source of organic carbon, and Cordruwisch [60] has estimated that each 1000 g of degradable organic material should support the generation of 630 g H_2S. The inorganic effects of hydrogen sulphide on corrosion of steels are discussed in Section 3.3.6.

Bacterial corrosion can affect a number of other materials in marine environments. For example, concrete may be susceptible to the acidic conditions brought about by the action of anaerobic bacteria. Hall [61] has identified *Thiobacillus concretivorous* in waste water as being harmful to concrete via the reaction of lime with sulphuric acid produced in the pore water as a result of metabolic processes.

Much of the recent work on biodeterioration of plastics has been aimed at making them more susceptible (for ecological purposes), and this is a measure of the characterisitic durability of most engineering plastics to microbial attack. Most of the reported biodeterioration mechanisms in plastics are physical and are associated with macro-organisms, which are dealt with later in this section.

Wood is the most susceptible of all materials to bio-effects, which is hardly surprising given its natural origins. In the literature more attention is paid to the macro-biological processes of timber degradation, although Boutelje and Goransson [62] suggest that micro-organisms (bacteria and fungi) are likely to play a part in the degradation of timber below the ground water table. Resistance to micro-organisms can, of course, be improved by various impregnation treatments. Jones and Irvine [63] indicate that a number of marine Ascomycetes and Deuteromycotina (Fungi Imperfecti) can cause soft rot in timber, but report that little is known of the tolerance of these organisms to preservatives.

The effects of macrofouling species on materials performance are rather more wide-ranging than those due to micro-organisms. Apart from the obscurement of surfaces which require regular inspection, the increased static, dynamic and fluid loading and the choking of filters, pipes and heat exchangers, some macro-organisms can cause direct physical damage to materials.

As might be expected, timber is susceptible to such attack. Perhaps the best known type of macro-organisms affecting timber in the marine environment is the ship-worm (*Teredo* sp.). These are in fact bivalves in which the shell is reduced to a boring instrument at the end of a worm-like body, and a schematic diagram of the effect of the worm is shown in Fig. 3.23. A number of other borers can affect timber, including *Limnoria* sp., which are isopods resembling woodlice [64]. Turner [65] classifies the wood borers into those which penetrate the solid simply for support or shelter and those which use the wood as a source of nutrient. As well as the two genera mentioned above, Turner mentions the *Xylophagainae* and *Sphaeromiidae* belonging in the former classification and *Martesia* and *Lignopholas* in the latter as being the most important organisms involved in the breakdown of wood in the marine environment. For a given set of protective measures, the danger of infestation is dependent upon the abundance of the invasive species in the specific locality or localities encountered. As an interesting example, Kuhl [66] quotes the grading system suggested by the Prussian Institute of Sea Bottom and Air Hygiene for *Teredo* infestation and notes that the grading for the Elbe estuary increased substantially between 1936 and 1951.

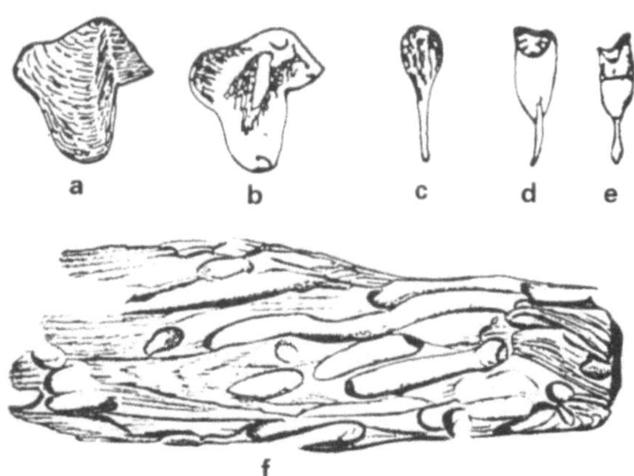

Fig. 3.23. Wood-boring shipworms (Teredinidae): **a**, **b** outer and inner views of shell of *Teredo norvegica*; **c**, **d**, **e** pallets of *T. norvegica*, *T. megatora* and *T. navalis* respectively; **f** wood which has been attacked by *Teredo* sp. (From Barrett and Yonge [64].)

Plastics can also suffer from the physical effects of marine macroflora and fauna. For example, Scott [67] reports that borers can attack many plastics in the marine environment, causing damage that varies from etching of the surface to total penetration. A number of field studies have been carried out at various sites and with various materials, one of the widest of these being by Connoly et al. [68, 69]. One of the major conclusions of this work was that no single class of polymer was completely immune to attack by borers (principally *Pholadidae*), although some were more resistant than others. It would appear that surface condition is an important factor influencing the likelihood of attack and the proximity of 'bait' materials such as wood and jute roving can have an important effect. Wolock [70] has concluded that most polymeric materials (including composites) can be attacked by borers, but that the damage is usually slight, even for lengthy exposures.

The obvious physical effects of fouling on offshore structures made of steel or concrete are of particular concern with regard to inspection requirements. However, the connection with corrosion is a rather more tenuous one. A number of arguments as to why macrofouling communities may enhance corrosion of structural steel have been advanced ranging from local chemical and biochemical conditions to interference with cathodic protection systems and removal of coatings by shell growth (e.g. Pipe [71]). However, little quantitative information is available which would allow designers to act on such suggestions, and the variability of the fouling community, even on adjacent platforms (e.g. Forteath et al. [72]), would seem to make this an unlikely prospect. Concrete does not appear to be susceptible to damage by marine fouling organisms. For example, surveys of structures [73] which have been in marine service for more than fifty years, although fouled, show no evidence of damage resulting from the fouling.

3.3.6 Hydrogen Sulphide

As discussed in Section 3.35, hydrogen sulphide can be produced in seawater by bacterial action, and a specific set of corrosion reactions is associated with this action when it occurs on the surface of metals and non-metals. This section deals only with the effects which hydrogen sulphide-containing solutions may have on metals and does not consider the means of production of the hydrogen sulphide. This is relevant to marine corrosion because it is possible that hydrogen sulphide may be produced remotely from the corrosion site (for example as is the case in sour crude).

As indicated in Section 3.3.2, corrosion in hydrogen sulphide environments can give rise to extremely high partial pressures of hydrogen gas, which can then dissolve in metals (and other materials). Because of their very small size, hydrogen atoms are quite soluble and highly mobile in many metals, but can often have a substantial effect on their properties. The effects of hydrogen in metals, particularly ferritic steels, have received an enormous amount of attention, principally because of the number of manufacturing and environmental processes which can give rise to dissolved hydrogen, for example welding, pickling, plating, corrosion and cathodic charging and, of course, because of the profound effect that even small amounts of dissolved hydrogen can have on the properties of ferritic steels.

Hydrogen sulphide, when dissolved in water or, more potently, in an electrolyte such as seawater, can react with steel in a way which can be described by the reaction

$$H_2S + Fe \rightarrow FeS + H_2$$

For this the equilibrium partial pressures of hydrogen and hydrogen sulphide (for Fe and FeS activities at unity) can be given (Opoku and Clark [25]) by

$$\lg \frac{P_{H_2}}{P_{H_2S}} = 11.3 \text{ at } 25\,^{\circ}C$$

which, even for a hydrogen sulphide partial pressure of 1 standard atmosphere (1.0133×10^5 Pa), results in a hydrogen partial pressure of around 2.5×10^{11} standard atmospheres (2.533×10^{16} Pa).

Mechanistically, the reaction is far more complex than suggested above, but the overall effect is that there is the potential to dissolve hydrogen in the steel at a concentration which is in equilibrium with this very high pressure. Since, at best, this is only a surface equilibrium, hydrogen will diffuse down the chemical potential gradient into the steel, producing an overall embrittlement. The response of the steel to the hydrogen depends very much on its strength and cleanliness in almost the same way as does the susceptibilty to hydrogen cracking during welding (see Chapter 5). There are three distinct, recognised responses of steels to hydrogen sulphide corrosion, which are largely related to their strength (Terasaki et al. [74]). These are blistering, hydrogen-induced (or stepwise) cracking (HIC) and sulphide stress corrosion cracking (SSCC or SSC), and these are illustrated schematically in Fig. 3.24.

Blistering is a phenomenon which is normally associated with steels with relatively low strengths. It occurs when the dissolved hydrogen encounters a heterogeneity in the steel at which its chemical potential can develop a sufficiently high hydrogen pressure to produce debonding at the interface between the heterogeneity and the steel matrix (Denham and Taylor [75]). The net effect is blistering within the steel, and individual blisters can become quite large and crack through to the surface of the component affected. Like lamellar tearing (see Chapter 5), this problem is largely associated with steels in which little control of inclusion content was practised and is now less commonly seen.

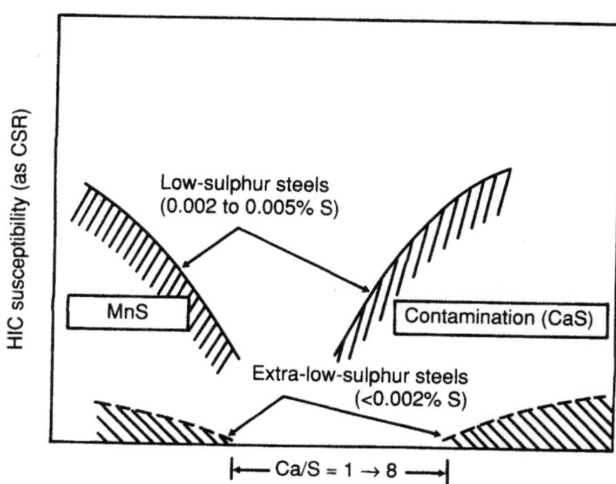

Fig. 3.25. Effect of inclusion shape control on susceptibility to hydrogen-induced cracking. (From Taira et al. [76].)

Hydrogen-induced cracking is a version of blistering which takes place in higher-strength, cleaner steels. Phenomenologically, it consists of gaseous hydrogen evolution at an inclusion but also an enhanced hydrogen solubility in the plastic stress concentrations at the edges of the developing delamination [74]. The combination of an intensified stress and a high level of dissolved hydrogen can lead to cracking between inclusions, which has the characteristic 'stepwise' appearance shown in Fig. 3.24. The susceptibility to HIC is therefore dependent upon steel chemical composition, heat treatment condition and the size and shape distribution of inclusions. This susceptibility is measured by a standard NACE test (see Section 3.5) and, for example, the plot of CSR value against Ca/S ratio shown in Fig. 3.25 illustrates the value of inclusion shape control on susceptibility (Taira et al. [76]).

Sulphide stress corrosion cracking is a particular example of hydrogen-induced cracking and therefore occurs only where there is an applied or inherent (e.g. residual) tensile stress at the corroding surface. The morphology of cracking varies with the strength of the steel, and low-strength steels tend to show short subsidiary cracks parallel to the surface, although a similar effect can also be seen in some high-strength steels (e.g. in wires along the drawing texture). Susceptibility to SSC is highly dependent upon steel heat treatment and chemistry and much less on inclusion content than is HIC, and can be assessed using the NACE test referred to in Section 3.5.

Other metals can also be affected by hydrogen sulphide, either by increased general corrosion or by embrittlement and/or cracking mechanisms, and some data on this are included in Chapter 4.

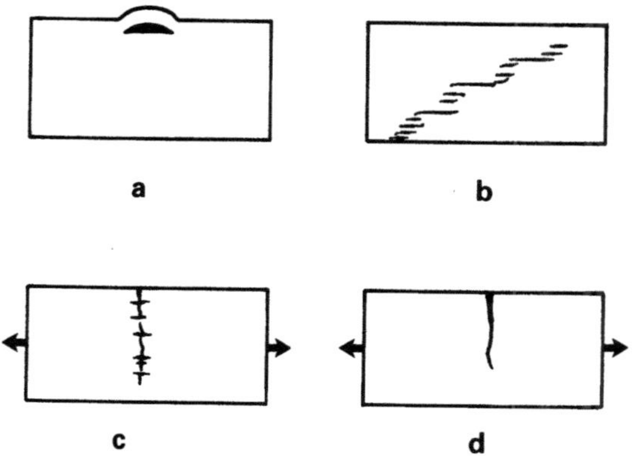

Fig. 3.24. The morphologies of hydrogen damage in steels: **a** blistering; **b**, HIC or stepwise cracking; **c**, sulphide stress corrosion cracking in low-strength steels; **d**, SSCC in high strength steels. (From Terasaki et al. [74].)

3.4 Corrosion Protection

In this section are included not only protection against corrosion of metals, but also the ways in which other materials such as polymers, concrete and timber can be protected against the degradative mechanisms dealt with in Section 3.3.

Clearly, the approaches which can be taken against degradation depend largely upon the mechanism involved. For example, the steps to be taken against corrosion of metals and against, say, UV degradation of polymers will be completely different. However, the rate of metallic corrosion, especially of steel in seawater, means that it is by far the most important and consequential of the processes, so that proportionately more effort has been put into its avoidance and the degree of detail in each of the following subsections reflects this.

3.4.1 Prevention of Metallic Corrosion

Before embarking upon a strategy for corrosion protection, it is first necessary to have an appreciation of the likely forms of corrosion expected and the likely extent to which they may occur. If the design cannot survive in an acceptable state with the expected corrosion (for any reason, be it engineering or aesthetic) then steps will be required to prevent the problem.

There are five basic methods for preventing or minimising corrosion in metallic systems:

1. to exclude the corrosive environment by use of a coating;
2. to depress the potential of the corroding surface hence opposing the thermodynamic tendency to corrode (cathodic protection);
3. to raise the potential of the corroding surface encouraging the formation of passive layers, if these will form on the material in question (anodic protection);
4. to alter the environment and render it non-corrosive (inhibition);
5. to select a material which is acceptably resistant to the environment in question.

Many corrosion protection schemes contain an element of more than one of the above, but it is convenient to discuss these separately. Item 5 is a matter of materials selection based on known data and experience and is therefore covered in Chapter 4.

3.4.1.1 Coatings

At their simplest, coatings can be regarded as barriers which prevent access of the corrosive environment to the surface to be protected. Whereas this is accurate in some cases, it should be realised that coatings are rarely totally impermeable and further that they may involve an element of inhibition or cathodic protection. Also, since coatings can never be chemically or mechanically indestructible, the possibility of local or general damage must also be considered as part of the coating selection process. In coating technology, small areas where the coating is not operating correctly, owing either to damage or to inadequate application, are known as 'holidays'.

Although much of the work on development of coatings has been aimed at the ability to apply to poorly prepared surfaces, it is still essential to ensure that adequate bonding is achieved at the metal–primer interface, and some type of cleaning or other preparatory surface treatment is usually required to do this. Johnson [77] has identified the major detrimental materials at paint–steel interfaces as greases and oils, moisture, chemical contaminants and oxides (mill scale or rust), and these need to be removed to achieve an acceptable coating lifetime. As well as surface preparation, it is also well recognised that the quality of application has an important influence on the performance of a coating in service and, like welding, coating application requires to be subjected to a rigorous inspection procedure. Whitehouse [78] has surveyed UK practice for the painting of offshore structures, and has stressed the importance of application and preparation to ensure the expected life of the coating system.

Coatings are most often used in the marine environment for the corrosion protection of steel in non-submerged applications where cathodic protection is inapplicable, or as a back-up to cathodic protection. Many different types of coating are available and these can be classified according to whether they are predominantly metallic, inorganic or organic. Table 3.5 summarises the different coating types and methods of application.

Metallic coatings can be applied by a variety of means including plating, hot dipping, spraying and cladding. The metal applied may be either anodic or cathodic to the substrate and different behaviour can be expected depending upon which of these is chosen.

Cathodic metallic coatings are normally chosen because of the inherently corrosion resistant nature of the cathodic material. For example cupronickels and monels (nickel–copper alloys) have been used (Carruthers [79]) as claddings on steel, particularly for splash-zone protection. An additional advantage of using a cupronickel for this purpose is the resistance of copper-base alloys to the settlement of fouling organisms. Designers need to be aware of

Table 3.5. Summary of coating types and means of application

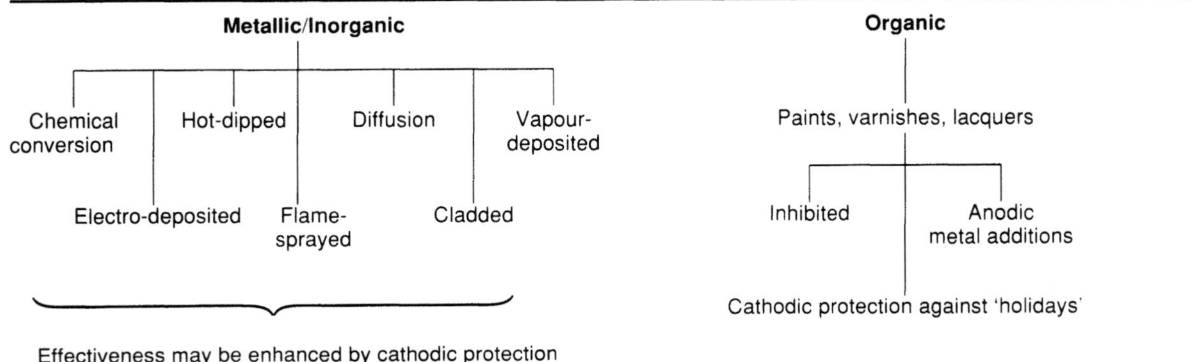

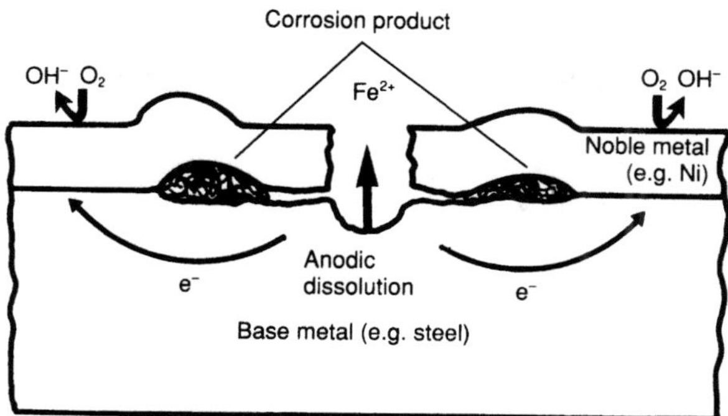

Fig. 3.26. Possible effect of disruption of a noble-metal cladding on a baser metal.

the possible effects of holidays in such claddings where the substrate can form the anode of a galvanic cell if the cladding is damaged (Fig. 3.26).

Such problems can be reduced by non-intimate contact between coating and substrate, thus increasing the anode to cathode area ratio and by the application of cathodic protection (a situation which has been considered by Astley and Rowlands [11] for copper pipes coupled to a steel hull). The claddings are also usually of substantial thickness, so that the likelihood of damage is low. Cathodic metallic coatings are also used widely as linings for corrosive service. For example, stainless steel linings for pipelines are used where the lining is a good fit and is often forged in place during pipe drawing (Yoshida et al. [80]).

The most familiar application of anodic metallic coatings is in galvanising, where a layer of zinc is applied to the substrate, usually by hot dipping. The advantage of an anodic coating is that the substrate will form the cathodic member of any galvanic couple produced by local disruption of the coating (Fig. 3.27).

Provided that the self-corrosion rate of the coating metal is reasonably low, this forms a stable coating system. Other examples of anodic metallic coatings include flame sprayed aluminium which has been applied, for example, as the protection of the high-strength tethers on a tension-leg platform [81]. The overcoating of flame-sprayed anodic metal with resin provides a considerable enhancement of life, especially in atmospheric service.

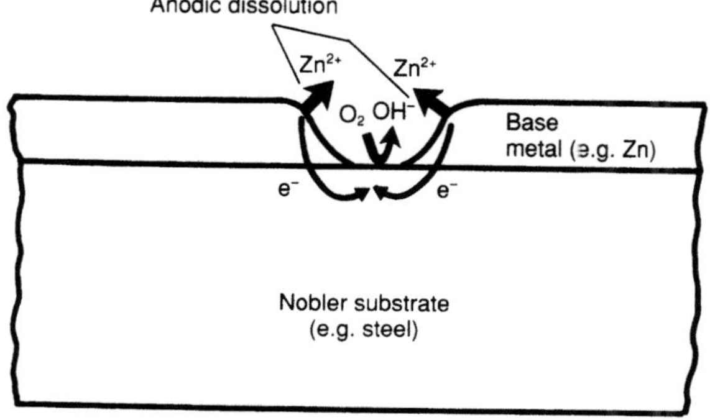

Fig. 3.27. Effect of disruption of a base-metal cladding on a nobler metal.

Inorganic coatings can also be applied by a variety of means including fusion, surface conversion and casting. The materials used are themselves usually oxides, and are therefore resistant to corrosion. Some examples include the artificially thickened oxide on aluminium produced by anodising, concrete and fused glass (vitrification). Inorganic materials are often part of a more general coating system, but do not usually form the only protection against corrosion.

Organic coatings form by far the greatest part of those used in the marine environment, owing mainly to their ease of application and, more particularly, maintenance. As mentioned previously, the preparation of the surface and the application are very important in determining coating life, but appropriate selection of type and thickness is also essential. Organic coatings are polymer-based and as such are rarely totally impermeable to oxygen and moisture as are metallic coatings. (Of course, concrete is also permeable to water and oxygen.) The requirements of a marine coating system (this usually is a composite of a number of composite coatings each composite coating laid on top of the next) can be summed up as follows [7]:

1. *Moisture.* The coating should be resistant to immersion and alternate wetting and drying. It should show a low permeability to water and also have a relatively low solubility for water.

2. *Other mass transfer.* The coating should also resist the passage of the ionic species which make seawater electrolytic, in particular chloride, sulphate and carbonate. If differences of solution concentration can be set up on either side of the paint film and the film is semipermeable (i.e. able to transmit water) an osmotic pressure can build up in the more concentrated solution (usually that under the paint), leading to blistering and possibly local failure. Thus, resistance to osmosis is also an important property of a coating system.

3. *Electrical properties.* First, coatings should be dielectric in nature so that they can resist the passage of electrons from a potentially large cathodic area to anodic sites on the substrate. It is also necessary for the coating to be resistant to electro-osmosis, a cross non-equilibrium thermodynamic phenomenon whereby mass (in this case water) is transported down a gradient of electrical potential (see e.g. Moore [1]). Since most paint coatings are of negative polarity and the metallic areas around a holiday are negatively charged, a phenomenon known as electroendosmosis can also occur, whereby moisture is forced through a membrane under a potential gradient in the direction of the pole which has the same sense of charge as the membrane.

4. *Environmental resistance.* Clearly the external surface of the coating must be resistant to the various degradative influences operative in the marine environment. For instance, it must be resistant to weathering without loss of thickness or continuity, and must be chemically resistant to seawater and other materials which may come into contact with it, including fungal, bacterial and other biological materials. The coating must also be resistant to shock and abrasion, depending upon the forseeable loadings that it might be expected to experience.

5. *Interfacial phenomena.* A coating cannot simply lie on its immediate substrate, whether this is the metallic surface or another coating. Firstly, adhesion is of primary importance to resist the transport effects described above,

and this adhesion must be uniformly good over large coated areas. If damaged, the coating should also be resistant to undercutting again by virtue of good adherence under aggressive conditions. It is also helpful for the coating to have inhibitive properties, so that corrosion is resisted even if bare metal is exposed.

6. *Application.* It is fundamentally important that the coating can be easily applied in the required thickness(es) without recourse to elaborate preparation or application methods. In particular, the 'hiding' properties of the coating are important, because this determines its ability to cover surface projections to the same thickness as flat surfaces. The ability to repair the coating locally to its original properties is also helpful.

The different specific chemical formulations of coatings and their properties are dealt with in Chapter 4. Organic coatings can be classified as paints, which consist basically of a dispersion of solid particles in a liquid which dries to leave a solid film on the surface (Chandler and Bayliss [82]), or as plastics which are cured on the surface, although this classification is not always distinct. The main components of a paint are binders, which are the final film formers in the dry film, pigments and extenders, which provide opacity and reinforcement against abrasion, solvents and diluents, which evaporate on drying, the former leaving the binder behind. Other additives are included, for example, to prevent skinning, accelerate drying or reduce sagging in thick films. Some pigments also perform special functions such as acting as inhibitors (useful only in the priming coat). The addition of zinc to primers also affords a certain degree of cathodic protection to the substrate if damaged and exposed to an electrolyte (Chandler and Bayliss [82]).

Thus it can be seen that a paint system is a very complex composite material with a number of simultaneous chemical and physical functions. Most coating systems consist of primers, undercoats and top coats, although some high-build coatings may be capable of protection with a single coat. As an example of the range of alternatives for offshore coating systems, Coke [83] has surveyed the fabrication and maintenance paint systems used for a number of platforms in the Gulf of Mexico comprising six different fabrication coating systems and seven different maintenance coating systems. Even with a given system, the effect of film thickness and surface preparation can be substantial.

3.4.1.2 Cathodic Protection

Cathodic protection consists of depressing the potential of a corroding surface to inhibit the anodic dissolution reaction. In simple thermodynamic terms, this can be understood as supplying an excess of electrons to the corrosion cell. For example, the corrosion of steel in seawater,

$$Fe \rightarrow Fe^{2+} + 2e^-$$

$$O_2 + 2H_2O + 4e^- \rightarrow 4OH^-$$

can be reduced by cathodic protection, where the supply of electrons to the surface can be seen as opposing the dissolution of iron and also aiding the reduction of oxygen. Care must be taken not to reduce the potential too far because this may allow cathodic evolution of hydrogen, which may lead to embrittlement of the steel.

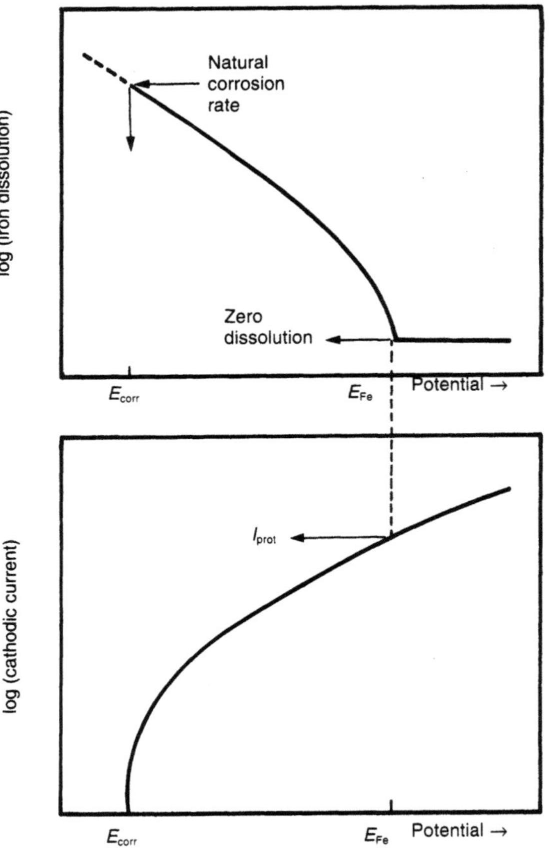

Fig. 3.28. Schematic diagram of the relationship between cathodic current and the corrosion of steel. (From Ashworth [84].)

Alternatively, cathodic protection can be understood in terms of the polarisation behaviour of the metal to be protected, and the appropriate current–potential diagram is shown in Fig. 3.28.

Figure 3.28 shows that once a sufficiently high cathodic current is applied (I_{prot}), the dissolution rate of iron drops to zero, and this corresponds to a potential reduction from E_{corr} to E_{Fe}. Further reduction in potential is not worthwhile, and the first increments of potential drop are the most effective because of the logarithmic relationship between potential and current in corroding systems.

The design of cathodic protection systems is aided by an electrical model for the corrosion cell, including a continuous metallic surface on which the anodic and cathodic reactions are taking place (not necessarily all on the same metal) in a continuous electrolyte. This behaviour can be modelled by the system of batteries and resistances shown in Fig. 3.29.

The corrosion current can now be given [85]:

$$I_{corr} = \frac{E_a - E_c}{R_m + R_a + R_c}$$

where the subscripts refer to the anode, cathode and metallic resistances. If an external circuit is now added, as in Fig. 3.30, with no metallic resistance, and if the impressed current, I_x, is sufficient to make the anodic current zero, then

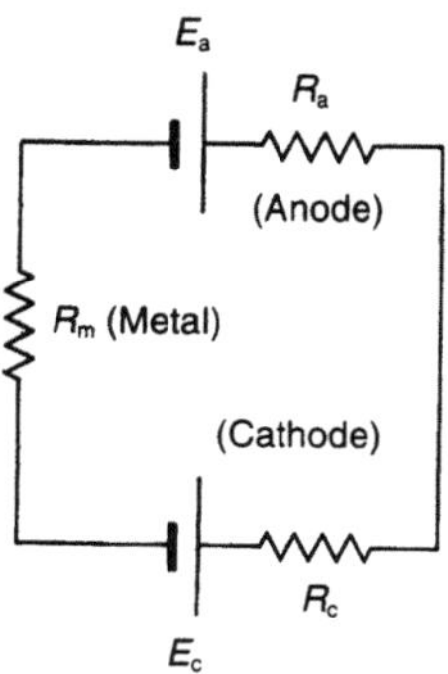

Fig. 3.29. The equivalent circuit of a simple electrochemical cell (From Morgan [85].)

we can see, using Kirchhoff's laws for networks (e.g. Morgan [85]), that

$$E_a = E_c + R_c I_x,$$

so that the application of a sufficiently large external current will result in reduction in the anodic current and an increase in the cathodic current in the corrosion cell. In the case of steel in seawater, this means that the iron dissolution reaction is impeded whereas the oxygen reduction reaction is enhanced. Of course, the anodic and cathodic reactions are subject to polarisation, so that the simple model above should be modified somewhat to account for this. The polarisation of anode and cathode can be expressed in terms of their open-circuit potentials, E_o, and polarization functions, $\phi(i)$, where the i are current densities:

$$E_a = E_{o,a} - \phi_a(i_a)$$
$$\text{and} \qquad E_c = E_{o,c} - \phi_c(i_c)$$

so that the condition of zero anode current is reached when

$$E_{o,a} = E_{o,c} - \phi_c\,(I_x/A_c) + R_c I_x$$

This indicates that protection is more easily achieved (lower required current) when cathode resistance is large and/or when the cathode is highly polarised.

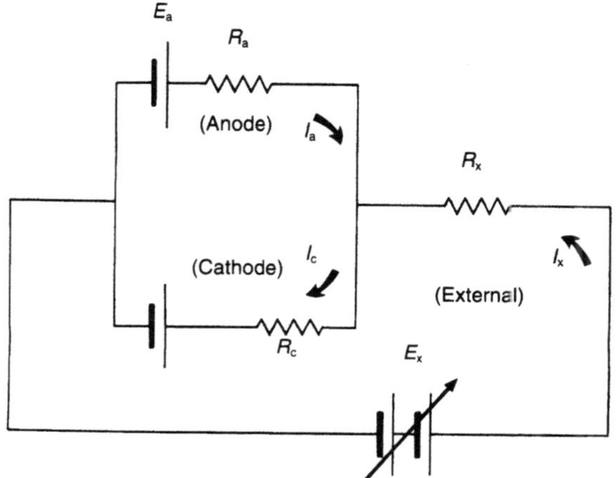

Fig. 3.30. The equivalent circuit of a simple electrochemical cell to which an external current has been applied. (From Morgan [85].)

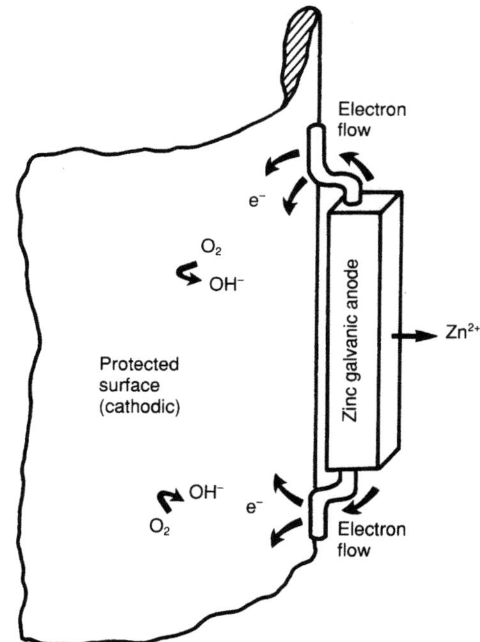

Fig. 3.31. Schematic diagram of cathodic protection using galvanic anodes.

In the practical case, different electrolytic resistances from external source to anode and cathode can be admitted, as well as an electrolytic resistance between anode and cathode. As before, the metallic resistance between anode and cathode is usually regarded as being negligible.

There are basically two ways of achieving cathodic protection, and these are normally referred to as 'protection by galvanic anodes' and 'protection by impressed current'.

In the first case (Fig. 3.31), the protective current is applied by welding blocks of a base metal to the surface to be protected. The lower potential of this base metal produces a similar effect to galvanic corrosion but in reverse, so that the corrosion rate of the anode is enhanced over that normally expected and corrosion is reduced on the nobler metal.

For protection of steel in seawater, suitable galvanic anode metals are zinc and aluminium with the possibility of magnesium, although this last has rather a low potential and is also consumed quite rapidly, so that the former two are generally preferred. The self-corrosion rate of the anode is also of importance in order to achieve long-term protection without the use of excessive amounts of galvanic material.

Protection using impressed current requires the use of an auxiliary anode which is made the positive pole of a d.c. supply where the structure to be protected is the negative pole (Fig. 3.32).

Impressed current anode materials can, in principle, be made from any conductor, although it is advisable to use a material with a low self-corrosion rate and high-current-density carrying capacity. Anode shapes are therefore often more elaborate than in the galvanic case, with variations on hemispheres, rods, rings and others. Whereas galvanic anodes typically operate at current densities of around 25 A/m² and consumption rates of around 10 kg/(A yr), some platinised surfaces for impressed current use can go as high as 700 A/m² at consumption rates as little as 10⁻⁵ kg/(A yr).

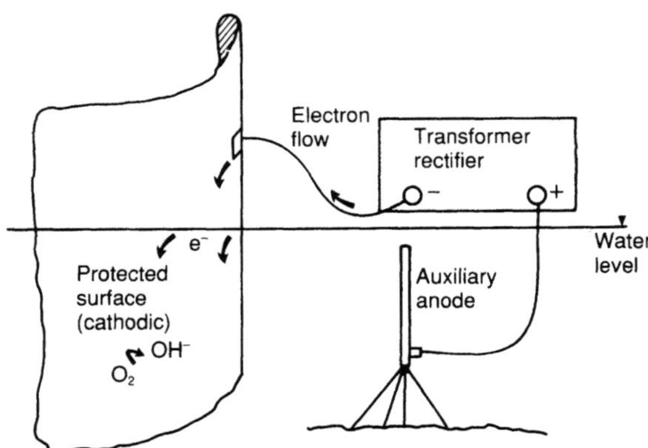

Fig. 3.32. Schematic diagram of cathodic protection using impressed current.

The advantages of galvanic anodes come principally with their passive nature and self-regulating behaviour. Their ease of installation (simply by welding) has ensured widespread use on ships [86], offshore structures [87], subsea pipelines [88] and even marine concrete [89], where the use of insulated leads and transformer and rectifier equipment could be difficult or inconvenient. Among the disadvantages of galvanic anodes are that they may induce an unacceptable degree of fluid dynamic drag, owing to the fact that anode sizes are larger than in impressed current systems and the smaller throw means that a larger number of anodes are normally required. Also, although designs are usually extremely conservative, damage or anode wastage is rather difficult to repair, usually requiring the use of underwater welding if the protected structure cannot be brought to a dry dock.

Impressed current systems normally require an operating platform and are therefore not suited, for example, to long submarine pipelines. However, they are frequently used for shore, ship and platform applications (Wyatt and Lothian [90], Tighe-Ford et al. [91] and Levings et al. [92] respectively). They are more versatile than galvanic systems, allowing a range of possible currents for a given configuration, but design has to be more carefully considered to avoid the danger of overprotection or underprotection, owing to the relatively small number of anodes involved and the consequent local high current densities. The problem of carrying current to the anodes without accelerated corrosion of leads also needs to be considered, and shielding is often necessary to avoid excessive potential gradients near the anodes, it sometimes even being necessary to switch the system off if divers are in the water. The enhanced throw of impressed current systems along with their variable driving potential makes them more suitable candidates in electrolytes of high resistivity such as concrete and in buried applications (Parker and Peattie [93]). The added advantage of relatively few points of contact with the surface to be protected can also be useful for buried and embedded structures. Monitoring of impressed current systems can also be carried out rather more conveniently than galvanic systems, it being sufficient to monitor the behaviour at the rectifier.

The design of a cathodic protection system revolves around ensuring that:

1. the potential of the surface to be protected is held within prescribed limits;

2. the potential is uniform over the metal surface;

3. anodes are capable of supplying the required external current for acceptable protection;

4. anodes have sufficient capacity to deliver current for the design life of the system.

Basic galvanic anode design is aimed principally at the last two of these criteria, the first two normally being considered to be taken care of by ensuring a reasonably uniform distribution of anodes over the surface to be protected and the low resistivity of seawater and saline mud. Furthermore, the natural potential of the anode materials is such that overprotection is unlikely even very close to attachment points. A typical submerged design procedure would consist first of finding anode resistance and hence likely anode current. From the total protective current density it is possible to calculate the total number of anodes required (e.g. Cochran [88]). The resistance of the anode to ground is usually considered to be the major contribution to the driving potential and this resistance is a function of geometry. For example the anode resistance for a cylinder of radius r and length L can be given by (Dwight [94]):

$$R_a = \frac{\rho}{2\pi L}\left[\ln\left(4L/r\right) - 1\right]$$

where ρ is the resistivity of the electrolyte. For shapes other than cylinders an effective circular radius can be used.

For most anode materials, the driving potentials to steel, E_d ($= E_{steel} - E_a$), are known and these can then be used to find the average anode current:

$$I_a = E_d/R_a$$

Again, for most seawater and saline mud conditions, the amount of cathodic current density required to achieve protection, i_{prot}, is known and this, along with the surface area, A, which requires to be protected, can be used to find the total number, N, of anodes required:

$$I_{tot} = Ai_{prot}$$

and

$$N = I_{tot}/I_a$$

These anodes are then spaced evenly over the surface to be protected. If distances to be covered per anode become excessive, or if there are geometric conditions which make it desirable, the number of anodes required can be increased by choosing an anode of smaller size.

Anodes are usually sold in convenient sizes for particular applications, for example pipelines, ships and offshore structures and are normally cast around a steel core which extends from the ends of the anode, thus providing a lug for welded attachment to the structure to be protected. This usually means that not all of the active material can be used and lifetime calculations normally account for this by multiplying the anode mass, M, by a utilisation factor, U, which varies from about 70 to 90 per cent depending upon the geometry. From Faraday's laws it is obvious that the current capacity, C, in $(A\,h)kg^{-1}$ is dependent upon the active constituent of the anode (about 2500 $(A\,h)kg^{-1}$ for aluminium anodes in seawater and about 750 $(A\,h)kg^{-1}$ for zinc anodes in seawater), but this can also vary with the environment. Thus, for an anode working at an average current of I_{ave} amperes, the lifetime (in hours) can be given by:

$$T = \frac{MCU}{I_{ave}}$$

If this lifetime is insufficient then it will be necessary to reiterate the calculation using a larger anode size. An example of a galvanic anode design is given in Case Study 7.4.

Impressed current cathodic protection design follows that for galvanic anodes to the extent that the resistivity of the electrolyte and the required protective current need to be known. Thereafter the design process differs somewhat. First, the groundbed location needs to be chosen so that cable lengths are not excessive, any hot spots can be protected and interaction or overprotection problems are not likely to occur. This is rather simpler for seawater than for buried applications, where a soil resistivity survey is usually necessary. The next stage involves the determination of circuit resistances for cathode (usually negligible), cable and anode (depending upon groundbed design). The effects of attenuation need also to be considered for remote areas of the structure in order to ensure adequate protection. A decision for protection voltage needs next to be made, in conjunction with anode material choice (related to capacity and consumption rate). The number, size and spacing of anodes can then be decided and the process iterated, if necessary, to produce an optimum design (Smith [95]). This process is compared with the galvanic anode method for pipeline cathodic protection in Case Study 7.4.

For critical or complex cases it may be necessary to calculate the potential distribution on the protected surface, in order to ensure that optimum conditions of protection are being achieved. This requires a solution, usually numerical, of the Laplace equation for potential distribution, $V(x, y, z)$, in a homogeneous conductor:

$$\frac{\delta^2 V}{\delta x^2} + \frac{\delta^2 V}{\delta y^2} + \frac{\delta^2 V}{\delta z^2} = 0$$

The solution is obtained subject to the polarisation behaviour of the anodic and cathodic boundaries, through which all current is constrained to flow. The anodic condition is usually relatively simple (although potential gradients along impressed current anodes may need to be accounted for), but the cathodic boundary condition may change with time and this may make a solution difficult to achieve (Sander [96]). Nevertheless, potential distributions for cathodically protected steel have been calculated with some success (e.g. Fu [97]).

3.4.1.3 Anodic Protection

Anodic protection is based upon the phenomenon of passivation, where the elevation of the potential of a metallic surface can, in some metals, result in a sudden drop in corrosion rate (see Section 3.1.2). Although the passive behaviour of some metals is sometimes exploited in marine corrosion-resistant materials selection, few if any applications of anodic protection have been documented, most records being confined to acid handling applications. Doubtless the reason for the reluctance to rely on active anodic protection against marine corrosion is the destabilising influence which chloride ions have upon passive films on metals, although anodised aluminium is of course an example of passive anodic protection.

3.4.1.4 Inhibition and Electrolyte Modification

Corrosion inhibitors can be broadly described as those chemicals which result, directly or indirectly, in the formation of films on the metal surface which render it resistant to corrosion. The inhibitor can be either added to the electrolyte or applied to the surface to be protected as part (or all) of a coating. This latter case has been dealt with in Section 3.4.1.1 and is not further considered here.

The way in which corrosion protection is achieved depends upon the type of inhibitor, some acting upon the solution and others acting at the metallic surface to interfere with the corrosion reaction. Most inhibiting films act by influencing polarisation effects (including concentration polarisation) or by increasing resistance at the metal–electrolyte interface.

The use of corrosion inhibitors usually has to be limited to closed systems, owing to the danger of pollution and also to the expense of such additives. In marine applications, one major use of inhibitors is in oil and gas production systems, where, for example , they can be used at the reservoir by squeezing or injection (Tischuk and Huber [98]), and also at various stages of processing and transportation, depending upon the corrosivity of the fluid and the amount of water present. Inhibitors are also used widely in seawater handling systems, for example in cooling applications, and are important constituents of many marine coatings.

The chemistry of corrosion inhibition is complex and difficult to study, and many proprietary formulations are available of whose components the designer is often unaware. Nevertheless, some idea of the types of action possible is necessary in order to properly select an inhibitor type and suitable application programme. Inhibitors can conveniently be classified in terms of their chemical nature (Jones [99]) as inorganic or organic, and anionic or cationic.

The inorganic inhibitors are often sodium salts where the cationic species formed on dissolution in water (e.g. chromate, silicate or phosphate) are the active constituents. Some of the organic inhibitors, for example sodium phosphonates, produce an active anion which contains an organic group [99]. These anions are thought to stabilise passivating films in near-neutral solutions and are therefore only effective against corrosion which is caused by dissolved oxygen. They are relatively ineffective against dissolved hydrogen sulphide or carbon dioxide.

Some inorganic inhibitors involve zinc (e.g. zinc silicate) and this cation (Zn^{2+}) can produce a protective insoluble salt film of $Zn(OH)_2$, restricting diffusion of oxygen to cathodic areas (where the hydroxyl ion concentration will have been at its highest) and also, by virtue of poor electronic conductivity, retarding the amount of reduction of oxygen on the film surface.

Organic cationic inhibitors usually have amine groups at one end of the organic molecule and, in solution, these pick up hydrogen ions, producing a long hydrocarbon (hydrophobic) tail on a hydrophilic polar head (Fig. 3.33).

The amine groups are able to adsorb onto metal surfaces and the oily tail acts as a water repellent. Long-chain amines are effective against carbon dioxide and hydrogen

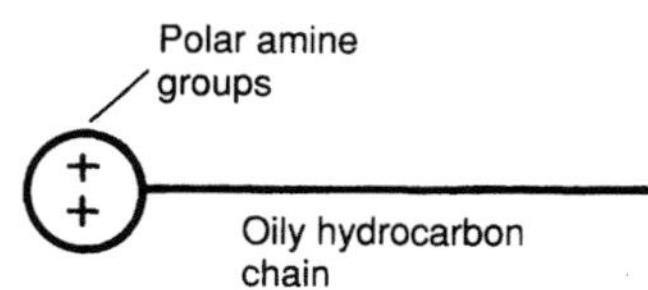

Fig. 3.33. Schematic structure of an organic cationic inhibitor molecule. (From Jones [99].)

sulphide corrosion, and the amine groups adsorb strongly onto iron sulphide. Furthermore, some amines are able to react chemically with hydrogen sulphide, producing complexes which add to the protective action.

Most oilfield inhibitors contain long-chain amines as their active constituent, and low-molecular-weight volatile amines are available for treating gas streams and vapour spaces. Water-soluble quaternary ammonium chlorides also find some oilfield applications and are also used as biocides. Most applications of the inorganic inhibitors are in cooling waters.

As well as inhibition, closed environments can be treated in other ways to reduce their corrosivity. For instance, aggressive gases can be stripped from water or process fluids by a variety of means, some chemical and some physical in nature. The principal gases usually considered in this context are carbon dioxide, hydrogen sulphide and oxygen.

Oxygen is normally removed physically by the use of counter-current towers (Fig. 3.34) or by vacuum deaeration, usually in packed towers.

Chemical removal of oxygen is achieved by addition of a scavenger such as hydrazine or a salt which produces sulphite ions, e.g.

$$SO_3^{2-} + \tfrac{1}{2}O_2 \rightarrow SO_4^{2-}$$

Unlike oxygen, dissolved hydrogen sulphide and carbon dioxide are present only in part in the molecular form. This means that physical stripping does not remove all of the gas from solution, and acid treatments are required to increase the molecular component of the dissolved gas. Such methods are not normally used in water treatment, because they may give rise to more problems than they solve, and other

solutions are normally sought to control corrosion such as inhibition. Low concentrations of hydrogen sulphide can be removed by chlorination, and scavenging with basic zinc carbonate to produce insoluble zinc sulphide is used in drilling operations (Jones [99]).

3.4.2 Protection Against Biological Effects

Apart from the corrosion prevention techniques mentioned above, protection against biodeterioration is achieved mainly by the use of biocides or by anti-fouling coatings, although cathodic protection may have an effect on some biocorrosion phenomena (e.g. Guezennec and Therene [100]).

Biocidal additions can be made to process streams in the same way as are inhibitors. However, great environmental care needs to be taken, particularly with once-through systems, to ensure that no unintentional damage is done. In water systems, it has been traditional to use chlorine or chlorine-releasing substances, and these continue to find wide use for a wide range of bacteria (e.g. Bessems [101]). Other chemicals used to control biological activity include phenolics, aldehydes and quaternary ammonium compounds, and Bessems has discussed the relative merits of these. Jones [99] indicates that the normal usage concentrations rarely produce complete sterilisation in oilfield water systems but that slugs of strong (0.5 to 2 weight per cent calcium hypochlorite) chlorine solutions are used to decompose slimes and effectively sterilise flowlines and tanks. The same author reports that commercial biocides (for example quaternaries or complex aldehydes) are often injected into oilfield water systems, or that batch treatment based on cell count data can be used.

The use of coatings against biodeterioration can range from the application of anti-fouling paints to the impregnation of timber against fungal decay or the effects of borers.

Whereas some timbers have a high resistance to marine biodeterioration, it is usual to consider some means of increasing durability. This may simply consist of selection of heartwoods of resistant species coupled with good control of moisture content, but the availability of such woods is limited. Because of the structure of timber, a useful method of protection is by impregnation with toxic chemicals. For high-duty service, such impregnation is usually carried out under pressure to ensure adequate penetration of the wood by the preservative. There is something of an irony here, in that it is much more difficult to permeate resistant woods with preservative, to the extent that this is not a viable method of protection for some species. In such cases, some physical protection can be given, especially against borers, using sheathing.

Anti-fouling agents normally act to discourage the initial settlement of the slime layer which provides the precursor to more extensive macrofouling. Such agents usually release a toxic substance at a slow, continuous rate, although other approaches which involve retaining a difficult-to-settle surface are also under active development. Since anti-fouling properties are rarely consistent with the other requirements of a marine material, it is usual to apply this property as a coating or cladding on the surface to be protected.

Perhaps the best known anti-fouling agent is metallic copper, whose benefits have been known for centuries. Its original use was initially as a metallic cladding (Chapter 1), and although it is still so employed copper is also incorporated as a compound in many anti-fouling paint formulations. In

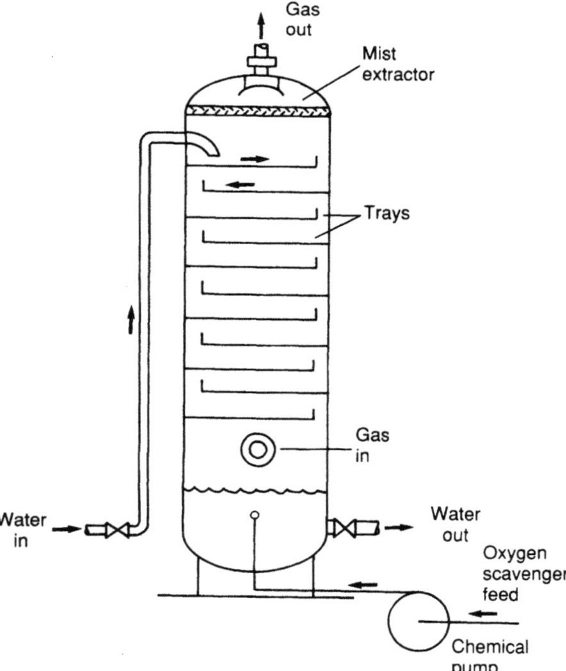

Fig. 3.34. Elements of a counter-current gas stripping tower for the removal of dissolved oxygen. (From Jones [99].)

some cases, the use of a copper alloy for corrosion protection can have the added benefit of anti-fouling properties [79]. The interaction of corroding copper and marine biofilms is a complex subject and is not yet entirely understood (Chamberlain and Garner [102]), although its effectiveness is almost certainly related to the corrosion rate of the alloy. Laque [7] indicates that a corrosion rate of about 2.5×10^{-2} mm/yr is required to suppress fouling and, further, the development of corrosion product layers can allow fouling to occur even in high-copper alloys. All other metals, with the exception of silver, can be expected to foul, so that there is a clear necessity for anti-fouling coatings.

Conventional anti-fouling paints also act to release toxic substances slowly into the water at the surface of the layer. Most of these substances have been metallic or organometallic compounds of copper, tin, lead, mercury or arsenic, and many are now prohibited for use in a number of countries of the world. Furthermore, the two most common types of heavy-metal toxins used in anti-fouling paints (based on copper and tin) are under increasing scrutiny as to their environmental effects, and it is not certain how long the use of such agents will continue to be permitted (Brady et al. [103]). Also, the surface layer of an anti-fouling paint eventually becomes exhausted and some physical surface treatment is usually required to reactivate it (Poretz [104]). Some self-polishing copolymeric vehicles have been developed which dissolve slowly with the toxin and reduce this requirement significantly (Miller [105]).

It has always been known that physical means can be used to discourage fouling from periodic cleaning to the use of high fluid flow rates. Laque [7] reports a critical velocity of about 1 m/s to inhibit fouling on smooth surfaces, but that this must be sustained. There is now considerable effort being expended to create an anti-fouling coating material which, by purely physical means, either completely prevents attachment of fouling organisms or which only permits a weak attachment resulting in drop-off under self-weight or detachment under normal hydrodynamic forces. This is of course an ideal, but there is some hope that coatings such as silicone-rubber-based materials to which an exuding fluid has been added (Miller [105]) or fluorinated polyurethanes (Brady et al. [103]) may provide a solution.

3.4.3 Protection of Polymers

Most of the effort in protection of polymers revolves around stabilisation against radiation damage, although autoxidation can also be protected against by the use of additives. Obviously, the use of an opaque surface film will protect polymers from radiation effects, but it is usually much more convenient to manufacture polymers with stabilisers which mitigate the effects of UV radiation on the polymer. Carbon black and titania are widely used for this purpose, the former being the more effective, and a number of non-coloured UV filters are also available. Additives which quench (reduce the mean lifetime of) the excited states, leading to photo-degradation, can also be added and some UV absorbers (e.g. benzotriazoles) fulfil this function in addition to their primary one [106]. It is also possible to inhibit the photo-oxidation cycle of degradation by the use of antioxidants. These act as scavengers for, for example, the peroxy radicals which are required for photo-oxidation.

Although it is not a process of environmental degradation, the protection of polymers and composites against fire is a matter of some concern in marine applications. Apart from fire-resistant cable sheathings, where the main concerns [107] are flame propagation along the cable, the generation of large volumes of smoke, and the evolution of corrosive or toxic by-products, there is also a requirement for retained structural integrity if composite materials are to compete effectively with steel. Marks [108], for example, has considered this latter point with reference to GRP pipes. Apart from a knowledge of inherent stability, which can be measured by standard fire propagation and flammability tests, along with those indicating toxicity and smoke evolution (e.g. Hunter [109]), it is necessary to consider the application of a fire protection system to combat the worst effects of loss of structural integrity under fire conditions. For example, glass-reinforced epoxy loses its structural integrity at about 350 °C [109], and one way of combating this is to use coatings with an ability to restrict heat flow and also to withstand high temperatures. The types of coatings capable of such function include cementitious coatings, inorganic foams, phenolic foams and intumescent coatings, the last of which operate by undergoing endothermic gas-evolving reactions during heating. The evolved gases absorb heat as they pass through the char layer and produce a stable foam from the char.

3.4.4 Protection of Concrete

Although concrete is a rather durable substance and can, in a loose way, be regarded as a coating itself in some instances, it is sometimes desirable to protect it against water penetration, salt intrusion, abrasion, frost damage and reinforcement corrosion. Some coatings are available which are resistant to the alkilinity of the concrete and are suitable for submerged or atmospheric service (Foscante and Kline [110]), but it has to be said that coating of concrete is not often considered in the marine environment.

Steel embedded in concrete can, in principle, be protected cathodically, although there are considerable difficulties in applying this in the critical splash zone area (Wilkins [89]). There is little evidence to suggest that protection of steel in underwater concrete is necessary [89].

3.5 Assessment of Materials Performance

This section considers the ways in which the performance (with regard to corrosion and other degradation mechanisms) of materials in the marine environment can be assessed so as to give engineers an insight into the ways in which such data can be interpreted and also into the necessary tests to specify for performance in particular environments. The question of corrosion monitoring (in-service corrosion measurement) is dealt with separately in Chapter 6.

The testing of materials for corrosion resistance is, with one or two notable exceptions, a laboratory exercise and is a matter for interpretation rather than specification by design engineers. The first point to note about corrosion tests is that they are usually carried out in synthetic (and often very severe) environments and therefore only give a comparative measure of performance. This is useful, of course, because most design engineers will have some experience of how existing materials perform in the environment concerned and will be concerned about the adoption of a new or modified material (or even the same material from a different batch or supplier).

Corrosion tests are generally carried out according to the mechanisms outlined in Section 3.2. For general corrosion under submerged conditions or in the marine atmosphere there is no substitute for direct exposure [7, 111], although the time scale involved means that this is not always practicable and there can be variations from site to site. Accelerated simulation of the marine atmosphere usually means a hot salt spray test, and Lee and Money [112] have discussed the applicability of measurements from such an apparatus. Simulated seawater [113] is occasionally used in the laboratory to avoid the biological difficulties in maintaining live seawater, and this is particularly useful if the effect of large doses of contaminants, such as sulphides [114] are to be examined. The interpretation of general corrosion results is relatively simple, except that it must be remembered that results quoted as weight loss are averaged and local variations across the surface may be substantial.

Pitting can be measured qualitatively or quantitatively [115], taking account of shape and size distribution as well as depth (Fig. 3.35). Such data can be extremely useful to

engineers, but are unfortunately rare, and most designers have to deal with relative qualitative data to assess pitting susceptibility. The pitting potential [116] can be measured during electrochemical experiments, and it may be useful to know the potential below which pitting will not occur. For direct use it is important to carry out quantitative tests under appropriate flow, as well as chemical conditions.

Crevice corrosion is usually measured by the application of a crevice-producing agent, usually in a rather aggressive environment [117]. The crevice may be as simple as the space underneath a rubber band, but more sophisticated and reproducible designs are also used. Obviously in either case the depth of crevicial attack is only a relative measure, but it is likely that this is all a design engineer requires to know. As with pitting, a protection potential below which crevice corrosion is observed not to occur can be helpful, but Hubbard et al. [116] warn that short-term tests of this quantity may not be conservative.

Velocity effects need to be very carefully interpreted, mainly owing to the conjoint effect and the need to have the correct relative magnitudes of the mechanical and chemical components. Also, different types of fluid erosion (turbulent, impingement and cavitation) will yield different results. As examples of the range of equipment used for such tests, rotating-wheel specimens have been used for erosion [118], and Venturi constrictions and vibratory methods for cavitation [119]. Matsumura et al. [120] have described a jet-in-slit impingement test used to measure velocity effects in copper alloys in seawater. Corrosion-erosion data can be reported quantitatively as rate of loss of thickness as a function of velocity but this is more difficult for cavitation. In this latter case, some success has been obtained [121] in correlating relative cavitation resistance (as weight loss relative to a standard material) against the resilience characteristic of the material (Fig. 3.36).

For environmental cracking, much use has been made of very severe tests for ranking specific material types, for example boiling magnesium chloride for stainless steels [122], and results are therefore comparative. Fracture mechanics tests, particularly the slow strain rate test referred to in Section 3.2.4, are of direct quantitative applicability provided that the test environment is appropriate. For sulphide stress corrosion (SSC), the NACE test TM01 77 [123] is widely used as an indicator of resistance and the related recommended practice for materials selection [124] is almost mandatory for those specifying materials for use in sulphide environments, although reference to other literature (e.g. [125,126]) may be necessary for specific material types.

Hydrogen-induced cracking tests are often specified for pipelines carrying sour fluids, and the NACE test [127] is again probably the most widely used, consisting of the immersion of a block of prepared material in sulphide-containing electrolyte. Terasaki et al. [74] show a method for quantifying the results of such a test in terms of the ratios of crack length to specimen length (RCL) and crack width to specimen width (RCW) (Fig. 3.37).

A number of tests are also required in relation to cathodic protection. For example, coatings may become disbonded when a cathodic current is applied to the surface to which they are attached owing to the evolution of hydroxyl ions when the oxygen reduction reaction is driven. The ASTM test to assess the susceptibility of coatings to such cathodic disbondment [128] consists of applying a small degree of overprotection to a sample of the coated

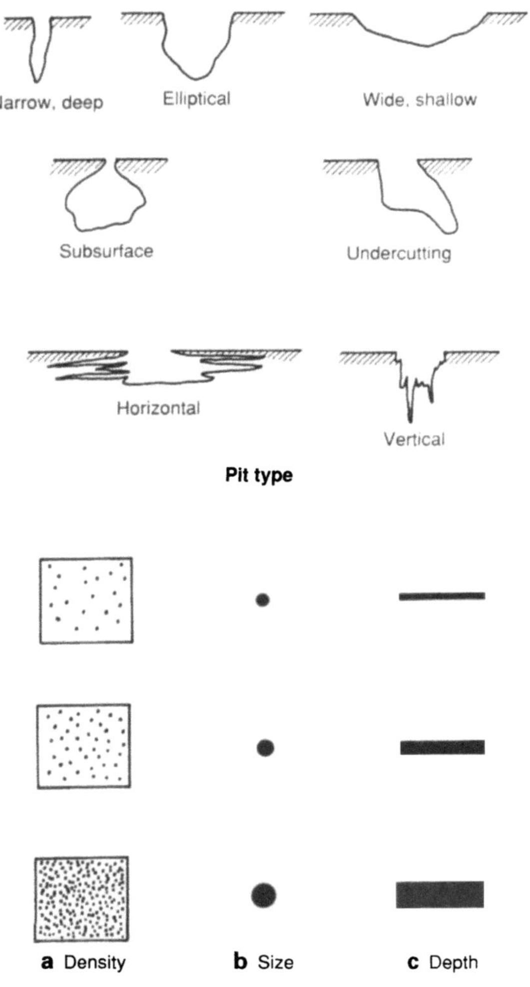

Fig. 3.35. Variables for use in the assessment of pitting corrosion. (From ASTM [115].)

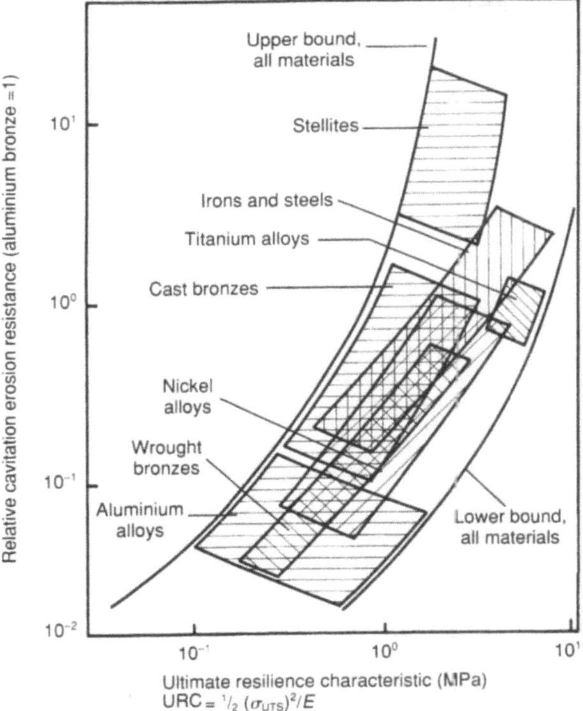

Fig. 3.36. Relationship between resilience characteristic and relative cavitation resistance for a range of metallic materials [121].

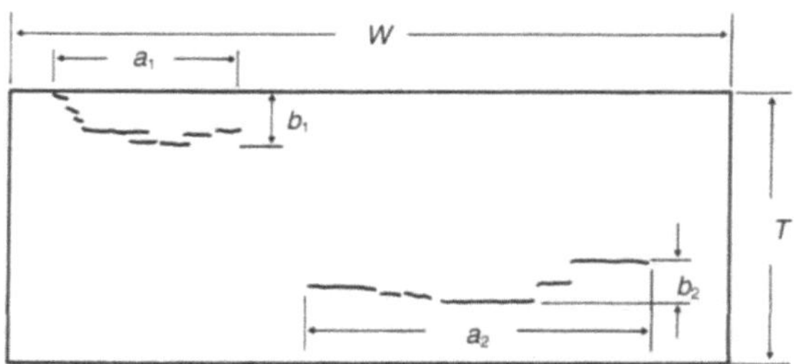

Fig. 3.37. Measurement of susceptibility to hydrogen-induced cracking by examining the cross-section of an exposed sample. RCL is the ratio of total crack length, Σa_i, to specimen width, W. RCW is the ratio of total crack width, Σb_i, to specimen thickness. (After Terasaki et al. [74].)

surface on which a holiday (local disruption in the coating) has been manufactured, and examining the remaining degree of adhesion after a specified period of time. Also, Nye et al. [129] describe a method for measuring the degree of polarisation (and hence drop in current demand) due to chalking (formation of calcareous deposit) on cathodically protected surfaces in seawater.

Environmental test methods for polymers are mostly associated with temperature and chemical resistance and weathering and radiation susceptibility. The assessment of temperature resistance usually involves thermal analysis, consisting of the measurement of some property change [130] with temperature. Quite often weight (thermogravimetric analysis) is quite sufficient, although another method

such as dimension change (thermomechanical analysis) is required for such phenomena as glass transition. Figure 3.38 shows typical TGA and TMA traces of a high-performance polymer.

The main aspects of chemical resistance of interest in marine technology for polymers are their behaviour in hydrocarbon fluids and their water absorption characteristics. These are generally measured by simple exposure tests coupled with observations of changes in volume and mechanical properties and even colour change.

Weathering factors include sunlight, oxidation, temperature variations and moisture [43], and these are usually tested simultaneously in so-called weatherometers. There are a variety of possible regimes of use of weatherometers, but the two commonly used to simulate accelerated natural weathering are:

1. continuous light exposure with an intermittent water spray;
2. alternating exposure to light and darkness with an intermittent water spray.

These, along with temperature and humidity control, allow the simulation of all of the natural weathering processes [43]. After exposure, specimens can be tested for retention of mechanical or other important properties. For coatings applied to metallic materials, an ASTM test [131] can be used to check for cracking and light fastness under ultraviolet and condensation exposure.

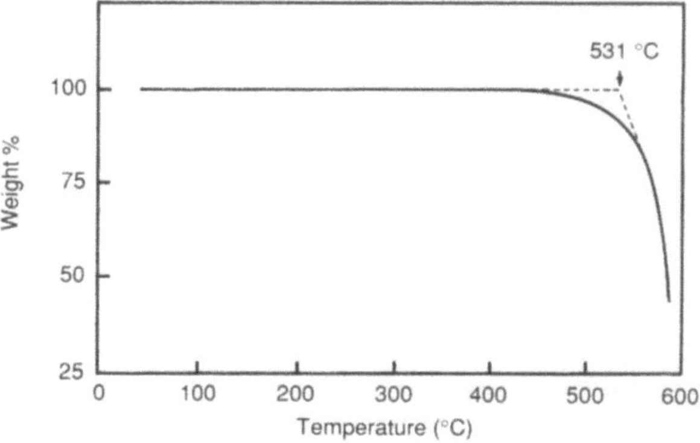

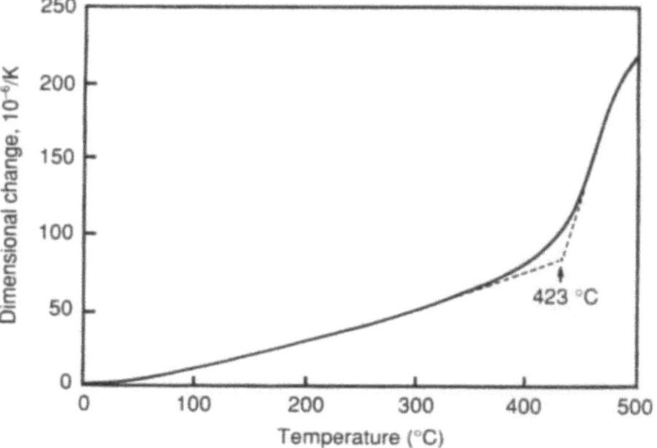

Fig. 3.38. Typical thermogravimetric (top) and thermomechanical (bottom) analysis traces for a polyimide material. (From Jones [130].)

References

1. Moore WJ. Physical chemistry. Longmans, London, 1972
2. Bard A, Parsons R, Jordan, J (eds). Standard potentials in aqueous solution. Marcel Dekker, New York, 1985
3. Pourbaix M. Atlas of electrochemical equilibria in aqueous solutions. NACE/CEBELCOR, Houston, 1974
4. Pourbaix A, Marquez-Jacome M. The enhancement of reaction kinetics by bacteria. In: Proceedings, 1st European Federation of Corrosion Workshop on Microbial Corrosion, Sintra, Portugal, 1988. Elsevier, London, 1988, pp 29–39
5. Fontana MG, Greene ND. Corrosion engineering. McGraw-Hill, New York, 1982
6. Bockris JO'M, Reddy AKN. Modern electrochemistry. Macdonald, London, 1970
7. LaQue FL. Marine corrosion – causes and prevention. Wiley, New York, 1975
8. Schumacher M (ed.). Seawater corrosion handbook. Noyes Data Corporation, Park Ridge, New Jersey, 1979
9. Oldfield JW. Electrochemical theory of galvanic corrosion. In: Hack HP (ed.) Galvanic corrosion, ASTP STP 978. ASTM, Philadelphia, 1988, pp 5–22
10. Munn RS. Materials performance. A mathematical model for a galvanic anode cathodic protection system. 1982; 21 (8); 29–32
11. Astley DG, Rowlands JC. Modelling of bimetallic corrosion in sea water systems. British Corrosion Journal 1985; 20 (2), 90–93
12. BS PD6484: Commentary on corrosion at bimetallic contacts and its alleviation. BSI, London, 1979
13. Fangteng, S. The influence of flow velocity on differential aeration corrosion of mild steel in seawater. In: Proceedings, International Conference on Corrosion Control for Offshore and Marine Construction, Xiamen, 1988. International Academic Publishers/ Pergamon, Oxford, 1989, pp 80–85
14. Wilkins NJM, Lawrence PF. Fundamental mechanisms of corrosion of steel reinforcements in concrete immersed in seawater. In: Concrete in the oceans, Technical Report, CIRIA/UEG. Cement and Concrete Research Association and the Department of Energy, London, 1980
15. Browne RD, Domone PLJ. The long-term performance of concrete in the marine environment. In: Proceedings, Conference on Offshore Structures, London, 1974. Institute of Civil Engineers, London, 1975, pp 49–59
16. Powers TC, Copeland LE, Hayes JC, Mann HM. Permeability of Portland cement paste. Journal of the American Concrete Institute 1954; 51(11): 285–298
17. Gjorv OE, Vennesland O, El-Busaidy AHS. Diffusion of dissolved oxygen through concrete. In: Proceedings, International Forum – Corrosion/76. NACE, Houston, 1976, pp 17/1–17/13
18. Hatch JE (ed.). Aluminium – properties and physical metallurgy. ASM, Metals Park, Ohio, 1984
19. Rothwell N, Turner MED. Corrosion problems associated with weldments. Materials Performance 1990; 29(2): 55–58
20. Wilhelm SM. Galvanic corrosion caused by corrosion products. In: Hack HP (ed.) Galvanic corrosion, ASTP STP 978. ASTM, Philadelphia, 1988, pp 23–34
21. Engel L, Klingele H. An atlas of metal damage. Wolfe Science Books and Carl Hanser Verlag, Munich, 1981
22. Staehle RW. Predictions and experimental verification of the slip dissolution model for stress corrosion cracking of low strength alloys. In: Proceedings, Conference on Stress Corrosion Cracking and Hydrogen Embrittlement of Iron Base Alloys. NACE, Creusot-Loire and Advanced Research Projects Agency, Unieux-Firminy 1973. NACE, Houston, 1977, pp 180–207
23. Brown BF. The application of fracture mechanics to stress corrosion cracking. Metallurgical Reviews 1968; 129: 171–183
24. Knott JF. Fundamentals of fracture mechanics. Butterworths, London, 1973
25. Opoku J, Clark WG. The effects of various hydrogen bearing environments on the K_{ISCC} of AISI Type 4340 and 3.5% NiCrMoV steels. Corrosion-NACE 1980; 36 (5): 251–258
26. Jones DA. Localised corrosion. In: Parkins RN (ed). Corrosion processes. Applied Science Publishers, London, 1982, pp 161–208
27. Jaske CE, Payer JH, Balint VS. Corrosion fatigue of metals in marine environments. Springer-Verlag and Battelle Press. Metals Park, Ohio, 1981
28. Nishoka K, Hirakawa K, Kitawa I. Low frequency corrosion fatigue strength of steel plate. Sumimoto Search 1976; (16): 40–54
29. UEG. design of tubular joints for offshore structures. UEG Publication UR33. UEG/CIRIA, London, 1985
30. Hooper WC, Hartt WH. Influence of cathodic polarisation upon fatigue of notched structural steel. In: Proceedings, Conference Corrosion '77, San Francisco. NACE, Houston, 1977
31. Congleton J, Hussein I and Parkins RN. Effect of applied potential on corrosion fatigue of wire ropes in seawater. British Corrosion Journal 1985; 20 (1): 15–18
32. Congleton J, Craig IH. Corrosion fatigue. In: Parkins RN (ed) Corrosion processes. Applied Science Publishers, London, 1982, pp 209–270
33. Mills NJ. Plastics – microstructure, properties and applications. Edward Arnold, London, 1986
34. Allen RC, Bauer RS. Moisture-related failure. In: Engineered materials, vol. 2: Engineering plastics. ASM International, Metals Park, Ohio, 1988, pp 761–769
35. Comyn J. Interaction of water with epoxy resins. In: Proceedings, Second International Conference Polymers in a Marine Environment. Marine Management (Holdings) Ltd, London, 1989, pp 153–160
36. Crawford RJ. Plastics engineering. Pergamon, Oxford, 1987
37. Lavers GM. The strength properties of timbers. In: The strength properties of timber. MTP Construction, Princes Risborough, 1974, pp 1–66
38. Illston JM, Dinwoodie JM, Smith AA. Concrete, timber and metals. Van Nostrand Reinhold, New York, 1979
39. Fookes PG, Simm JD, Barr JM. Marine concrete performance in different climatic conditions. In: Proceedings, Conference on Concrete in the Marine Environment. Concrete Society, London, 1986, pp 115–130
40. Griffin RB. Marine corrosion – marine atmospheres. In: Metals handbook, vol. 13: Corrosion. ASM, Metals Park, Ohio, 1987, pp 902–905
41. Brown PW, Masters LW. Factors affecting the corrosion of metals in the atmosphere. In: Ailor, WH (ed.) Atmospheric corrosion. Wiley, New York, 1982, pp 31–50
42. Sereda PJ, Croll SG and Slade HF. Measurement of the time-of-wetness by moisture sensors and their calibration. In: Dean SW, Rhea EC (eds) Atmospheric corrosion of metals. ASTM STP 767. American Society for Testing and Materials, Philadelphia, 1982, pp 267–285
43. Delre LC and Miller RW. Characterisation of weather aging and radiation susceptibility. In: Engineered materials, vol. 2: Engineering plastics. ASM International, Metals Park, Ohio, 1988, pp 575–580
44. Ostroff AG. Introduction to oilfield water technology. NACE, Houston, 1979
45. Beccaria AM, Poggi G. Effect of some surface treatments on kinetics of aluminium corrosion in NaCl solutions at various hydrostatic pressures. British Corrosion Journal 1987; 21 (1): 19–22
46. Oldfield JW, Todd B. Corrosion of metals in deaerated seawater. Desalination 1979; 31: 365–83
47. Goodman PD. Effect of chlorination on materials for sea water cooling systems – a review of chemical reactions. British Corrosion Journal 1987; 22 (1): 56–62
48. Amann H, Oebius HV, Gehbauer F, Schwarz W, Weber, R. Soft ocean mining. In: Proceedings, 23rd Offshore Technology Conference, Houston, Texas, paper no. OTC 6553, 1991, pp 469–480

49. Fischer KP, Bue, B. Corrosion and corrosivity of steel in Norwegian marine sediments. In: Escalante E (ed.) Underground corrosion, ASTM STP 741. ASTM, Philadelphia, 1981, pp 24–32

50. King, RA. Prediction of corrosiveness of sea-bed sediments. In: Proceedings Symposium Marine Corrosion on Offshore Structures. Society of Chemical Industry, London, 1979, pp 23–32

51. Tonyali K. Organic chemical related failure. In: Engineered materials, vol. 2: Engineering plastics. ASM International, Metals Park, Ohio, 1988, pp 770–775

52. Dexter SC. Marine corrosion – seawater. In: Metals handbook, vol. 13: Corrosion. ASM, Metals Park, Ohio, 1987, pp 893–901

53. Eriksrud E, Sontvedt T. Effect of flow on CO_2 corrosion rates in real and synthetic formation water. In: Proceedings, Conference Corrosion '83. NACE, Houston, 1983, pp 44/1–44/40

54. deWaard C, Milliams DE. Prediction of carbonic acid corrosion in natural gas pipelines. In: Proceedings, 1st International Conference on Internal and External Protection of Pipes. British Hydraulics Research Association, Cranfield, 1975, F1-1–F1-7

55. Tiller AK. Electrochemical aspects of microbial corrosion – an overview. In: Proceedings, Conference on Microbial Corrosion. Metals Society, London, 1983, pp 54–65

56. King RA, Miller JDA. Corrosion of ferrous metals by bacterially produced iron sulphide and its control by cathodic protection. In: Proceedings, 1st International Conference on Internal and External Protection of Pipes. British Hydraulics Research Association, Cranfield, 1975, pp F2-9–F2-16

57. International critical tables, vol. 3. McGraw-Hill, New York, 1928

58. Wilkinson TG. Biological mechanisms leading to potential corrosion problems. In: Lewis JR, Mercer AD (eds.) Corrosion and marine growth on offshore structures. Ellis Horwood, Chichester, 1984, pp 31–37

59. Gooch TG. Cathodic protection and steel properties. In: Lewis JR, Mercer AD (eds.) Corrosion and marine growth on offshore structures. Ellis Horwood, Chichester, 1984, pp 81–94

60. Cordruwisch R, Kleinitz W, Widdel F. Sulfate-reducing bacteria and their economic activities. In: Proceedings, International Symposium Oilfield and Geothermal Chemistry, Richardson, Texas, SPE paper SPE 13554, 1985, pp 55–64

61. Hall GR. Control of microbiologically induced corrosion of concrete in waste water collection and treatment systems. Materials Performance 1989; 28 (10): 45–50

62. Boutelje JB, Goransson B. Decay in wood constructions below the ground water table. In: Proceedings, 2nd International Biodeterioration Symposium. Applied Science Publishers, London, 1972, pp 311–318

63. Jones EBG, Irvine J. The role of marine fungi in the biodeterioration of materials. In: Proceedings, 2nd International Biodeterioration Symposium. Applied Science Publishers, London, 1972, pp 422–431

64. Barrett J, Yonge CM. Collins pocket guide to the sea shore. Collins, London, 1984

65. Turner RD. An overview of research on marine borers. In: Costlow JD, Tipper, RC (eds.) Proceedings, Symposium on Marine Biodeterioration. Spon, London, 1984, pp 3–17

66. Kuhl H. Teredos and fouling. In: Traurig J-O (ed.) Fishing boats of the world. Fishing News (Books). London, 1975, pp 227–229

67. Scott PJB. Biodeterioration of plastics in the sea. Materials Performance 1989; 28 (10): 52–55

68. Connolly RA. Effect of seven year marine exposure on organic materials. Materials Research and Standards 1963; 3: 193–201

69. Connolly RA, DeCoste JB, Gaupp HL. Marine exposure of polymeric materials and cables after fifteen years. Journal of Materials 1970; 5: 339–362

70. Wolock I. Polymeric materials and composites. In: Schumacher M (ed.) Seawater corrosion handbook. Noyes Data Corporation, N. J., 1979, pp 455–463

71. Pipe A. North Sea fouling organisms and their potential effects on the corrosion of North Sea structures. Proceedings, Symposium Marine Corrosion on Offshore Structures. Society of Chemical Industry, London, 1979, pp 13–22

72. Forteath GNR, Dicken GB, Ralph R. Patterns of macrofouling on steel platforms in the central and northern North Sea. In: Lewis JR, Mercer AD (eds.) Corrosion and marine growth on offshore structures. Ellis Horwood, Chichester, 1984, pp 10–22

73. Taylor Woodrow Research Laboratories. Surveys of existing concrete marine structures. UK Department of Energy Offshore Technology Report OTH 87 244. HMSO, London, 1991

74. Terasaki F, Ohtani Y, Ikeda A, Nakanishi M. Steel plates for pressure vessels in sour environment applications. Proceedings, International Conference on Fusion Welded Steel Pressure Vessels. Mechanical Engineering Publications, London, 1984, pp 45–64

75. Denham JB, Taylor RE. Pressure vessels in the oil industry. Proceedings, International Conference on Fusion Welded Steel Pressure Vessels. Mechanical Engineering Publications, London, 1984, pp 123–134

76. Taira T, Koyabashi Y, Matsumoto K, Tsukad K. Resistance of line-pipe steels to wet sour gas. Corrosion-NACE 1984; 40 (9): 478–486

77. Johnson WC. Detrimental materials at the paint–steel interface. In: Berger DM, Wint RF (eds) New concepts for coating protection of steel structures, ASTM STP 841. ASTM, Philadelphia, 1984, pp 28–43

78. Whitehouse NR. Survey of painting practice for protection of offshore structures. Paint Research Association, Teddington, 1983

79. Carruthers R. The use of 90/10 cupro-nickel as a splash zone cladding. Proceedings, Conference on Copper Alloys in Marine Environments. Copper Development Association, Potter's Bar, 1985, paper 6

80. Yoshida T, Matsui S, Atsuta T. Corrosion problems of pipeline and a solution. Proceedings, 12th Offshore Technology Conference, Houston, paper no. OTC 3891, 1980, pp 361–370

81. Salama MM, Tetlow, JH. Selection and evaluation of high strength steel for Hutton TLP tension leg elements. Proceedings, 15th Offshore Technology Conference, Houston, paper no. OTC 4449, 1983, pp 57–63

82. Chandler KA, Bayliss DA. Corrosion protection of steel structures. Elsevier, London, 1985

83. Coke, JR. Protective coatings for offshore equipment and structures. Materials Performance 1990; 29 (5): 35–38

84. Ashworth V. The theory of cathodic protection and its relation to the electrochemical theory of corrosion. In: Ashworth V, Booker CJL (eds) Cathodic protection. Ellis Horwood, Chichester, 1986, pp 13–30

85. Morgan J. Cathodic protections. NACE, Houston, 1987

86. Jensen JO. Ships and semi-submersibles. In: Ashworth V, Booker CJL (eds) Cathodic protection. Ellis Horwood, Chichester, 1986, pp 128–142

87. Wyatt BS. Cathodic protection of fixed offshore structures. In: Ashworth V, Booker CJL (eds) Cathodic protection. Ellis Horwood, Chichester, 1986, pp 143–171

88. Cochran JC. New mathematical models for designing offshore sacrificial cathodic protection systems. Proceedings, 12th Offshore Technology Conference, Houston, paper no. OTC 3858, 1980, pp 27–35

89. Wilkins NJM. Cathodic protection of concrete structures. In: Ashworth V, Booker, CJL (eds) Cathodic protection. Ellis Horwood, Chichester, 1986, pp 172–182

90. Wyatt BS and Lothian AM. Cathodic protection of a roll-on/roll-off facility. Materials Performance 1990; 29 (4): 17–23

91. Tighe-Ford DJ, McGrath JN and Wareham MP. Evaluation of warship cathodic protection systems. Institute of Marine Engineers Paper, March 1988. Marine Management Holdings, London, 1988

92. Levings EA, Finnegan JE, McKie WM and Strømmen RD. Proceedings, 15th Offshore Technology Conference, Houston, paper no. OTC 4565, 1983, pp 445–450

93. Parker ME, Peattie EG. Pipe line corrosion and cathodic protection. Gulf Publishing Company, Houston, 1984

94. Dwight HB. Calculation of resistances to ground. Electrical Engineering 1936; 55: 1319–1328

95. Smith CA. Designing a cathodic protection system. Electrical Review 1980; 207 (18): 35–41

96. Sander KF. The problem of predicting potential and current distributions. In: Ashworth V, Booker CJL (eds) Cathodic protection. Ellis Horwood, Chichester, 1986, pp 31–37

97. Fu JW, Chow JSK. Calculation of cathodic protection potential and current distributions using an integral equation numerical method. In: Collected papers on cathodic protection current distribution. NACE, Houston, 1989, pp 15–18

98. Tischuk JL, Huber DS. Use of economic analysis to select the most cost effective method of downhole corrosion control. Proceedings, Conference UK Corrosion 1984. Institution of Corrosion Science and Technology, London, 1984, pp 13–18

99. Jones LW. Corrosion and water technology for petroleum producers. OGCI Publications, Tulsa, 1988

100. Guezennec J, Therene M. A study of the influence of cathodic protection on the growth of SRB and corrosion in marine sediments by electrochemical techniques. Proceedings, 1st European Federation of Corrosion Workshop on Microbial Corrosion, Sintra, Portugal, 1988. Elsevier, London, 1988, pp 256–265

101. Bessems E. Biological aspects of the assessment of biocides. Proceedings, Conference on Microbial Corrosion. Metals Society, London, 1983, pp 84–89

102. Chamberlain AHL, Garner BJ. Microbial fouling and corrosion of 90/10 copper–nickel alloys. Proceedings, 1st European Federation of Corrosion Workshop on Microbial Corrosion, Sintra, Portugal, 1988. Elsevier, London, 1988, pp 431–442

103. Brady RF, Griffith JR, Love KS, Field DE. Non-toxic alternatives to antifouling coatings. Proceedings, 2nd International Conference Polymers in a Marine Environment. Marine Management (Holdings), London, 1989, pp 191–196

104. Poretz I. Precision reactivation of antifouling paints. In: Berger DM, Wint RF (eds) New concepts for coating protection of steel structures. ASTM STP 841. ASTM, Philadelphia, 1984, 79–94

105. Miller D, Cameron AM, Shone EB. Applications of novel anti-fouling coatings on offshore structures. In: Lewis JR, Mercer AD (eds) Corrosion and marine growth on offshore structures. Ellis Horwood, Chichester, 1984, pp 139–148

106. Schnabel W. Polymer degradation. Hanser International, Munich, 1981

107. Manley TR, Harrison DA. The fire performance of cables on warships. Proceedings, Conference Polymers in a Marine Environment. Institution of Marine Engineers, London, 1989, pp 103–108

108. Marks PR. The fire endurance of glass-reinforced epoxy pipes. Proceedings, Conference Polymers in a Marine Environment. Institution of Marine Engineers, London, 1989, pp 69–78

109. Hunter, J. The advantages of glass-reinforced phenolics in demanding construction applications. Proceedings, Conference Polymers in Offshore Engineering. Plastics and Rubber Institute, London, 1988, pp 25/1–25/15

110. Foscante RE, Kline HH. Coating concrete – an overview. Materials Performance 1988; 27 (9): 34–36

111. Ailor WH (ed.). Atmospheric corrosion. John Wiley, New York, 1982

112. Lee TS, Money KL. Difficulties in developing tests to simulate corrosion in marine environments. Materials Performance 1984; 24 (8): 28–33

113. Castle JE, Chamberlain AHL, Garner B, Parvizi MS, Aladjem A. The use of synthetic or natural seawater in studies of the corrosion of copper alloys. In: Francis PE, Lee TS (eds) The use of synthetic environments for corrosion testing. ASTM STP 970. ASTM, Philadelphia, 1988, pp 174–189

114. Mack RD, Wilhelm SM, Steinberg BG. Laboratory corrosion testing of metals and alloys in environments containing hydrogen sulphide. In: Haynes GS, Baboian R (eds) Laboratory corrosion tests and standards. ASTM STP 866. ASTM, Philadelphia, 1985, pp 246–259

115. ASTM G46-76. Standard recommended practice for examination and evaluation of pitting corrosion. ASTM, Philadelphia, 1980

116. Hubbel M, Price C, Heidersback R. Crevice and pitting corrosion tests for stainless steels. In: Haynes GS, Baboian R (eds) Laboratory corrosion tests and standards. ASTM STP 866. ASTM, Philadelphia, 1985, pp 324–336

117. ASTM G48-76. Standard test method for pitting and crevice corrosion resistance of stainless steels and related alloys by use of ferric chloride solution. ASTM, Philadelphia, 1980

118. Vyas B. Erosion-corrosion. In: Preece CM (ed.) Treatise on materials science and technology, vol. 16: Erosion. Academic Press, New York, 1979, pp 357–390

119. Preece CM. Cavitation erosion. In: Preece CM (ed.) Treatise on materials science and technology, vol. 16: Erosion. Academic Press, New York, 1979, pp 249–305

120. Matsumura M, Oka Y, Okumoto S, Furaya H. Jet-in-slit test for studying erosion-corrosion. In: Haynes GS, Baboian R (eds) Laboratory corrosion tests and standards. ASTM STP 866. ASTM, Philadelphia, 1985, pp 358–370

121. Fulmer Research Institute. Fulmer materials optimiser. Fulmer Research Institute, London, 1980

122. ASTM G-36 73. Standard practice for the performance of stress corrosion cracking tests in a boiling magnesium chloride solution. ASTM, Philadelphia, 1981

123. NACE TM-01-77. Testing of metals for resistance to sulphide stress cracking at ambient temperatures. NACE, Houston, 1977

124. NACE MR-01-75. Sulphide stress cracking resistant metallic materials for oil field equipment. NACE, Houston, 1984

125. Whitehead T, Baloun CH. Evaluating the suitability of the NACE standard test TM-01-77 for testing 13% chromium martensitic stainless steels for sulphide stress cracking resistance. In: Haynes GS, Baboian R (eds) Laboratory corrosion tests and standards. ASTM STP 866. ASTM, Philadelphia, 1985, pp 400–414

126. Treseder RS, Kachik EA. MTT corrosion tests for iron- and nickel-base corrosion resistance alloys. In: Haynes GS, Baboian R (eds) Laboratory corrosion tests and standards. ASTM STP 866. ASTM, Philadelphia, 1985, pp 373–399

127. NACE TM-02-84. Test method for the evaluation of pipeline materials for resistance to stepwise cracking. NACE, Houston, 1984

128. ASTM-G8-79. Standard test for cathodic disbonding of pipeline coatings. ASTM, Philadelphia, 1982.

129. Nye TL, Smith SW, Hartt WH. Once through vs. recirculated seawater testing for calcareous deposit polarisation of cathodically protected steel. In: Haynes GS, Baboian R (eds). Laboratory corrosion tests and standards. ASTM STP 866. ASTM, Philadelphia, 1985, pp 207–214

130. Jones RJ. Characterisation of temperature resistance. In: Engineered materials handbook, vol. 2: Plastics. ASM International, Metals Park, Ohio, 1987, pp 559–567

131. ASTM G53. Recommended practice for operating light and water exposure apparatus for exposure of non-metallic materials. ASTM, Philadelphia, 1980

4 Marine Materials

There is a tendency to believe that only a few materials are suitable for the marine environment, and that marine materials constitute a special class developed entirely for that application. This is not quite true and, although many of the advances that have been made in marine engineering have become possible due to the realisation or development of classes of materials previously considered unsuitable, it is also true that many design advances have been made to allow the use of well-studied materials, particularly structural ones. Classic examples are, of the former, the iron-hulled ships referred to in Chapter 1 and, of the latter, the development of conventional structural steel for offshore applications.

This chapter is concerned primarily with an overview of the types of materials in current marine use, and those likely to become important in the immediate future. The discussion also includes the types of materials in use for particular applications. This chapter is designed to be self-contained in that most of the design data in the book can be found here. The remaining chapters are provided as a guide to the meaning of the particular quantities referred to, for information on how the data may be used, and for examples of such use.

Each material or group of materials has a separate section, in which its nature and general properties are discussed and compared with those of others. Each section then continues with detailed properties and references to the literature. A coarse selection guide is included in Chapter 1 as an overview of the range of materials covered in this chapter. For the purposes of this discussion the broad materials categories are taken as

Metals

Polymers

Inorganic materials

Composites

Cement and concrete

and these are discussed in turn below.

4.1 Marine Alloys

The decision as to whether to use a metal or alloy in a marine application is becoming a rather less clear-cut one as the traditional metallic advantages of high strength, toughness and stiffness are matched and, in some cases,

exceeded by other classes of material such as ceramics or resin-based composites, particularly when these latter are normalised for density. Nevertheless, alloys, particularly steel, remain extremely important materials in marine applications and it is unlikely that any material will overtake steel and concrete for very large-scale constructions in the foreseeable future.

Because of the very wide range in specification for what may be very similar materials, the text and tables in this section employ principally British Standard and AISI designations. The reader is referred to the Equivalence Guides [1, 2] to cross-reference metals and alloys using other designatory systems. It is, however, far more useful to be able to identify a material by its generic type and its properties and to use standards simply to ensure that a specification is met.

Alloys can be defined in terms of their microstructures as well as composition and heat treatment, and the interrelationship of these parameters is fundamental to determining the final properties, one single parameter not normally being sufficient.

The following is a brief review of the way in which alloys are constituted and how this constitution (the microstructure) can be controlled by heat treatment and alloying. For a description of the meaning of the metallurgical terms the reader is referred to, for example, Askeland [3].

4.1.1 Alloy 'Architecture'

Most metals are crystalline in nature, as are most compounds with a metallic component. Each metallic element and compound has a unique crystal structure (or range of crystal structures) into which it will arrange its atoms in a certain temperature range. Each of these structures is a distinct phase, and alloys may contain many different phases (as is the case in complex bronzes, for example) or simply one (for example copper–nickel alloys). If a pure metal (single phase) is alloyed with a second one, this alloying element may dissolve (as a solid solution), react (to form an intermetallic compound with its own distinct crystal structure) or retain its own identity as a second phase. In a given alloy system more than one such phenomenon may take place over the temperature and composition space, and exactly how an alloy series will behave is a matter for observation rather than calculation, these observations being embodied in phase diagrams. Such diagrams are plots of temperature–composition space showing the phases which are present under equilibrium conditions. Of course, alloying is normally carried out in the liquid phase, so that an alloy must solidify and cool to develop its microstructure, and the way in which metals solidify is normally that a number of very small seed crystals grow until they impinge upon each other, at which point the alloy is solid. Further changes of phase in the solid state must take place by diffusion or by specialised types of transformation involving very short-range movement of atoms (martensitic transformations). Clearly the cooling conditions (particularly the cooling rate) will influence the final phase constitution (whether or not it reaches equilibrium) and the phase distribution (size, shape and morphology).

The individual crystals in engineering alloys (at least for the matrix phase) are often of polygonal shape and this is particularly true for wrought alloys. The cross-section of a piece of metal therefore usually has the appearance of a tiling pattern where the 'grain size' is typically of the order of a few tens of microns, as is shown schematically in Fig. 4.1a. In as-cast structures, the primary phase may take on a more dendritic form, which in cross-section looks 'cored' (Fig. 4.1b) unless some homogenisation takes place during cooling.

In a pure metal or solid solution each grain will be of the same chemical composition, but in multiphase mixtures different grains will have different compositions, crystal structures and hence properties, so that many alloys can be regarded as metal–metal or metal–ceramic composites, although the control over morphology is more limited than in the case of true composites.

Phase distributions can take a wide range of different morphologies in metals, but two very common ones are the lamellar eutectic, of which pearlite in steel is a typical example (Fig. 4.2a), and the spheroidal or lath precipitate, of which temper carbides in steel are typical examples (Fig. 4.2b).

The microstructures of metals and alloys are crucial in determining their properties, and microstructures can be changed quite radically by thermal and other treatments, even with a single alloy composition. The remainder of this section is aimed at indicating how metallic properties can be altered by alloying and treatment.

4.1.1.1 Control of Strength

Most alloy developments over the years have been aimed at increasing strength without compromise to other properties such as toughness and weldability. In general, increases in strength are measured as yield strength or ultimate tensile strength. Young's modulus is a property primarily of the matrix crystal structure and cannot be changed (short of changing the crystal structure or including a substantial fraction of second phase), so that for example an ultra-high-strength steel with a yield strength of 1500 MPa will have substantially the same Young's modulus as an ordinary mild steel with a yield strength of 250 MPa.

The individual components which go to make up the strength of a metal are fundamentally its crystal structure and, then, a series of modifications which are in the control of metal processors from primary processes such as smelting down to final processes such as welding. The following brief descriptions indicate how each of these strengthening mechanisms operates and hence how the properties might be modified or compromised during processing.

1. *Inherent strength.* Each crystal structure has an inherent strength related to the ease with which dislocations can move in the lattice. This strength is, however, much less than the theoretical strength for rupture of atomic bonds right across the section of a perfect crystal, and it

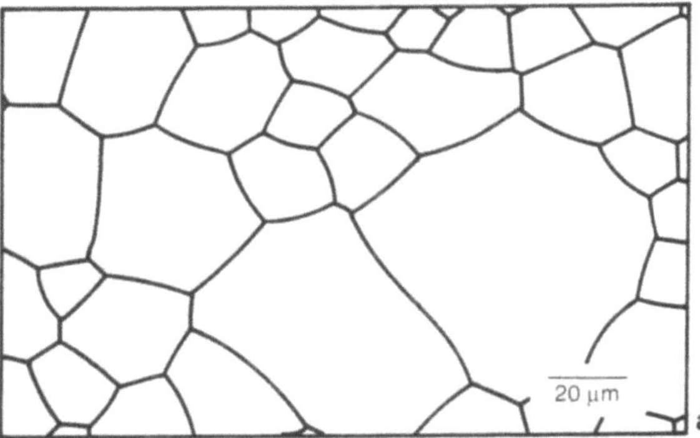

Fig. 4.1. Sketches of typical **a** wrought and **b** cast grain structures in metals.

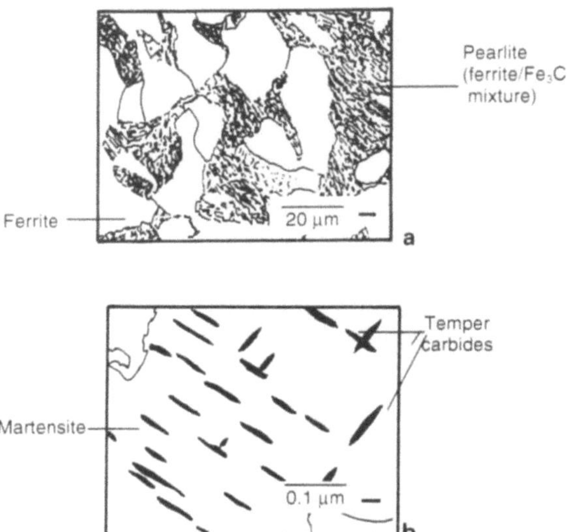

Fig. 4.2. Examples of the range of possible second-phase morphologies in metals. **a** Lamellar eutectic or eutectoid as typified by pearlite in steel. **b** Uniform lath as typified by temper carbides in steel.

was this observation which led to the postulation of the dislocation as the main mode of metallic plastic deformation, the yield strength essentially being the stress which, when averaged over all grains and resolved as a shear stress on the slip planes, will result in dislocation movement. The inherent strength of a crystal lattice containing some (non-interacting) dislocations is in the region of $E/300$ and any increase over this has to be achieved by alloying and/or thermal and/or mechanical treatment. Thus, an aluminium alloy with a yield strength of 500 MPa would be regarded as a high-strength aluminium alloy, whereas the same strength in a heat-treated steel would be quite unremarkable. The theoretical strength of metals can be more closely approached by special manufacturing techniques to reduce the dislocation density to very low levels, and this is how very high-strength levels are reached in 'whiskers', which are beginning to find applications in advanced metal matrix composites.

2. *Cold work.* The effect of cold working of an alloy is to multiply dislocations, which then interact with each other making further deformation more difficult. Although such treatments (e.g. cold rolling) will result in a stronger product, there is also a consequent loss of ductility and toughness. While this can be tolerated in some products (e.g. wires), it is normally an unacceptable method of strengthening in others (e.g. pipelines). Cold work can be reversed by simple thermal treatments (annealing), and this reversal must be considered in welding cold-work-strengthened alloys.

3. *Grain size.* Grain boundaries also act as obstacles to dislocation movement and to the propagation of yield from one grain to the next, and thus finer-grained alloys are stronger. The refinement of grains can be achieved by accelerated cooling (almost the only way in casting alloys), by careful cold-work–anneal cycles, and by special thermo-mechanical treatments involving thermally stable precipitates (e.g. niobium carbides in steel). This is one of the most satisfactory ways of strengthening, since fine-grained alloys are tougher as well as stronger.

4. *Solid solution.* The lattice strain field around solute atoms reacts with the strain fields of any approaching dislocations, so that strengthening is achieved simply by virtue of there being 'foreign' atoms in the lattice, the larger the solute strain field the greater the strengthening. Atoms can dissolve in a crystal lattice, either as interstitials (like carbon in iron) or substitutionally (like copper in nickel).

Generally, the interstitial solutes in metals are the smaller atoms (such as carbon, nitrogen, oxygen, boron, phosphorus and sulphur) and produce greater distortion (since interstices are usually much smaller than the atoms) and hence greater strength increases per atom. The potency of an interstitial strengthener is dependent upon the relative sizes of the atoms and the interstices in the lattice into which it fits. The difference between austenitic and ferritic steels in their tolerance to dissolved carbon is a good example of this and is, in fact, the basis of the heat-treatability of steels, carbon having to be rejected from interstitial solid solution in the austenite phase during cooling. Substitutional solid solution strengthening also shows a relative size effect, but additionally in this case large negative size differences can be just as effective as large positive ones.

5. *Second phase.* Solid solutions, like liquid solutions, show a solubility limit, and in the case of solid solutions this limit depends upon the chemical affinities of the components, size differences and differences in crystal structure. Once the solubility limit has been exceeded a second phase will be present (according to the phase diagram), and this may be an intermediate compound (e.g. Mg_2Al_3 in aluminium–magnesium alloys) or another solid solution (e.g. β-Si in aluminium–silicon alloys). The effect of this second phase on strength depends upon:

The size of the second-phase grains/particles;

The amount of second phase (volume fraction);

The properties of the second phase;

The distribution and shape of grains/particles;

The coherency (lattice matching) between second phase and matrix (Fig. 4.3);

The contiguity of second phase paths in the solid.

Since most intermetallic compounds are hard, precipitation of these types of compounds as second phase leads to useful strengthening of otherwise weak alloys. Coherency is a very important issue for some precipitates, particularly aluminium alloys, where very small precipitates only are coherent and care must be taken not to over-age. Strengthening using coherent precipitates is known as 'age-hardening' (or 'precipitation hardening') to distinguish it from more general dispersion strengthening by second phase.

6. *Martensitic transformation.* Many alloys show martensitic transformations, and only some of these increase strength, so that this category generally applies

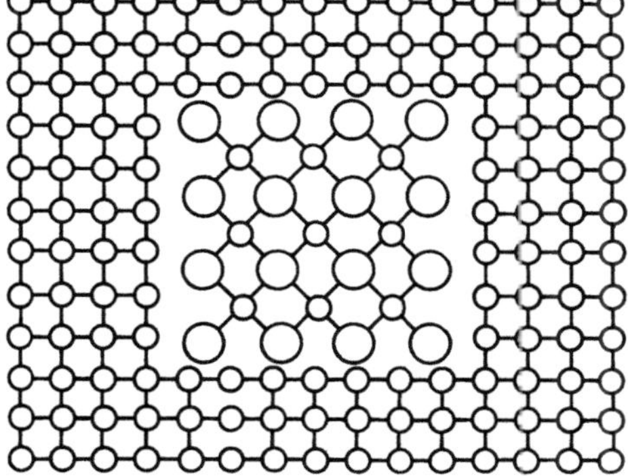
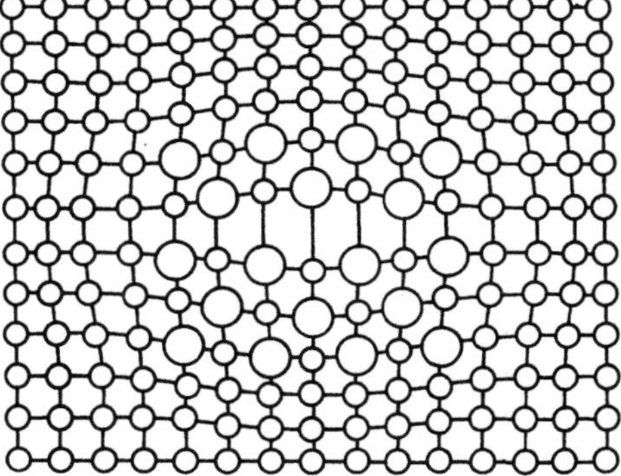

Fig. 4.3. Schematic view of the effect of coherency of second phase. For equal size, shape, volume fraction and distribution, coherent precipitates (right) provide much greater strengthening than incoherent (left).

only to the particular martensitic transformation which occurs in steel. This is described in more detail in Section 4.1.2 and is mentioned here for completeness because it represents an important means of strengthening low-alloy steels.

The strengthening mechanisms outlined above are summarised in Table 4.1, along with some examples of alloy systems which employ them.

4.1.1.2 Control of Toughness

Like inherent strength, inherent toughness is a property of the crystal structure which depends upon the relative ease of fracture by cleavage of atomic planes and yield through dislocation movement. Obviously, any change in strength within a given crystal structure will alter this relative ease and carries with it the danger of embrittlement. As a general statement, materials which do not yield easily are likely to behave in a brittle fashion, and most of the development of engineering materials has been based on increasing strength (i.e. making yield more difficult) without compromising toughness. Among the metals, face-centred cubic structures (copper, nickel, austenite) are more likely to yield than fracture, whereas body-centred cubic structures show more resistance to yielding (e.g. chromium, ferrite), and more care has to be taken over embrittlement.

Table 4.1. Summary of the various compositional and constitutional factors which affect the strength of metals and alloys

Factor	Examples in commercial alloys
Inherent strength	None specifically, but metals with higher Young's modulus (e.g. iron) have a higher basic strength level than those with lower moduli (e.g. aluminium).
Cold work	More commonly used for alloys with FCC basis metal structure (e.g. copper alloys, aluminium alloys) or for development of spring properties (e.g. steel wire).
Grain size	Especially structural steels, but fine grain size preferred in most wrought and cast alloys.
Solid solution	Substitutional used in alloys where there is high mutual solid solubility (e.g. cupronickels) but not an important factor in, for example, steels. Interstitial strengthening used notably in steels (including some stainless steels).
Second phase	Dispersion with large particles used in multi-phase copper alloys (e.g. α–β brasses) and C–Mn steels, also some titanium alloys. Dispersion with smaller particles in some aluminium alloys, tempered steels and mechanical alloys (e.g. some superalloys and metal matrix composites). Precipitation strengthening used in aluminium alloys, some copper and nickel alloys and some stainless steels.
Martensitic transformation	Most notably in heat-treatable steels (low-alloy and some high-alloy). Also used in maraging steels, some cast irons.

It is for the above reason that cold work strengthening is a rather unsatisfactory method of increasing strength and is most useful for metals of FCC structure. On the other hand, reduction of grain size is one of the few methods of strengthening which also increases toughness.

The need to avoid grain boundary or other continuous networks of brittle phase is obvious, so that, from the point of view of toughness as well as strength, the best precipitate morphology is a fine, discontinuous dispersion of spheroidal particles, and the difference can be appreciated by considering the properties of tempered martensite and a coarse ferrite–pearlite structure in a heat-treatable steel. Examples of embrittlement by continuous second phase can be seen, for example, in overheated brazements or sigma-phase precipitation in stainless steels.

Some interstitial solutes can embrittle metals (e.g. hydrogen in ferrite), the effect being greater if other interstitials are already present (e.g. hydrogen in high-strength ferritic steels).

Finally, the effect of temperature on the toughness of some materials needs to be considered, the most important case being the ductile–brittle transition in ferritic steels.

4.1.1.3 Fatigue Resistance

Since fatigue is a fracture-mechanical phenomenon, its progress is related very much to geometry and loading condition. Within an alloy system, there is little which can be done to alter fatigue resistance, and fatigue properties are rarely used as a selection criterion. Most design against fatigue is therefore aimed at reducing the chance of fatigue initiation by careful surface contour control and surface preparation, as well as the induction of compressive residual stresses on the surfaces, for example by shot-peening and nitriding. If cyclic loading can be avoided or its effects reduced (for example by preloading of bolts), this is also useful.

4.1.1.4 Corrosion Resistance

As seen in Chapter 3, there are two types of inherent corrosion resistance which can be shown by metals.

The first of these is nobility, which is shown by metals which are high in the electrochemical series (e.g. gold, platinum) and do not oxidise easily. Such metals are relatively inactive electrochemically; they are usually prized for this property and are consequently expensive.

The ability to form a passive film, shown by, for example, titanium, aluminium and chromium, is a rather more realistic selection aim, but passivity is a property which is usually only shown over a particular range of temperature, chemistry and electrochemical potential, so that passivating alloys may show localised corrosion effects such as pitting and crevice corrosion in some environments. Passivating properties can be conferred by making (usually substantial) additions of a strong passivator, the best-known example being the addition of chromium (in amounts greater than about 13 per cent) to produce stainless steels.

The purpose of the foregoing discussion has been to bring out some of the essential elements of alloy design so that the reader can be aware of the driving force behind the developments seen in the alloys described in the following subsections.

4.1.2 Carbon–Manganese and Low-Alloy Steels

Carbon–manganese and low-alloy steels are by far the most widely used alloys in marine technology, as they are in most other areas of engineering. Much of the reason for this is the versatility of the alloys, where very small compositional changes can produce very large changes in strength, but also their availability, which has led to the commoner grades of steels being relatively cheap.

The major characteristics of steel which have made it such a successful engineering material are its high strength and high Young's modulus. In fact, these properties combine to outweigh the significant disadvantages of poor marine corrosion resistance and (relative) technical difficulty in welding. The susceptibility of the ferritic structure to embrittlement by hydrogen or by low temperatures is a further factor which must be borne in mind by the designer, although, as with the other disadvantages, established (standard) routes are available for overcoming these.

Between the two types of steels considered in this section, and normally referred to as 'carbon–manganese steels' and 'low-alloy steels', the distinction is not always clear. For the purposes of this discussion, 'low-alloy' refers to those steels which have alloying additions made to them for the purpose of rendering them suitable for quenching and tempering, so that the 'microalloyed' steels used for pipelines and offshore structures only fall marginally into this category.

Uses of carbon–manganese steels in the marine environment are almost unlimited, but they are in bulk use for items such as ships' hulls, offshore structures and pipelines; it is for such structures that most of the marine-technology-led developments in steels have taken place and these applications are discussed below.

Low-alloy steels find more limited use, owing to the difficulties with welding arising from their hardenability (see Chapter 5). However, for components which do not require welding, or which can be re-heat-treated in their entirety, such steels find a number of applications, particularly as petroleum engineering tubulars, although it must be said that NACE requirements for sulphide resistance (see Section 3.5) often mean that their full strength potential cannot be reached. The use of high-strength steels is of considerable interest in offshore and marine engineering, because it permits substantial weight savings in designs where strength is the limiting criterion. Of course, Young's modulus is not any higher in a 'high-strength' steel than in any other, so that such savings have to be made without compromising stiffness, buckling or dynamic properties. At present such high-strength steel applications occupy a halfway stage between carbon–manganese and low-allow steels, and their use is unlikely to extend to the genuine low-alloy steels in the foreseeable future.

These two classes of ferritic steel cover the majority of that produced and, indeed, the majority of metallic materials produced. The range of available alloys is vast, and it is essential to break these down into different groups by virtue of their composition and treatment and, in some cases, their intended application.

Steels of the carbon–manganese type are generally classified in terms of their carbon content, which is the most important alloying element. The manganese content is usually between 0.5 per cent (in the poorer grades) and 1.5 per cent, and no other deliberate alloying additions are made. These steels are not usually heat-treated and are used in the ferrite–pearlite condition (for a very readable account of the transformation characteristics of steels, see Honeycombe [4]), so that their grain structure has the characteristic appearance shown in schematic form in Fig. 4.2a, and the amount of the pearlitic phase increases with carbon content.

This has a number of implications as regards mechanical properties. First, the size and distribution of the two phases (ferrite and cementite, Fe_3C) or, more practically, the ferrite and pearlite, has a profound effect upon mechanical properties. It is widely known that increasing carbon content increases the amount of pearlite in the structure and thus the strength of the steel. However, the corresponding increased hardenability (ease with which the alloy can be quenched to produce a martensitic structure) and increased brittleness of the equilibrium structure make this an undesirable approach for most structural purposes, particularly where welding is required.

Almost all of the developments in structural steel over recent years have been aimed at increasing strength without altering the chemistry, particularly the carbon content, of the steel. The most effective way of doing this while retaining or even enhancing the toughness is to reduce the grain size.

The traditional method of reducing grain size in metals is to cold work and then raise the temperature to a point at which recrystallisation takes place, to soak to allow complete or partial recrystallisation and to cool, 'freezing in' the recrystallised grain size. With steels this is slightly complicated by the fact that a phase change also takes place on heating, but the new austenite grain size is also affected by cold work and the final transformed ferrite–pearlite structure reflects this. Such a method can only produce a limited degree of grain refinement, and modern grain refining practice is based on dynamic recrystallisation.

Dynamic recrystallisation results from hot working at a temperature where recrystallisation takes place simultaneously with working. In steels, this usually occurs in the austenite phase and 'microalloying additions' of aluminium, vanadium, titanium or niobium [5, 6] are added to impede the growth of austenite grains as illustrated in Fig. 4.4. It should also be pointed out that some alloying elements, notably aluminium, have been used successfully to refine grains during normal processing [7].

As can be seen from the above discussion, the interplay of the various contributions to steel strengthening and toughening by the precipitate size and distribution, the final ferrite grain size, the amount of pearlite and the dislocation density is complex (Platts et al. [8]). If the element of a certain degree of accelerated cooling is added, it is not surprising that a wide range of subtly different processes for producing such high-strength–low-alloy (HSLA) steels have evolved.

Aside from the never-ending search for increased strength, a number of other recent developments in steelmaking practice have been driven by the particular requirements for offshore structures and pipelines. One of these, control of sulphur, is considered in Chapter 5 and secondarily in Chapter 3, because it affects both the weldability and susceptibility to some forms of corrosion. Baker [9] has discussed recent trends in sulphur reduction for the two main bulk steel types, finding a drop of over 50 per cent in specified sulphur content for the period 1970 to 1980.

De Koenig [10] has indicated more generally the effect of composition of modern structural steels on weldability

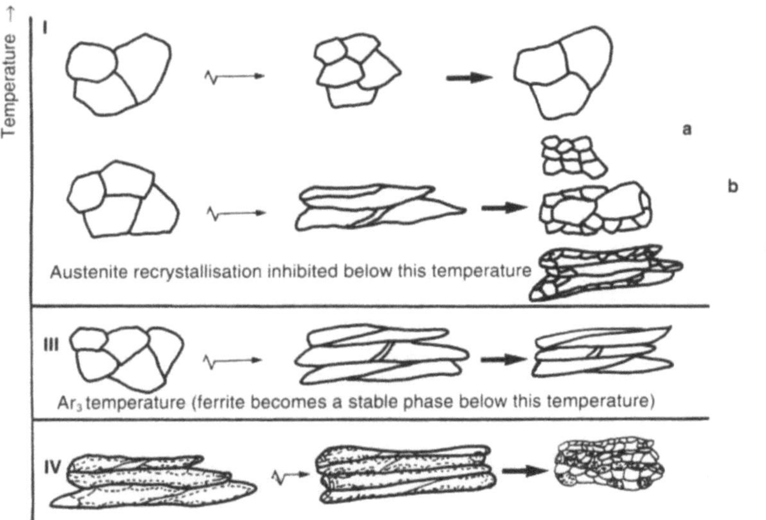

Fig. 4.4. Schematic diagram of the thermomechanical processing of high-strength low-alloy (HSLA) steels. I, rapid dynamic or static recrystallisation and grain growth of austenite; II, slower recrystallisation kinetics or secondary grain growth may lead to mixed austenite grain structures, **b**, **c**; III, recrystallisation retarded, subsequent passes are cumulative; IV, rolling of two-phase austenite–ferrite producing strain-hardening in the ferrite. (Adapted from Cohen and Hansen [5].)

and properties, and Table 4.2 summarises some of these effects, although a more detailed quantitative description of composition effects on weldability is given in Chapter 5.

The specific steel requirements for offshore structures are partly dependent upon their mode of fabrication and partly upon operational conditions. A typical steel jacket structure for offshore use is fabricated mostly from plate material of various thicknesses, the tubular morphology being obtained by rolling the plates into cylinders and seam welding these into the basic building blocks known as cans. The structure is then made from cans of various sizes and wall thicknesses by a series of butt and intersection welds [11].

The pattern of steel requirement for offshore platforms has therefore been to provide good shop- (for cans) and site-weldability for very large section thicknesses coupled with high toughness, fatigue resistance and strength. In Britain, these requirements have traditionally been met from steels specified in BS 4360, *Weldable Structural Steels* [12], and grades available from this standard include modifications for, for example, improving through-thick-

Table 4.2. Effect of chemical composition of modern structural steels on properties and weldability

Performance	C	Si	Mn	P	S	Cu	Ni	Cr	Mo	Nb	V	Al	Ti	B	O	N	Rem. Ca	Sn	Sb
Steel																			
Strength	↑ 1	2	1			1	1	1	1	1	1			Free B 1					
Toughness	↑ 3		3			2	2					AlN 2	TiN 2		3	4			
HAZ																			
Hardness	↓ 4	3	4	3	?	3	3	4	4	3	3			Free B 3	Oxide 2	2			
Cold cracking	↓ 4		4	3	2	3		4	3	3	3			Free B 3	Oxide 2				
Lamellar tearing	↓		3	4								Al_2O_3 4			4		1	1	
Reheat cracking	↓ 3		2					4	4	3	4								
Charpy value	↑ 3		3			2	2		2	3		AlN 2	TiN 1	BN 2	Oxide 3 or 2	4	Oxide 2		
COD	↑ 3		3	3		2	1			3		AlN 2	TiN 1	BN 2	Oxide 3 or 2	4	Oxide 2		
SSCC	↓ 4	3	4	3		3	3	4	4	3	3			Free B 3					
Temper embrittlement	↓	3	3	4														4	4
Creep embrittlement	↓					4													4
Irradiation embrittlement	↓			4		4													

Source: De Koenig [9].
1, very beneficial; 2, beneficial; 3, detrimental; 4, very detrimental.

ness properties as described in Chapters 2 and 5. The microalloying additions niobium (columbium) and vanadium feature in most of the BS 4360 compositions.

For UK offshore structures the general approach has been to use Grade 43 in 'secondary' structural areas (such as module walls and decks), Grade 50 in 'primary' areas (such as jacket legs and module support frames) and Hyzed Grade 50 in critical areas such as nodal joints (Billingham [13]). The 'D' toughness designation was normally used as this accorded with the requirements of BS 6235 [14], the *Code of Practice for Offshore Structures* (now withdrawn), although modern practice is rather less restrictive than this.

Often, however, platform designers add their own modifications to standards in the light of experience gained in other structures. For example, Salama et al. [15] indicate that the material used for the Hutton tension leg platform (TLP) had modifications made to restrict chemical composition and an additional mechanical testing requirement was introduced to ensure better fracture toughness, improved weldability and improved through-thickness properties over BS 4360 Grade 50E.

The UEG/CIRIA tubular joints design guide [11] contains a collection of a wider range of structural steel specifications, and this is shown in Table 4.3. As can be seen from this table, the band of mechanical and chemical variables is remarkably narrow, considering the range available to carbon–manganese steels. A lot of this conservatism can be understood in terms of the sheer bulk of steel involved in such platforms and the need to be confident about final properties, but there is a growing tendency to become involved with higher-strength steels, although, understandably, this progress is cautious.

A survey by the UK Offshore Supplies Office of the Department of Energy [16] found that the use of higher-strength structural steels has a potential to reduce weight in offshore structures and cites BS 4360 Grade 55F as having the ability to to save 10 to 15 per cent in weight over more conventional structural steels, provided that the stiffness and buckling criteria can be met in the resulting reduced sections. To this might be added the observation (Gurney [17]) that strength has little influence on fatigue life, although this statement is a general one and requires to be validated on a case-by-case basis. Table 4.4 shows some potential areas where high-strength steels might be used to save weight in offshore structures. These low-carbon, quenched and tempered steels are slowly gaining acceptance for use as offshore structural materials, and an idea of their potential can be seen by comparing the BS 4360 Grade 55 steels with the roller-quenched and tempered steels shown in Table 4.5.

Billingham [13] has pointed out that Grade 55 steel was used in the Magnus platform in UK offshore waters and, further, that there is some evidence that some higher-strength steels may be more fatigue-resistant than their lower-strength counterparts for the same level of stress intensity (Lieurade and Lecoq [19]). This observation, however, applies only to initiation in the base metal, and the effect in offshore structural welded joints is complicated by the fact that life is propagation-dominated, and Lieurade and Lecoq conclude that, since propagation is only marginally affected by yield strength, both high and traditional yield strength steels will show approximately the same fatigue life in offshore structural applications.

The development of offshore structural steels continues,

although acceptance is a slow process, which is not surprising given the amount of research effort which has been expended on the existing structural types and the crucial importance of very careful validation procedures.

The process developments required to achieve these developments in strength are no less involved, and one example is presented by Suzuki et al. [20] for Arctic-grade structural steel of high strength with good weldability and extreme low-temperature toughness. This steel is produced by the controlled-rolled and on-line accelerated cooling process (CR + OLAC), and the effect of just one of the processing variables on mechanical properties is illustrated in Fig. 4.5.

These authors also note that, whereas substantial grain refinement and hence high yield strength and excellent low-temperature toughness can be obtained by the CR + OLAC process in a variety of steel chemistries, a lot of this benefit can be lost in the weld-heat-affected zone. It is suggested that the interplay of Al, Ti and N precipitation kinetics in the heat-affected zone can be exploited to enhance heat-affected zone toughness, and that a high-aluminium, low-nitrogen steel microalloyed with titanium could find application for structural work in Arctic waters even after welding with high-energy input.

Steel for shipbuilding use has a rather longer history than for offshore structures and the requirements which have driven its development are slightly different. As pointed out in Chapter 1, ships with welded steel hulls began to form an important fraction of those manufactured in the 1930s and most of the early materials developments were concerned with the eradication of the problems of brittle fracture.

From the structural point of view, the ship's hull, including decks, stiffeners and bulkheads, is its most important and fundamental part and, as such, is the part that is most often made from a structural material such as steel or concrete.

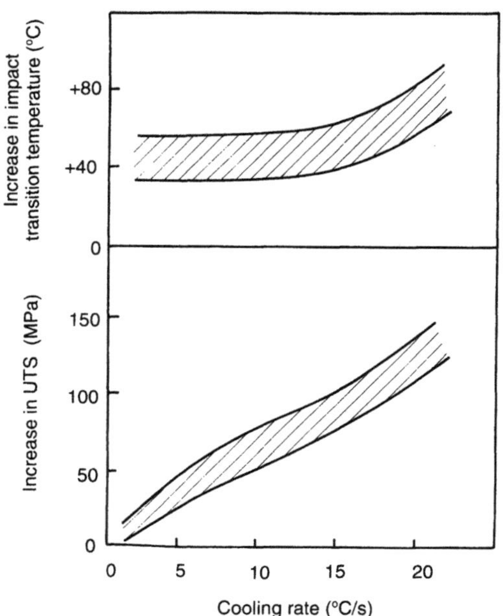

Fig. 4.5. Effect of cooling rate on the strength and toughness of controlled-rolled and on-line accelerated cooled (CR+OLAC) steels as compared with conventional controlled-rolled (CR) plates. (From Suzuki et al. [20].)

Table 4.3. Selection of structural steels from different specifications

Specification	Chemical composition: maximum of element (%)														Mechanical properties				Condition
	C	Si	Mn	P	S	Ni	Cr	Mo	Nb	V	Ti	Al	Cu	N	$\sigma_{y,min}$ (MN m^{-2})	UTS (MN m^{-2})	Elongation, min. (%)	Charpy energy, min. (J)	
BS 4360: 43D	0.16	0.50	1.50	0.04	0.04				0.003/0.1	0.003/0.1					240/280	430/510	20	27 at −20°C	Normalised
BS 4360: 43E	0.16	0.1/0.5	1.50	0.04	0.04										240/280	430/510	20	27 at −50°C	Normalised
BS 4360: 50D	0.18	0.1/0.5	1.50	0.04	0.04				0.003/0.1	0.003/0.1					340/355	490/620	18	27 at −30°C	Normalised
BS 4360: 50E	0.18	0.1/0.5	1.50	0.04	0.04				0.003/0.1	0.003/0.1					340/355	490/620	18	27 at −50°C	Normalised
BS 4360: 50F	0.16	0.1/0.5	1.50	0.025	0.025				0.003/0.08	0.003/0.1					390	490/620	18	27 at −60°C	Quenched and tempered
BS 4360: 55E	0.22	0.1/0.6	1.60	0.04	0.04				0.003/0.01	0.003/0.2					400/450	550/700	17	27 at −50°C	Normalised
BS 4360: 55F	0.16	0.1/0.5	1.60	0.025	0.025				0.003/0.08	0.003/0.08					415/450	550/700	17	27 at −60°C	Quenched and tempered
API 5L Grade A	0.22		0.90	0.04	0.04										205	330			Hot finished/ normalised
API 5LX 52	0.30		1.35	0.04	0.05										358	455			Hot finished/ normalised
API 2H	0.18	0.15/0.30	0.9/1.60	0.04	0.035									0.015	289	427/565	20	34 at −40°C	Normalised
ASTM AI31 Grade C5	0.16	0.1/1.35	1.0/1.35	0.04	0.04										235	400–490	21	19 at −40°C	Normalised
AI31 Grade EH36	0.18	0.1/0.5	0.90/1.60	0.04	0.04	0.40	0.25	0.08	0.05	0.10		0.10	0.35		360	490–620	22	23 at −40°C	Normalised
Euronorm Fe E 255	0.15	0.35	1.4	0.03	0.03	0.3	0.25	0.1	0.05	0.10		0.015	0.35		235–255	370–490		28 at −50°C	Normalised
Euronorm Fe E 355	0.18	0.5	1.6	0.03	0.03	0.3	0.25	0.1	0.06	0.20		0.015	0.35		325–355	490–630		28 at − 50°C	Normalised

Source: UEG/CIRIA [11].

Table 4.4. Assessment (at 1987) of the technical feasibility of replacing 'standard' structural steels with high-strength steel for offshore and marine structural applications

Application	High-strength steels RQT 501 (55F)	RQT 701	Comments
Structural			
Trusses	S	L	High strength and the need for welding predominate
Transverse plate girders	S	L	High strength and the need for welding predominate
Deck plate	X	X	Deflection critical
Stringers	X	X	Deflection critical
Walkways	X	X	
Handrails	X	X	
Stairways	X	X	
Caissons	X	X	
External cladding	X	X	
Equipment support	S	L	Simple construction
Crane pedestals	S	L	Suited to high-strength steels
Drilling derricks	S	L	Fatigue and strength critical designs
Mud tanks	X	X	
Flare stack	S	L	
Piping			
Open drains	X	X	
Water injection	X	X	
Firewater	X	X	Opportunities exist for a greater use of duplex stainless steels
Seawater	X	X	
Cooling water	X	X	
Hydrocarbon processing (liquid and vapour)	X	X	
Pipe support	X	X	
Equipment			
Pressure vessels	X	X	High operating pressures and temperatures
Low-pressure tanks	X	X	
Heat exchangers	X	X	
Skids	X	X	
Electrical			
Cables	X	X	Specialist technology
Rack and tray	X	X	
Lighting equipment	X	X	Specialist technology

Source: Adapted from Offshore Supplies Office [16].
Technical feasibilities of material/application combinations:
S, short-term development (0 to 3 years)
M, medium-term development (3 to 7 years)
L, long-term development (> 5 to 10 years)
X, indicates impossible, inappropriate or unsuitable combinations.

Table 4.5. Comparison of some of the properties and chemical components of high-yield weldable structural steel plates with BS 4360 Grade 55
(Figures are for comparison only – some compositional details depend on sub-grade or manufacturer and some specified compositional constraints are not included)

Alloy	%C, max.	%Mn, max.	%Si, max.	Other alloying elements	Min. yield (MPa)
Gr 55	0.22	1.60	0.5	0.1%Nb, 0.2%V max.	400 to 450
RQT 501	0.16	1.50	0.5	0.08%V, 0.2%Mo max.	470
RQT 601	0.20	1.50	0.50	0.06%Nb, 0.04%Ti, 0.003%B max.	620
RQT 701	0.20	1.50	0.50	0.06%Nb, 0.04% Ti, 0.003%B max.	690

Source: British Steel Corporation [18].

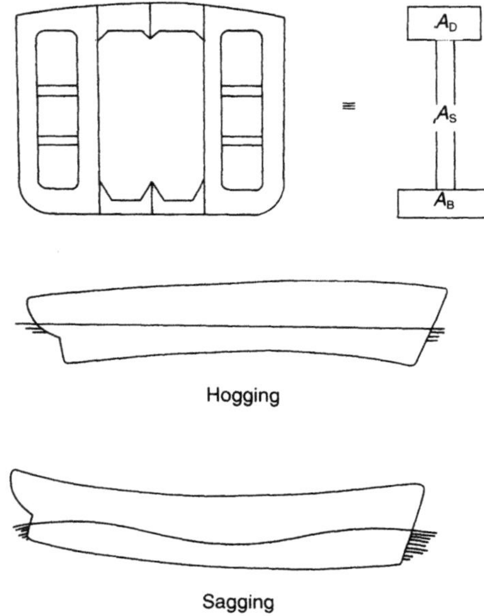

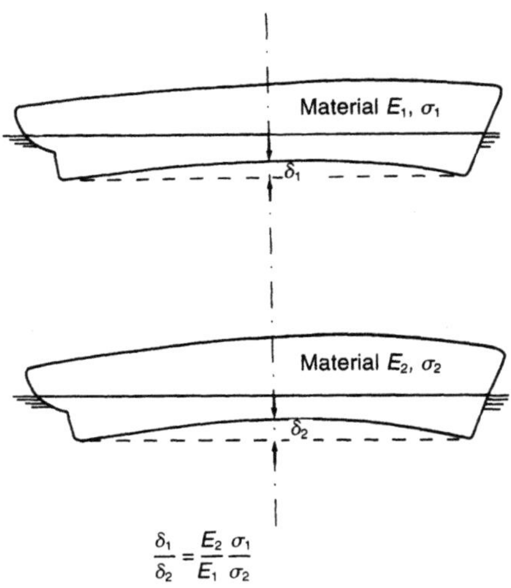

$$\frac{\delta_1}{\delta_2} = \frac{E_2}{E_1}\frac{\sigma_1}{\sigma_2}$$

Fig. 4.7. Deflection ratio for comparing different materials for ship construction. (Adapted from Crane and Charles [21].)

Fig. 4.6. Schematic loading of the ship beam. Top diagram shows how the ship's section can be idealised as an I-beam, with A_D the area of deck plating, deck longitudinals and other longitudinal material above 90 per cent of the section height, A_S the area of the sheet plus longitudinal bulkheads and other longitudinal material between 10 and 90 per cent of section height and A_B the area of longitudinal material below 10 per cent of section height. Bottom two diagrams show loading of the beam in a seaway. (Adapted from Crane and Charles [21].)

(An exception is discussed in Case Study 7.3.) This hull must be fabricated by welding and, in use, is most often subjected to overall reversed bending as shown in Fig. 4.6.

This is, of course, a simplified view, but a major factor in the selection of ship materials is deflection of the full shape under static and dynamic wave loading (Crane and Charles [21]). These authors also indicate that the substitution of higher-strength steel (according to the deflection ratio illustrated in Fig. 4.7) for a ship of identical dimensions and load distribution (transferring any weight savings to the cargo) leads to a deflection which is increased in the ratio of the increase in allowable stress. Using the limits on deflections set by *Lloyd's Classification Rules* [22], they were able to show that no benefit is obtained by using steels of higher yield stress than 340 MPa. This suggestion is, however, likely to break down for 'composite' beams (i.e. ships which use one or more steel or material in the structure). Muckle [23] has considered this problem from the point of view of two materials with the same elastic modulus (e.g. high- and low-strength steels) and two materials

Table 4.6. Some typical ship structural materials

Steel	Approximate equivalent specifications	Yield strength (MN/m^2)	Tensile strength (MN/m^2)	Limit of proportionality (MN/m^2)	Young's modulus (GN/m^2)	Poisson's ratio	Charpy temp./energy (°C/J)	Elongation (%)	Cost, relative to mild steel
Mild steel	NES 791 pt 1 BS 4360 43 A Lloyd's 'A'	245($\geqslant$63 mm) to 275 ($\leqslant$16 mm)	430 to 510	216 to 263	201 to 208.5	0.3	–/–	22	1
Notch tough MS	NES 791 pt 2 BS 4360 43 D Lloyd's 'D'	245($\geqslant$63 mm) to 280($\leqslant$16 mm)	430 to 540	216 to 263	201 to 208.5	0.3	–20/27	22	1.1
B quality (Crack arresting steel)	NES 791 pt 3 BS 4360 50 DD Lloyd's EH32	280($\geqslant$32 mm) to 310($\leqslant$22 mm)	450 to 590	263 to 355	199 to 207	0.3	–30/40	20	1.15
Q1N (HY 80)	NES 736 pt 1	550–655 (0.2% proof)	$\nless$1.4 x proof stress	480 to 585	197 to 207	0.3	–84/70 to 100	20	6
Q2N (HY 100)	NES 826 pt 1 (to be issued)	690–790 (0.2% proof)	$\nless$1.11 x proof stress	–	197 to 207	0.3	(not yet specified)	20	–
HSLA-80	MIL-S-24645(SH)	550–690 (0.2% proof)	–	–	207	0.3	–84/35	20	–

Source: Betts and Clayton [24].

Table 4.7. Range of steels used in the UK Type 23 warship

Specification	Composition (%)	Yield strength	Uses
BS 43A	0.25 C, 0.5–1.6 Si, 0.05 P, 0.05 S	225–275 MN m^{-2}	Frames, deck stiffeners, superstructure
BS 43D	0.25 C, 0.5–1.6 Si, 0.003 Nb, 0.003 V, 0.05 P, 0.05 S	225-275 MN m^{-2}	Hull shell
'B' quality mild steel	0.19 C, 0.1–0.35 Si, 1.2–1.7 Mn	280–310 MN m^{-2}	Keel, bilge & keel strakes, No. 1 deck
BS 3602 HFS, 360	0.17 C, 0.35 Si, 0.4–0.8 Mn, 0.05 P, 0.05 S	215 MN m^{-2}	Pipework – water, air, oil

Source: Hadden [27].

of different elastic moduli (e.g. aluminium and steel) (Fig. 4.8). This author gives an example of a large liner where a reduction in weight of about 10% could be achieved by the use of higher-strength steel at the outer fibres of the longitudinal beam. The steel ultimate strength was increased from the value for normal shipbuilding mild steel (402 to 494 MPa) to 540 MPa, corresponding to 'higher-tensile' steel. The reduction in steel section, of course, results in a reduction in stiffness, and the increased deflection is calculated by Muckle to be a factor of 1.2 times the original. The buckling stress was also found to be well above yield stress, and buckling was therefore not a problem introduced by this alteration. The composite beam problem is most relevant to hull and superstructure considerations, and Muckle gives some examples of this.

Betts and Clayton [24] have reviewed the range of steels used in shipbuilding, commenting that mild steel is still commonest, but for cold regions where toughness could be a problem, impact energies usually need to be specified. There is a suggestion that the HSLA steels referred to above for offshore structures are beginning to be used also in shipbuilding, particularly in the USA and Japan. Table 4.6 shows some data on steels used for ships' hulls [24] and other compilations (e.g. [25], [26]) might be consulted for US and military applications.

Betts and Clayton suggest that acceptable bending stresses in ships' hulls depend upon a multiplicity of factors, but that the maximum total (wave plus still water) stress usually amounts to a factor of safety of about 2 to 2.5 on yield stress, at which level fatigue is probably implicitly covered for the low-frequency, high-strain cycles encountered by most surface ships, although it is quite possible to design the hull girder for a given fatigue life.

Hadden [27] has discussed the materials used in modern warships, and indicates the use of the hull steels shown in Table 4.7. He also indicates that, for plates up to 10 mm in thickness, mild steel is usually adequate. For thicker sections (up to 25 mm) a lower inclusion count and reduced porosity are preferred to obviate the danger of brittle fracture. This is met by the Ministry of Defence 'B' Grade structural steel, which is fully killed (by adding aluminium to the melt to reduce inclusion and porosity contents).

The design of submarine hulls calls for rather different types of steels. Because of the different structural loading (predominantly hydrostatic external pressure), submarines and other submersible structures containing atmospheric internal pressure have yield strength as one of the more important selection variables. Masubuchi [26] describes how US combat submarines have evolved from using steel of yield strength 220 MPa prior to 1940 up to the current use of the HY80 and HY100 quenched and tempered steels, and Hadden [27] has indicated the use of quenched and tempered steels of yield strengths up to 690 MPa in British submarines. Table 4.8 summarises some of the grades of steels in use for such applications.

The development of pipeline steels has some elements in common with that for submarines, in that the pressure vessel element of the design lays much emphasis upon yield strength as a selection criterion, although the requirement here is primarily to resist internal rather than external pressure, so that failure by buckling is only important under non-operational external loading such as during laying. The

Table 4.8. Some steels used for submarine pressure hulls

Application [Reference]	Steel
Submarines [26]	HY80 (typ. 0.18%C/1.5%Cr/3%Ni/0.5%Mo, quenched and tempered to 550 MPa yield strength)
Submarines [26]	HY100 (typ. 0.18%C/1.5%Cr/3%Ni/0.5%Mo, quenched and tempered to 690 MPa yield strength)
Submarines [27]	Quenched and tempered steel to 690 MPa yield strength
Deep-submergence rescue vehicle [26]	HY130 (typ. 0.12%C/0.55%Cr/5%Ni/0.5%Mo/0.07%V/0.15%Cu – yield strength 895 MPa)

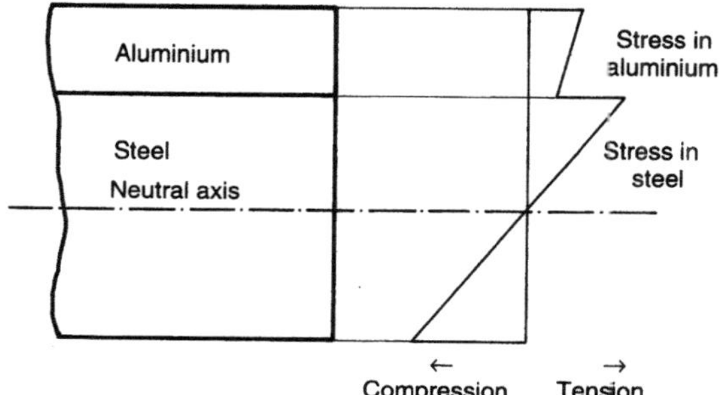

Fig. 4.8. Stress distribution in a composite beam, for example steel–aluminium. (From Muckle [23].)

development required for an increase in yield strength in pipeline materials is set against the same background as for the structural steels described above, in that it is essential to retain weldability and toughness, both in the parent plate and in the heat-affected zone. These difficulties have been solved in largely the same ways: first by the use of reduced grain sizes and subsequently by thermomechanical treatment coupled with accelerated cooling.

The specification of steels for pipelines is achieved almost universally by reference to the American Petroleum Institute Standard API-5L [28], although most users append some other requirements, particularly in regard to low-temperature toughness. API-5L designates yield strength as the primary specification variable, and grades are referred to directly in these terms, where the two digits following the 'X' are the yield stress in thousands of pounds-force per

Table 4.9. The development of linepipe steels (based on data from two producers)

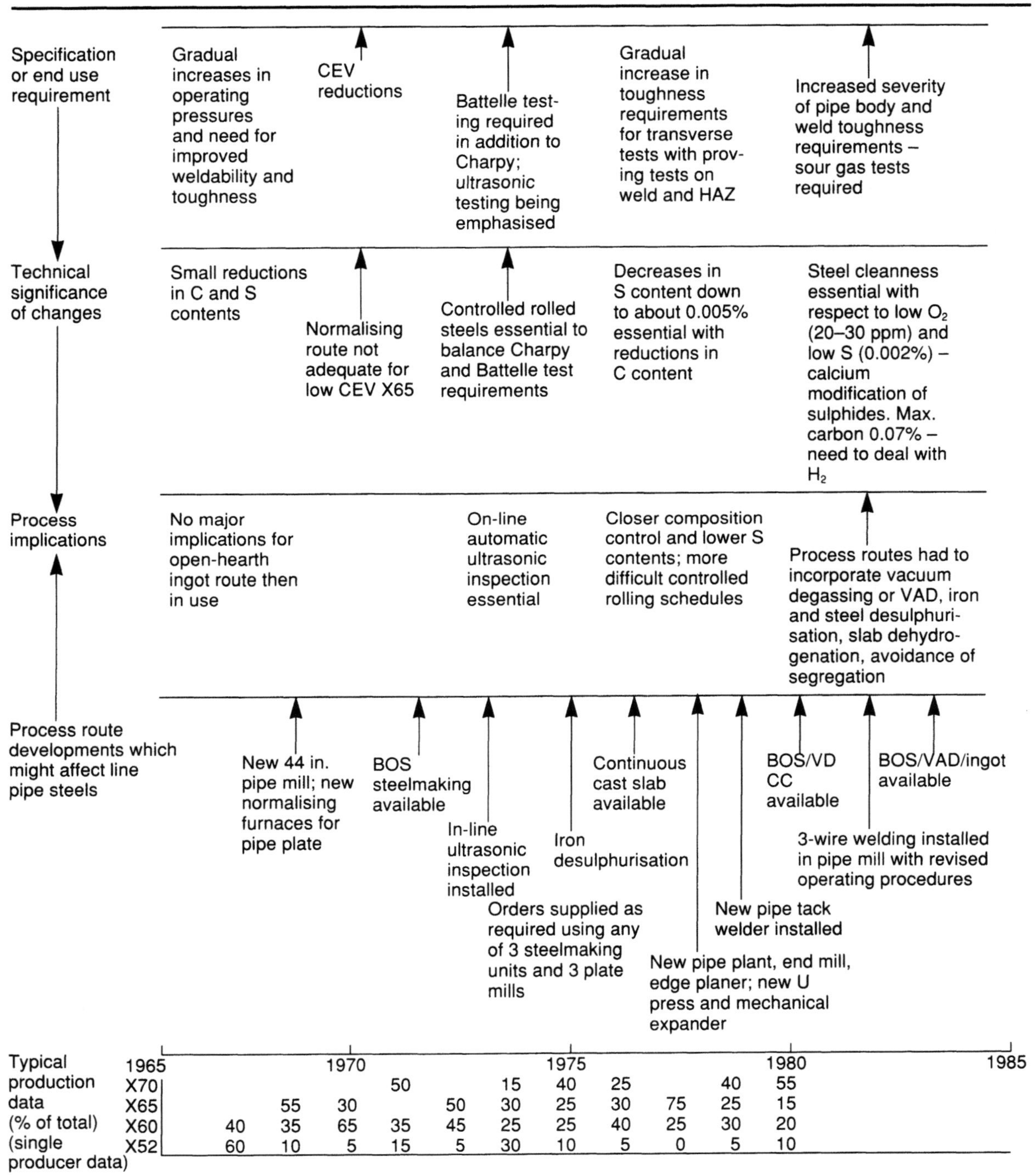

Specification or end use requirement
- Gradual increases in operating pressures and need for improved weldability and toughness
- CEV reductions
- Battelle testing required in addition to Charpy; ultrasonic testing being emphasised
- Gradual increase in toughness requirements for transverse tests with proving tests on weld and HAZ
- Increased severity of pipe body and weld toughness requirements – sour gas tests required

Technical significance of changes
- Small reductions in C and S contents
- Normalising route not adequate for low CEV X65
- Controlled rolled steels essential to balance Charpy and Battelle test requirements
- Decreases in S content down to about 0.005% essential with reductions in C content
- Steel cleanness essential with respect to low O_2 (20–30 ppm) and low S (0.002%) – calcium modification of sulphides. Max. carbon 0.07% – need to deal with H_2

Process implications
- No major implications for open-hearth ingot route then in use
- On-line automatic ultrasonic inspection essential
- Closer composition control and lower S contents; more difficult controlled rolling schedules
- Process routes had to incorporate vacuum degassing or VAD, iron and steel desulphurisation, slab dehydrogenation, avoidance of segregation

Process route developments which might affect line pipe steels
- New 44 in. pipe mill; new normalising furnaces for pipe plate
- BOS steelmaking available
- In-line ultrasonic inspection installed
- Orders supplied as required using any of 3 steelmaking units and 3 plate mills
- Continuous cast slab available
- Iron desulphurisation
- New pipe plant, end mill, edge planer; new U press and mechanical expander
- New pipe tack welder installed
- BOS/VD CC available
- 3-wire welding installed in pipe mill with revised operating procedures
- BOS/VAD/ingot available

Typical production data (% of total) (single producer data), across the period 1965–1985:

Grade											
X70				50		15	40	25		40	55
X65		55	30		50	30	25	30	75	25	15
X60	40	35	65	35	45	25	25	40	25	30	20
X52	60	10	5	15	5	30	10	5	0	5	10

Source: Cavaghan et al. [30], Coolen et al. [31].

Test temperature T_t from minimum design temperature T_D (°C)

Nominal wall thickness (mm)	Risers gas and liquid	Pipelines	
		Gas or gas and liquid	Liquid
$t \leqslant 20$	$T_t = T_D - 10$	$T_t = T_D - 10$	$T_t = T_D$
$20 < t \leqslant 30$	$T_t = T_D - 20$	$T_t = T_D - 10$	$T_t = T_D$
$t > 30$	to be decided on a case-by-case basis		

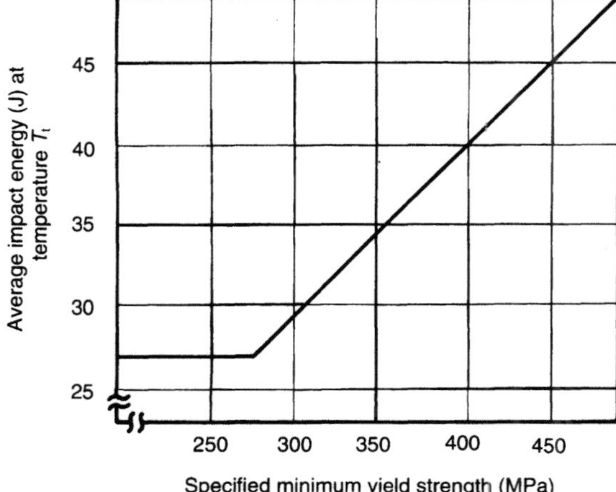

Fig. 4.9. Low-temperature toughness criteria for pipeline materials suggested by DnV [31].

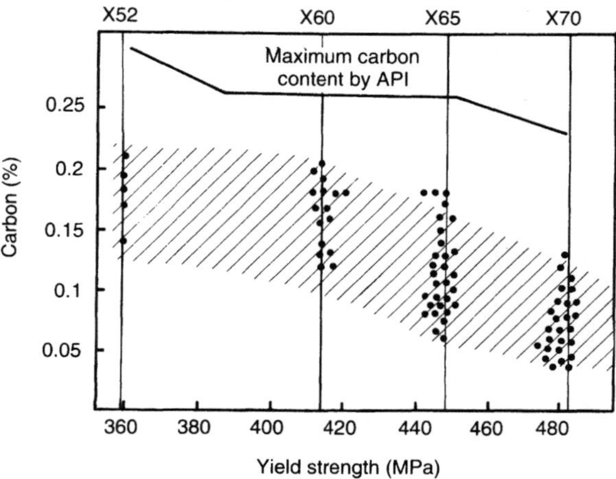

Fig. 4.10. Changing carbon content specification with increasing strength grade from 28 manufacturers. (From Billingham [13].)

square inch. Thus for example the American Petroleum Institute (API) in its publication API 5L X52 refers to a pipeline steel of specified minimum yield stress (SMYS) 52 000 lbf/in² (358 MPa). The commonly used API pipeline grades are X52, X60 and X65.

One important mechanical aspect not addressed in detail by the API specification is toughness. This is usually achieved by defining an impact energy requirement and, for example, DnV *Rules for Submarine Pipeline Systems* [29] recommend the adoption of low-temperature toughness criteria as outlined in Fig. 4.9.

As with structural steels, the metallurgical developments and engineering uses of pipeline steels have run in parallel, and this can be seen from production data from two steel-makers combined in Table 4.9. The potential savings in wall thickness for a given internal pressure and diameter by the use of higher strength steel are obvious.

These improvements in strength have been brought about in the background of a general decrease in carbon content and increased requirement for low-temperature toughness as shown in Fig. 4.10, and this has largely been achieved by the use of microalloying elements and special thermo-mechanical treatments (e.g. Fig. 4.11) not unlike those for advanced structural steels. Currently pipeline steels are used up to X70 strength, and these are met with relatively low carbon contents [32–34].

There are some indications that steels with a duplex ferrite–fine-bainite microstructure (Shiga et al. [35]) can attain X80 properties with good low-temperature toughness by a combination of controlled rolling and accelerated cooling. The strength–toughness relationship for such high-strength linepipe steels has been summarised by Lander et al. [36] (Fig. 4.12).

Of course, all such developments can only be carried out subject to the requirement to retain weldability and

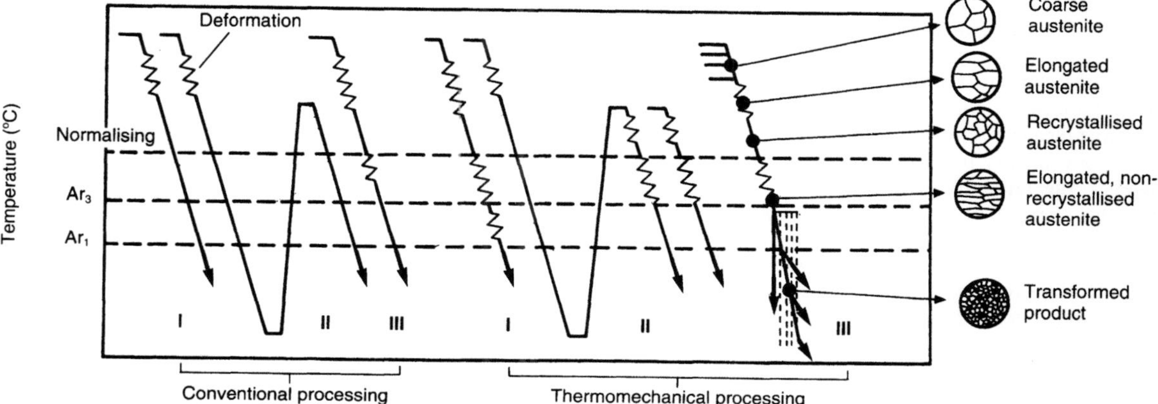

Fig. 4.11. Comparison of conventional and thermomechanical processing for steel plate. (From de Koenig [9].)

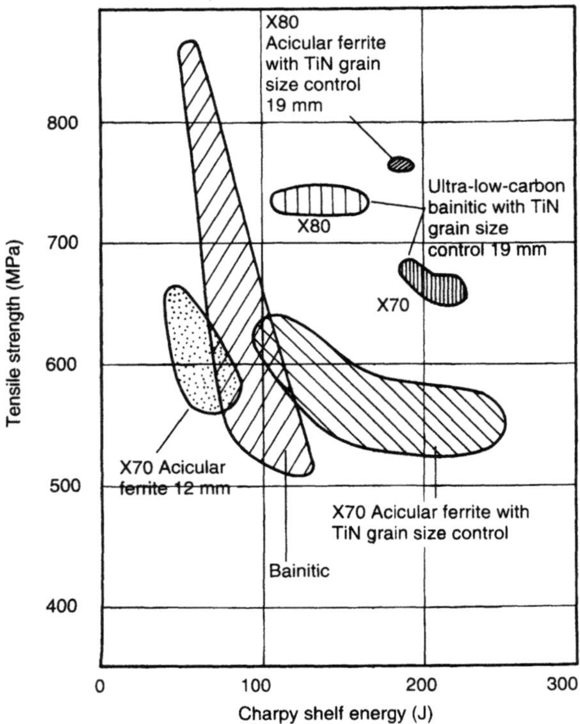

Fig. 4.12. Strength vs. toughness diagram for a range of high-strength plates. (From Lander et al. [36].)

corrosion resistance and these factors are discussed in Chapters 3 and 5. Toughness and defect assessment in pipelines are discussed further in Case Study 7.1.

Pipeline fittings (such as valves, flanges, bends, tees) utilise a different manufacturing procedure from the linepipe, requiring a secondary forming operation (usually hot). Furthermore, section sizes are generally larger than the pipeline, so that the steel properties are more difficult to achieve. In general, it is possible to produce X65 fittings in a normalised condition, but X70 properties would require a quench-and-temper heat treatment (Blondeau et al. [37]). ASTM requirements [38] identify 8 grades of chemical requirements and 4 classes of mechanical requirements. Wada et al. [39] have reviewed these and have indicated how fittings can be specified.

Cast steels have been seen as viable alternatives for structural steels in critical fatigue-prone areas of offshore structures, owing principally to the ability with which sections can be contoured, drastically reducing stress concentration factors. Also, casting may represent a viable alternative means of producing more intricate shapes which cannot conveniently be made from plate or tube material. However, the radically different manufacturing process means that some of the conventional strengthening methods cannot be used, and materials requirements are less clear. Ohba et al. [40] have described a design for leg nodes in a jack-up drilling rig, and Ohtake et al. [41] have described the use of high-strength casting alloys. Ellis et al. [42] have described the structural aspects, and Wood et al. [43] the metallurgical aspects of the castings used as transition pieces between the pontoon corners and columns of a tension leg platform. The composition and properties of this

structural cast steel are compared with those of the nearest wrought counterpart in Table 4.10.

For cast steels, some elements of mechanical design have to be altered. In particular, different residual stress distributions and defect size distributions make it desirable to obtain specific fatigue data rather than simply to apply an S–N curve for welded structural steel with an appropriate reduction in S to account for the reduction in SCF.

Apart from these three main applications of carbon–manganese steels, a number of other uses are found for ferritic steels (carbon–manganese and low-alloy types) in marine environments. These applications depend greatly upon whether or not toughness and/or weldability are of prime concern.

Depending upon the application, it is normal to increase strength either by increasing carbon content and using in the normalised (or occasionally the quenched-and-tempered) condition or by increasing the hardenability and using quenching and tempering as the production route. The former method reduces toughness, but the relatively high quantity of pearlite (and hence Fe_3C) increases hardness, so that carbon–manganese steels with up to 0.7 per cent carbon may be found in wear-resistant applications where stresses are relatively low. Table 4.11 illustrates a selection of both types from AISI and BS specifications.

For quenched and tempered steels it is important to use steel of appropriate hardenability, so that uniform properties can be obtained through the entire section. The hardenability of a low-alloy steel is dependent upon both its carbon content and its alloy contents, and is usually expressed in terms of the maximum section size (ruling section) where martensite (or a specified condition) can be reliably produced. This is often presented in terms of a continuous cooling transformation (CCT) diagram, which shows the structures expected under various cooling conditions for a given type of steel [44]. These diagrams (e.g. Fig. 4.13) have cooling rate as the horizontal axis, so that

Table 4.10. Comparison of a cast steel for a particular offshore fabrication with the nearest wrought counterpart

	Rolled plate specification	Cast steel specification
C	0.18 max.	0.18 max.
Mn	1.0 to 1.6	1.2 to 1.6
Si	0.50 max.	0.2 to 0.50
P	0.025 max.	0.020 max.
S	0.010 max.	0.008 max.
Cr	0.10 max.	0.15 max.
Mo	0.08 max.	0.20 max.
Nb	0.05 max.	0.03 max.
V	0.08 max.	0.08 max.
Ti	0.06 max.	0.050 max.
Nb + V	0.12 max.	0.10 max.
Nb + V + Ti	—	0.12 max.
Ni	0.25 max.	0.60 max.
Cu	0.25 max.	0.25 max.
Al (soluble)	0.06 max.	0.015 to 0.030
N	0.01 max.	0.012 max.
C equivalent	0.42 max.	0.48 max.
Yield strength (MPa)	Approx.	340
UTS (MPa)	BS 4360	460
Charpy Impact	Gr 50D	30 J at −40°C

Source: Ellis et al. [42].

Table 4.11. Composition of a selection of AISI and BS 970 carbon, carbon–manganese and low-alloy steels (compositions and properties are only intended to illustrate the range and do not encompass the specification)

Type	Principal alloying elements (wt%)	Heat treatment/ yield strength (MPa)
BS 970 080M40 (En8)	0.4%C/0.8%Mn 0.4%C/1.7%Mn	Normalised/ 250 to 280 (depending on section size) Annealed/435
AISI 1340		Normalised/560 Quenched and tempered/ 620 to 1600 (depending on tempering temperature)
BS 970 817M40 (En24)	0.4%C/0.6%Mn/1.2%Cr/ 0.3%Mo/1.5%Ni	650 to 1235 (depending on heat treatment from softened to quenched and lightly tempered)
AISI 4140	0.4%C/1%Mn/1%Cr/0.2%Mo	Annealed/415 Normalised/650 Quenched and tempered/650 to 1650 (depending on tempering temperature)
AISI 1060	0.6%C/0.7%Mn	Annealed/370 Normalised/420 Quenched and tempered/520 to 780 (depending on tempering temperature)

the transformation behaviour at a given cooling rate can be traced by following a vertical line downwards from the austenitising temperature. Note that these diagrams are not reversible (like some of the transformations), and so only apply to cooling. Quite often the cooling rate is given in practical terms as the section size and cooling mode. Thus the point 10 mm on the 'air-cooled' scale represents the average rate at which the centre of a circular bar of diameter 10 mm will cool in still air. Some diagrams are provided with a hardness scale at the bottom, which can be read by a direct extension downwards of the vertical line onto the 'as-cooled' curve or that for any appropriate subsequent tempering temperature. The examples of CCT diagrams in

Fig. 4.13 illustrate the difference between a steel which is essentially intended for use in the 'normalised' (i.e. air-cooled) condition and one to be used in the quenched and tempered condition. As can be seen, the carbon content determines the final hardness of the untempered martensite, whereas the ease with which this martensite can be produced (i.e. hardenability) is dependent on the carbon and alloy contents. Alloying elements (particularly chromium and molybdenum) are also selected for their behaviour during the tempering operation, especially with regard to high-temperature stability of carbides.

It should also be appreciated that a given alloy can be used in a variety of conditions, depending upon its heat

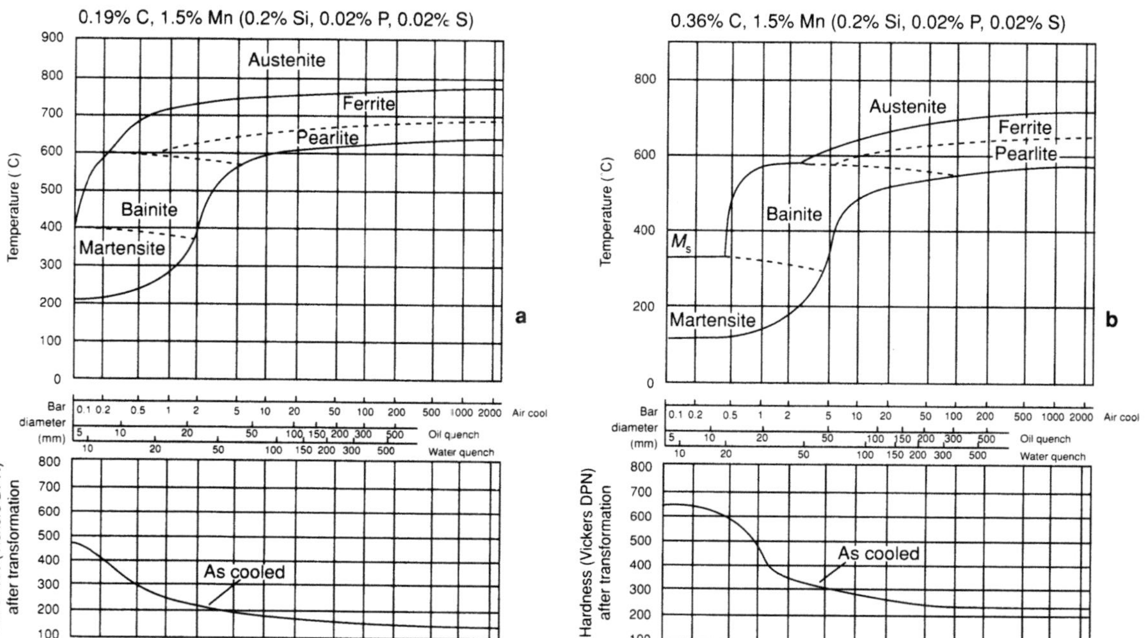

Fig. 4.13. Continuous cooling transformation (CCT) diagrams for the heat treatment of steels, illustrating the effects of carbon and alloy content on the hardenability and as-quenched hardness. (From Atkins [44].)

(*continued overleaf*)

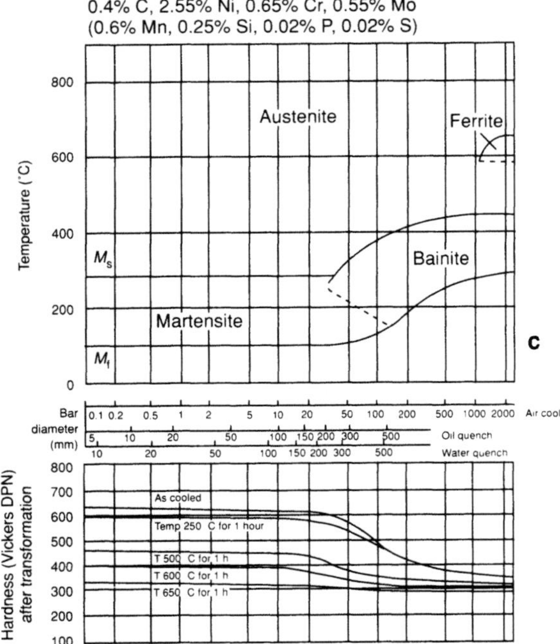

Fig. 4.13. (continued)

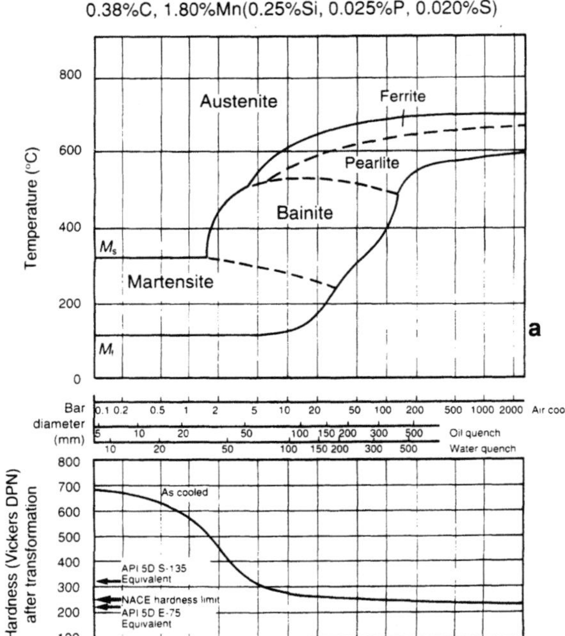

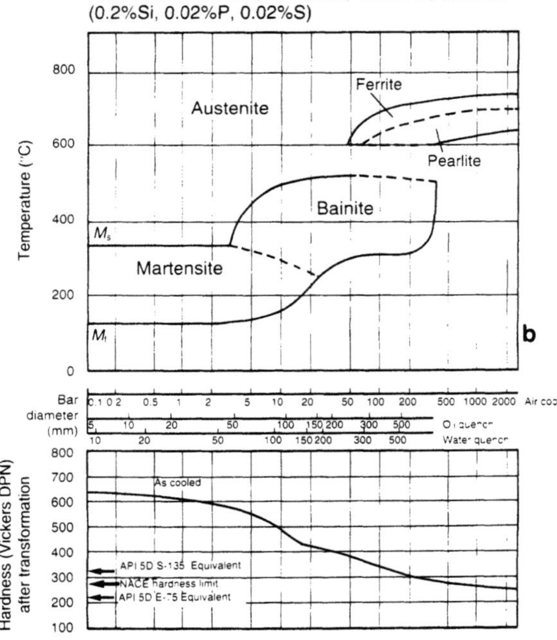

Fig. 4.14. CCT diagrams close to the specifications AISI 1340 and 4140 respectively. (From Atkins [44].)

treatment and section size. For example, the widely used BS 970 high-strength steel 817M40 (formerly En 24) can be supplied in a variety of heat treatment conditions, each with a specified ruling section, as can be seen from an examination of the standard or other compilation (e.g. [18]). The hardest conditions (e.g. 'Z') correspond to the martensitic structure with varying degrees of tempering, whereas the softer conditions (e.g. 'T') with the larger ruling sections could be achieved by slacker cooling.

One example of the use of high-strength steels is for application as petroleum engineering tubular goods such as drillpipe, well-casing and production tubing. Such materials are almost invariably specified with reference to API standards. For example, the drill-pipe specification API-5D [45] contains four strength grades, the chemical composition of which is only weakly specified, as shown in Table 4.12.

These requirements can be met by a number of different steels and treatments, but the AISI specifications 1340 and 4140 shown along with their CCT diagrams in Fig. 4.14 (and their British equivalents) are widely used. The API specifications are also shown superimposed on these diagrams (using an accepted hardness–tensile strength relationship [46]), as are the NACE requirements for these steels. This illustrates the complex interrelationship of the various constraints on high-strength steels in marine technology. Since the drill string is made by screwing together lengths of drillpipe, the problem of welding only arises during the manufacture of the pipes when the possibility of heat treatment of the entire weldment exists (refer to Chapter 5).

Another example of high-strength tubular application is given by Salama and Tetlow [47] for the tethers used in the tension leg platform installed in NW Hutton in the UK sector of the North Sea. These were of a 3½% Ni, CrMoV steel whose properties are shown in Table 4.13. This example is quoted because it is one of the relatively rare examples of such a structural element and required careful fracture

mechanics and fatigue testing before its adoption.

Bolts, by their nature, usually need to be made from high-strength alloys. The ISO strength specifications for bolts involve two digits, one of which is related to the ultimate tensile strength and the second of which relates to the ratio of yield strength to ultimate tensile strength. For example, the strength grade ISO 8.8 specifies material of

Table 4.12. Summary of the API-5D specification for drill pipe (not a specification)

Grade	Composition (weight %) (maxima unless stated)							Treatment	Tensile strength (MPa)	Yield strength (MPa)
	C	Si	Mn	P & S	Cr	Mo	Others			
E-75	—	—	—	P 0.040 S 0.060	—	—	—	N or N & T or Q & T	690	517/724
X-95	—	—	—	P 0.040 S 0.060	—	—	—	Q & T or N & T	724	655/862
G-105	—	—	—	P 0.040 S 0.060	—	—	—	Q & T or N & T	793	724/931
S-135	—	—	—	P 0.040 S 0.060	—	—	—	Q & T or N & T	1000	931/1138

Source: British Steel Corporation [18].

tensile strength 80 kgf/mm^2 (758 MPa) from the first digit, and yield strength of 80 per cent of the UTS (second digit), i.e. 630 MPa. The standard grades vary from 4.6 to 14.9.

For applications to steel, it is normally necessary, from the point of view of strength of the bolted connections, to use grades of 8.8 or better, and this usually requires quenched and tempered steels. Since it is rarely necessary to weld onto bolted materials, the usual limitations on quenching and tempering are relaxed. Bolts can, of course, be made from any material, but alloy steels are usually chosen from BS 1506 or the ASTM standards A193 or A320. A number of different grades are specified in each of these standards, but all are low-alloy or carbon–manganese steels which could, in principle, be found in a general bar standard such as BS 970.

Table 4.13. Properties of an alloy steel used for tension leg platform tethers

	Specifications	Production
Mechanical properties		
0.2 % proof stress (N/mm^2)	795 to 950	835 to 895
Ultimate strength (N/mm^2)	900 min.	935 to 990
Elongation (%)	15 min.	15 to 22
Hardness (HV$_{30}$)	280 to 335	294 to 323
FATT (°C)	—	−80 to −110
CTOD for 92.5 mm		
thick specimen at 0 °C	0.50 mm	0.50 to 0.81
Metallurgical data		
ASTM grain size number	5 min.	5 to 7
Inclusion length (mm), max/ave	—	0.29/0.04
Chemical composition		
C	0.35 max.	0.27 max.
Si	0.10 max.	0.06 max.
Mn	0.50 max.	0.24 max.
P	0.01 max.	0.006 max.
S	0.008 max.	0.002 max.
Ni	3.5 min.	3.66 to 3.74
Cr	1.25 min.	1.79 to 1.85
Mo	0.30 min.	0.39 to 0.42
V	0.15 max.	0.13 max.
N	0.01 max.	0.004 max.
Sn	0.02 max.	0.006 max.
Al	0.01 max.	0.005 max.
Cu	0.15 max.	0.05 max.
As	0.015 max.	0.01 max.
Sb	0.005 max.	0.005 max.

Source: Salama and Tetlow [47].

For example, Scott [48] indicates that steel bolts for structural marine use (grades 8.8 or 10.9) are often of grade L7 (ASTM A320) or B7 (ASTM A193) or BS 1506 Gr621A. There is no reason why an equivalent cannot be chosen from, for example BS 970, or the AISI series with appropriate specification of the other requirements. Table 4.14 illustrates some of these specifications and includes some equivalents from the more general standards.

For pressure containment purposes, the pressure vessel codes normally refer to specific compatible standards for materials (e.g. BS 5500 refers to BS 4882, which in turn refers to specific grades of BS 1506 or BS 970), but the principles of design for strength and toughness are paramount and similar in many ways to the considerations for pipelines, especially if hydrocarbons are to be contained.

Steels for the manufacture of wire are generally of a rather high carbon content, and are cold worked to further increase their yield strength. This produces a steel of relatively low toughness, but for applications such as wire ropes strength is the primary consideration. Wire ropes find widespread use in marine technology, particularly for mooring or lifting applications, and much work has been done on their marine performance, particularly with regard to degradation by corrosion and/or fatigue [49]. Wire rope materials and geometries are also used elsewhere in marine engineering applications, particularly in the armouring layers of flexible components (such as umbilicals and pipes, covered in Case Study 7.5).

Rope wire is usually used in the hard-drawn condition and a number of different standards are used to cover treatment of wires and manufacture of ropes. Typically, UTS is the specified variable with chemistry and degree of drawing having an effect on strength, which varies with the diameter of the individual wires. Typical compositions and properties of rope wires are given in Table 4.15.

One series of low-alloy steels finds widespread use in oil refineries at high temperatures because of their creep resistance and because of their low carbon level and hence weldability. These so-called 'refinery grade' steels use chromium and molybdenum as the main alloying elements, the chromium being useful in the improvement of corrosion resistance. Breen [50] and Glendinning and Vincent [51] have reviewed the compositions and properties of such alloys, and further details are given in ASTM A204 and A387.

Other applications of carbon–manganese and low-alloy steels include chains for mooring and gearing for marine engines. Chains, like ropes, are usually purchased in a

Table 4.14. Comparison of ISO bolt-grade mechanical specifications with those achievable using alloyed and carbon–manganese steels

| Mechanical property | ISO BS 3692 | | BS 1506 | Gr 621A | BS 4486 | BS 970 817 M40(U) | BS 970 817 M40(V) | Ferralium alloy |
| | | | L7 | B7 | | | | |
	Gr 8.8	Gr 10.9	ASTM A320	A193	Macalloy	(EN24U)	(EN24V)	255
Tensile strength (N/mm^2)	785	981	860	860	1011	925 to 1075	1000 to 1150	740 to 830
0.2% proof stress (N/mm^2)	628	883	730	730	893	740 to 860	835 to 960	490 to 590
Rockwell hardness (HRC)	18 to 31	27 to 38				27 to 35	31 to 38	
Brinell hardness (HB)	225 to 300	280 to 365	248 to 335	248 to 335		269 to 331	293 to 352	230 to 270
Vickers hardness (HV)	225 to 300	280 to 370	261 to 353	248 to 335				
Elongation after fracture (%)	12	9	14	14	8	12	12	246

Source: Scott [48].

specific strength class, having been manufactured by flash-butt welding and heat-treated. A variety of material grades and conditions are used for marine gearing, and some of these used by the Royal Navy are reviewed by Cooper [52]. As with non-marine gearing applications, surface-hardening treatments are applied to some components.

Steels with still higher strengths than the carbon–manganese and low-alloy types can be produced by maraging. These steels are highly alloyed with nickel, cobalt and molybdenum and have a very low carbon content. Maraging consists of producing martensite and age-hardening this with a variety of intermetallic compounds including Ni_3Ti, Fe_2Mo and Ni_3Mo. Table 4.16 shows a selection of compositions and properties for maraging steels. These alloys may be used for pressure hulls and are weldable with care (e.g. Dexter [53]), but the possibilities of stress corrosion and hydrogen embrittlement need to be considered [53].

The properties of steels can be broadly categorised as those which are specified and those which, although crucial to the designer, are not covered in the standards to which the steel is purchased. In the first category are primarily strength and chemical composition, and these are the primary specification variables for steels. Weldability can be inferred from the composition by calculating the carbon equivalent or other weldability parameter (see Chapter 5), and is in fact specified separately in some standards. Toughness is specified only in so far as some standards include impact energy specifications and ductility, in the form of percentage elongation, is also normally specified.

Table 4.15. Some of the carbon–manganese and low-alloy steel compositions and cold-drawn properties which might be used for wire ropes (not specifications – original standards should be consulted)

Specification	Composition	UTS (MPa)
AFNOR FM15	0.15%C/0.5%Mn	785
AFNOR FM72	0.7%C/0.8%Mn	1500
BS 2763	Not specified	Varies with diameter
BS 1408 (BS 970 En49)	0.7%C/1%Mn	1550
ASTM A648/11	0.65%C/0.8%Mn	1500
ASTM A679	0.7%C/0.8%Mn	2000

More quantitative toughness parameters require to be the subject of test by the user or by some arrangement between the user and manufacturer. Corrosion resistance and fatigue resistance are so much a function of specific application as to be impracticable to specify.

The general data on steels are categorised on the above basis, with Table 4.17 covering the 'specifiable' data. Table 4.18 covering fatigue and fracture data and Table 4.19 covering corrosion data. Great care should be taken in using Tables 4.18 and 4.19, and the reader is referred to Case Studies 7.1 and 7.2 and Chapters 2 and 3 to obtain an appreciation of how the application affects the interpretation of such data.

Table 4.16. A selection of compositions and mechanical properties of maraging steels

| Grade | Composition (%) | | | | | Heat treatment | Tensile strength | Yield strength | Elongation in 50 mm | Reduction in area | Fracture toughness |
	Ni	Mo	Co	Ti	Al	(a)	(MPa)	(MPa)	(%)	(%)	(MPa$\sqrt{m}$)
18Ni(200)	18	3.3	8.5	0.2	0.1	A	1500	1400	10	60	155 to 200
18Ni(250)	18	5.0	8.5	0.4	0.1	A	1800	1700	8	55	120
18Ni(300)	18	5.0	9.0	0.7	0.1	A	2050	2000	7	40	80
18Ni(350)	18	4.2	12.5	1.6	0.1	B	2450	2400	6	25	35 to 50
18Ni(Cast)	17	4.6	10.0	0.3	0.1	C	1750	1650	8	35	105

Source: American Society for Metals [46].
(a) Treatment A: solution treat 1 h at 820 °C (1500 °F), then age 3 h at 480°C (900 °F). Treatment B: solution treat 1 h at 820 °C (1500 °F), then age 12 h at 480°C (900°F). Treatment C: anneal 1 h at 1150°C (2100 °F), age 1 h at 595 °C (1100 °F), solution treat 1 h at 820 °C (1500 °F) and age 3 h at 480 °C (900 °F).

Table 4.17. Composition and strength summary of carbon–manganese and low-alloy steels

Composition and example from BS, AISI or ASTM standards	Mechanical properties (strengths in MPa)
Low-carbon (around 0.2%C/ 0.8%Mn); e.g. BS 970 080M15	Yield strength in region of 200 and ultimate around 500, depending upon carbon content, section thickness, hardenability and whether or not cold working has been applied.
Low-carbon, grain-refined or microalloyed (around 0.2%C/1.5%Mn with microallaying additions such as V, Nb and Ti); e.g. BS 4360 Gr50D	With appropriate thermomechanical treatment and accelerated cooling yield strengths of around 450 are achievable and some treatments produce even higher strength.
Low-carbon martensitic (0.15%C with Cr and Mo); e.g. ASTM A387	Yield strengths of around 300 in normalised and tempered condition.
Medium-carbon (0.25 to 0.5%C/0.5 to 1.5%Mn); e.g. AISI 1050	Yield strength around 300 to 400 in non-heat-treated (higher carbon and manganese contents may be quenched and tempered) or cold-worked condition. Also depends on carbon content.
Heat-treatable low-alloy (0.4%C/>1% Cr, Ni, Mo total); e.g. BS 970 817M40	Up to 2000 yield in the higher-carbon steels in the lightly tempered condition. More often used at around 700 to 1000 yield. Alloy content chosen primarily on hardenability for section in question.
High-carbon (> 0.5%C)	Mainly used for wear resistance or in cold-worked condition for spring applications. In latter case yield is usually around 1500.

Table 4.18. Fracture and fatigue data for carbon–manganese and low-alloy steels

Steel type (see Table 4.17)	Fracture properties	Fatigue properties
Low and medium carbon	Critically dependent on carbon content, heat treatment and temperature.	Endurance limit around 100 MPa, but critically dependent on factors mentioned in text.
Low carbon, grain-refined and microalloyed	Usually measured on case-by-case basis using impact, but δ_c is expected to be greater than around 0.1 mm. Actual value depends upon a variety of factors, but weld region is usually design-critical.	See Figs 7.8, 7.9.
Heat-treatable, low-alloy	Typical value of K_{Ic} 30 to 100 MPa $\sqrt{m}$ at 20 °C depending upon heat treatment.	Typical fatigue strength 400 MPa or greater, but depends on heat treatment and inclusion content as well as other factors.
High-carbon	—	Fatigue limit around 400 MPa.

Table 4.19. Corrosion data for carbon–manganese and low-alloy steels except low-carbon Cr–Mo steels (Information is for all steels in the class unless stated otherwise)

Environment	Corrosion comments and rates
Seawater (quiescent surface)	Typically 75 to 125 μm/yr, although pits can be as deep as 4 mm after 16 years' immersion. Initial rate (up to 2 years) is usually higher, up to 400 μm/yr.
Seawater (deep and bottom sediment)	Usually settling to a value of 25 to 75 μm/yr after one year. Sediment rates subject to bioactivity.
Marine atmosphere	Typically 30 to 100 μm/yr. Small alloying additions (particularly of copper, although Cr and Ni also have an effect) can reduce corrosion rates substantially.
Splash zone	Up to 40 μm/yr, more if a substantial velocity effect.
Other	All corrosion rates aggravated by fluid velocity and/or abrasion. Crevice effect relatively mild. Susceptible to hydrogen sulphide and carbon dioxide, the former producing cracking in some steels and blistering in others. Substantial effects of oxygen and temperature.

4.1.3 Stainless Steels

The difference between the stainless steels and the carbon–manganese and low-alloy steels considered in the previous section is primarily in the use of chromium as a substantial alloying element. The use of chromium in stainless steels is associated with the enhancement of the atmospheric corrosion resistance which it imparts to the the steel and, as chromium content is increased, the transition to 'stainless' behaviour is a gradual one [54] but, at 13 per cent chromium, the steel can definitely be regarded as stainless, and most commercial alloys contain at least this amount of chromium. The reason for their corrosion resistance lies in the ability of the dissolved chromium to form a passive layer of a chromium-rich oxide on the surface of the steel and the ability of this layer to 'self-heal' if disrupted (provided that the environment is conducive to this). It is important to understand that stainless steels have limitations as to their corrosion resistance but, provided that these limitations are known, they represent one of the cheapest alternatives to carbon–manganese and low-alloy steels where unprotected corrosion resistance is required.

As their name implies, stainless steels are primarily ferrous alloys, and the main alloying elements are chromium and nickel, leading to a rather complex metallurgical constitution best understood by examining the iron–chromium–nickel ternary alloy series. The functions of the two alloying elements can immediately be appreciated, once it is known that chromium normally forms a body-centred cubic (BCC) crystal structure whereas nickel forms a face-centred cubic (FCC) structure. The significance of this is the interaction of these (substitutional) solutes on the behaviour of the iron crystal structure, which, if unalloyed, will show the following allotropic transformations:

Phase (structure)	Stability range (pure iron)
δ-delta ferrite (BCC)	melting to 1400 °C
γ-austenite (FCC)	1400 °C to 910 °C
α-ferrite (BCC)	910 °C and below

The structure of the delta and alpha forms of ferrite is the same, and so austenite can be seen as an interruption to the stability of iron's crystal structure. The γ–α transition is the basis of the heat treatment of steels, owing mainly to the differential solubility of carbon in the two phases, as has been described briefly in Section 4.1.2. The effects of the substitutional elements chromium and nickel in stainless steels are that chromium extends the range of stability of ferrite and nickel extends the range of stability of austenite.

The iron–chromium system shown in Fig. 4.15 illustrates the effect of chromium in confining the stability range of austenite, about 17 per cent chromium being sufficient to ensure that austenite will not form at any temperature. Another feature of the iron–chromium system of importance in stainless steels is the presence of σ-phase, which is a brittle intermetallic compound of iron and chromium detrimental to the properties of the steel. Fortunately, the formation of σ-phase requires the diffusion of the substitutional solute chromium at relatively low temperatures (for a substitutional solute), and it can therefore be avoided, provided that steels of susceptible composition are not exposed to temperatures of 500 to 800 °C for a prolonged period.

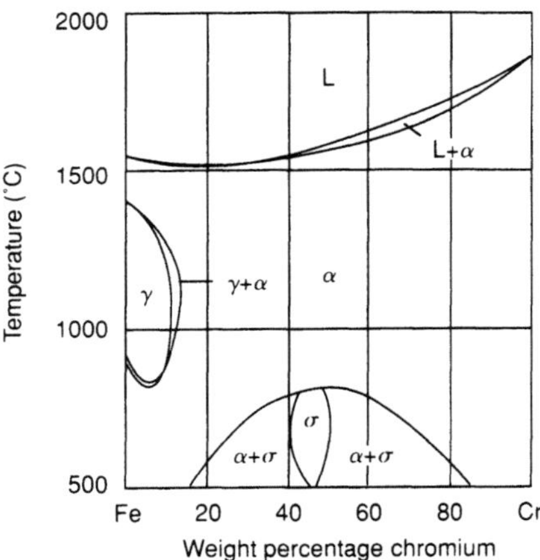

Fig. 4.15. The iron–chromium phase diagram showing the limited range of austenite stability and the occurrence of sigma phase. (From Askeland [3].)

The behaviour of ternary alloys can be appreciated from the two isothermal sections of the iron–chromium–nickel system illustrated in Fig. 4.16. Both of these sections show clearly the ferrite-stabilising influence of chromium and the austenite-stabilising influence of nickel. These diagrams should, however, only be used qualitatively, because the σ-phase forms even more sluggishly in Fe–Cr–Ni alloys than it does in Fe–Cr alloys, and there is a pronounced reluctance of metastable austenite to transform once established at high temperatures (Speich [55]). Schneider [56] has modified the diagram of Schaeffler (see Fig. 5.23) to indicate the phases present in stainless steels as a function of their chromium and nickel equivalents:

$$\text{Ni equivalent} = \%\text{Ni} + \%\text{Co} + 30(\%\text{C}) + 25(\%\text{N}) + 0.5(\%\text{Mn}) + 0.3(\%\text{Cu})$$

$$\text{Cr equivalent} = \%\text{Cr} + 2(\%\text{Si}) + 1.5(\%\text{Mo}) + 5(\%\text{V}) + 5.5(\%\text{Al}) + 1.7(\%\text{Nb}) + 1.5(\%\text{Ti}) + 0.75(\%\text{W})$$

As well as identifying the three major matrix phases possible in stainless steels (i.e. austenite, ferrite and martensite), such diagrams allow the effect of alloying elements other than chromium and nickel to be assessed, the major limitation being that they are normally formulated for weld metals.

Stainless steels are normally classified in terms of the matrix phase(s) with one exception. The five main forms are

Austenitic

Martensitic

Ferritic

Precipitation hardening

Duplex

and these are discussed in turn below. Table 4.20 summarises the stainless steel range, referring to some of the main alloys discussed below.

4.1.3.1 *Austenitic Stainless Steels*

Austenitic stainless steels form the major proportion of those produced and, although relatively weak, have the

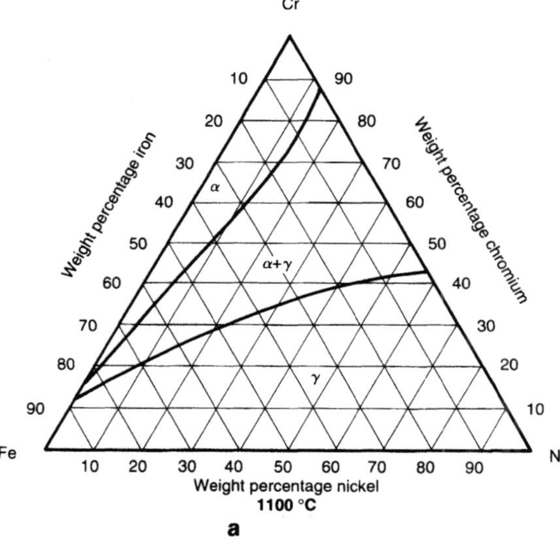

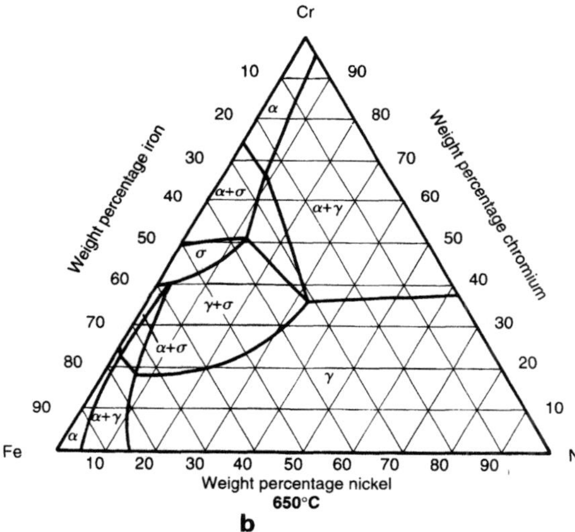

Fig. 4.16. Isotherms at 1100 °C and 650 °C in the iron–chromium–nickel system. (From Speich [55].)

Table 4.20. Broad classification of the stainless steels using AISI designations
(The last two categories do not generally have AISI designations)

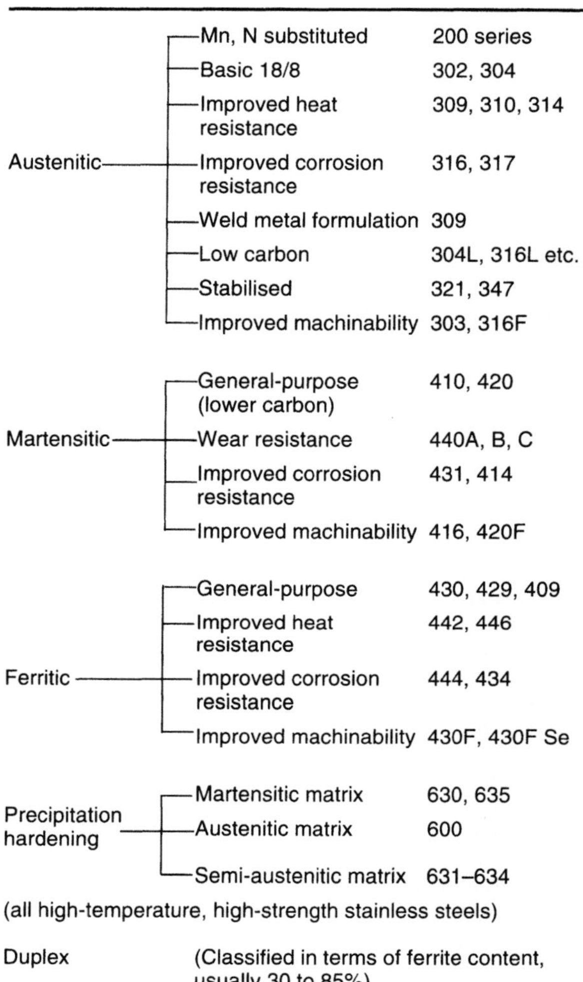

Austenitic	Mn, N substituted	200 series
	Basic 18/8	302, 304
	Improved heat resistance	309, 310, 314
	Improved corrosion resistance	316, 317
	Weld metal formulation	309
	Low carbon	304L, 316L etc.
	Stabilised	321, 347
	Improved machinability	303, 316F
Martensitic	General-purpose (lower carbon)	410, 420
	Wear resistance	440A, B, C
	Improved corrosion resistance	431, 414
	Improved machinability	416, 420F
Ferritic	General-purpose	430, 429, 409
	Improved heat resistance	442, 446
	Improved corrosion resistance	444, 434
	Improved machinability	430F, 430F Se
Precipitation hardening	Martensitic matrix	630, 635
	Austenitic matrix	600
	Semi-austenitic matrix	631–634

(all high-temperature, high-strength stainless steels)

Duplex	(Classified in terms of ferrite content, usually 30 to 85%)

advantage of cryogenic toughness and a high degree of homogeneity, which improves their general corrosion resistance. Austenitic stainless steels are based around two compositions, the so-called '18/8' and '25/20' steels. The 18/8 composition contains 18 per cent chromium, and a consideration of Figs 4.16 and 5.23 indicates that about 8 per cent nickel will be needed to stabilise austenite to room temperature with this amount of chromium. Likewise, 25 per cent chromium requires the presence of about 20 per cent nickel to stabilise austenite.

The most widely used designation system for stainless steels is the AISI system, and austenitic stainless steels are specified by three-digit numbers in the 200 and 300 series. In addition, there are a few important austenitic stainless steels without designations in the AISI series. There are a great many austenitic stainless steels, but it is sufficient to examine a few of these in order to appreciate the important compositional constituents.

AISI 304 steel is the archetypal austenitic stainless, showing properties typical of such steels. These are a rela-

tively modest strength coupled with a good response to work hardening (shown by the yield/UTS ratio and also by the difference between annealed and cold-worked properties). This steel is available in a number of minor compositional variants which serve to illustrate the effects of the interstitials carbon and nitrogen on the properties of this steel and, more generally, in the austenitic series; Table 4.21 shows some of these variants.

Carbon is usually considered to be a nuisance in austenitic steels, because it shows an affinity for chromium and will preferentially form chromium carbides at grain boundaries when exposed to an appropriate temperature range (usually about 500 to 800 °C). Because of the relatively slow diffusion rate of chromium in austenite compared with that of carbon, such precipitation results in chromium-depleted zones, with consequent loss of corrosion resistance, usually along grain boundaries. To avoid this problem during welding or other processing, a low-carbon version of 304 is available for steels which are likely to be exposed to the above-mentioned

Table 4.21. Compositional variants on the AISI 304 and 316 designations

AISI designation	Composition (%)							
	C	Mn	Si	Cr	Ni	P	S	Others
304	0.08	2.00	1.00	18.0 to 20.0	8.0 to 10.5	0.045	0.03	0.10 N
304L	0.03	2.00	1.00	18.0 to 20.0	8.0 to 12.0	0.045	0.03	0.10 N
304H	0.04–0.10	2.00	1.00	18.0 to 20.0	8.0 to 10.5	0.045	0.03	
304N	0.08	2.00	1.00	18.0 to 20.0	8.0 to 10.5	0.045	0.03	0.10 to 0.16 N
304HN	0.08	2.00	1.00	18.0 to 20.0	8.0 to 10.5	0.045	0.03	0.16 to 0.30 N
304LN	0.03	2.00	1.00	18.0 to 20.0	8.0 to 12.0	0.045	0.03	0.10 to 0.16 N
316	0.08	2.00	1.00	16.0 to 18.0	10.0 to 14.0	0.045	0.03	2.0 to 3.0 Mo; 0.10 N
316L	0.03	2.00	1.00	16.0 to 18.0	10.0 to 14.0	0.045	0.03	2.0 to 3.0 Mo; 0.10 N
316H	0.04–0.10	2.00	1.00	16.0 to 18.0	10.0 to 14.0	0.045	0.03	2.0 to 3.0 Mo
316F	0.08	2.00	1.00	17.0 to 19.0	10.0 to 14.0	0.20	0.10 min	1.75 to 2.5 Mo
316Ti	0.08	2.00	1.00	16.0 to 18.0	10.0 to 14.0	0.045	0.030	5(C+N) min, 0.70 max Ti; 0.10 max N; 2.0 to 3.0 Mo
316Cb	0.08	2.00	1.00	16.0 to 18.0	10.0 to 14.0	0.045	0.030	10 x %C min, 1.10 max (Nb+Ta); 0.10 max N; 2.0 to 3.0 Mo
316N	0.08	2.00	1.00	16.0 to 18.0	10.0 to 14.0	0.045	0.03	2.0 to 3.0 Mo; 0.10 to 0.16 N
316LN	0.03	2.00	1.00	16.0 to 18.0	10.0 to 14.0	0.045	0.03	2.0 to 3.0 Mo; 0.10 to 0.16 N

temperature range (L Grade). However, carbon, being an interstitial solute, is a potent strengthener of austenite (although not as much as in ferrite), and the reduction in specified carbon leads to a slight reduction in design allowables. Thus, some nitrogen strengthening can be used to replace this without danger of sensitisation. In fact, nitrogen is also believed to act as a retarder of sensitisation (Lula [57]).

An alternative method of resolving the difficulty of sensitisation is to add an element which has a greater affinity for carbon than has chromium. Elements which do this include niobium (columbium), tantalum and titanium, and additions of these at about 5 to 10 times the amount of carbon form the basis of the stabilised grades such as AISI 321, although many other grades have stabilised versions. Higher chromium contents, such as in the 25/20 steel AISI 310, are usually only required for high-temperature oxidation resistance, but molybdenum (to about 4 per cent) enhances corrosion resistance, especially that to pitting and crevice corrosion and is used, for example, in 316. Additional austenite stabilisation (in the form of increased nickel and manganese) is required to counteract the ferrite-forming power of molybdenum in such steels. Free-machining austenitics (e.g. 303) contain more than the usual amount of sulphur; this precipitates as manganese or selenium sulphides, which act as chip-breakers during the metal cutting process, although the inhomogeneity resulting from such precipitation slightly compromises the corrosion resistance. The interstitial strengthening effects of carbon and nitrogen are exploited in the 'Nitronic' grades of austenitics. These steels do not have AISI designations, but have developed from the manganese-substituted austenitics of the 200 series. The original use for manganese was as a substitute for nickel, but the nitronics exploit the effect that manganese has on increasing the solubility for nitrogen. This results in substantially increased yield stress over the conventional austenitics and also improved wear resistance. Nitronic 60, in particular, shows good wear and anti-galling properties in unlubricated metal-to-metal contact.

4.1.3.2 *Ferritic and Martensitic Stainless Steels*

These steels are essentially iron–chromium–carbon alloys, and are therefore of low nickel equivalent. Their behaviour is best understood by means of isopleths through the iron–chromium–carbon systems, and these isopleths can be interpreted as pseudo-binary diagrams. The isopleths at 5 and 17 per cent chromium over a range of carbon content and at 0.1 and 1.5 per cent carbon shown in Fig. 4.17 serve to illustrate the effects of alloying with chromium and carbon in all the alloys of commercial interest.

The first thing to notice about the Fe–Cr–C system is the effect of chromium on confining the range of stability of the austenite phase. As can be seen, the effect of chromium on iron–carbon alloys is to shrink the 'γ-loop', making it

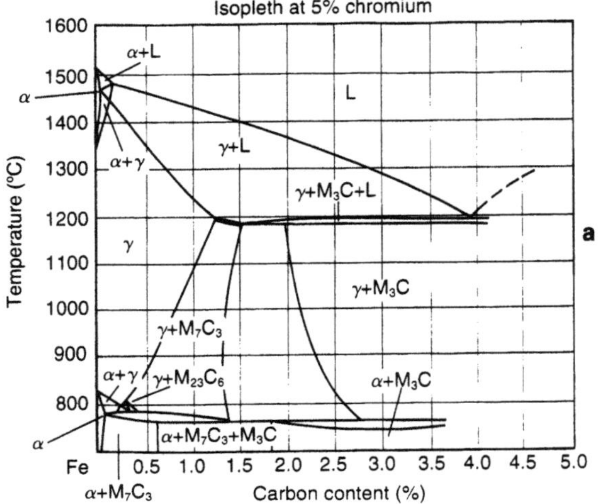

Fig. 4.17. Isopleths in the iron–chromium–carbon system [46]. *(continued on next page)*

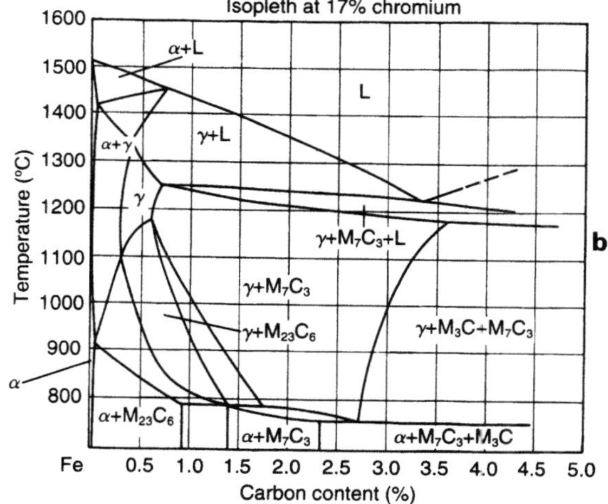

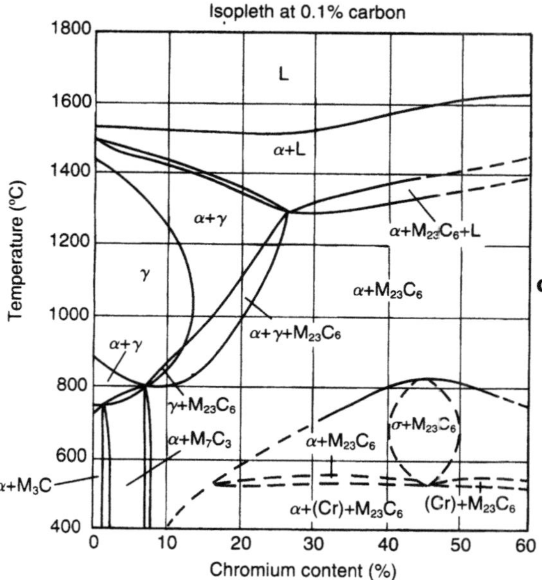

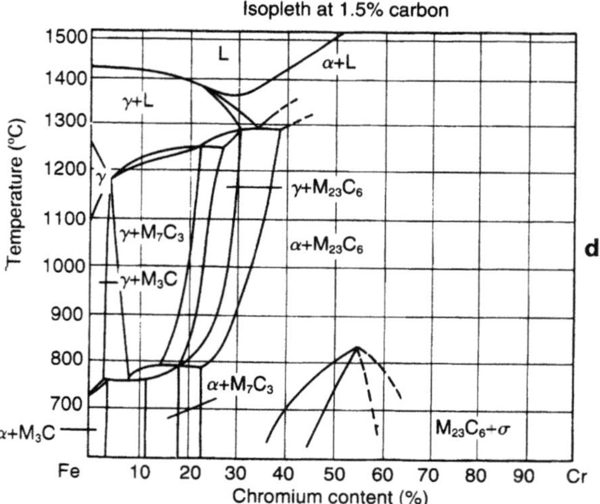

Fig. 4.17. (*continued*)

difficult and eventually impossible to austenitise the alloy. The consequences of this on the heat-treatability of the alloy depend upon the carbon and chromium contents of the steel.

At lower carbon contents and higher chromium contents the formation of austenite is completely inhibited, leading to the ferritic stainless steels, and a summary of the main alloys in this range is given, for example, by Lula [57].

The so-called first-generation ferritics are simply Fe–Cr–C alloys. Clearly, the total inhibition of austenite in such alloys requires quite a high chromium level, and it is possible to form austenite at high temperatures in many first-generation ferritics and hence heat-treat them to produce a duplex ferrite–martensite structure. They are, however, rarely used in this condition, because they would be very brittle.

The second-generation ferritics have a somewhat lower chromium content in favour of other ferrite formers such as aluminium and titanium (e.g. 405 and 409). Recent steel-making developments have allowed the interstitial (carbon and nitrogen) content of ferritic stainless steels to be lowered and this, coupled with stabilisation by titanium and niobium, has led to the development of the third-generation ferritic stainless steels, which include types 444 and 26-1, with improved weldability and corrosion resistance. The selection of ferritic stainless steels should be carried out in the light of possible difficulties with grain growth, toughness and embrittlement and sensitisation (Lula [57]).

At higher carbon contents and lower chromium contents, it becomes possible to render the steel austenitic or into a state of austenite plus carbides, thus producing the martensitic stainless steels which can be broadly divided into two groups, depending on their carbon content and the mechanical properties are highly dependent upon this. The lower carbon grades also have a lower chromium level to avoid them showing behaviour characteristic of the ferritic grades. Clearly, provided that they are austenitisable, the very high hardenability of these steels (associated with the chromium content) ensures that they produce martensite under most conditions of cooling. All of the martensitic stainless steels are used in the quenched (quenching being achieved by air cooling) and tempered state, although softening treatments are sometimes applied prior to forming, for example.

The low-carbon grades, for example 410, have structures similar to low-alloy steels, with a tough, strong-tempered martensite structure in which the main temper carbide is a chromium one. The susceptibility of martensitic stainless steels to temper brittleness (e.g. Lula [57]) means that tempering in the range 425 to 565 °C must be avoided. Thus, tempering is usually carried out above the upper end of the embrittlement range, but excessive carbide precipitation must be avoided, from the points of view both of properties and of corrosion resistance.

The higher carbon grades are typified by the 440 series. These steels contain a substantial amount of carbon and, at least in some cases, cannot be fully austenitised, at best going into a two-phase austenite + carbide phase field. Subsequent quenching and tempering produce a bimodal distribution of carbides with fine-temper carbides and coarser carbides which did not dissolve during austenitisation, this process being illustrated schematically in Fig. 4.18. Because of their high carbon content, these steels would suffer severely reduced corrosion resistance if tempered to give useful ductility, and are therefore used in

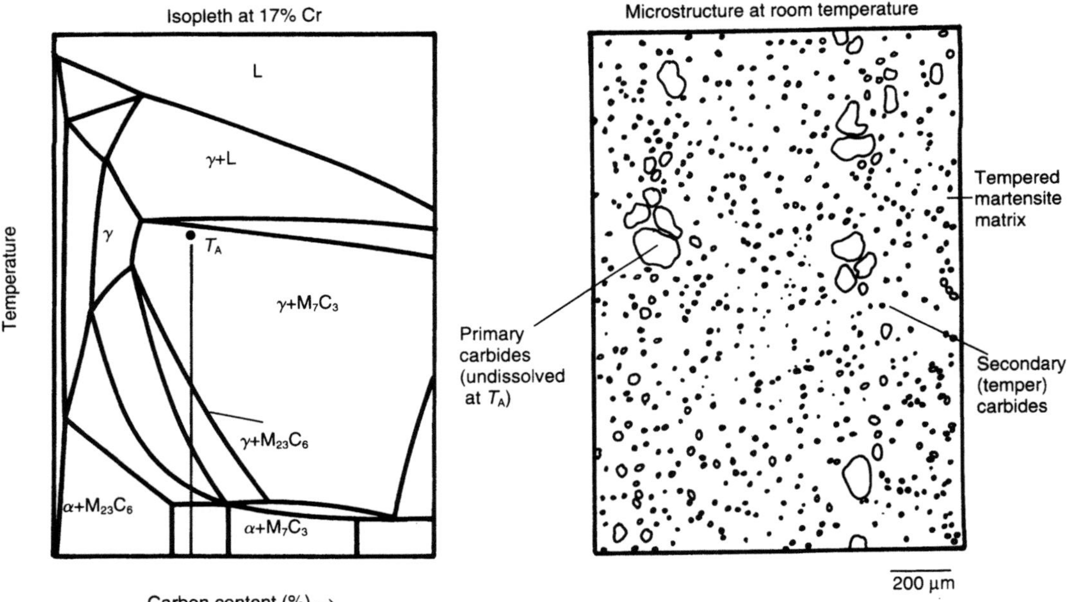

Fig. 4.18. Illustration of the development of the microstructure of an AISI 440C stainless steel (1.2%C, 17%Cr).

applications where the combination of corrosion resistance and wear resistance is required and where limited toughness can be tolerated.

In brief, the martensitic stainless steels provide a high-strength, corrosion-resistant option, the strength being usable (i.e. coupled with adequate toughness) in the low carbon grades and being exploitable only as increased wear resistance (i.e. hardness) in the higher carbon grades.

4.1.3.3 Duplex and Precipitation Hardening Stainless Steels

Duplex stainless steels are a relatively recent addition to the range of engineering alloys. Although the principle of controlling the balance of ferrite and austenite content by alloying and heat treatment has been known for some time, it was not until the advantages of these steels for petroleum service were realised that duplex stainless steels found a major industrial application.

According to Solomon and Devine [58], the metallurgy of duplex stainless steels is best understood using pseudo-binary sections of the Fe–Cr–Ni system, shown in Fig. 4.19. The main effects to be seen as iron content increases in the Fe–Cr–Ni system are the introduction of the sigma phase and the eventual formation of an 'austenite loop' at higher iron contents. When the iron content exceeds about 60 per cent, austenite becomes unstable at room temperature and may transform to martensite.

Most duplex stainless steels contain about 70 per cent iron and, according to Fig. 4.19, austenite–ferrite structures are likely. However, the metastable nature of the phase relationship at room temperature means that different structures will be developed during solidification to those developed during hot working.

The casting duplex alloys can be treated very much like weld metal, since their structure is developed during solidification. Most commercial cast duplex alloys have a composition of around 19 per cent Cr and 9 per cent Ni,

which with a wrought alloy would be associated with an austenitic structure. The effect of casting can be seen using a Schaeffler or DeLong constitution diagram (see Chapter 5), although these are really only applicable to relatively rapidly cooled castings (since they were developed for weld metal).

Wrought duplex stainless steels are usually produced by hot working, and it is important to ensure that some ferrite

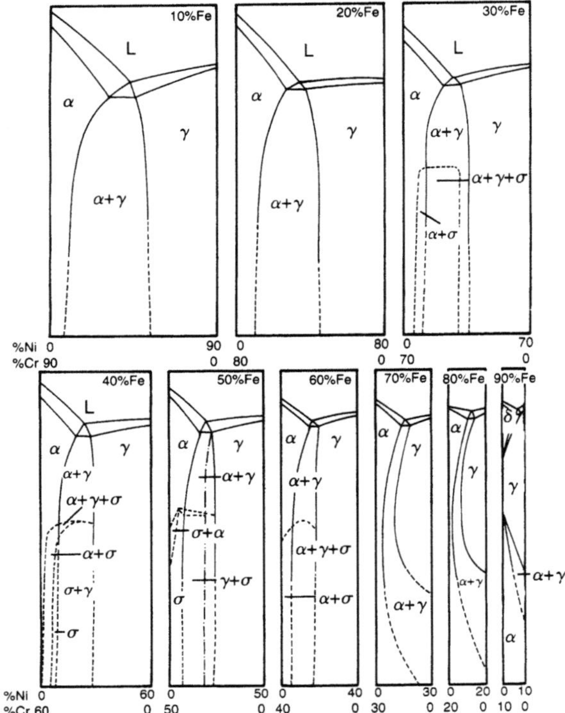

Fig. 4.19. Pseudo-binary sections in the iron–chromium–nickel system. (From Solomon and Devine [58].)

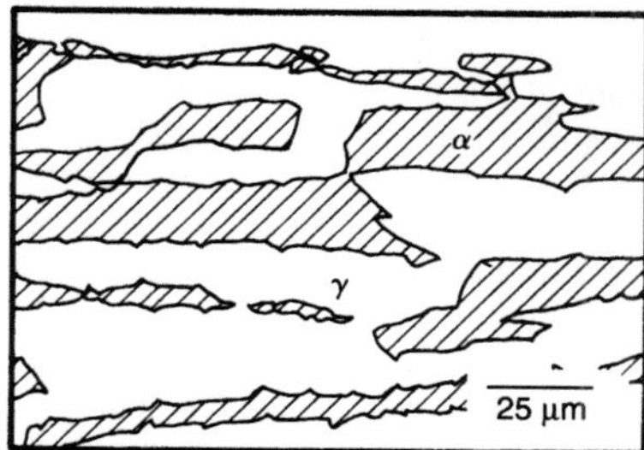

Fig. 4.20. Schematic microstructure of a duplex stainless steel hot-worked in the temperature range 1000 to 1200 °C. The illustrated amount of ferrite (about 20%) corresponds roughly to a 20%Cr, 8%Ni, 3%Mo alloy.

is present at the hot-working temperature (typically 1000 to 1200 °C) to avoid hot cracking. This usually produces a structure consisting of bands of ferrite in an austenitic matrix, as illustrated schematically in Fig. 4.20.

Modified Schaeffler diagrams (e.g. Fig. 4.21) have been developed to quantify the ferrite present in wrought duplex stainless steels. These diagrams are usually specific to a particular hot-working temperature and the hot-worked steels generally have a lower ferrite content than would a casting of the same chromium and nickel equivalent. It is also possible to produce microduplex structures where the ferrite and austenite grain sizes are very small. This is generally achieved by special thermomechanical treatments, occasionally involving cold work. These microduplex steels are superplastic at about 1000 °C, and typical thermomechanical treatments are discussed by Solomon and Devine [58].

Wrought duplex stainless steels usually contain around 20 to 25 per cent Cr, 5 per cent Ni and a few per cent Mo and have rather higher yield strengths and lower ductilities than the austenitics, although the UTS values are similar. This makes their general mechanical behaviour intermediate between ferritic and austenitic steels as might be

expected, and means that any precipitation behaviour depends upon whether this precipitation is in the austenitic or ferritic phase (discussed below) and also that low temperature toughness is an issue which needs to be considered. The effect of ferrite content on UTS, yield strength and toughness is shown in the schematic diagram (Fig. 4.22), although it should be realised that this is rather simplistic, particularly from the point of view of toughness where a number of compositional parameters (i.e. not just ferrite content) need to be considered. However, the principal point is that designers should be aware of the need to specify low-temperature toughness in duplex stainless steels for some types of service.

Precipitation is usually to be avoided in duplex stainless steels. The (mainly) chromium carbides $M_{23}C_6$ and M_7C_6 and various other undesirable phases can form if the alloy is cooled insufficiently slowly or held in inappropriate temperature ranges. However, precipitation can be used to strengthen other types of stainless steels, these normally being called precipitation-hardened stainless steels. Like the duplex stainless steels these do not generally have systematic AISI numbers, but some appear under the AISI 600 series for high-strength, high-temperature stainless steels. They are often known by trade names, although the Unified Numbering System [1] gives these as designations. The broad classes of precipitation hardening stainless steels are those with martensitic, austenitic or semi-austenitic matrices.

The conventional way of strengthening martensitic steels (i.e. by using a higher carbon content) is not always helpful, in that it leads to reduced ductility and toughness, so that precipitation provides a useful way of providing additional strengthening. The martensitic P-H stainless steels therefore have relatively low carbon contents (compared with the 400 series, for example) and have small additions of the precipitation-hardening elements. Their compositions are based around 16 per cent chromium (higher than in martensitics, owing to the lower carbon content), with additions of Cu, Al, Mo and others.

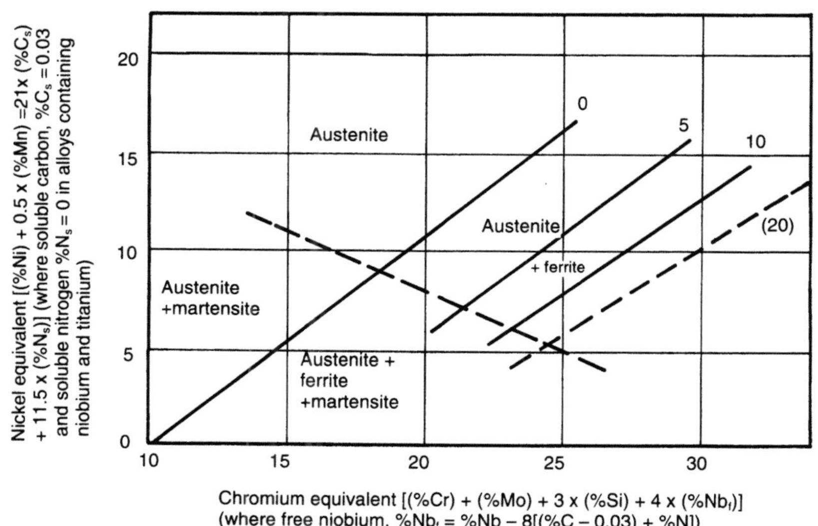

Fig. 4.21. Schaeffler-type diagram for the prediction of ferrite content in wrought duplex stainless steels hot-worked at 1150 °C. (From Solomon and Devine [58].)

The semi-austenitic grades are still austenitic after quenching, and must be further treated (by sub-zero quenching or by tempering out some of the carbon in the form of chromium carbides) to produce martensite. Once this has been done, the steels show similar mechanical properties to the martensitic grades, but have higher chromium and nickel equivalents.

Whereas precipitation hardening is used in the martensitic grades to provide strength with low carbon content, austenitic stainless steels are in greater need of strengthening. The precipitation sequence is rather simpler in the austenitic P-H steels, and consists mainly of strengthening by the coherent Ni_3Ti precipitate.

4.1.3.4 Properties of Stainless Steels

The mechanical properties of stainless steels are dependent largely upon the matrix and thence upon the compositional details which have been outlined above. In broad terms, the austenitics are weakest but toughest, while the martensitics show highest strength but require care to be taken over their toughness. There are two major ways to compromise between these opposing sets of properties; the first is to use a mixture of the two phases (duplex stainless), the second to employ precipitation strengthening. Table 4.22 illustrates in a very general way how the mechanical properties vary across the stainless steel grades, but a more detailed

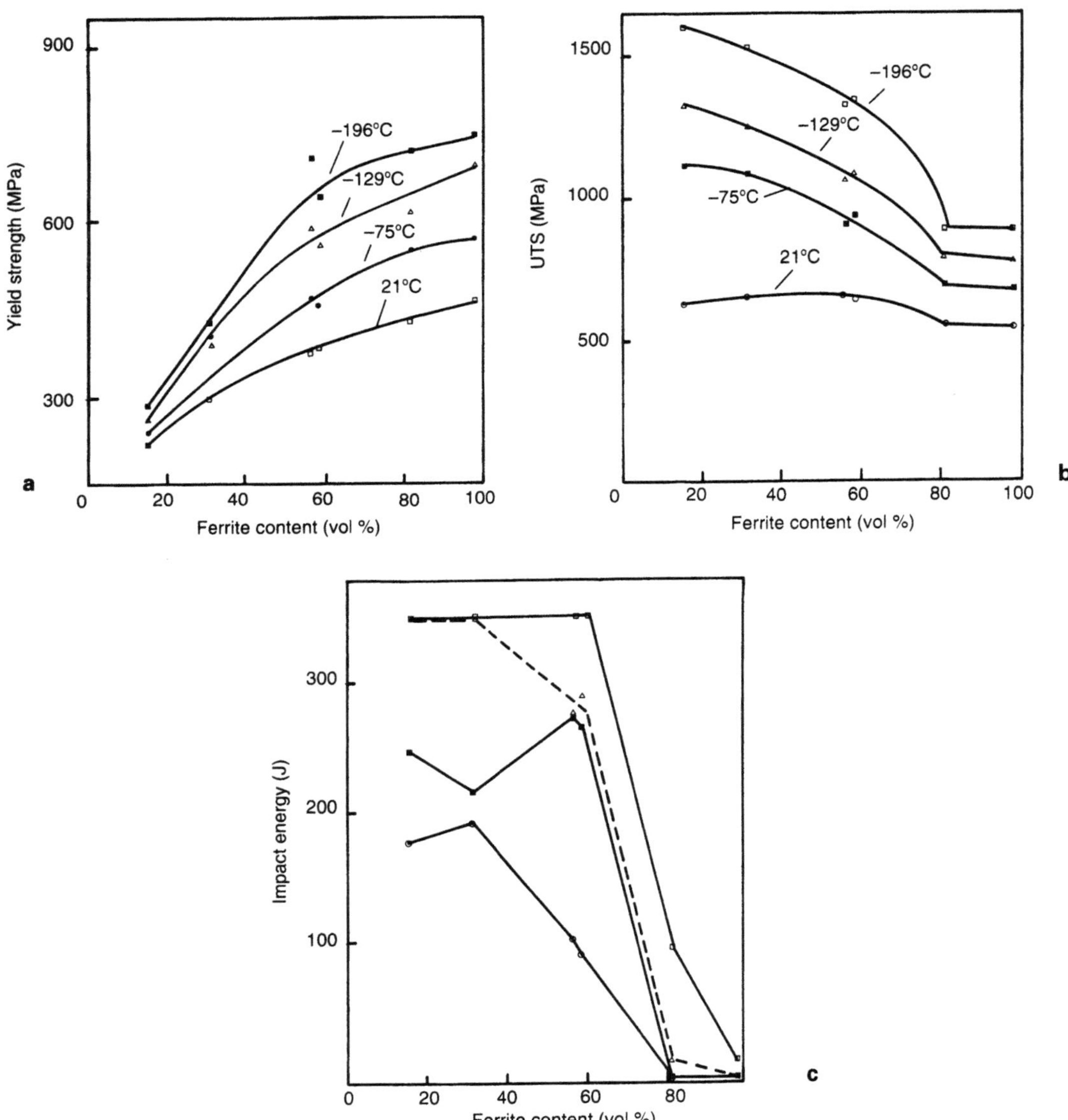

Fig. 4.22.   Effect of ferrite content on mechanical properties of duplex stainless steels. **a** Yield strength. **b** Ultimate tensile strength. **c** Impact energy. (From Solomon and Devine [58].)

Table 4.22. Selection of strength and ductility data of stainless steels

Designation	Condition and type	0.2% Proof stress (MPa)	UTS (MPa)	Elongation (%)
304	Austenitic annealed	205	515	40
304	Austenitic cold-worked	→1035	→1240	55 to 10
Nitronic 60	Austenitic nitrogen-strengthened, annealed	345	655	35
410	Martensitic lower carbon, annealed	275	515	30
410	Martensitic lower carbon, quenched and tempered	415 to 1035	620 to 1310	30 to 15
440C	Martensitic higher carbon, quenched and tempered	620 to 1860	895 to 1930	10 to 2
430	Ferritic general purpose, annealed	240 to 380	415 to 585	20 to 35
430	Ferritic general purpose, cold-worked	620 to 860	690 to 1000	5 to 2
600	Precipitation hardening austenitic, aged	690	1000	24
631 (17-7 PH)	Precipitation hardening semi-austenitic, aged	1310 to 1795	1415 to 1825	9 to 2
630 (17-4 PH)	Precipitation hardening martensitic, aged	930 to 1275	1035 to 1380	17 to 14
SAF 2205	Duplex ≈ 45% ferrite annealed	410 to 450	680 to 900	25

compilation (e.g. [46]) should be consulted for specific applications.

The corrosion of stainless steels is a complex subject but, as might be expected for a passivating alloy, the major points to be considered are the possibility of local effects such as pitting and crevice corrosion and the dangers of stress corrosion.

The general corrosion rates of stainless steels in seawater and marine atmospheres are of course lower than for non-stainless steels, but some staining and pitting does occur, depending upon the grade. As can be seen from the data in Table 4.23, it is pointless to consider weight loss as a useful measure of penetration, since pitting is the predominant mode of metal wastage. In general, atmospheric performance is rather good, but pitting may be a problem in submerged service, the higher grades (higher chromium, austenitics and molybdenum-containing) showing superior performance.

As far as pitting and crevice corrosion are concerned, the effect of chloride ion on passivation behaviour (see Chapter 3) means that stainless steels are generally more susceptible in those marine environments where the supply of oxygen is low.

The propensity of a stainless steel to pit is usually measured by its pitting potential, although various measurements of number and size of pits are also used. Also, a number of tests are available for measuring susceptibility to crevice corrosion, but these data are not normally used directly by designers; the important point is whether or not the steel will show acceptable performance in service, although the term 'acceptable' is rarely quantified. Since both pitting and crevice corrosion are caused in stainless steels by essentially the same phenomenon (Chapter 3), those steels which are resistant to one are normally resistant to the other, and the best performance under pitting or crevice conditions is offered by steels with higher chromium contents (> 20 per cent) and with a few per cent of molybdenum. Of course, other grades may perform perfectly adequately under conditions of good agitation and aeration (unless oxygen can be eliminated entirely) and where crevices can be avoided. The use of inhibitors and also of cathodic protection can be effective.

The principal interest in stress corrosion in stainless steels centres around chloride SCC, although polythionic acids, hydrogen sulphide and hydrogen embrittlement need also to be considered for some grades and in some applications. A large number of variables affect the susceptibility of stainless steels to chloride stress corrosion. These can be conveniently classified as environmental (including chloride concentration, cation type and concentration, temperature, pH, stress level and oxygen concentration) and compositional (including matrix type, nickel content and degree of sensitisation). It is broadly the case that a temperature of at least 70 °C is required, and that ferritic stainless

Table 4.23. Some marine corrosion data for stainless steels after eight years' exposure

a Immersed

	AISI type			
	410	302	316	321
Weight loss (g/dm^2)				
1 yr	5.99	2.93	1.18	2.32
2 yr	9.56	4.43	0.65	3.44
4 yr	15.87	7.06	0.54	6.55
8 yr	28.09	10.89	4.08	10.22
Average depth of deepest pits (mm)				
1 yr	1.55	1.78	1.12	1.63
2 yr	1.63	1.88	1.32	3.05
4 yr	3.76	2.72	1.22	4.45
8 yr	4.09	3.56	3.91	4.90
Deepest pits (mm)				
1 yr	6.6	6.63	6.22	6.86
2 yr	6.71	6.05	2.08	5.23
4 yr	6.6	7.26	2.36	6.93
8 yr	6.58	5.99	6.22	6.91
Type of corrosion attack	Localised concentration-cell pitting	Random pitting	Random pitting	Random pitting

b Marine atmosphere (coastal)

AISI type	Weight loss (g/dm^2)				Type of corrosion attack
	1 yr	2 yr	4 yr	8 yr	
410	0.08	0.11	0.11	0.08	Some localised pitting
430	0.05	0.07	0.07	0.07	Some localised pitting
301	0.00	0.00	0.01	0.01	No visible attack
316	0.00	0.00	0.00	0.00	No visible attack
321	0.01	0.00	0.01	0.01	No visible attack

Source: Lula [57].

steels are rather less susceptible to chloride SCC. The effects of temperature and nickel content on chloride SCC of austenitic stainless steels are shown in Fig. 4.23. Dissolved oxygen is known to aggravate chloride SCC, and a critical applied stress is required, these two factors being illustrated in Fig. 4.24, again for austenitics.

Chloride stress corrosion in non-austentitic grades is much less likely, although, for example, stress corrosion of a 430F ferritic stainless in a marine atmosphere has been reported [59].

In ferritic and martensitic grades, particularly those of high strength, the stress corrosion problem is more one of hydrogen embrittlement. NACE [60] have limitations on the grades and conditions of stainless steels for sulphide service and these are summarised in Table 4.24.

One of the stated advantages of duplex stainless steels for marine service is that they show better chloride SCC resistance than do their austenitic counterparts. For example, Bernhardsson et al. [61] and Rhodes et al. [62] have investigated the suitability of duplex stainless steels for aggressive petroleum applications where sulphides, chlorides and dissolved carbon dioxide may be present, finding a synergic effect of chloride and sulphide. An example of the operating envelope for a high-strength duplex stainless steel in hydrogen sulphide is shown in Fig. 4.25. Whitehead and Baloun [64] have reported a passivatory effect of carbon dioxide in the presence of hydrogen sulphide for martensitic stainless steels.

Because of their reliance on passivating behaviour, the corrosion of stainless steels can be aggravated by some mechanical effects. For example, fatigue can be enhanced by the presence of seawater but, on the other hand, mildly corrosive environments (such as seawater) generally do not cause a great deal of corrosion–erosion in stainless steels, even in the presence of solid particles. This is particularly so for the martensitic grades, which are used in a variety of wear- and corrosion-resistant applications, for example as valves and trim in oil wells producing fluids with abrasive particles. Austenitic and P-H stainless steels are also known to be resistant to cavitation damage (Sedriks [54]).

Despite their useful properties, stainless steels do not find their major application in marine environments. This is probably due to the (not insurmountable) complication of the existence of chloride ions, which can render one of the most obvious advantages of using a stainless steel inoperative unless sufficient care is taken over design and selection. Stainless steels are most likely to be found in atmospheric rather than submerged marine service, although castings can be used in condensers, marine exhausts, propellers and pump impellers and usually show good performance where flow is maintained, although problems may arise if they are left idle filled with seawater.

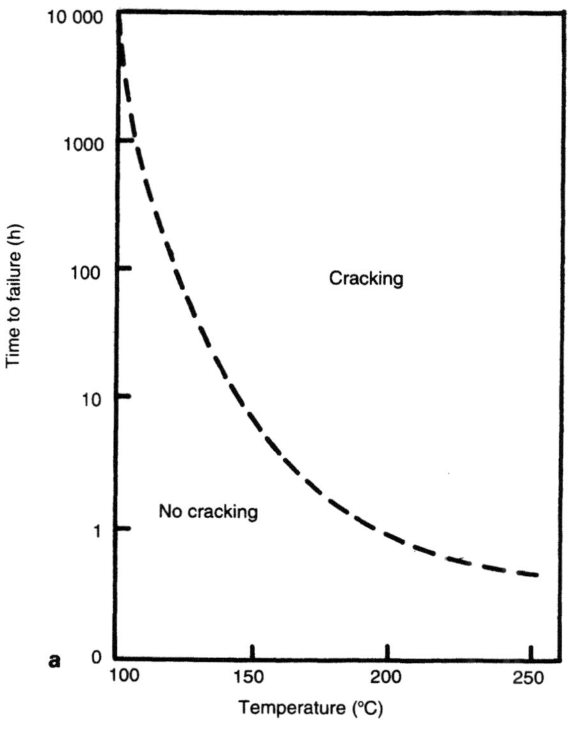

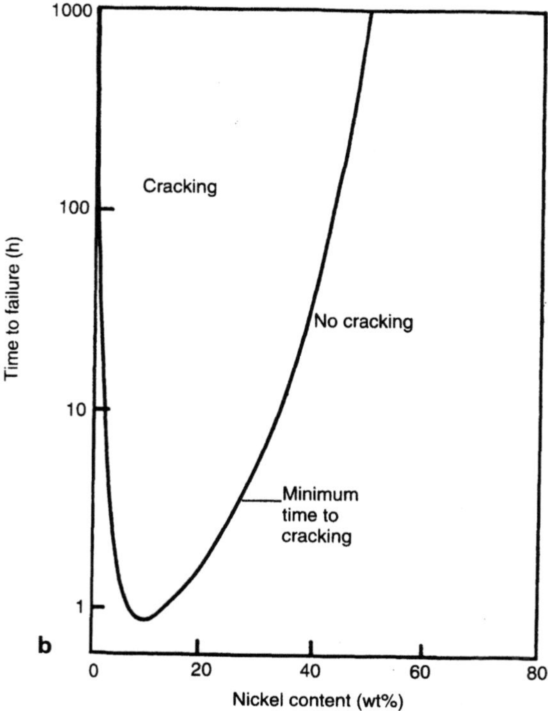

Fig. 4.23. Effects of temperature and nickel content on the susceptibility of stainless steels to stress corrosion cracking: **a** types 304 and 347 in oxygen-containing sodium chloride solutions; **b** wires containing 18 to 20% chromium in boiling magnesium chloride solution at 154 °C. (From Sedriks [54].)

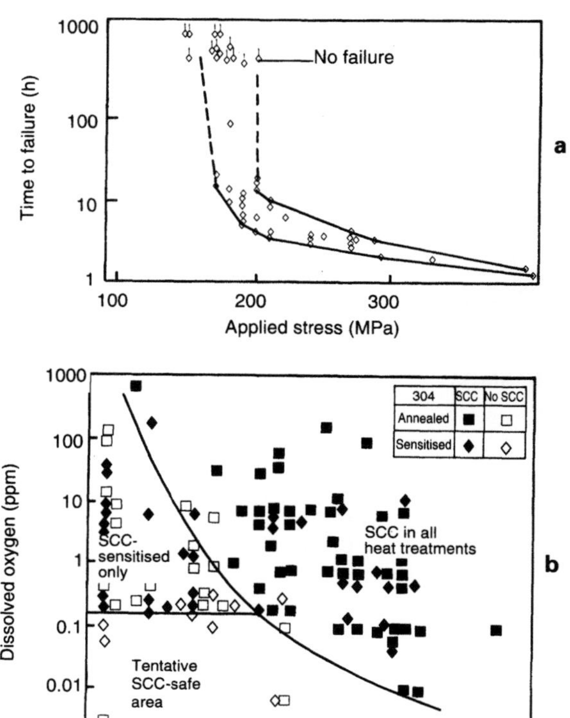

Fig. 4.24. Effects of applied stress and electrolyte composition on the susceptibility of stainless steels to stress corrosion cracking: **a** electropolished type 347 in boiling magnesium chloride solution at 154 °C, **b** type 304 at temperatures in the range 260 to 300 °C, with applied stress in excess of yield and exposures in excess of 1000 h. (From Sedriks [54].)

Atmospheric applications include deck fittings and coastal architecture, although with some grades (e.g. 304) periodic cleaning is required to avoid pitting. A number of petroleum applications are found for many grades, for example martensitics in reservoir fluids for pump and valve components and duplexes for pipeline applications. Applications for seawater handling have been discussed by Shone [65], the broad conclusion being that seawater handling systems could be constructed from stainless steels but that these require to be carefully selected. In particular, it was noted that steels of the AISI 316L type are unlikely to be suitable for this application unless cathodically protected. Jasner [66] has discussed applications in reverse osmosis seawater desalination plants with particular regard to pitting performance, as measured by a pitting resistance equivalent:

$$PRE_N = \%Cr + 3.3 \times (\%Mo) + 30 \times (\%N)$$

A 6%Mo austenitic stainless steel was recommended for this purpose, although it was expected that ferritic and duplex stainless steels could prove suitable under some circumstances.

4.1.4 Cast Irons

Cast irons are metallurgically rather sophisticated alloys, and it is only their long history which makes them seem

Table 4.24. Stainless steels listed by NACE as being acceptable for direct exposure to sour environments
(The original standard should be consulted for other details)

Ferritic	Martensitic	Precipitation hardening	Austenitic	Duplex (austenitic/ferritic)
AISI	AISI	ASTM	AISI	UNS S31803
405	410	A 453 Gr 660	302	
430	501	A 638 Gr 660	304	UNS S32550, wrought condition only
			304L	
ASTM	ASTM	UNS S17400	305	
A 268	A 217 Gr		308	Cast duplex (austenitic/
GR TP 405,	CA 15	UNS S45000	309	ferritic) stainless steel
TP 430,	A 268 Gr TP		310	Z 6CNDU20.08M, NF A 320-55
TP XM 27,	410		316	French National Standard
TP XM 33	A 743 Gr	UNS S66286	316L	
	CA 15M		317	
	A 487 Cl		321	UNS S32404
	CA 15M		347	
	A 487			
	Cl CA6NM		ASTM	
	A 182		A 182	
	Cl F6NM		A 193	
			Gr B8R, B8RA	
			B8, B8M	
			B8MA	
			A 194	
			Gr 8R, 8RA,	
			8A, 8MA	
			A 320	
			Gr B8, B8M	
			A 351	
			Gr CF3, CF8,	
			CF3M, CF8M	
			B 463	
			B 473	
			UNS S20910	
			UNS N08020	

old-fashioned. In fact, there are a number of modern applications of cast irons in marine environments and these will be reviewed in the following. In addition, cast iron is often encountered as existing structures in marine (particularly coastal) environments, and some comments are included pertinent to repair or strengthening. Engineers should always bear in mind that, unlike steels, where a wide range of strength is possible with little change in elastic modulus, cast irons may contain substantial amounts of graphite, and hence their modulus may vary considerably.

The metallurgy of irons can be understood initially by reference to the iron–carbon and also the iron–cementite phase diagrams shown in Fig. 4.26. Cast irons occupy the region around the eutectic point in these diagrams, and thus contain around 3 per cent carbon. As can be seen, this results in a rather lower solidification temperature than in steels, and this has been the main reason for the development of these compositions for casting applications and, unlike almost all other alloy types, cast-iron compositions are never available in wrought form. Cast irons also normally contain a substantial amount of silicon, and this has an effect on the position of the eutectic, which can be summarised by the carbon-equivalent formula:

$$CE = \%C + (\%Si + \%P)/3$$

The eutectic composition occurs at a carbon-equivalent value of 4.3 per cent and hypereutectic irons normally pecipitate primary graphite (kish graphite), following the Fe–C diagram at high temperatures, rather than primary cementite (as would be indicated by the Fe–Fe$_3$C diagram).

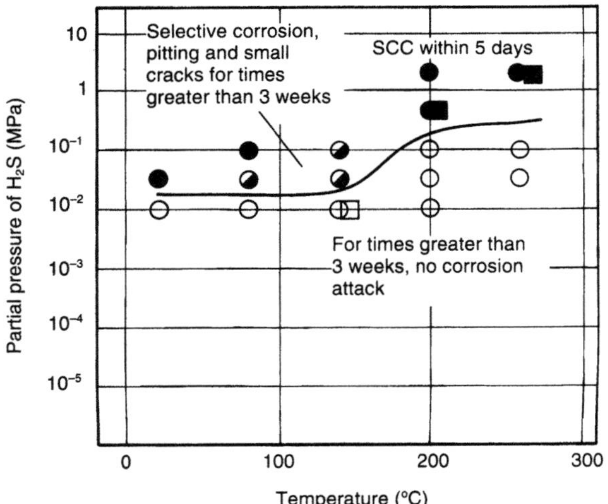

Fig. 4.25. Engineering corrosion diagram for a 22%Cr–5.5%Ni–3%Mo duplex stainless steel cold-worked tube material in hydrogen sulphide. (From Tynell [63].)

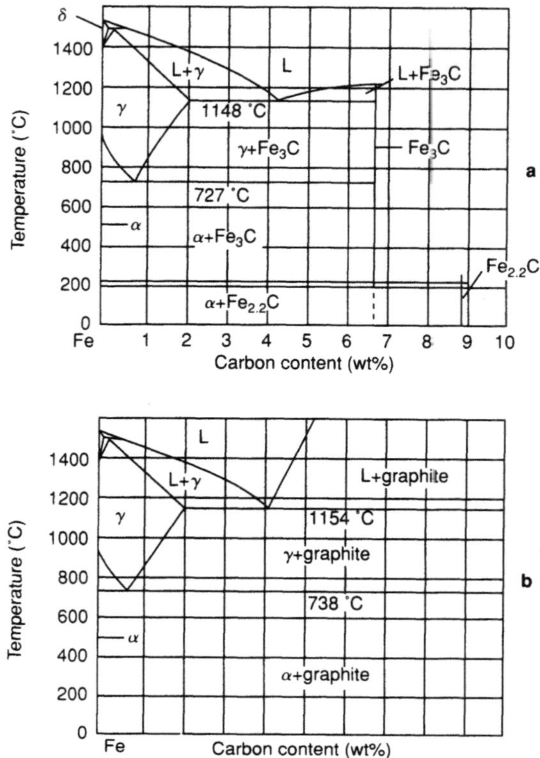

Fig. 4.26. The **a** iron–cementite and **b** iron–carbon phase diagrams [46].

Cast irons are classified in terms of their microstructure, which can contain any of the phases normally found in steels (i.e. ferrite, austenite, martensite and cementite), along with a phase not found in steels, namely graphite. Given that cementite contains 6.67 per cent carbon, a cast iron in which all the carbon exists as cementite will contain a very substantial proportion of that phase, with consequent extreme hardness and brittleness. The precipitation of graphite (100 per cent carbon) can thus be seen as a way to allow the carbon to be out of solution without excessive embrittlement. Nevertheless, cast irons are brittle alloys and are rarely used in components where the general tensile stress level is high.

The microstructural range available in cast iron can be classified in terms of: the existence or otherwise of graphite (white, graphite-free or grey, graphitic irons), the form of the graphite (flake, compacted, spheroidal or nodular), and the matrix (ferritic, pearlitic, martensitic or austenitic). The microstructure is controlled by alloying and heat treatment, and the various types are summarised in Table 4.25.

The primary means of classification of cast irons, and that which produces the most fundamental distinction in properties, is between white and grey irons, the grey irons being produced by trying to promote equilibrium (Fe–C) conditions, particularly at higher temperatures. Special heat treatments after solidification give rise to the nodular (or malleable) irons.

White irons are produced by relatively rapid cooling in irons of low carbon equivalent. The formation of white iron is further promoted by alloying with chromium, molybdenum and vanadium, which incidentally are also useful carbide formers. Metallurgically, white irons produce primary austenite on cooling (since they are normally of low carbon equivalent) and the remaining liquid transforms to the eutectic mixture of austenite plus cementite, normally referred to as 'ledeburite'. As the eutectoid line is crossed, all austenite transforms to cementite plus ferrite, so that the final structure is a complex mixture of ferrite and cementite. These alloys are hard and hence wear-resistant, making them rather difficult to machine. The use of unalloyed white irons is often in parts of castings by selective chilling, but a wide range of alloy white irons is also available for bulk form. The addition of nickel and chromium encourages the transformation of austenite to martensite rather than pearlite, so that these irons are even harder than unalloyed white irons. The major use of alloy white irons is in the handling of abrasive materials such as minerals and slurries, particularly in the mining industry. Even with a single composition, there is quite a range of hardnesses available by varying heat treatments or cooling rate of castings.

Grey irons do not generally develop the same degree of hardness as white irons; however, they are not necessarily less brittle, because they contain graphite, which is usually present as rather large particles. Their main uses are where high stiffness is required at a relatively low stress, for example in machine beds and engines, and this normally involves fairly massive structures. Graphite can be formed from the liquid phase in two distinct ways, depending on the cooling rate and on the use of small alloying additions of magnesium (normally referred to as inoculation). The graphite forms are known as 'flake' and 'spheroidal', although intermediate forms, produced by control of rare earth and other additions, are also possible, being referred to variously as 'compacted', 'vermicular', 'quasi-flake' and 'semi-nodular' graphite. (A third form known as 'nodular' or 'temper' carbon is produced from the solid state in malleable irons.)

As grey irons are cooled, either kish graphite or austenite is produced as the primary precipitate (depending upon whether the iron is hyper- or hypo-eutectic). At the eutectic temperature the remaining liquid solidifies by co-precipitation of austenite and graphite, irrespective of

Table 4.25.   Summary of the microstructural variations possible in cast irons

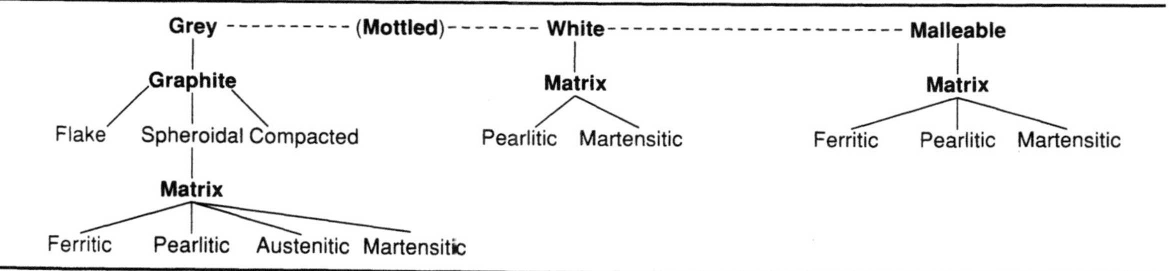

whether the iron is hyper- or hypo-eutectic. Depending upon cooling rate and composition, the remaining austenite can transform at the eutectoid temperature to pearlite, to ferrite plus graphite, or to all three, leaving a final structure which can vary between ferrite plus graphite to pearlite plus graphite. A useful range of alloy irons based on the grey structure with flake or spheroidal graphite is also possible. These are based on large additions of nickel or of silicon. The nickel-alloyed group usually contain from 13 to 35 per cent nickel, with chromium and occasionally copper as secondary additions. The nickel content prevents the transformation of the austenite, and these irons consist of graphite dispersed in austenite at room temperature. The corrosion resistance of these alloys ensures their wide use for seawater and petroleum handling. The silicon irons contain about 14 per cent silicon, and have rather low (less than 1 per cent) carbon contents. Their structure is ferrite plus graphite, and main uses are in the handling of mineral acids. They do, however, have one major marine use, and this is as impressed current cathodic protection anodes. The normal grades perform adequately as anodes in fresh water or in soil, although additions of noble metals such as chromium or nickel are required to ensure good performance in seawater. Figure 4.27 shows schematically a selection of grey iron structures.

Malleable or nodular irons are produced by heat treatment of white irons. As with quenching and tempering of steels, the reheating of white irons allows a greater degree of microstructural control than is possible by controlling cooling. Heat treatment is fairly complex and involves first the decomposition of cementite to austenite plus carbon at a temperature above the eutectoid (first-stage graphitisation). After this, the iron may be cooled at a moderate or an accelerated rate to room temperature, producing a pearlitic or martensitic matrix. This may then be further tempered in the same way as a martensitic steel would be. The pearlite may also be reheated to spheroidise the cementite. Cooling from first-stage graphitisation may also be controlled to allow graphitisation at the eutectoid transformation, leading to a ferritic matrix. The graphite form produced from such a solid-state processing is always of nodular form. The range of possible heat treatments and microstructures involved in the production of malleable iron is summarised in Fig. 4.28.

Table 4.26. Marine corrosion properties of cast irons

Environment	Corrosion rates and comments
Seawater (quiescent surface)	Initially 250 to 600 µm/yr for grey and 25 to 75 µm/yr for high nickel irons. Rates fall to about 100 µm/yr and about 20 µm/yr respectively after one or two years. Both types may pit as much as 2.5 mm over about 10 years, austenitic types being less susceptible.
Seawater (deep ocean)	Uniform rate of 12 to 120 µm/yr for grey and 10 to 35 µm/yr for high-alloy irons.
Marine atmosphere	About 10 µm/yr for grey and usually <3 µm/yr for high-nickel irons.
Other	Both types form a graphitic layer which helps to prevent further corrosion, so that velocity effects are important, more so for the unalloyed irons. Graphitic layer also has galvanic effect if old iron is connected to new construction.

Tables 4.26 and 4.27 summarise the corrosion properties and the fatigue and fracture properties respectively of a range of cast irons [67–72]. As before, care should be taken in interpreting such data in design applications. The range of available mechanical properties with type and composition is rather complex, and compilations such as Walton and Opar [67] and Angus [68] should be consulted for a detailed discussion on how composition, processing and section thickness can affect properties. Only some standards specify strength for cast irons, hardness often being the mechanical property yardstick. One exception to this is the family of ductile irons, which normally yield before fracture of a test bar.

Apart from their well-known uses in machinery and engine blocks, cast irons find a number of applications in pumping, piping and valves for seawater systems where the increased corrosion resistance of high-nickel grey irons may be of additional benefit.

Table 4.27. Fatigue and fracture data for cast irons

Type	Fracture	Fatigue
Grey	Impact tests often carried out on unnotched specimens. For CVN, impact energies are around 1.5 to 7 J and K_{Ic} is in the range 10 to 25 MPa$\sqrt{m}$ (both at room temperature).	Fatigue limit in the range 65 to 250 MPa.
Ductile	CVN usually 5 to 20 J, K_{Ic} around 50 MPa$\sqrt{m}$ (both at room temperature). K_{Ic} for austempered ductile irons can be as much as 100 MPa$\sqrt{m}$.	Fatigue limit in the range 200 to 350 MPa for unnotched specimens, 125 to 200 for notched specimens.
Austenitic	Depends on graphite form. K_{Ic} around 60 MPa$\sqrt{m}$ at room temperature in SG forms.	Fatigue limit about 50% of ultimate tensile strength in softer alloys, less (about 30%UTS) in higher-strength alloys.
White	Alloys only used in wear-resistant applications where global fatigue and fracture are not major design considerations.	

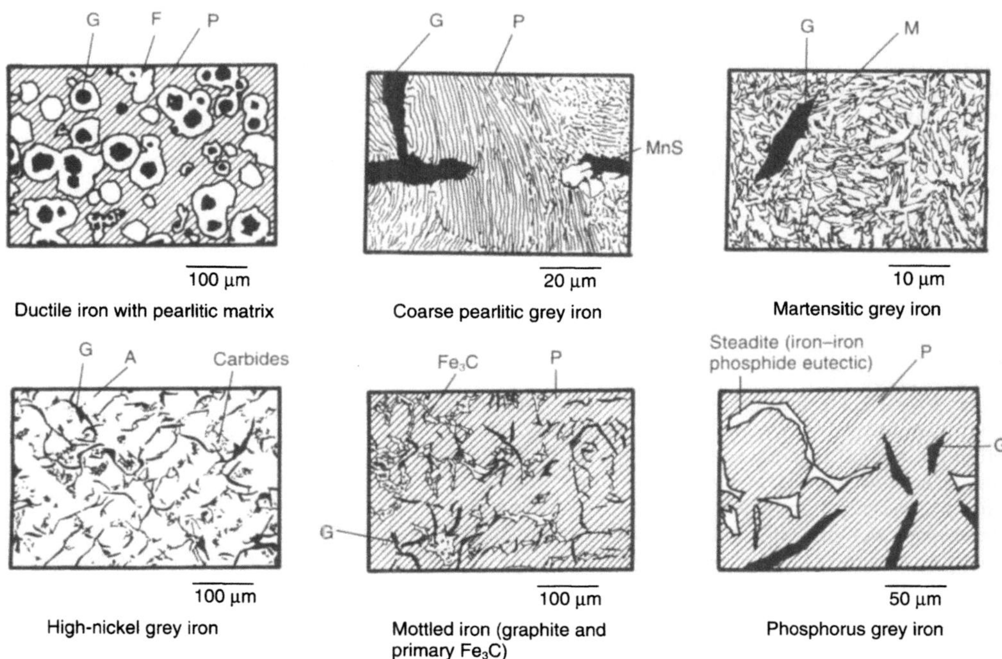

Fig. 4.27. Some of the range of possible grey iron structures. (Adapted from Walton and Opar [67].) G, graphite; P, pearlite (ferrite + cementite); F, ferrite; M, martensite; A, austenite.

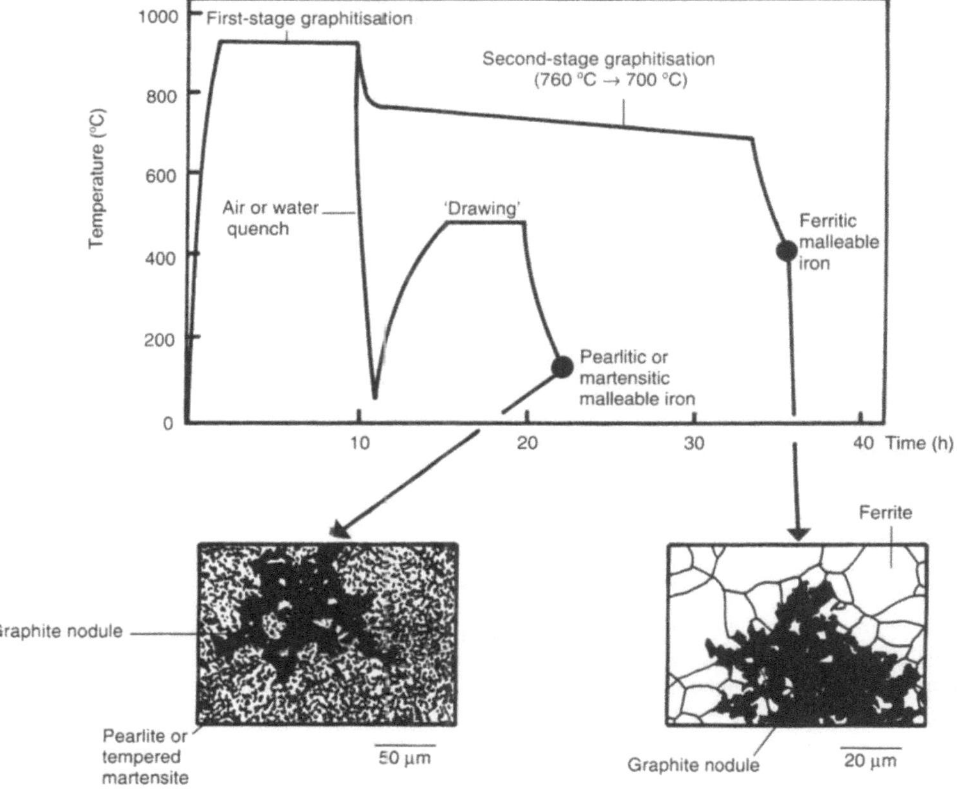

Fig. 4.28. Heat treatments for the production of malleable iron from white iron and the resulting microstructures. (Adapted from Askeland [3].)

4.1.5 Aluminium Alloys

Although aluminium is itself a base metal, its protective oxide means that corrosion rates in the marine environment are, under some circumstances and for some alloys, quite acceptably low. As with its land-based applications, the other main impetus for the use of aluminium alloys is their relatively high modulus to density and yield strength to density ratios. The possibility of stress corrosion cracking in seawater needs to be addressed when considering the use of some alloys, and great care needs also to be taken to avoid coupling with noble metals (such as copper), owing to the highly electronegative nature of aluminium.

Pure aluminium is a relatively weak metal, and were it not for its substantial response to various strengthening mechanisms, particularly precipitation hardening, it would find very few engineering applications. All aluminium alloys are strengthened by a combination of cold work, precipitation hardening, dispersion strengthening and solid solution strengthening, and it is important to know which of these has been used for any alloy in question, as there are implications for welding and any other thermal processing which may be undertaken.

The designation of aluminium alloys has in the past been a complex matter, with several different systems being in use at once, but, for wrought alloys, the major US aluminium producers and the British Standards Institution have now adopted a common system. This system involves a four-digit composition designation and a temper designation, which is indicative of the condition of the alloy.

The leading digit of the alloy designation indicates the principal alloying element:

1xxx   commercially pure alloys (>99 per cent Al)

2xxx   copper

3xxx   manganese

4xxx   silicon

5xxx   magnesium

6xxx   magnesium and silicon

7xxx   zinc

8xxx   others

The temper designations differ between the American Aluminium Association and the BSI. The British system (as summarised by Ray [73]) uses the following letters:

M   as manufactured

O   annealed

H   strain-hardened; degree indicated by an additional digit from 1 to 8

T   heat-treated, then

    B   solution-treated and aged at room temperature

    D   solution-treated, cold-worked and aged at room temperature

    E   cooled from an elevated temperature shaping process and precipitation-treated (by reheating)

    F   solution-treated, then precipitation-treated

    H   solution-treated, cold-worked and then precipitation-treated

The AA system differs slightly (Hatch [74]):

F   as manufactured

O   annealed

H   strain-hardened followed by two digits: first digit indicates whether partial annealing or stabilisation (for Al–Mg alloys) and second digit indicates degree of work hardening

T   heat-treated, then

    1   cooled from elevated temperature shaping and aged (at room temperature)

    2   cooled from elevated temperature shaping, cold-worked and aged

    3   solution-treated, cold worked and aged

    4   solution-treated and aged

    5   cooled from elevated temperature shaping and precipitation treated

    6   solution-treated and precipitation-treated

    7   solution-treated and stabilised

    8   solution-treated, cold-worked and precipitation-treated

    9   solution-treated, precipitation-treated and then cold-worked

    10   cooled from elevated temperature shaping, cold-worked and then precipitation-treated

(The second digit after T and the third digit after H indicate specific treatments.)

(A 'W' temper designation indicates an unstable heat-treated condition.)

Most aluminium alloys contain only a few per cent of alloying additions, but the alloys may become quite complex with additions of several different elements. The principal age-hardening sequences exploited for aluminium alloys are:

Al–Cu: coherent and semi-coherent precipitates are formed on the way to the equilibrium $CuAl_2$ precipitate (which is associated with overageing).

Al–Zn–Mg: coherent and semi-coherent precipitates are formed based on magnesium zinc compounds with occasional substitution of Al for Zn. Equilibrium precipitate is $MgZn_2$.

Al–Cu–Mg: at high Cu–Mg ratios, precipitates from the Al–Cu system can form, but sequences of coherent and semi-coherent precipitates involving all three elements leading to the equilibrium Al_2CuMg also form.

Al–Mg–Si: coherent and semi-coherent precipitates based on magnesium and silicon leading to equilibrium precipitate Mg_2Si.

In all the above cases, precipitation treatments need to be optimised to ensure the maximum strength development, but many non-ageing aluminium alloys are in widespread use, especially where stiffness, and not strength, is the governing mechanical property.

The 1xxx alloys are non-heat-treatable, because they contain no deliberate alloying additions. The major impurity in commercially pure aluminium is iron, so that a small amount of dispersion strengthening is possible from small $FeAl_3$ particles. The alloys are generally suitable for submerged or atmospheric marine service because of their resistance to corrosion, although good aeration is required and coatings, such as anodising, may be of help. However, their limited strength obviates use as structural marine parts.

The 2xxx alloys are the traditional, heat-treatable aluminium alloys containing typically a few per cent of copper with magnesium as the commonest second element, although silicon, iron, manganese, zinc, chromium, titanium and a few others are also used. Despite the moderate strength levels attained by these alloys and their considerable history in aircraft use, it is unlikely that much application will be found in marine environments, owing to the poor corrosion resistance caused by the presence of copper-bearing precipitates. This can be overcome by cladding with pure aluminium, but a distinct technical advantage over other aluminium alloys would be required to make this worthwhile.

The 3xxx alloys contain manganese as their main alloying element, with lesser additions of magnesium being used occasionally. Manganese provides some dispersion strengthening via the precipitates $(Mn, Fe)Al_6$ and $(Fe, Mn)_3SiAl_{12}$, and magnesium adds some solid solution strengthening. Nevertheless, the alloys are of modest strength, and cold working is often used to enhance this. Relatively few alloys are available in this series and the corrosion properties, although acceptable in some alloys, may make protection necessary. The alloys are therefore of limited marine application.

The 4xxx alloys are also non-heat-treatable, strengthening being provided by the precipitation of silicon. These alloys are not widely available in wrought form, the principal uses being in welding electrodes, claddings and castings.

The 5xxx alloys are also non-heat-treatable, containing up to about 6 per cent magnesium, with occasional minor additions of manganese and chromium and a few others. The main mode of strengthening is by solid solution, although cold working and dispersion strengthening by Mg_2Al_3 are also used. Strength levels applicable for structural use can be developed, and the marine corrosion resistance of many of the alloys and treatments in this series have ensured their continued usage for marine structural applications, particularly alloy 5086. Atmospheric corrosion resistance is normally acceptable, although protection is usually advised for submerged applications. Splash zone corrosion rates tend to be higher than marine atmospheric rates, and crevices should be avoided.

The 6xxx alloys are age-hardenable using the Mg–Si intermetallics, and contain up to about 1.5 per cent of each of these elements, with minor additions of manganese, chromium or copper. Although these are not generally the highest-strength precipitation-hardening aluminium alloys, their seawater compatibility is quite good and there is a proven record of their successful use in marine structural applications.

The 7xxx alloys are also age-hardenable and use magnesium and zinc in relatively high quantities (up to about 8 per cent Zn and 4 per cent Mg), some alloys also containing substantial amounts of copper. These alloys are of the highest strength in the aluminium alloy range, and find major application in aerospace, where their toughness and resistance to stress corrosion has, however, always been a matter for careful consideration. It is reasonable to say that the same is true of marine applications, although field experience is not as widespread, and the additional possible problems of pitting, crevice, intergranular and exfoliation corrosion need to be considered.

Aluminium casting alloys are often rather different from their wrought counterparts, owing to the fact that some of the processing routes (for example cold working to produce enhanced precipitation) are not available for castings. The designation system is also different and not as standardised as for the wrought alloys. As one example, the American Aluminium Association's three-digit system is probably the most logical, since it is closest to the four-digit wrought alloy system:

1xx	> 99% Al
2xx	Al–Cu
3xx	Al–Si–Mg, Al–Si–Cu and Al–Si–Cu–Mg
4xx	Al–Si
5xx	Al–Mg
8xx	Al–Sn

As can be seen from the above, no commercial use is made of the Mg–Zn precipitation hardening system in castings, although some of the casting alloys are heat-treatable. Wide use is also made of silicon, often in quite high concentrations, which provides dispersion strengthening using the Al–Si eutectic. The major marine use of aluminium casting alloys is in the provision of galvanic anodes for cathodic protection. Several alloys are used [75], including binary Al–Zn alloys with 3 to 7 per cent zinc and ternary alloys with zinc and indium. Some alloys with zinc and tin have been reported [75] to suffer from intergranular corrosion at low temperatures, resulting in a drop in efficiency. Table 4.28 summarises the electrochemical behaviour of some aluminium-based galvanic anodes [76].

Table 4.28. Electrochemical data for some aluminium alloys used for galvanic anodes

Alloy	Environment	Driving potential (mV)	Current capacity (Ah/kg)	Consumption rate (kg/A year)
Al–Zn–Hg	Seawater (5 to 30 °C)	200 to 250	2600 to 2800	3.1 to 3.4
Al–Zn–In	Seawater (5 to 30 °C)	250 to 300	2500 to 2700	3.2 to 3.5
Al–Zn–In	Saline mud (5 to 30 °C)	150 to 250	1300 to 2300	3.85 to 6.7
Al–Zn–In	Saline mud (30 to 90 °C)	100 to 200	400 to 1300	6.7 to 22

Source: Det norske Veritas [76].

Table 4.29. Guide to the strengthening mechanisms used in aluminium alloys

Wrought alloys

Alloy	Condition	Strengthening mechanism	UTS (MPa)	Elongation (%)
99.999%Al	Annealed	No strengthening	45	60
99%Al	Annealed	'Impurity'-strengthened	90	45
99%Al	Cold-worked to 75% reduction	Strain-hardened	165	15
1.2%Mn	Annealed	Dispersion-strengthened	110	35
5%Mg	Annealed	Dispersion- and solid-solution strengthened	290	35
5.6%Zn–2.5%Mg	Precipitation treated	Precipitation-hardened	570	11

Casting alloys

Alloy	Treatment	Strengthening	UTS (MPa)
4.5%Cu–0.8%Si	Sand-cast and precipitation treated	Precipitation-hardened	248
5.2%Si	Sand-cast	Dispersion	131
	Die-cast	Dispersion and grain size	228
12%Si	Die-cast	Dispersion (eutectic)	297

Table 4.29 gives a rough guide to aluminium alloy strengthening mechanisms, although it is a summary and, for example, Polmear [77] should be consulted for more detailed information. Tables 4.30 and 4.31 contain some guidance on fracture, fatigue and corrosion data [78].

Aluminium alloys find a substantial number of uses in marine and offshore engineering, not least for structural applications. A few examples of large structures fabricated from aluminium alloys are described in [16], and it will be noted that many of these are concerned with weight saving or transportation applications. It should also be noted that there is no evidence (see, for example, West [79]) that aluminium or its alloys burn or produce fuel during fires. Indeed, the melting of aluminium (at between 550 and 650 °C) can even be of benefit in fires on board ships, by providing a vent for the smoke and toxic gases produced, for example, by the burning of some polymers [80].

4.1.6 Copper Alloys

Alloys based on copper are the traditional marine metallic materials, owing principally to the corrosion and fouling-resistant properties of copper, both of which are largely transferred to its alloys. In fact fouling and corrosion are related in copper alloys, and it is recognised (see Dexter [53]) that the application of cathodic protection, for example, will curtail the antifouling properties of copper-based alloys. The other major advantage of copper and its alloys in marine service is that the form of corrosion is usually very uniform, and can be designed against satisfactorily (see Chapter 3).

Because of its thermal and electrical conductivity, copper and its alloys find a number of other non-marine applications, and consequently not all copper alloys have been developed for marine use. Mechanically, since copper is a face-centred cubic metal, it is inherently ductile and is not particularly strong, so that some alloying or strain hardening is usually required to improve its performance for structural use. The balance of wrought and casting alloys is more even for copper than for many other alloy-basis metals, and this is reflected in the range of available alloys.

There are four principal types of copper alloys, the high coppers, cupro-nickels, brasses and bronzes, although a

Table 4.30. Fracture and fatigue data for some aluminium alloys

Series	Fracture	Fatigue
2xxx	Depends upon alloy and heat treatment. K_{Ic} usually in range 20 to 40 MPa\m.	Notch-sensitive but typically 120 MPa for 10^7 cycles.
5xxx	Toughness not normally a design consideration owing to low strength of alloys.	Dependent upon degree of cold work and alloy. Usually 80 to 140 MPa for 5×10^8 cycles.
7xxx	Depends on alloy and heat treatment, but K_{Ic} usually around 20 to 55 MPa\m.	Around 150 MPa for 10^7 cycles, depending on alloy and heat treatment.

Table 4.31.   Marine corrosion data for some aluminium alloys

Series	Marine corrosion
All	Most alloys susceptible to corrosion fatigue when immersed in seawater, fatigue strengths at 10^8 cycles in NaCl solutions being about 25 to 30% of those in air. All alloys galvanically active. Alloys tend to pit rather than show general corrosion, so assessment of rate is usually qualitative. Pitting tendency generally increases with depth.
2xxx	Not recommended for use in marine environments in unprotected condition. Maximum pit depths of 350 µm have been found in aluminium-clad 2014 after 10 years' submerged exposure.
5xxx	In well-aerated surface water pitting penetration is about 25 to 75 µm/yr over about 10 yr although not linear. In deeper water and in mud rates become larger and values between 125 µm/yr and 4 mm/yr have been reported. General weight loss in marine atmospheres is usually less than 1 µm/yr but this will again be distributed as pits.
6xxx	Rather similar to 5xxx alloys. Some temper conditions (e.g. T6) seem to show poorer performance.
7xxx	In surface water, penetration rate 280 to 500 µm/yr and increasing with depth. Higher temper conditions generally more susceptible. Atmospheric rates appear to be similar to 5xxx and 6xxx alloys, although data are somewhat scarce. Some alloys and tempers are susceptible to SCC in seawater (e.g. 7079 T6 at above 70 and 420 MPa in transverse and longitudinal directions respectively).

number of other more complex alloys occupy a space between these last two categories and they will therefore be considered together.

4.1.6.1   The High-copper Alloys

These alloys consist substantially of copper with only a few per cent of alloying additions, and therefore include the commercially pure coppers. Most of the high coppers are conductivity alloys with chemical specifications based on those impurities which affect thermal or electrical properties. These classes of alloys, if strengthened at all, are usually strengthened by cold working so as to produce minimum disruption to the conductivity.

Small additions of cadmium, beryllium and chromium can be made to copper to produce age-hardenable alloys, and mechanical properties can also be improved by small additions of iron or phosphorus.

The high coppers do find some applications in marine environments, but these are usually as small parts.

4.1.6.2   The Cupro-nickels

The copper–nickel alloy system shows unlimited solid solubility, so the cupro-nickels are single-phase alloys employing only solid solution strengthening. Because there comes a point at which the alloy ceases to be a copper alloy strengthened by nickel and becomes a nickel alloy strengthened by copper (see Section 4.1.7), there will be a maximum in the curve of strength vs. nickel content.

Because of their homogeneity and hence improved corrosion resistance, their excellent weldabililty and their improved corrosion–erosion resistance, the cupro-nickels have supplanted brasses from a number of the latter's more traditional applications.

The basic cupro-nickel compositions are based on the 90% Cu–10% Ni and 70% Cu–30% Ni alloys, although other nickel contents up to 40 per cent are used occasionally. Addition of up to 2 per cent iron can usually be made to increase strength without compromising corrosion resist-

ance. Two important alloys for condenser tubes and seawater piping are the 90/10 (with 10% Ni and 1.5% Fe) and the 70/30 (with 30% Ni and 0.5% Fe), and these offer good corrosion and fouling resistance with easy fabricability. The higher cost of these alloys over other competitors is usually justified by higher permissible water speeds and improved reliability. A range of high-strength cupro-nickels is available using precipitation hardening with the intermetallic compound $Ni_3(Al, Ti)$ achieved by small additions of aluminium or titanium, usually to 1 or 2 per cent and further additions of iron and manganese can also be made for added solid-solution strengthening. Relatively few casting alloys are available in this class, but there are some alloys under development for naval applications. One such reported alloy based on 30% Ni and 2% Cr with other minor additions is being developed as a replacement for nickel–aluminium bronze (Powell [80]) in submarine seawater handling systems. Although some of the mechanical properties were superior to the competitor nickel–aluminium bronze, the main advantage was in the corrosion and crevice corrosion resistance, allowing for components which will last for the lifetime of the vessel. Copper–nickel alloys are also used as cladding for marine structural steels, and one notable case in UK offshore waters uses cupro-nickel sheathing for the splash zone area of the legs of an offshore structure (Carruthers [81]).

4.1.6.3   Brasses and Bronzes

Traditionally, brasses and bronzes have been considered to be alloys based on the copper–zinc and copper–tin alloy systems respectively. The addition of the term 'aluminium bronzes' to describe the copper–aluminium alloy series complicated this somewhat, and now terms such as manganese brasses, manganese bronzes and nickel aluminium bronzes can also be heard. Furthermore, zinc can be used to replace tin in copper–tin alloys, for instance, so it is probably simplest to describe such alloys in terms of their principal alloying element and, if there is more than one, to refer to these as 'complex' alloys. The following discussion therefore deals first with alloys which are mainly copper–zinc, copper–tin and copper–aluminium and then the more complex alloys together.

4.1.6.3.1 Copper–Zinc Alloys

Copper–zinc alloys, or true brasses, have been in use for many years in marine environments, particularly in naval applications. Like many copper alloys, the phase diagram is a complex one, showing a number of solid–solid reactions (e.g. [46]).

However, the relative sizes of copper and zinc (affecting diffusion coefficients), combined with the relatively low temperatures of these solid–solid reactions, usually means that they rarely take place unless under prolonged heating or very slow cooling. Most useful copper–zinc alloys are based on the 70 per cent copper, 30 per cent zinc or 60/40 alloys.

70/30 alloys, commonly called 'yellow brasses', are single-phase alpha alloys. Additional alloying elements include lead for improved machinability or fluidity for castings, and those required to eliminate or mitigate the effects of dezincification (see Chapter 3). The addition of 1 per cent tin provides a slight improvement in dezincification resistance (previously thought to be more significant [83]), hence the name 'Admiralty brass' for this alloy. A much more effective solution to the problem of dezincification is the addition of 0.02 to 0.06 per cent arsenic. 'Aluminium brass', produced by adding 2 per cent aluminium to a nominal 76/24 alloy, with a small arsenic addition as above, provides a corrosion-resistant alloy which is also resistant to velocity effects, making it particularly useful for condenser tubes.

60/40 brass, sometimes known as 'Muntz metal', is two-phase and is thus dispersion-strengthened by the strong, but brittle, beta phase. It also provides the basis of many of the more complex high-tensile brasses. Lead is quite often used in these alloys to improve machinability. Because of their duplex nature, 60/40 brasses are, if anything, more prone to dezincification than 70/30 alloys, but the addition of 1 per cent tin to produce 'Naval brass' provides some improvement. This alloy finds some use in marine hardware and tube plates in condensers, but should be exposed to seawater only with caution. Dezincification resistance for some applications can be conferred by adding arsenic and carefully controlling composition [82].

4.1.6.3.2 Copper–Tin Alloys

As with the copper–zinc system, the copper–tin alloy phase diagram shows a number of solid–solid reactions, and again these are often inhibited at low temperatures. This means that alloys up to about 10 per cent tin consist at room temperature of single-phase alpha. Phosphor bronze is based on this composition and contains an addition of 0.2 per cent phosphorus to improve bearing properties. Since zinc is rather cheaper than tin, replacement on the basis of 2 per cent zinc for every 1 per cent of tin is often made.

Many of the commercial copper–tin alloys are cast and are dispersion-strengthened by delta-phase. These casting bronzes are often called 'gunmetals' or 'G-bronzes'. The two most important compositions are 5% Sn–5% Zn–5% Pb and 10% Sn–2% Zn, and these find widespread marine and non-marine uses for pump and valve bodies and parts.

4.1.6.3.3 Copper–Aluminium Alloys

The copper–aluminium alloys are almost universally known as 'aluminium bronzes' and, unlike the previous two alloy series, some exploitation is made of the phase changes in this system. The relevant phase diagram is shown in Fig. 4.29.

The low-temperature reaction is rare, but the $\beta \rightarrow \alpha + \gamma_2$ eutectoid can be used in heat treatments. Useful alloys normally have less than about 11 per cent aluminium. The main group of wrought alloys contain about 5% Al and are single-phase alloys, often homogenised by cold working

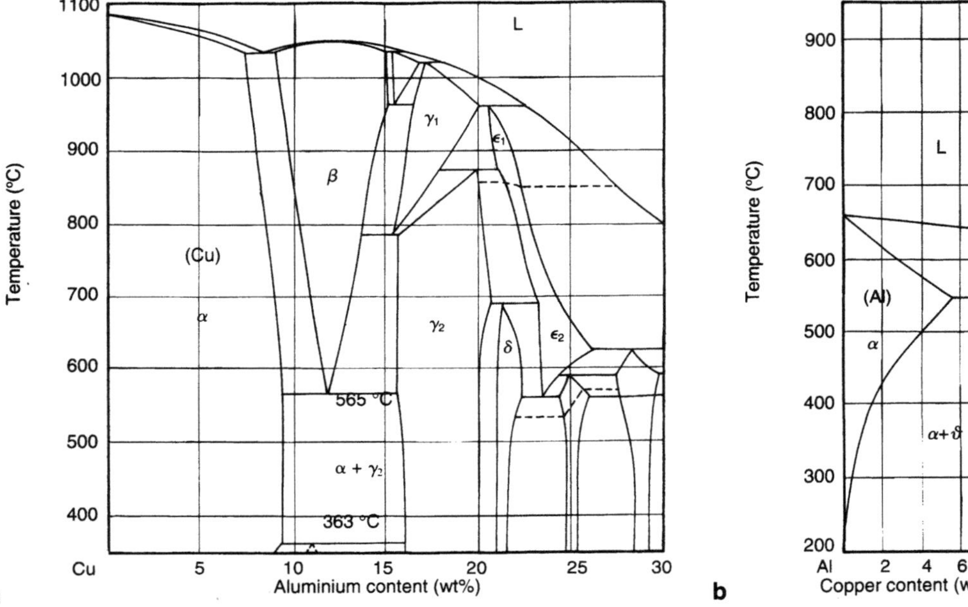

Fig. 4.29. The copper–aluminium phase diagram: **a** aluminium bronzes and **b** aluminium–copper alloys [46].

Table 4.32. Broad description of the strengthening effects in copper alloys

Alloy	UTS (MPa)	Elongation (%)
Pure copper, annealed	210	60
Commercially pure copper, cold worked	390	4
2% beryllium, precipitation-treated	1310	4
10%Ni–1.5%Fe	385	40
30%Ni–0.5%Fe	415	40
Precipitation-strengthened cupro-nickel	740 (typ)	25
30% zinc, annealed	300	70
30% zinc, cold-worked	450 to 600	15 to 10
40% zinc, hot-worked	500	17
10%Sn–2%Zn, cast	309	20
5%Sn–5%Zn–5%Pb, cast	215	15
7% aluminium, cold-worked	620	18
10% aluminium, quenched and tempered	630	48
Complex manganese bronze, cast	650 (typ)	18
Complex aluminium bronze, hot-worked	750 (typ)	20

and annealing. The two-phase alloys contain about 10% Al, and these can be cast or wrought, in which latter case they are normally hot-worked. At elevated temperatures (about 600 °C) these alloys consist of an alpha–beta mixture, and the beta transforms to an $\alpha + \gamma_2$ eutectoid on moderate cooling rates. Quenching suppresses this reaction, and produces an acicular structure like martensite in steel. This can be tempered to precipitate finely dispersed γ_2 in a similar way to that found in the tempering of martensite. The presence of aluminium provides very good corrosion and also erosion resistance, and the heat-treatability of these alloys ensures a wide range of applications including tanks and tubing for the wrought alloys and pumps, impellers and propellers for the casting alloys [83].

4.1.6.3.4 More Complex Alloys

Almost all of the developments into more complex copper alloys have been aimed at increasing strength and reducing the quantity of expensive alloying elements, without substantially compromising corrosion resistance. Also, many of the complex alloys are casting alloys, mainly because of the fact that strengthening options are more limited than for wrought alloys, leaving alloying as one of the most important methods.

The principal alloying elements besides those already discussed are manganese, nickel, silicon and iron, although many others are used in varying quantities.

Starting with a copper–zinc alloy, a number of variants are possible. The term 'manganese bronzes' has been used for copper-zinc alloys (brasses) which have been strengthened by adding up to 4 per cent manganese, and more complex alloys can be ranked using a zinc equivalent which quantifies their susceptibility to form the β-phase:

Silicon	10
Aluminium	4 to 6
Tin	2
Magnesium	2
Lead	1
Iron	0.9
Manganese	0.5
Nickel	– 1.2

In general, such 'high-tensile brasses' are strong, but can be susceptible to stress corrosion. Large additions of nickel (about 20 per cent) produce the so-called 'nickel silvers'.

Apart from the additions of lead and phosphorus mentioned above and the replacement of some tin with zinc, there are relatively few other variants on the copper–tin alloys. A series of bearing bronzes is available by adding between 8 and 30 per cent lead, and these sometimes have a small addition of nickel to improve distribution of lead globules.

A range of complex aluminium bronzes containing manganese iron and nickel is available in both wrought and casting forms. The cast nickel–aluminium bronzes and nickel–aluminium–manganese bronzes are useful for applications where improved erosion resistance over manganese bronze is required.

Table 4.32 gives a rough guide to the strengthening effects in copper alloys, illustrating the effects discussed above; more details on mechanical properties and compositions are given in, for example, [46]. Tables 4.33 and 4.34 summarise fatigue and corrosion data respectively [84].

Table 4.33. Fatigue properties of copper alloys (Fracture is not generally a major design consideration so data are uncommon)

Alloy type	Fatigue considerations
High copper	For 'pure' copper generally around 100 MPa for 10^8 cycles. Susceptible to water vapour. Beryllium copper 220 to 290 MPa for 10^8 cycles.
Brass	Wrought alpha brasses 70 to 280 MPa for 10^8 cycles depending on grain size, degree of cold work and alloy.
G-bronze	Around 90 MPa for 10^8 cycles.
Nickel silver	140 to 170 MPa for 3×10^8 cycles depending upon degree of cold work.
Complex alloys	Around 120 MPa at 10^8 cycles for manganese bronze. Silicon brasses and bronzes around 150 MPa at 10^8 cycles. Complex aluminium bronzes 170 to 220 MPa at 10^8 cycles.

Table 4.34. Marine corrosion data for a selection of copper alloys

Alloy type	Marine corrosion considerations
High coppers	Submerged corrosion generally uniform and relatively independent of depth. Rates in the region of 2.5 to 15 μm/yr may initially be somewhat larger. Susceptible to velocity effects and crevice corrosion. Atmospheric metal loss 1 to 40 μm/yr.
Cupro-nickels	Submerged rate 2.5 to 40 μm/yr. Susceptible to pollution but relatively resistant to velocity effects.
Admiralty brass	Submerged rate 12.5 to 50 μm/yr. More resistant than many other copper alloys to hot-spot attack in cooling applications using seawater containing hydrogen sulphide. Atmospheric rate in the region of 2 μm/yr or less.
Aluminium brass	Most interest in heat exchanger applications. Submerged rate 12.5 to 50 μm/yr. Relatively resistant to hot-spot attack in seawater containing hydrogen sulphide.
G-bronze	Submerged rate 25 to 50 μm/yr. Some resistance to velocity effects.
Aluminium bronze	Submerged rate 50 to 60 μm/yr. Good resistance to velocity and cavitation effects. Atmospheric rate 1 to 16 μm/yr depending upon pollution.

4.1.7 Nickel Alloys

The single outstanding feature of nickel and its alloys is corrosion resistance, even at high temperatures and in marine environments. On the other hand, nickel is a relatively expensive metal, so that a sound reason is normally required before contemplating its use as a basis metal over other corrosion resistant alloys.

Alloying of nickel is a relatively simple matter. Since toughness is rarely a problem and additions to improve corrosion resistance are not normally required, the impetus for alloying is to increase strength and, in some cases, to develop specific properties as in the nickel-based superalloys. Nickel alloys are normally strengthened by solid solution or occasionally by precipitation hardening, although a special range of high-temperature alloys, often referred to as the 'superalloys', are strengthened by a mechanical suspension of refractory oxides. Nickel work hardens very effectively and this can make it very difficult to machine, but it can also be used to advantage as in burnished surface coatings.

The principal alloy series are provided by copper and chromium, the latter often including iron or molybdenum.

The 'Monels', or nickel–copper alloys, are a continuation of the cupro-nickel alloy series described in Section 4.1.6. There are two principal alloys, known as 400 and K500. The 400 alloy contains 30 per cent copper with some minor additions of iron, manganese and carbon for strengthening. The K500 alloy is also based on 30 per cent copper, but with aluminium and titanium additions to allow precipitation hardening with the phase $Ni_3(Al, Ti)$. Cast forms of both of these alloys are available and both wrought and cast forms find marine uses in shafts and pump bodies, impellers, fasteners, marine exhausts and even for splash zone protection as a cladding material. The galvanic compatibility with copper alloys is such that the nickel alloy will be cathodically protected by any copper alloy connected to it. This can be exploited by taking advantage of galvanic size effect in pump body–impeller combinations where the (nickel-based) impeller is the critical component and is relatively small compared with the body, so that wall thickness loss in the (copper-based) body is insignificant.

Nickel–chromium alloys were specifically developed for high-temperature applications where retention of strength, creep and oxidation properties are essential. Aluminium–titanium precipitation hardening and/or refractory oxide dispersion are widely used in these alloys. One such alloy, 625, containing about 20 per cent chromium and 10 per cent molybdenum, has found some marine applications owing to its excellent resistance to pitting and crevice corrosion. The most significant of these are in high-speed propellers, in critical heat exchanger components and also as a cladding deposited by weld overlaying. A still more resistant range of alloys, the 'Hastelloys', with slightly less chromium and a large addition of molybdenum (typically around 15 per cent), with significant additions of cobalt and tungsten, provide the ultimate in corrosion resistance for critical marine and petrochemical applications, where extreme reliability is required with regard to corrosion resistance.

The somewhat less expensive alloy 825, with about 30 per cent iron and 20 per cent chromium, with copper and molybdenum additions, was developed to replace stainless steels where there was a difficulty of pitting and stress corrosion cracking. Typical stainless steel replacement

Table 4.35. General description of strengthening mechanisms in nickel alloys

Alloy	Strengthening mechanism	UTS (MPa)
Ni (commercially pure, annealed)	—	350
Ni (commercially pure, cold-worked)	Work hardening	650
Ni–32%Cu/1%Fe/ 1%Mn (annealed)	Solid solution	550
Ni–29%Cu/3%Al/ 0.6%Ti (aged)	Solid solution and precipitation	1050
Ni–15%Cr/2%Mo/ 2.5%Ti/4.5%Al/ 4%W/2%Ta/ 1.1%Y_2O_3	Dispersion with refractory particles	Developed for creep resistance

Table 4.36. Fracture and fatigue properties of nickel alloys

Alloy type	Fracture	Fatigue
Ni–Cu	CVN values for solid solution alloys in region of 270 J, even at very low temperatures. For aged alloys values from 40 to 100 J.	140 to 280 MPa for 10^8 cycles, depending upon heat treatment and degree of cold work.
Ni–Cr–(Fe)	For lower-iron alloys 200 to 300 J, depending upon cold work. For higher-iron alloys values are lower.	220 to 350 MPa for 10^8 cycles, depending mainly upon whether hot-worked, cold-worked or annealed.
Ni–Cr–Mo	70 to 40 J CVN, depending a little on temperature.	480 to 690 MPa for 10^8 cycles in air. About half these values in seawater.

applications include expansion bellows for piping, exhausts and valves.

Table 4.35 summarises the behaviour of nickel alloys with regard to strengthening, and Tables 4.36 and 4.37 provide some data on corrosion resistance and fatigue and fracture. As before, the reader is referred to Chapters 2 and 3 before making use of these latter data.

4.1.8 Other Alloys

A number of other alloy types and series find minor and not-so-minor uses in the marine environment, and some of these are summarised in the following.

4.1.8.1 Titanium Alloys

Titanium is an expensive metal and, as with nickel alloys, the use of titanium alloys normally requires there to be a distinct technical advantage. Titanium is typified by very high stiffness:density and strength:density ratios and hence is a major candidate material for airframes. Its marine uses are due only partly to this, and more to its outstanding cor-

Table 4.37. Marine corrosion properties of nickel alloys

Alloy type	Marine corrosion considerations
Ni–Cu	Atmospheric rates very low, 0.1 to 1 μm/yr, depending upon degree of pollution. In immersed conditions rates less than 250 μm/yr maintained to moderate velocities. In stagnant conditions fouling may cause some pitting at a rate of penetration of about 200 μm/yr.
Ni–Cr–(Fe)	Atmospheric rate negligible. General corrosion rate low in flowing seawater, but crevicial and subfouling pitting penetration may be rapid (up to 1.5 mm/yr) under stagnant conditions. Rates in hydrogen sulphide solutions low even at elevated temperatures.
Ni–Cr–Mo	Virtually inert under all marine conditions.

rosion resistance, although there are occasional reports of crevice corrosion and stress corrosion [77]. A wide range of strength is available by control of microstructure through alloying and heat treatment, although the strength of commercially pure or lightly alloyed titanium is quite good, so that few applications for the higher alloys are found outside aerospace.

Titanium is metallurgically quite an interesting metal, since it shows an allotropic transformation from hexagonal close-packed alpha phase (a relatively uncommon structure among engineering alloys) to body-centred cubic beta phase at 882 °C. This means that titanium alloys can consist of single-phase alpha (which can be strengthened by solid solution), duplex alpha–beta, or (less commonly) single-phase beta. Also, martensitic structures can be formed and age hardening is possible in some alloys.

Typically, marine applications include heat exchangers on offshore oil production facilities for aggressive conditions and in power station condenser tubes, for example in polluted estuarine sites (Heaton et al. [85]). An interesting example is given by Dexter [53] of the use of a titanium 6% Al–2% Nb–1% Ta–0.8% Mo alloy for the main personnel pressure spheres for the deep-submergence three-man scientific submersible *Alvin*. This alloy was chosen as a moderate-strength alloy over the higher-strength Ti, 6% Al–4% V alloy owing to the few reports of SCC of the latter in seawater. The submersible also used unalloyed titanium for its frame and the 6% Al–4% V alloy for much of the piping.

Although annealed, unalloyed titanium has a yield stress of about 170 MPa, alloys are available with yields as high as 1300 MPa, with 900 MPa being relatively easily obtained. Details of compositions and properties of titanium alloys can be found in Polmear [77].

4.1.8.2 Magnesium Alloys

Magnesium is the third (along with aluminium and titanium) of the 'light-alloy' basis metals and is in fact the one of lowest density. Magnesium is, however, a rather weak metal, does not respond particularly well to strengthening and also has a low Young's modulus (45 GPa) for a metal.

The corrosion resistance of magnesium is generally poor, approaching that of mild steel in marine atmospheres for the most susceptible alloys. Impurity content is a very important factor in determining resistance, and some alloys are susceptible to stress corrosion cracking in dilute chlorides. Some form of protection is required in immersed conditions and, of course, cathodic protection by galvanic anodes is not an option, owing to the highly electronegative behaviour of magnesium.

When all the above factors are considered there is little impetus to develop magnesium alloys for any other use than as a galvanic anode material. For this application, magnesium metal can be used, but the alloys AZ63 and M1A are more likely to be used, since they offer higher relative potential drops. Even so, some care to avoid overprotection or excessive consumption is required. In fact, it is relatively uncommon to find magnesium alloys in use as galvanic anodes in seawater, although they can be advantageous in other environments.

Magnesium alloys rarely develop strengths (0.2% proof) of more than 200 MPa, and 100 MPa is a more common figure. Further details can be found, for example, in Polmear [77].

4.1.8.3   Zinc Alloys

Zinc is also a highly electronegative metal, a property which it shares with aluminium and magnesium. Unlike magnesium, zinc is roughly as dense as steel, but like magnesium it is weak, although it responds rather better to strengthening. Its main general engineering uses as a basis metal revolve round its low melting point and hence ease of casting. However, corrosion resistance is fairly limited in seawater and, like magnesium, the main marine use of zinc is as galvanic anode material.

Zinc is, in fact, the commonest cathodic protection anode material and has the longest history. The main problem is to ensure an adequate, though not excessive, corrosion rate, in particular avoiding intergranular corrosion, which may cause deterioration of the integrity of the anode or the formation of adherent oxide which may prevent its proper operation. Commercial-quality zinc has iron as its main impurity, and this causes excessive polarisation [75]. The avoidance of this problem usually requires the control of the iron content to very low levels or the use of a controlled iron content along with small amounts of alloying additions, typically 0.5 per cent aluminium with up to 0.1 per cent silicon and/or cadmium. Mercury is added as an activator to some alloys to further reduce polarisation. DnV [76] suggest that zinc alloys for cathodic protection in seawater should conform to the compositional limits given in Table 4.38, and this table also compares the electrochemical behaviour of zinc alloys with aluminium alloys for galvanic anode applications.

Zinc alloy ultimate tensile strengths are usually in the region of 200 MPa, with higher strengths only being obtained with substantial alloying additions (e.g. 25 per cent aluminium) or with cold working.

4.1.8.4   Others

A number of other specialised applications can be noted, such as noble and refractory metals for electrical applications and lead as ballast.

4.2 Polymers for Marine Use

Polymeric materials have been used for many years in marine environments, although often in a semi-disposable way. Their great advantage over metals is that they do not suffer from corrosion in the classic sense, although they are weak and compliant and may be susceptible to other degradation processes, as indicated in Chapter 3. The low density, corrosion resistance and insulating properties of polymers assure them an important place in engineering use. Their main disadvantage (lack of strength and stiffness) means that polymers find few structural applications, but when combined with stronger, stiffer components in composites they can provide structural properties approaching the metals and concrete. Such materials are, however, dealt with separately in Section 4.4.

Polymers do not have microstructures in the same way as do metals, or indeed concrete and ceramics, and their classification is normally carried out on the basis of molecular structure and molecular interaction because these are the primary determinants of the mechanical properties.

The broad types of structural classifiers (e.g. Mills [86]) which can be used for polymers are thermoplastics, thermosets and elastomers, each of which imples a specific type of molecular structure, although other sub-classifiers to do with chemical composition and manufacture such as chain length, degree of branching, polymerisation mechanism and degree of crystallinity are all important. The following detailed discussion is divided among the three former categories, but for the general discussion below each of these will be defined.

Because of the structural differences between polymers and the (strictly) crystalline solids, polymers are able in general to dissolve molecules (rather than atoms), and this without substantial activation. This means that properties such as the absorption of gas and water and the permeability of the polymer become important, and these will be indicated in the ensuing detailed discussion where appro-

Table 4.38.   Zinc alloys for cathodic protection

a   Compositional requirements

	Max. %	Min. %
Aluminium	0.2	0.1
Cadmium	0.06	0.03
Iron	0.002	
Copper	0.005	
Silicon	0.125	
Lead	0.006	
Zinc	Remainder	

b   Electrochemical performance (see also Table 4.28)

Alloy	Environment	Driving potential (mV)	Current capacity (A h/kg)	Consumption rate (kg/A yr)
Al-based	Seawater	200 to 300	2500 to 2800	3.1 to 3.5
Al-based	Saline mud	100 to 250	400 to 2300	3.85 to 22
Zinc	Seawater	200 to 250	760 to 780	11.2 to 11.5
Zinc	Saline mud (0 to 60°C)	150 to 200	760 to 780	11.2 to 11.5

Source: Det norske Veritas [76].

priate. In particular, Case Study 7.5 includes some discussion of barrier properties of polymers.

Thermoplastics are perhaps the simplest of polymers in that their structure consists of unconnected long-chain molecules. The defining feature of thermoplastics is that, once made, the polymer can be melted and moulded to shape. Processing is therefore relatively simple in that the polymer can, in principle, be recycled *ad infinitum* by simply heating it to softening temperature and subjecting it to the appropriate moulding process.

Thermosetting plastics have a rigid three-dimensional structure which is formed by the mixing of two components and/or by the application of heat and pressure. The defining feature is not, however, that the structure is three-dimensional (albeit associated with this), but that, once formed, this structure cannot be modified, so that the polymer is set into its final shape, which can only be changed by cutting. Reheating of such polymers will only result in irreversible degradation.

Elastomers, or rubbers, consist fundamentally of coiled (mostly linear) chains resulting from the asymmetry introduced usually by a *cis-* bond, so that their structure allows for very large elastic deformations. Viscous sliding of the chains can also occur, with consequent time-dependent properties, and most modern elastomers are cross-linked so that this sliding can be inhibited, giving a semi-three-dimensional structure.

In terms of application, the thermosets can bond well to other materials (owing to cross-linking) and are therefore to be preferred as coatings and in composites whereas the unique mechanical properties of the elastomers secure their place in a specific set of circumstances. The major failings of thermoplastics, namely their limited temperature resistance and low strength and stiffness compared with thermosets, are gradually being overcome and new manufacturing methods are helping to develop the use of continuous-fibre reinforcement for thermoplastics. Thus, the market in thermoplastics is gradually increasing at the expense of that of thermosets, and the ease of recycling of thermoplastics makes this a desirable situation.

Polymerisation, whether of thermoplastics, thermosets or elastomers, can be achieved by addition, where individual monomer segments join together without the production of any other compound, or by condensation (or step growth) where another molecule is eliminated as a by-product. Important characterisers of the polymerisation reaction are chain length, degree of branching and stereoregularity.

Also, it is sometimes possible to produce copolymers where two or more different monomers combine to produce a polymer. The disposition of the two components along the chain can be random or in blocks, and branching both with respect to copolymer and randomly are also possible. Furthermore, another dimension of variability can be introduced, associated with the disposition of pendant groups and this, along with stereoregularity, means that the range in properties of copolymers can be very wide indeed.

Polymer blends, or alloys, should be seen as distinct from copolymers. These former are simply mixtures of two or more polymers, although there will be some bonding across the interfaces of the individual particles.

Polymers can also be classified in terms of their physical state, which can be semicrystalline or amorphous.

In the semicrystalline state, chains conform into groups which are lamellar crystals. The individual crystal lamellae are around 15 nm thick and about a micrometre long, but these do not grow together as in metal crystals; rather, they are linked with amorphous material into a spherulite, and it is these spherulites which impinge upon each other, producing polyhedral spherulite boundaries. The spherulites will usually fill the entire space of the solid, and the overall degree of crystallinity (i.e. amount of crystalline material) can be in the range of 40 to 90 per cent.

The amorphous state is characterised in the solid by the existence of a glass transition. This transition can be regarded as a further change in the mobility of the polymer chains after solidification into the rubbery state. Just after solidification, the amorphous polymer has sufficient rigidity to maintain its shape, but the degree of entwinement of the chains means that it can be very easily deformed by relative movement of the chains, as well as by elastic deformation. As the temperature becomes lower, the amount of viscous deformation at a given stress becomes less and the polymer behaves in a rubbery manner. At the glass transition temperature, a degree of local ordering is thought to take place; the main feature of the glassy state, however, is that the molecules have little capacity to change shape, and therefore deformation is inhibited, leading to glass-like properties.

Although most thermoplastics are used in the rubbery (or, of course, the crystalline) state, most thermosets have glassy properties over their entire operating range. On the other hand the glass transition and the temperature at which viscous deformation becomes unacceptably high normally define the operating temperature range of a thermoplastic or an elastomer. There is a tendency for polymers of higher melting temperature to have higher glass transition temperatures, so that the useful temperature range is not necessarily extended by having a high-melting polymer, and it is important for designers to remember this. Thus, key temperatures for polymers are the softening, crystalline melting or decomposition temperatures at the upper end and glass transition at the lower end. Figure 4.30 illustrates important temperature considerations for a typical thermoset, elastomer and thermoplastic.

The following discussion is divided between thermoplastics, thermosets and finally elastomers, concentrating on those members of each group which are widely used in marine environments but excluding those which are used

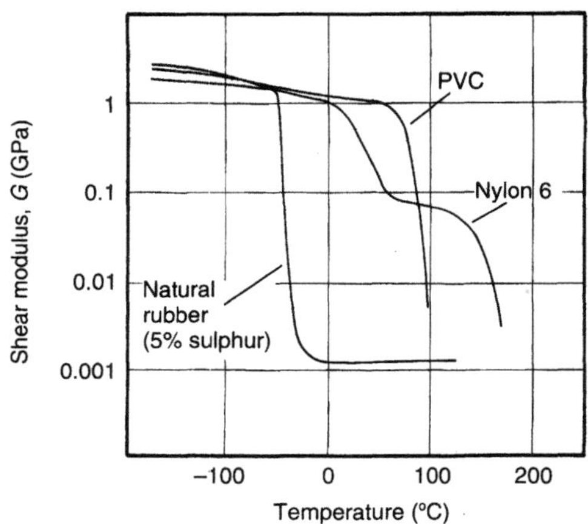

Fig. 4.30. Thermal behaviour of three important engineering polymers. (From McCrum et al. [87].)

Table 4.39. Relative consumptions of the three classes of plastics materials

Plastic class	UK consumption per annum ('000 tonnes)
Thermoplastics	1954
Thermosets	632
Elastomers	940

Source: Mills [86].

as coatings (dealt with in Section 4.5.3). In addition, many polymers contain additives other than those which provide reinforcement (composites are dealt with in Section 4.4), and these are mentioned briefly in the discussion for each polymer, but only as to function.

The final complication is that in the polymer industry (far more than in metals or other materials) trade names abound, but these are avoided here except where the trade name (e.g. nylon) is so widely accepted that the chemical name is almost unrecognisable. Apart from these obvious cases, the engineer is advised to enquire of manufacturers using unfamiliar trade names as to what chemical class any polymer under consideration belongs.

Table 4.39 indicates the relative usages of thermoplastics, thermosets and elastomers in the UK (1985 data, [86]) and it can be seen that thermoplastics dominate the market. However, some care should be taken in interpreting these figures in the context of the subject of this book, owing to the very high usage of plastics (particularly thermoplastics) for 'non-engineering' applications such as food packaging and clothing. Nevertheless, the three types of polymers will be discussed in the order in which they appear in this table.

4.2.1 Thermoplastics

The most convenient way to classify the thermoplastics is by their chemical make-up, although it must be realised that degree of crystallinity, degree of polymerisation, amount of branching and other factors may affect the properties of an individual polymer type, and the existence of polymer mixtures and copolymers further complicates the picture. These factors will be discussed in the following in so far as they affect each of the individual polymer types.

Table 4.40 shows data for thermoplastics use in the UK at 1985, and it can be seen that there are four thermoplastics for which there is a very high demand, and six so-called 'engineering thermoplastics' which are in relatively high demand for their superior engineering properties (mostly mechanical). Finally, there is a range of other thermoplastics with specific properties. The prices of such materials largely reflect the demand, so that the 'commodity' thermoplastics tend to be cheapest, with the 'speciality' thermoplastics being the most expensive. There can be a factor of ten or more difference in cost between the cheapest and most expensive thermoplastics [86].

The thermoplastics are discussed below according to usage, although many of the applications of the 'big four' are in such areas as packaging. The 'engineering' thermoplastics are discussed as a group and other important thermoplastics are dealt with in this group. Finally, polymer blends and copolymers are briefly discussed.

4.2.1.1 Polyethylenes

These polymers have probably the highest volume consumption of any plastics in the world, and they are also probably the simplest. They are made up of the simple repeating unit C_2H_4 and the polymerisation reaction can be written:

$$n(CH_2{=}CH_2) \rightarrow [-CH_2-CH_2-]_n$$

The polymer is usually categorised as 'high-density' (HDPE), 'low-density' (LDPE) or 'ultra-high molecular weight' (UHMWPE), reflecting the influence of molecular weight (i.e. degree of polymerisation) and degree of branching on the properties of this polymer. In fact there are hundreds of grades of polyethylene available based mainly on the following variables:

Degree of short-chain branching

Degree of long-chain branching

Variation in average molecular weight

Variation in molecular weight distribution

Presence of small amounts of comonomer or polymerisation residues and impurities

Compounding and cross-linking

Table 4.40. Consumption of significant thermoplastics in the UK (1985)

Plastics	Abbreviation	Physical state	Consumption (1000 tonnes)
Polyethylene	LDPE	Semicrystalline	570
	HDPE	Semicrystalline	250
Polypropylene	PP	Semicrystalline	325
Polyvinylchloride	PVC	Glassy	450
Polystyrene	PS	Glassy	180
Acrylonitrile butadiene styrene terpolymer	ABS	Glassy	55
Polyamides (nylon)	PA	Semicrystalline	21
Polyethylene terephthalate	PET	Semicrystalline	
Polybutylene terephthalate	PBT	Semicrystalline	65
Polymethylmethacrylate (acrylic)	PMMA	Glassy	27
Polyoxymethylene (acetal)	POM	Semicrystalline	11

Source: Mills [86].

The phenomenon of branching in polyethylene can occur during some of the manufacturing methods, and it is thought that these branches result from the formation of extra methyl groups by a 'back-biting' mechanism.

Such short-chain branches inhibit the close packing of molecules associated with crystallisation, and polyethylene containing significant branching is usually termed 'low-density' because of the effect of branching on the density of the polymer. Other effects of short-chain branching are reduced opacity, lower melting point and lower strength, lower Young's modulus and lower hardness.

Longer branches are also possible (usually only a few per chain, but these can be as long as the original chain), probably arising from a transfer reaction, and these give a large distribution in molecular weight. The so-called linear low-density polyethylenes are free of such branches, and the advantage of this is a reduction in melt viscosity, allowing a higher degree of processability.

The principal effect of increased molecular weight is on strength properties at high strain, for example the UTS, since the higher degree of entanglement only makes itself felt at such high molecular displacements.

Molecular weight distribution is normally characterised by M_w/M_n which is the ratio of weight average molecular weight to number average molecular weight. Aside from the possible effect of long-chain branching, a decrease in M_w/M_n is found to increase toughness, tensile strength and resistance to softening.

Ethylene can be copolymerised with, for example, propylene and vinyl acetate, but these will be dealt with under the section on mixtures and copolymers. Smaller additions of other alkenes can be made during the polymerisation reaction to produce controlled amounts of branching, and others can be used to impede crystallisation.

Cross-linking can be achieved in polyethylene by radiation, by peroxide curing or by the use of vinyl silane. The effect of cross-linking of the chains is to inhibit crystallisation with the corresponding mechanical penalty which accrues from that. However, the cross-linking ensures that a measure of strength is maintained above the crystalline melting point and hence the major application in thermal or electrical sheathing materials. Cross-linked HDPE foam has found a number of applications, notably as a flotation medium for oil-carrying and dredging hose.

According to Brydson [88], the T_G in polyethylene has little physical significance owing to the crystalline nature of the polymer (values from −130 °C to +60 °C can be found in the literature). In fact, the major temperature consideration for polyethylene is the crystalline melting temperature, which, depending upon the detailed structure, is usually between 108 and 132 °C. This low melting temperature is to be expected for a polymer with no strong intermolecular forces, and most of the strength results from the molecular packing associated with the crystalline state.

The resistance of polyethylene to the environment is good, as one might expect from what is essentially a paraffin of high molecular weight, and this points to potential marine applications; a collection of relevant properties of polyethylene is shown in Table 4.41.

4.2.1.2 Polyvinyl Chloride (PVC)

PVC is another simple addition polymer produced by the reaction

$$n(C_2H_3Cl) \rightarrow [—C_2H_3Cl—]_n$$

although, of course, the raw feedstock of vinyl chloride monomer (VCM) needs first to be prepared, unlike that of ethylene, which is obtained directly by cracking of natural-gas products. In fact, the commonest way to produce VCM is by the chlorination of ethylene.

The molecular structure of PVC suggests that it might be similar in properties and structure to PE. This is not so; the major feature of PVC is its inherent instability, and it is virtually useless without additives.

Apart from its effects on stability, the replacement by chlorine of one of the hydrogen atoms in the mer results in an increased attraction between chains and hence increased hardness and stiffness. The question of tacticity does not arise in PVC, since it appears that polymerisation takes place through a head-to-tail linkage as opposed to head-to-head and tail-to-tail linkage.

However, the most fundamental difference between PVC and PE is that the former is amorphous, with a glass transition temperature of about 80 °C.

The main difference between the individual grades is in the average molecular weight, the weight distribution and the particle size, shape and shape distribution. Chains are normally only slightly branched. The particle structure is important because of the low degree of crystallinity of the as-polymerised material. This consists of particles which are

Table 4.41. Some properties of polyethylene

	Low density	Medium density	High density	High molecular weight
Elastic modulus (MPa) – tension	140 to 186	170 to 380	414 to 1240	135 to 6900
Elastic modulus (MPa) – flexure	69 to 207	240 to 790	620 to 1050	520 to 970
Maximum service temperature (°C)	80	100	110	—
Water absorption (% in 24 h)	< 0.1	< 0.01	< 0.01	< 0.01
Specific gravity	0.91 to 0.925	0.93 to 0.94	0.95 to 0.96	0.94
Tensile strength (MPa)	6 to 17	14 to 16.5	20 to 30	17 to 38
Elongation (% at UTS)	80 to 725	200 to 425	50 to 1000	200 to 500
Environmental	All types of PE will craze rapidly in sunlight although weather-resistant grades are available. Resistant to marine exposure.			

Source: Dexter [53].

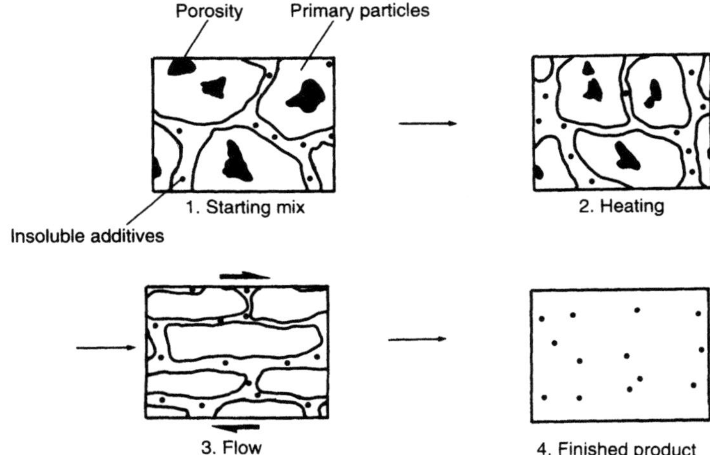

Fig. 4.31. Schematic diagram of the fusion of PVC particles during extrusion. (From Mills [86].)

blended with additives and heated to form the final product. During this heating, the crystalline material melts and binds the particles together, as shown schematically in Fig. 4.31.

Clearly, plasticisers and fillers are of vital importance in the processing of the polymer, and final properties are rather dependent upon these and other additives. Generalisations regarding PVC or, more accurately, PVC compounds are therefore quite difficult, although Table 4.42 is an attempt to summarise the important properties.

4.2.1.3 Polypropylene (PP)

Polypropylene is the next highest polyolefine (polyalkene) from polyethylene and is produced by a simple addition polymerisation reaction, giving the following repeating unit:

$$-\overset{\displaystyle |}{\underset{\displaystyle |}{C}}-\overset{\displaystyle |}{\underset{\displaystyle |}{C}}-$$
$$\overset{}{H}\quad CH_3$$

This may seem like a simple matter, but it was only relatively recently (around 1955) that polypropylene became available commercially.

Not surprisingly, the properties of polypropylene are rather similar to those of polyethylene, its major advantage being an increased crystalline melting temperature. Because of the asymmetry of the monomer, three orientations are possible in the polymer, namely 'isotactic', where all the pendant groups are on the same side of the polymer chain, 'syndiotactic', where they alternate, and 'atactic', where they are randomly oriented on either side of the chain (Fig. 4.32).

The main consequence of tacticity is on the ability of the polymer to crystallise. In the isotactic form, the planar zig-zag crystallinity of polyethylene is not possible, and a helical form results which can permit small amounts of the other forms. At the other end of the scale, the atactic form is amorphous, has rubbery properties and is not of significant commercial use. Most commercial forms aim to be at least 90 per cent isotactic, and the effect of tacticity on properties can be seen in Fig. 4.32.

Unlike those of polyethylene the strength and stiffness of polypropylene tend to decrease with increasing molecular weight, and this is usually attributed to the fact that the degree of crystallinity is less in the higher molecular weights. However, one of the major limiting factors of

Table 4.42. Property data for PVC compounds

	Plasticised PVC	Unplasticised UPVC
Elastic modulus (MPa) – tension	2.7–21	24–40
Elastic modulus (MPa) – flexure	—	20–40
Maximum service temperature (°C)	60–105	65–75
Water absorption (% in 24 h)	0.15–1	0.03–0.04
Specific gravity	1.16–1.55	1.3–1.45
Tensile strength (MPa)	7–27	38–62
Elongation (% at UTS)	4.5–65	2–40
Izod impact energy per unit width (J/m)	—	21
Environmental		Slight effects due to UV exposure, which depend on additives. Reduction in hardness (5 to 10% over 5 years) due to water absorption, otherwise no substantial effect of seawater exposure. May be attacked by marine borers.

Source: Brydson [88] and Dexter [53].

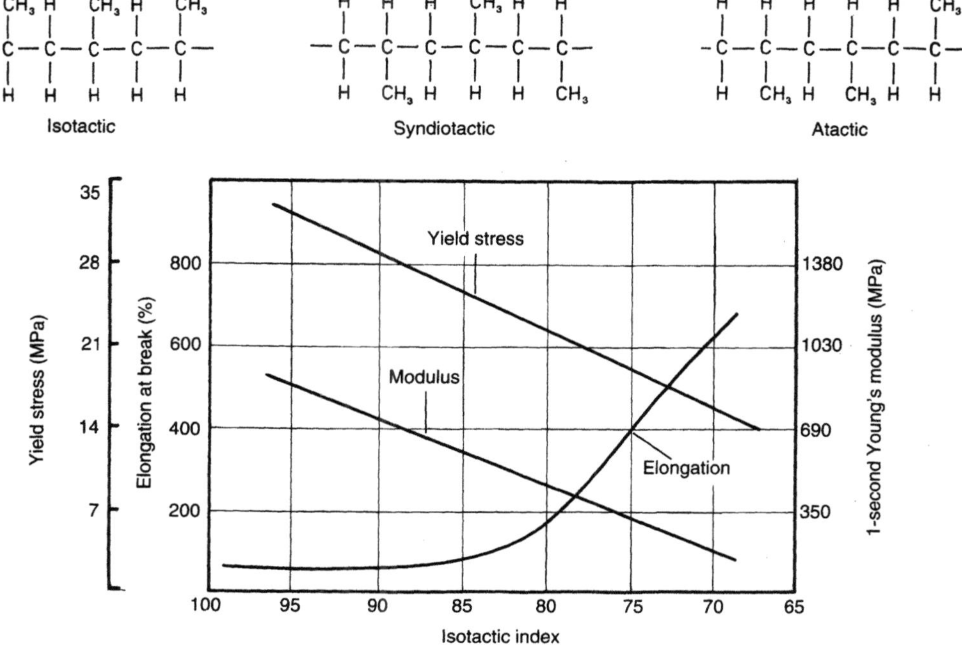

Fig. 4.32. The effect of tacticity on the properties of polypropylene. (Adapted from Brydson [88].)

polypropylene is its brittleness. There is a transition at about 0 °C but this can be overcome by copolymerisation (about 10 per cent) with ethylene.

Although they find broadly similar applications, polypropylene has a slight disadvantage over polyethylene in its lower resistance to oxidation and to ultra-violet radiation. It does, however, have an excellent inherent fatigue resistance.

Atactic polypropylene does find some applications, particularly in coatings, adhesives and sealants.

Table 4.43 provides a summary of some property data for polypropylene.

4.2.1.4 Polystyrene (PS)

Polystyrene is the last of the four 'commodity' thermoplastics, but still finds very extensive application. It again is a relatively simple polymer, produced by addition of styrene monomer (Fig. 4.33).

Of course, as with VCM, the styrene monomer needs first to be prepared, usually by reaction of ethylene and benzene and dehydrogenation of the resulting ethylbenzene.

Most polystyrene produced is a mixture of the syndiotactic and atactic forms and is amorphous with a glass transition temperature of about 100 °C. The molecular weight needs to be over about 50 000 before significant strength is developed, and this continues to increase up to molecular weights of 140 000 [88]. The presence of the benzene ring makes the polymer rather more reactive than polyethylene.

Because of its glassy nature, polystyrene is hard, rigid and transparent. It is not particularly tough, has poor temperature resistance and is not very resistant to oil or organic solvents. A number of so-called high-impact polystyrenes (HIPS) have been developed, but these mainly involve blending, copolymerisation and the use of additives.

Table 4.44 summarises some of the properties of polystyrene, including those of its high-impact versions.

Table 4.43. Properties of polypropylene

	Homopolymer	Block copolymer
Elastic modulus (MPa) – tension	689 to 1520	
Elastic modulus (MPa) – flexure	1100 to 1310	1030 to 1290
Softening temperature (°C)	145 to 150	148
Water absorption (% in 24 h)	<0.01 to 0.03	
Specific gravity	0.89 to 0.91	
Tensile strength (MPa)	29 to 34	25 to 29
Elongation (% at UTS)	115 to 350	40 to 240
Izod impact energy per unit width (J/m)	27 to 107 (for general purpose)	
	53 to 800 (for high-impact grades)	
Environmental	All types of PP will craze rapidly in sunlight, although weather-resistant grades are available (using carbon black). Resistant to marine exposure with low water absorption.	

Source: Brydson [88] and Dexter [53].

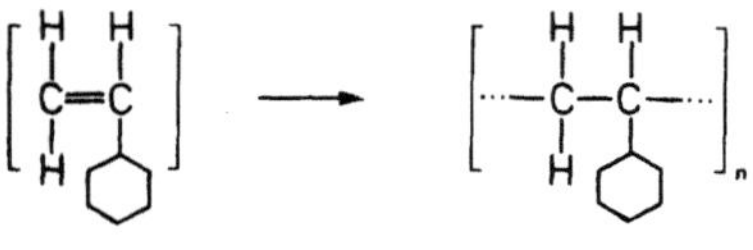

Fig. 4.33. The polymerisation of styrene to produce polystyrene.

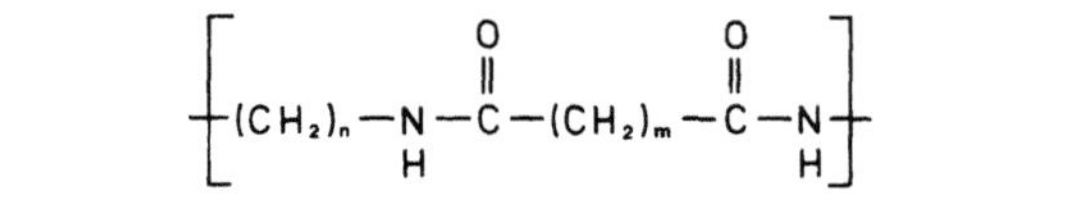

Fig. 4.34. The general structure of linear polyamides. For example, nylon-11 when $m = 10$ and $n = 0$ (excluding middle N—H and C=O groups) and nylon-6,6 when $n = 6$ and $m = 4$.

4.2.1.5 The Engineering Thermoplastics

These plastics are of fairly wide application, but not for the same reasons as the commodity thermoplastics, which have wide non-structural applications. The 'engineering' thermoplastics have been developed for their superior strength, stiffness and temperature resistance. The main plastics under consideration here are polyamides (PA), polyethylene and polybutylene terepthalates (PET and PBT), polymethylmethacrylate (PMMA) and polyoxymethylene (POM). (Of the six main 'engineering' thermoplastics, one (ABS) is a copolymer and is considered in the next section.)

Polyamides (nylons) are condensation polymers formed by one of three possible main reactions:

Diamine + carboxylic acid

Self-condensation of ω-amino acid

From a lactam

and, as with many other polymers, these reactants need first to be prepared. Other reactions are possible, particularly diamines with diacid chlorides, to produce aromatic polyamides; these however are rather complex to deal with in a text of this scope, and Brydson [88], for example, can be consulted for further details. Figure 4.34 shows the general nylon structure, which depends upon whether preparation is by the first or one of the latter two routes. The description of a polyamide usually consists of one number indicative of the 'R' group in either the lactam or ω-amino acid or of two numbers indicative of the 'R' group in the diamine followed by the 'R1' group in the carboxylic acid. For example, nylon 66 is prepared from hexamethylene diamine and adipic acid and the structure of this is shown in Fig. 4.35. Copolymers are named according to the components, e.g. nylon 66/6.10 (60: 40) is a copolymer of nylon 66 and nylon 6.10 in the ratio of 60: 40.

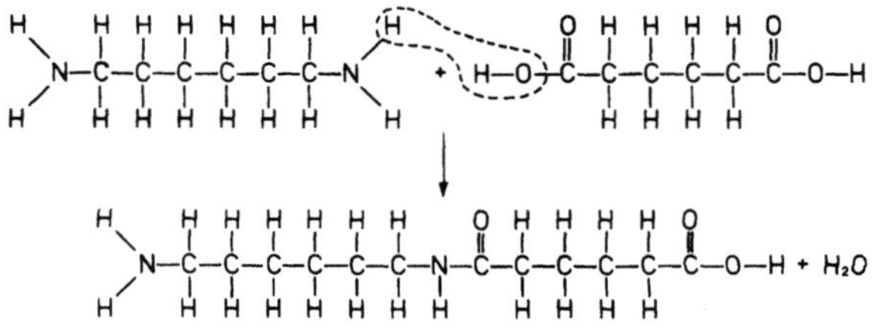

Fig. 4.35. The preparation of a nylon-6,6 unit from the reaction of hexamethylenediamine and adipic acid. (From Askeland [3].)

The polyamides have traditionally been used as fibres (mainly nylon 6 and nylon 66), but more recently they have found use as engineering plastics, particularly 6, 66, 11 and 12 and some copolymers. Aromatic polyamides also find some engineering uses, the best known being as reinforcing fibres.

The aliphatic polyamides are semi-crystalline thermoplastics owing to the high degree of intermolecular attraction afforded by the regularly spaced —CONH— groups, but the aliphatic chains provide a degree of flexibility and toughness. These complementary properties are affected by the structure of the nylon, in particular the distance between repeating —CONH— groups, number of methylene groups in the intermediates, molecular weight and nitrogen substitution. Of course, copolymerisation between two or more nylons will also affect properties. These factors are summarised in Fig. 4.36.

The other major structural parameter is degree of crystallisation, which is very sensitive to processing, particularly cooling rate, where higher cooling rates result in a lower degree of crystallisation. Resistance to water absorption by the polymer is reduced by having a higher degree of crystallinity. The spherulite size is also

Table 4.44. Properties of polystyrene

	General-purpose	High-impact
Elastic modulus (MPa) – tension	3100 to 3500	2500 to 3500
Elastic modulus (MPa) – flexure	2750 to 3500	1030 to 3500
Maximum service temperature (°C)	70	50 to 80
Water absorption (% in 24 h)	0.03 to 0.2	0.03 to 0.6
Specific gravity	1.04 to 1.09	1.04 to 1.10
Tensile strength (MPa)	40 to 48	21 to 48
Elongation (% at UTS)	1 to 2.5	2 to 90
Izod impact energy per unit width (J/m)	10 to 20	25 to 60
Environmental	Slight yellowing and slight loss of strength on UV exposure. Good resistance to seawater immersion and low water absorption.	

Source: Brydson [88] and Dexter [53].

sensitive to processing, and has an effect on mechanical properties.

Water absorption is important in nylons, particularly for marine use, and alternate wetting and drying may lead to damage. However, their toughness, rigidity and resistance to abrasion and to hydrocarbons has led to a number of marine applications, for example as seals and even as propellers for small craft.

Table 4.45 summarises property data for aliphatic polyamides.

The aromatic polyamides are also worthy of some mention at this point. An almost bewildering array of such materials exist, and a number of these are glassy and transparent. The best-known aromatic polyamides are the fibre-forming aramids, defined as long-chain synthetic polyamides in which at least 85 per cent of the amide linkages are connected to two aromatic rings (Fig. 4.37).

Perhaps the best-known of such polyamides is poly-(p-phenyleneterephthalamide), better known under its trade name of 'Kevlar' [89, 90]. It is produced by the reaction of p-phenylenediamine with terephthalyl chloride, and the fibres are widely known for their exceptional strength. A number of other fibre-forming aramids are available and the properties of these are discussed in Section 4.3.3.

The polyethylene and polybutylene terephthalates are of the family of polyesters which are defined as having ester linkages in the main chain (rather than in side groups such as in polyvinyl acetate), and they cover a wide range of materials, only some of which are suitable for moulding (as opposed to for surface coatings, rubbers, etc.). Those (thermosetting) polyesters suitable for laminating resins are dealt with in Section 4.4.

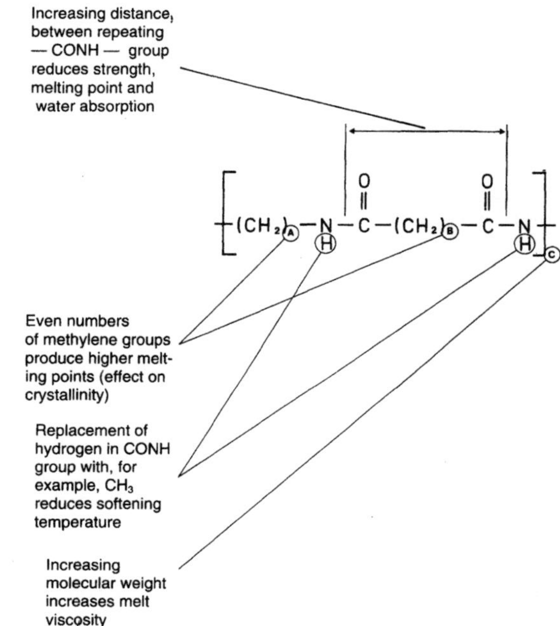

Fig. 4.36. Schematic of the relationship between structure and properties of aliphatic polyamides.

Brydson [8] describes five different commercially important methods of preparation of polyesters, but the PET and PBT polymers are usually prepared by condensation from dimethyl terephthalate either with 1,4-butanediol or ethylene

Table 4.45.   Properties of some aliphatic polyamides (nylons)

Property	66	6	610	612	11	12	66/610 (35:65)	66/610/6 (40:30:30)
Tensile stress at yield (MPa)	80	76	55	60	38	45	38	—
Tensile stress at break (MPa)	—	—	—	60	52	54	—	51
Water absorption (% in 24 h)	0.3 to 0.5	1.3 to 1.9	0.4	—	0.4	0.25	—	—
Elongation at break (%)	80 to 100	100 to 200	100 to 150	100 to 250	30 to 300	200	>200	~300
Tension modulus (MPa)	3000	2800	2100	—	1400	1400	1400	1400
Impact strength (dry) (IZOD) (J/m)	70	53	59	35 to 45	—	—	—	—
Rockwell hardness	R118	R112	R111	R114	R108	R107	—	R83
Specific gravity	1.14	1.13	1.09	1.07	1.04	1.02	1.08	1.09
Environmental	All grades embrittled by prolonged exposure to sunlight, although stabilisation with carbon black is possible. All grades susceptible to dimensional change and loss of mechanical properties on prolonged exposure to water. Types 11 and 12 are least susceptible and Type 6 is most susceptible.							

Source: Brydson [88] and Dexter [53].

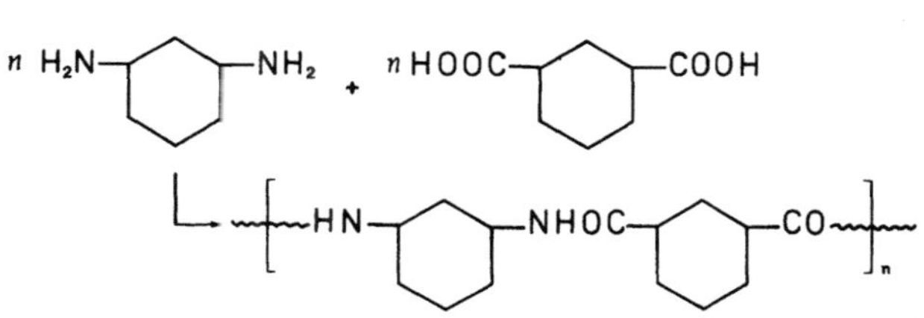

Fig. 4.37.   The preparation of a poly-(m-phenyleneisophthalamide) fibre-forming aramid by condensation of 1,3-phenylenediamine with isophthalic acid.

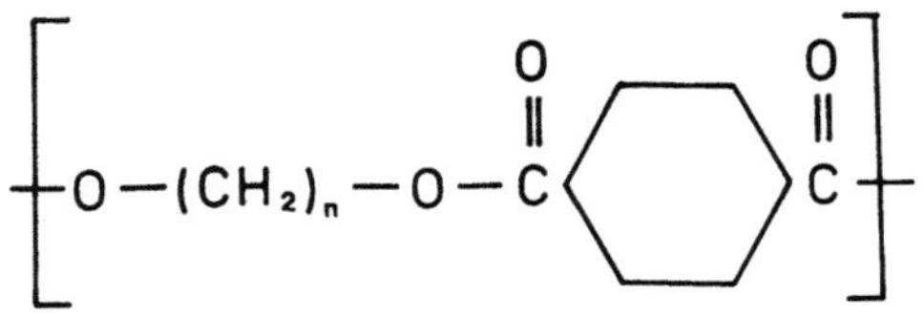

Fig. 4.38. The polyesters polyethylene terephthalate (PET), $n = 2$, and polybutylene terephthalate (PBT), $n = 4$.

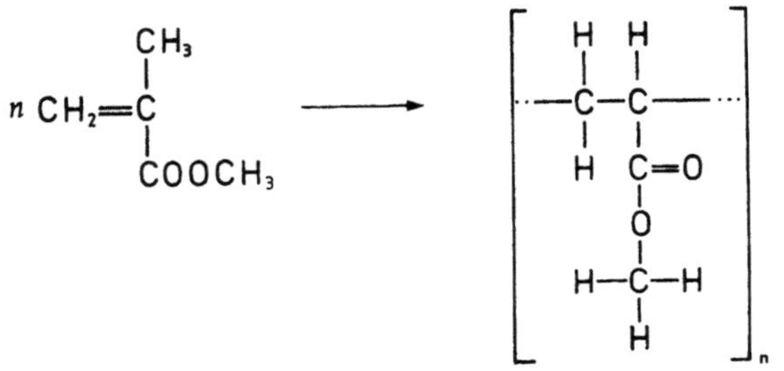

Fig. 4.39. The polymerisation of methylmethacrylate to produce PMMA.

Table 4.47. Some properties of PMMA acrylic plastics (Variation includes that between cast and mouldin compositions)

	General-purpose	High-impact
Elastic modulus (MPa) – tension	2400 to 3000	1600 to 2300
Elastic modulus (MPa) – flexure	2750	1900 to 2500
Maximum service temperature (°C)	60 to 95	
Water absorption (% in 24 h)	0.1 to 0.5, reducing for extended immersion	
Specific gravity	1.12 to 1.28	
Tensile strength (MPa)	41 to 86	38 to 55
Elongation (% at UTS)	2 to 7	25 to 200
Izod impact energy per unit width (J/m)	10 to 21	26 to 122
Environmental	Generally resistant (apart from some discolouration) to UV and weathering. In seawater some reduction in properties might be expected due to water absorption (for example, up to 30% reduction in tensile strength).	

Source: Brydson [88] and Dexter [53].

glycol to give the final structures illustrated in Fig. 4.38.

These polymers have good strength, stiffness and chemical stability. They are crystalline and are often used with glass reinforcement. Some mechanical properties are given in Table 4.46.

Polymethylmethacrylate, better known as 'acrylic', 'Perspex' or 'Plexiglas', is one of a range of polymers derived from acrylic acid. It is prepared by addition from the monomer in the reaction shown in Fig. 4.39.

The two main features of PMMA are that it is amorphous, with a glass transition at between 90 and 105 °C (and is therefore glassy), and that it is transparent. The polymer is unbranched, and therefore the main difference in commercial grades (excluding compounds) is in the molecular weight, a higher weight mainly affecting the processability of the material but having a minor effect on mechanical properties. The weathering resistance of PMMA is generally good compared with other polymers.

Impact-resistant versions of PMMA can be made by copolymerisation and blending. The major use of the polymer is as a (less-abrasion-resistant) glass replacement, and it is rare that it is used when transparency is not required due to its high cost compared with the larger bulk production plastics. An interesting application is in windows for pressure vessels for human occupation and, for example, DnV [91] give design rules for calculating minimum window thickness and for specifying acrylic materials for such a purpose.

Table 4.47 summarises some properties of acrylics.

The last of the major engineering thermoplastics is polyoxymethylene (POM), or acetal as it is more widely known. As with PMMA and acrylics, POM is only one of a range of acetal resins characterised by polymerisation from formaldehyde, although they are all part of a more general family of aliphatic polyethers. The polymerisation of formaldehyde is rather a complex phenomenon, producing, among other possible products, a trimer trioxane. The two principal commercial acetal resins are the homopolymer of formaldehyde and a copolymer of formaldehyde and

Table 4.46. Some properties of the thermoplastic polyethylene and polybutylene terephthalates (unfilled compounds)

	PET	PBT
Elastic modulus (MPa) – flexure	2350	2350
Water absorption (% in 24 h)	0.08 to 0.09	0.08
Specific gravity	1.31	1.31 to 1.32
Tensile strength (MPa)	55	56
Elongation (% at UTS)	250 to 300	—
Izod impact energy per unit width (J/m)	53	43 to 53
Environmental	Generally resistant (apart from some discolouration) to UV and to immersion.	

Source: Brydson [88] and Dexter [53].

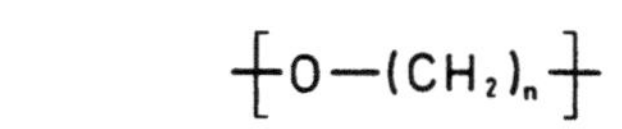

$$+O-(CH_2)_n+$$

Fig. 4.40. The acrylic polymers polyoxymethylene ($n = 1$) and polyethylene oxide ($n = 2$).

trioxane. Figure 4.40 shows the structure of POM and the rather similar polyether, poly(ethylene oxide).

The homopolymer structure can be compared to polyethylene, and it can be seen that the main difference is in the shorter —C—O— bond length, leading to a harder, higher-melting polymer. The presence of copolymerisation in POM leads to a reduction in ability to crystallise in the same way as does branching in PE.

The main competitors of POM are the polyamides, and the main advantages of POM are in superior fatigue resistance, creep resistance, stiffness and water endurance. The latter of these properties is of significance for immersed use. In addition, the resistance of POM to petroleum is good, and friction properties are also excellent.

Properties of POM and poly(ethylene oxide) are shown in Table 4.48.

A number of other thermoplastic polymers are also in engineering use. Most notable among these are polytetrafluoroethylene (PTFE), polyvinyl acetate (PVA) and a range of polymers containing *p*-phenylene groups in the main chain (as opposed to side groups). Some of these have already been considered (e.g. aromatic polyesters and polyamides), but there are some others of significance such as polycarbonates, polysulphones and aromatic polyether ketones (such as polyether ether ketone, or PEEK). These are listed in Table 4.49, along with some of the key properties; to summarise the discussion of thermoplastics, some data on those already covered are included also in this table.

Table 4.48. Some properties of the polyethers polyoxymethylene (POM or acetal) and poly(ethylene oxide)

	POM	Poly(ethylene oxide)
Elastic modulus (MPa) – tension	2800 to 3500	—
Elastic modulus (MPa) – flexure	2500 to 2800	—
Maximum service temperature (°C)	90 to 105	66 (melts)
Water absorption (% in 24 h)	0.25	Water-soluble
Specific gravity	1.41 to 1.43	—
Tensile strength (MPa)	61 to 70	7 to 11
Elongation (% at UTS)	25 to 75	700 to 1200
Izod impact energy per unit width (J/m)	64 to 120	—
Environmental	POM will chalk slightly owing to UV and weathering. In seawater a minor reduction in hardness might be expected for long exposures. POM also has very high fatigue endurance for a thermoplastic. Poly(ethylene oxide) will suffer a severe drop in strength at relative humidities of greater than 80%.	

Table 4.49. Summary of thermoplastic data (blends and copolymers not included)

Polymer	Strength	Toughness	Environmental
Polyethylene	3 to 4	1	2 to 3
PVC	2 to 3	2 to 4	2
Polypropylene	3	2 to 3	3
Polystyrene	2 to 3	3 to 4	2
Nylons	1	1 to 3	2 to 4
PET and PBT	2	3	1
Acrylic (PMMA)	1 to 3	3 to 4	3
POM (acetal)	1 to 2	3	2
Poly(ethylene oxide)	4	2?	4
PTFE	3 to 4	3	1
Polyvinylidenefluoride (PVdF)	2 to 3	3	1?
Polycarbonate (PC)	1 to 2	2 to 3	2
Polysulphone (PS)	1	3	2?
Poly(etherketones) (e.g. PEEK)	1	3	1
Polyimide	1	3	2 to 3

Values are relative only, a rating of 1 representing a desirable value and 4 representing a poor value. ('Environmental' represents a combination of UV and water resistance. Polymers with poor UV resistance can usually be modified to alleviate this and are therefore not given a rating of 4 unless they also deteriorate unacceptably in seawater.)

4.2.1.6 Thermoplastic Copolymers and Polymer Blends

Copolymerisation and blending are most often used to enhance mechanical properties such as strength, stiffness and toughness. As described previously, copolymerisation occurs at the molecular level, where either blocks or individual groups of the comonomer appear in the basic chain. Individual groups can be regular or randomly spaced and blocks can be in the chain or grafted. Blending, on the other hand, implies the mixture of two different polymers, usually of quite different properties, and the only bonding between the types occurs between the particles of the individual components.

The best illustration of blending is provided by the remaining undiscussed engineering thermoplastic, acrylonitrile butadiene styrene (ABS). This range of materials is based on the three polymers mentioned in the name, the second two of which are discussed in Sections 4.2.3.3 and 4.2.1.4 respectively while the first, acrylonitrile, finds only limited use as the homopolymer polyacrylonitrile (Fig. 4.41).

There are two main types of ABS, the more important being interpolymers of polybutadiene with styrene and acrylonitrile, the others being polymer blends of acrylonitrile-styrene copolymers with butadiene–acrylonitrile rubber. The difference between these is illustrated in schematic form in Fig. 4.42.

The blends can be altered in terms of the proportion of the two components to vary processability, toughness and

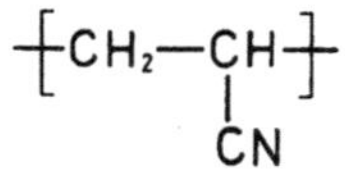

$$+CH_2-CH+$$
$$|$$
$$CN$$

Fig. 4.41. The homopolymer polyacrylonitrile.

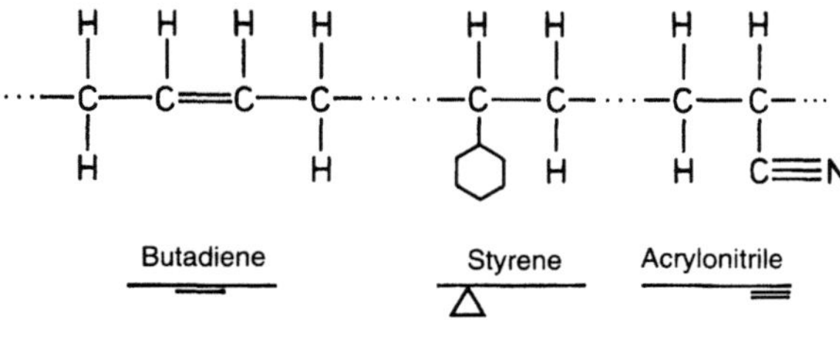

a

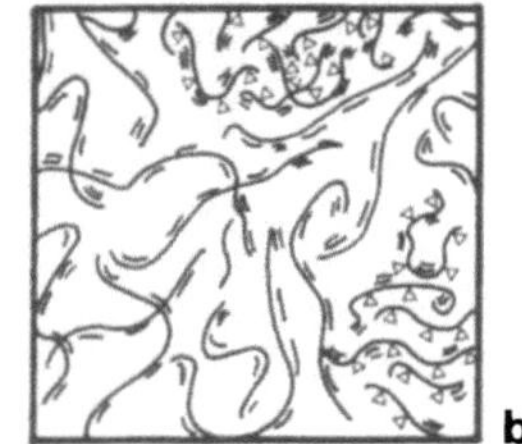

b

c

Fig. 4.42. Schematic illustration of the two main types of ABS: **a** interpolymers of polybutadiene with styrene and acrylonitrile, **b** blends of acrylonitrile–styrene with butadiene–acrylonitrile rubber; **c** shows high-impact polystyrene composed of polystyrene in a matrix of styrene–butadiene rubber (SBR).

The interpolymers are produced by mixing styrene and acrylonitrile with polybutadiene latex and heating to about 50 °C, giving a mixture of polybutadiene, polybutadiene grafted with acrylonitrile and styrene, and styrene acrylonitrile copolymer. Clearly again, a wide range of mixtures and degrees of grafting and colpolymerisation is possible.

The introduction of other monomers and the use of proprietary names means that data on ABS are necessarily product-specific, but the general properties are good impact resistance and a softening temperature similar to that of general-purpose polystyrene. Flame-retardant variations have recently been introduced by blending with PVC or by additives. Blends with polycarbonate and polysulphones have been used to give approximately intermediate properties, particularly to improve heat distortion temperature. Blending with PMMA can give reasonable transparency.

Table 4.50 illustrates something of the variability in properties available in the styrene–butadiene–acrylonitrile system, and also indicates the effect of blending polystyrene with SBR, one of the principal ways of producing the so-called 'high-impact' polystyrenes (HIPS).

Apart from the cross-linked form of PE mentioned in Section 4.2.1.1, there are a number of copolymers involving ethylene, and chlorination can be used to produce elastomers with useful oil, heat and weathering resistance. Copolymerisation of PE with small amounts of the higher olefins (e.g. propylene and but-l-ene) tends to produce slightly inferior mechanical properties, but nevertheless these find applications for blow and injection moulding and show a marked improvement in environmental stress cracking resistance. Another set of copolymers with vinyl acetate (EVA) find uses mainly as hot melt adhesives, but can also be used as cable insulation.

As mentioned above, PVC is used as part of a compound material rather more than is the case with other commodity thermoplastics. The two main classes of compounds are plasticised PVC and unplasticised (UPVC). The plasticisers are non-volatile PVC solvents which, at a temperature of about 150 °C, allow sufficient molecular mixing to allow the plasticiser to gel to a degree determined by the ratio of plasticiser to polymer and also by the type of plasticiser used. A wide number of such plasticisers are used for PVC and this, along with the variation of fillers, extenders, impact modifiers and other additives, makes PVC materials properties rather difficult to define. Formulations are avail-

heat resistance. Of course, the variation of the proportions of the components within the two copolymers introduces two further dimensions for variability, and this can affect the toughness if the nitrile content of the rubber is varied; cross-linking of the rubber can be used to further modify properties.

Table 4.50. Range of properties available in styrene–butadiene–acrylonitrile blends and copolymers

Property	GP polystyrene	Medium-impact PS-SBR blend	Very high impact PS-SBR blend	Styrene–acrylonitrile	Medium-impact ABS	High-impact ABS	MBS
Specific gravity	1.05	1.05	1.02	1.06	1.04 to 1.07	1.01 to 1.04	1.07 to 1.10
Tensile strength (MPa)	52	35	14	40 to 70	38 to 48	24 to 40	40 to 55
Tensile modulus (MPa)	3900	3250	1750	3700	2000 to 3000	1500 to 2000	1700 to 2500
Impact strength per unit width (J/m)	13	32	31	16 to 32	53 to 360	374	—
Vicat softening point (°C)	100	94	94	106	100 to 108	104	96 to 99

Source: Brydson [88].
Abbreviations: GP, general purpose; PS, polystyrene; SBR, styrene–butadiene rubber; ABS, acrylonitride–butadiene–styrene; MBS, methylmethacrylate–butadiene–styrene.

able for applications as diverse as general-purpose insulation compounds and rigid opaque formulations suitable for pipes [89]. A number of PVC copolymers are used, probably the best known being that with vinyl acetate for record player discs. A more important product for engineering applications is the copolymer with 2–10 per cent propylene, which has improved heat stability and hence has made processing without plasticiser a lot easier.

Two of the main methods of improving the impact resistance of polystyrene involve the use of rubbery additives (described above) and copolymerisation. The first of these is effectively a blend, and the main processes for toughening polystyrene are combinations of the two methods. For example, it is now common practice to dissolve the styrene in the rubber (e.g. butadiene) and to polymerise the styrene, producing a blend but also some graft polymer chains where short styrene side chains are attached to the rubber molecules (Fig. 4.42). SBS and triblock EPDM rubbers are also used. In SBR–polystyrene blends the amount of SBR used is crucial in determining the impact resistance, but reduces tensile strength and softening point (Table 4.51). Copolymers of styrene with acrylonitrile (SAN) are of course a component of ABS mentioned above, but SAN copolymers can be used alone as a rigid transparent product, although their transparency and weathering properties are not as good as those of PMMA. Two of the major limitations of ABS are the limited light resistance and opacity. The first of these has led to the replacement of butadiene with more oxidation-resistant rubbers, notably ethylene propylene termonomer rubbers, to produce AES or acrylic ester rubbers to produce ACS. By grafting methyl methacrylate and styrene onto the polybutadiene backbone in ABS, the alternative, high-clarity MBS polymer blend can be made, although this has rather lower strength than ABS, as Table 4.50 shows.

A number of important blends and copolymers are produced based on the engineering thermoplastics. For polyamides, one of the major objectives is toughening, and the toughened nylons usually contain an elastomer (for example EPDM rubber) as the dispersant rather than the matrix. The exceptional toughness of these materials is usually bought at the expense of a slight reduction in stiffness,

Table 4.51. Specifications for three types of toughened polystyrene according to BS 3126

	Type 1	Type 2	Type 3
Impact strength			
(J per m of notch) (min.)	53	27	27
Softening point (°C) (min.)	80	85	95
Tensile strength			
(MPa) (min.)	21	24	28
Elongation at break			
(%) (min.)	15	10	7.5
Volatile matter (%) (max.)	3.0	2.5	2.5
Water absorption			
(mg) (max.)	20	15	20

Source: Brydson [88].

strength and creep resistance. PET- and PBT-based copolymers have been introduced to improve toughness, for example by substituting a longer-chain aliphatic dicarboxylic acid for some of the dicarboxylic acids. As with polystyrene and PVC, most of the modifications to PMMA are made for the purpose of toughening and result in some loss of strength and stiffness. As with PS, both copolymerisation and blending have been used. Butyl acrylate and acrylonitrile have been used as copolymers and blending has been with rubbers. The HI PMMAs generally show better weathering resistance than ABS and even ASA (acrylonitrile–styrene–acrylic ester rubber blend), but strength and stiffness are generally lower, as is impact strength. Some of these effects are summarised in Table 4.52.

One copolymer of POM has already been dealt with in Section 4.2.1.5. Also, blends with PTFE (20–25 per cent) have been used to provide very low-friction materials.

4.2.2 Thermosetting Plastics

The properties of thermosets are generally very different from those of thermoplastics. In particular, the rigid three-dimensional network results in high strength and stiffness but also limited toughness. The widest uses of thermosets

Table 4.52. Comparison of methylmethacrylate–acrylonitrile copolymer, poly(butylacrylate) blended with PMMA, ABS and ASA rubber modified polymers and unmodified PMMA

Property	PMMA	MMA–ACN	ABS	ASA	HI-PMMA I	HI-PMMA II
Specific gravity	1.18	—	1.06	1.07	1.12	1.15
Vicat softening temperature (°C)	110	80	98	92	76	95
Impact strength, (mJ mm^{-2}),			No	No	No	
NKS small standard rod	18	40	fracture	fracture	fracture	45
Notched impact strength (mJ mm^{-2}),						
NKS small standard rod	2	3	12	14	8	3.5
Tensile strength (MPa)	75	85	45	44	20	40
Elongation at break (%)	3.5	60	20	20	50	30
E-modulus (MPa)	3300	4500	2400	2300	800	1800
Degree of transmission ($d = 3$ mm)	92	90	Opaque	Opaque	approx. 70	88

Source: Adapted from Brydson [88].
Abbreviations: PMMA, unmodified poly(methylmethacrylate) moulding compound; MMA-ACN, copolymer of methylmethacrylate and acrylonitrile; ABS, acrylonitrile–styrene–butadiene blend; ASA, acrylonitrite–styrene–acrylic ester rubber blend; HI-PMMA I, rubber-modified PMMA with 30% poly(butylacrylate); HI-PMMA II, rubber-modified PMMA with 15% poly(butylacrylate).

Table 4.53. Consumption of thermosets in the UK

Thermoset type	Consumption ('000 t/yr)
Aminos	125
Urethanes	94
Polyesters	51
Phenolics	50
Epoxies	18
Others	294

Source: Mills [86].

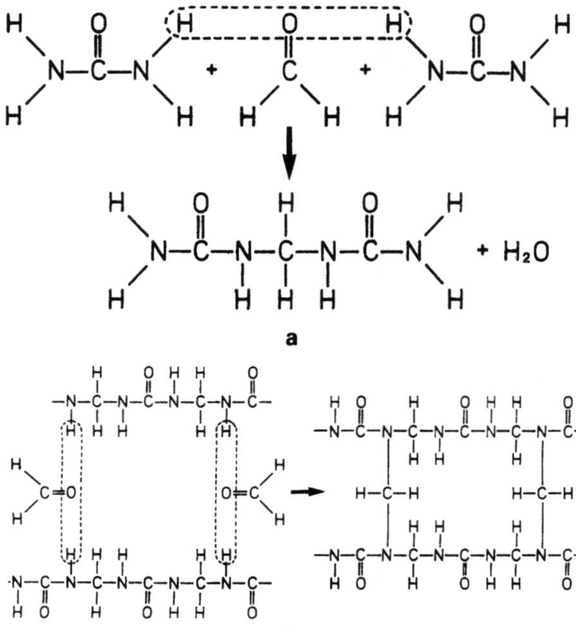

Fig. 4.43. Schematic illustration of the buildup of an aminoplastic from urea and formaldehyde: **a** formation of urea–formaldehyde chain; **b** cross-linking of chains to produce a framework. (From Askeland [3].)

are as the base materials for fibre-reinforced composites and as coatings in both their reinforced and unreinforced conditions.

The normal method of producing thermosets is to produce a linear chain polymer which is then cross-linked in a second reaction. This second reaction can be achieved through the application of heat and/or pressure, or can come about by mixing two parts, often as liquids.

Table 4.53 shows the ranking order of usage of thermosets, and comparison with thermoplastics shows that the usage is more widely spread and also that the total quantity of thermosets in use is rather less. In fact the use of thermosets has been in decline for some years, because their major advantage of temperature resistance has been eroded by advanced thermoplastics, although increased use of resin matrix composites may cause this trend to reverse. The major thermosetting resins used for glass-reinforced plastics are unsaturated polyesters and (to a lesser extent) epoxies.

The thermosets are discussed below according to the rank order of their production as indicated in Table 4.53. The only thermosets discussed under the category 'other' in this table are the polyimides because of their importance as matrix materials for composites.

4.2.2.1 The Aminoplastics

These plastics are produced by the combination of amines or amides with aldehydes. The two principal plastics of this type are those formed from urea or melamine with formaldehyde. The formaldehyde provides links between the amine molecules to give linear chains, and any excess provides cross-linking into a three-dimensional network. For example, the urea–formadehyde network can be built up as shown in Fig. 4.43. As can be seen, the functionality of the urea molecule can be as much as four. In the case of melamine, similar reactions condensing water can take place, and this time the monomer functionality can be as much as six.

The U–F and M–F resins can be filled with a variety of substances for moulding purposes and the U–Fs can be plasticised, mainly to allow the use of more highly condensed resins, which reduces curing shrinkage while maintaining good flow properties.

Melamine can also be copolymerised with phenol because of the similar nature of the condensation reaction with formaldehyde.

It has to be said that the wide use of the aminoplastics comes about because of their cheapness, ease of colouration and resistance to heat, at least up to common domestic requirements, and in fact they are mainly used domestically. The unreinforced moulding materials are not often employed in engineering.

Table 4.54. Some properties of urea–formaldehyde and melamine–formaldehyde mouldings

Property	Urea–formaldehyde				Melamine–formaldehyde		
	α-cellulose-filled	Woodflour-filled	Plasticised	Translucent	Cellulose-filled	Glass-filled	Mineral-filled
Specific gravity	1.5 to 1.6	1.5 to 1.6	1.5 to 1.6	~ 1.5	1.5 to 1.55	~ 2.0	~ 1.8
Tensile strength (MPa)	52 to 80	52 to 80	48 to 66	48 to 69	55 to 83	41 to 69	28 to 41
Impact strength (J)	0.3 to 0.5	0.2 to 0.5	0.2 to 0.3	0.2 to 0.3	0.2 to 0.3	0.2 to 0.3	0.15 to 0.3
Cross-breaking strength (MPa)	76 to 117	76 to 114	93 to 107	89 to 117	89 to 144	62 to 96	41 to 76
Water absorption (mg)							
24 h at 20 °C	50 to 130	40 to 170	50 to 90	50 to 100	10 to 50	10 to 20	7 to 14
30 min at 100 °C	180 to 460	250 to 600	300 to 450	300 to 600	40 to 110	20 to 35	15 to 40

Source: Brydson [88].

Table 4.53 summarises some properties of U–F and M–F mouldings.

4.2.2.2 Polyurethanes

Polyurethanes are produced by the reaction of polyhydroxy materials with polyisocyanates, as illustrated in Fig. 4.44, which shows the combination of 1,4-butanediol (a polyhydroxy material) with the polyisocyanate, hexamethylenedi-isocyanate. No condensation is involved here, but there is a transfer of hydrogen atoms from the polyol to the polyisocyanate. Under some conditions polyurethanes can be treated as thermoplastics or elastomers, depending upon the details of the processing and composition. Although fibre-forming (as is the polymer illustrated in Fig. 4.44) and rubbery polyurethanes are known, it is as foams and coatings that polyurethanes find their widest use.

Foams may be flexible or rigid, depending upon the degree of cross-linking achieved which in turn depends on the functionality of the polyol. The immediate advantage of PU foams is that they can be formed *in situ* and act as an adhesive to the skin material. The flexible foams are mainly of use in upholstery and packaging, but the rigid foams have more engineering applications such as in buoyancy and insulation applications, and even structurally for some of the denser foams.

The early foams were based on polyesters, which contain carboxyl groups, and on reacting these with isocyanates, the rigidity of the foam being again dependent upon the functionality of the polyol. By using polyesters of low carboxyl values, gas evolution (for blowing the foam) can be produced by the isocyanate–water reaction.

Polyisocyanurates differ slightly from the polyurethanes, but are also based on the chemistry of isocyanates, employing a trimerised isocyanate and, in some cases, polymeric isocyanates. Foams made from these materials show rather better flame and fire resistance than the polyurethanes, but can be brittle, so that combinations have been used. Blends of polyurethane with other polymers are also used. For example, polyurethane intermediates methylmethacrylate monomer and unsaturated polyester resin can be blended to give interpenetrating polymer networks formed by a mixture of addition and rearrangement reactions.

4.2.2.3 Polyesters

Like polyurethanes, polyesters can be thermosetting or thermoplastic polymers. As mentioned in Section 4.2.1.5, the thermoplastics PET and PBT belong to the broad family of polyesters. They are formed by one of five possible reactions [89].

The only common feature of polyesters is the (—COO—) link, which may only be a small part of the molecule, so

that this is an enormously diverse group of polymers. The thermosetting polyesters are mainly of use as laminating resins or for surface coatings with the moulding applications being filled mainly by the thermoplastics.

The unsaturated polyester laminating resins are produced by condensing a glycol with both a saturated and an unsaturated dicarboxylic acid. The unsaturated acid provides the sites for later cross-linking, so that the ratio of saturated to unsaturated acids can be used to control the degree of cross-linking in the cured product. The commonest glycol is 1,2-propylene glycol, although others such as di- and triethylene glycols are used.

The most widely used unsaturated dicarboxylic acids are maleic or fumaric acids, the former usually as its anhydride, and the commonest saturated acid is phthalic acid, again normally used as its anhydride.

The finished thermoset is made by mixing the resin (condensed glycol plus acids) with an active diluent such as styrene and a catalyst, and the process is illustrated in Fig. 4.45.

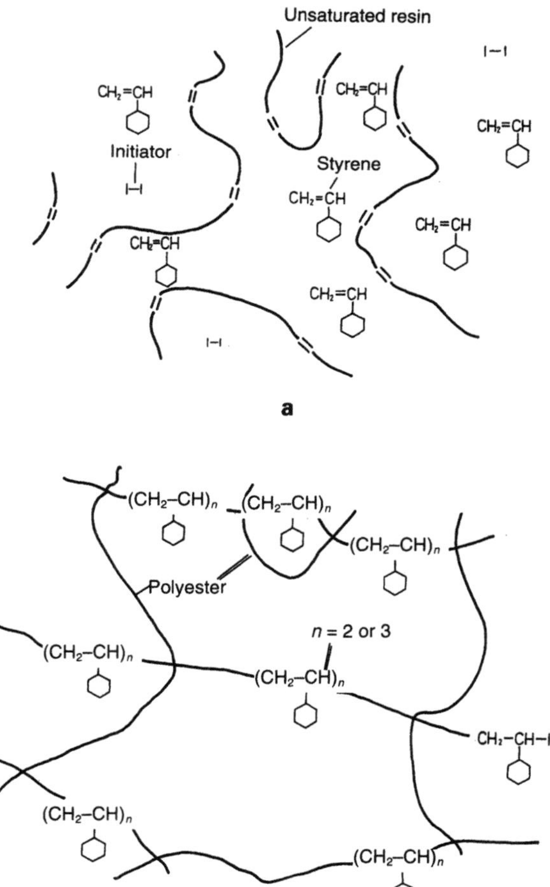

Fig. 4.45. The curing reaction in polyester laminating resins: **a** ready for laminating and consisting of low molecular weight unsaturated resin molecules, reactive diluent molecules (shown here as styrene) and some initiator molecules and **b** cured resin cross-linked by addition polymerisation and incorporating an average of 2 or 3 styrene molecules in a typical general purpose resin. (Adapted from Brydson [8].)

Fig. 4.44. The production of a polyurethane from a polyisocyanate and a poly-hydroxy material. In this specific case, 1,4-butanediol and hexamethylenedi-isocyanate react to produce a fibre-forming polyurethane.

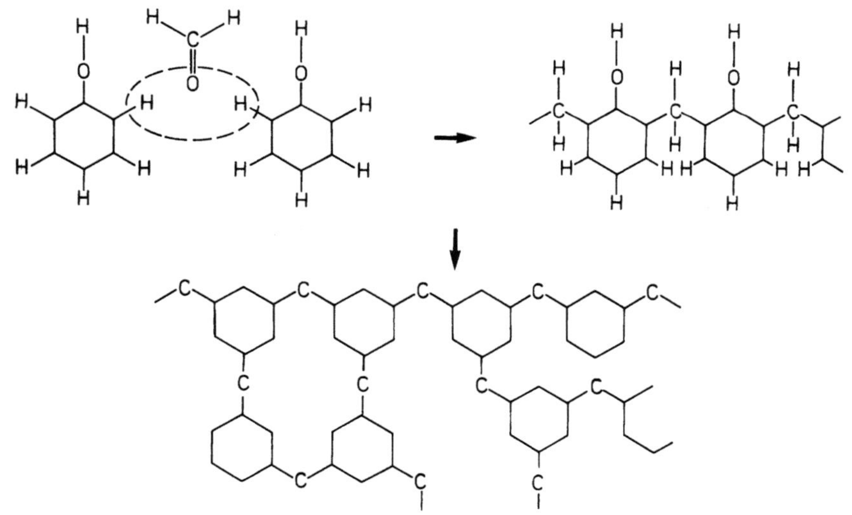

Fig. 4.46. Example of the production of a phenolic resin by condensation with formaldehyde into a linear chain and finally by cross-linking with excess formaldehyde. (From Askeland [3].)

Depending upon the mixture, curing may be hot or cold and may take from minutes to hours; also, because cross-linking takes place via an addition mechanism, no volatiles are given off, which makes the resin particularly suitable for larger structures.

There are some polyester thermosets used for moulding, but these are unlikely to take over from phenolics and aminos for this application. The moulding polyesters can be based simply on similar compositions to the laminating resins containing glass reinforcement, and these can be moulded as doughs or sheet (sometimes referred to as dough moulding and sheet moulding compounds, DMCs and SMCs). In addition to these, alkyd, diallyl phthalate and diallyl isophthalate compositions are available.

4.2.2.4 Phenolics

Phenolic resins are produced by the condensation of phenols with an aldehyde, the aldehyde providing three-dimensional linkage, as illustrated for the example of phenol–formaldehyde in Fig. 4.46. Phenol and the cresols are the most commonly used phenols and formaldehyde (and occasionally furfural) is the most commonly used aldehyde. The phenol-to-formaldehyde ratio is crucial to the final properties of the resin. For mouldings, a low-molecular-weight resin is first produced and further cross-linking is provided at the moulding stage. The initial resins are either novolaks or resols. A typical novolak has 5 or 6 benzene rings per molecule and a typical resol 2 to 4 such rings.

Whereas the resols have reactive methylol groups and can be induced to cross-link under heating, the novolaks require the addition of materials capable of forming methylene bridges (e.g. hexamethylenetetramine). The mouldings are hard and heat-resistant, but mechanical properties are rather dependent on amount and type of filler, as shown in Table 4.55.

Phenolics are also used for laminates, and for this application it is necessary to use resols with enough methylol

Table 4.55. Components and properties of four grades of phenolic resin

	GP grade	Electrical grade	Medium shock-resisting grade	High shock-resisting grade
Component				
Novolak resin	100	100	100	100
Hexa (hardener)	12.5	14	12.5	17
Magnesium oxide	3	2	2	2
Magnesium stearate	2	2	2	3.3
Nigrosine dye	4	3	3	3
Woodflour	100	—	—	—
Mica	—	120	—	—
Cotton flock	—	—	110	—
Textile shreds	—	—	—	150
Asbestos	—	40	—	—
Property				
Specific gravity	1.35	1.85	1.37	1.40
Impact strength (J)	0.22	0.18	0.39	1.08 to 1.9
Cross-breaking strength (MPa)	80	76	76	83
Tensile strength (MPa)	55	58	48	45
Blister temperature (°C)	175	190	170	175
Water absorption (24 h at 23 °C)(mg)	45–65	2–6	30–50	50–100

Source: Brydson [88].

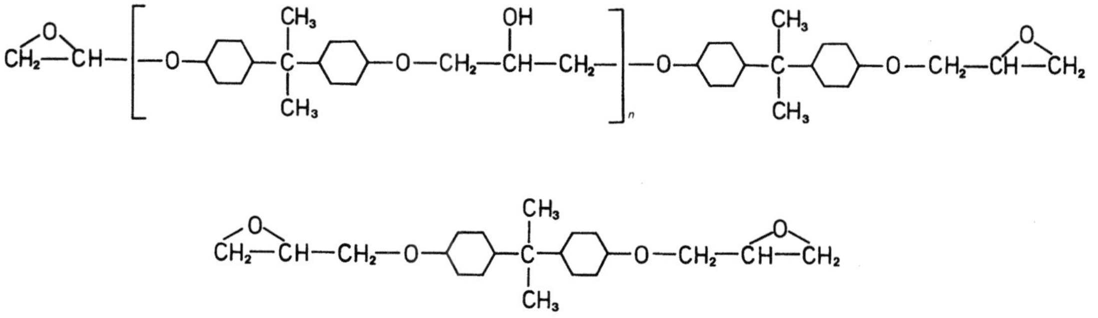

Fig. 4.47. Epoxy resins resulting from the reaction of bis-phenol A and epichlorhydrin.

groups, so that cross-linking can be achieved without the use of formaldehyde donors. The best strength is obtained from resins based on phenol or phenol–cresol mixtures.

4.2.2.5 Epoxies

Epoxy polymers are based on rather a complex range of monomers whose common feature is that they contain a tight C—O—C ring whose scission and rearrangement forms part of the polymerisation reaction. Although this definition allows for a very wide range of possible compounds, most epoxy resins are based on products of reactions between bis-phenol A and epichlorhydrin. These products vary from the diglycidal ether of bisphenol A to the higher-molecular-weight products (with n up to about 10), both of which are shown in Fig. 4.47. Curing of these glycidal ethers can be carried out using the epoxy- or hydroxy-groups, and this can be achieved either catalytically or by using a polyfunctional agent such as an amine.

It is possible to group the very wide range of epoxy resins in terms of their uncured structures, but much of the distinction disappears when these are cross-linked, so that it is difficult to make statements about properties in general terms, although the main skeleton of the resins usually has good chemical resistance. Epoxies generally show good strength and toughness for thermoset materials; hence their use as resin matrices for composites.

The main applications of epoxy resins are in coatings and adhesives, although they are often used in laminates.

4.2.2.6 Polyimides

Polyimides are characterised by the functional group shown in Fig. 4.48, which allows the production of polymers which have a backbone consisting predominantly of ring structures, leading to high temperature capability (typically to around 300 °C).

Polyimides may be either thermoplastics, derived from a condensation reaction between anhydrides and diamines, or thermosets, produced by addition polymerisation of pre-formed imide monomers or oligomers (which have been

produced by the condensation reaction). An example of each of these is shown in Fig. 4.49.

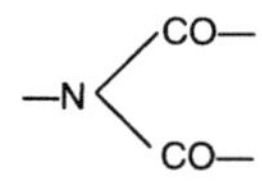

The main applications of polyimides are as high-performance composite matrices for aerospace applications, and for this reason their mechanical properties have been intensively studied. There is fairly limited interest in the engineering properties (other than as coatings or adhesives) of the neat resins of the thermosetting plastics, so that much of the property discussion is deferred until Section 4.4, where their properties in composites are described.

4.2.3 Elastomers

Apart from the coiling of chains, there is no molecular structural difference between the elastomers and the thermoplastics, the distinction normally being made upon the degree of elastic deformation which can be undergone by these materials. This extensibility is related to the coiling of the polymer molecules brought about by the *cis*-arrange-

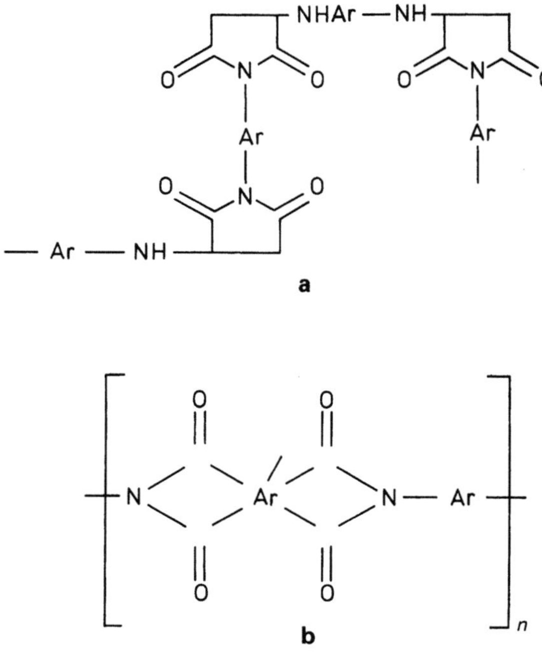

Fig. 4.48. The functional group which characterises the polyimides.

Fig. 4.49. Some polyimides: **a** an amine cross-linked poly-bismaleimide produced by addition polymerisation; **b** a condensation polyimide [92].

Table 4.56. Approximate world rubber supply data by type

Rubber	Usage (%)			
	1977	1980	1983	1986
Natural rubber (NR)	34.2	35.7	38.6	43.2
Styrene–butadiene rubber (SBR)	38.4	34.6	31.6	27.7
Butadiene rubber (BR)	10.3	10.9	11.2	11.0
Ethylene–propylene rubber (EPDM)	3.3	3.8	4.7	5.2
Butyl rubber (IIR)	4.5	5.0	5.1	4.5
Chloroprene rubber (CR)	3.7	3.7	3.7	3.3
Nitrile rubber (NBR)	2.2	2.5	2.3	2.4
Synthetic isoprene rubber (IR)	2.4	2.6	1.6	1.4
Acrylic rubbers (ACM and AEM)	0.3	0.4	0.4	0.3
Other rubbers	0.7	0.8	0.8	1.0
	Usage ('000 000 t)			
SBR supply (solid + latex)	3.33	2.89	2.56	2.56

Source: Brydson [93].

ment of the bonds. Mitigation of this behaviour can be achieved by cross-linking, and this is usually necessary to provide some mechanical integrity to structures containing elastomers. Many rubbers are hydrocarbons and therefore dissolve or swell (depending upon whether or not vulcanised) in hydrocarbon solvents. The unsaturated rubbers are also fairly reactive to oxygen.

Brydson [93] has ranked the rubbers according to annual supply. His 1986 data are reproduced in Table 4.56 and will be discussed in the order in which they are given there, although, as with the other polymeric materials, these data are distorted, in this case by the use of rubber in the tyre industry.

4.2.3.1 Natural Rubber (NR)

Natural rubber remains the most widely used polymer of the elastomeric class, although it is not oil-resistant and does not have particularly good temperature resistance. Its resistance to fire, oxygen and ultra-violet radiation are not remarkable either, but NR has high strength (for an elastomer and when appropriately treated), toughness and reversible elasticity.

Botanically, most commercial natural rubber comes from the tree *Hevea brasiliensis*; chemically, in its raw form, it

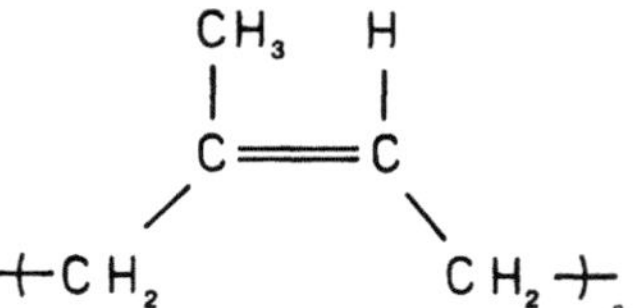

Fig. 4.50. The structure of *cis*-1,4-polyisoprene, the principal component of natural rubber.

consists of about 60 per cent water, 33 per cent dry rubber and various resinous, proteinous substances and sugars. The rubber itself consists almost entirely of *cis*-1,4-polyisoprene (Fig. 4.50).

Natural rubber owes its properties to the fact that it is an unsaturated aliphatic hydrocarbon polymer with a very regular structure. The molecular weight of the natural product is very high, and this makes plastic forming very difficult. It is normal to subject the rubber to mastication, a process of intense mechanical working which reduces the molecular weight and improves the plasticity of the material. Natural rubber is relatively reactive, owing to the presence of the unsaturated bond, and this has implications for halogenation, vulcanisation, ageing and epoxidation.

The principle of vulcanisation is quite simple, crosslinking of the rubber molecules being achieved by the use of sulphur (Fig. 4.51). However, as well as the sulphur, an accelerator, an activator and a fatty acid are also used, and a typical vulcanisate has a variety of different types of cross-link and the chemical groupings formed and the cross-links are not as simple (or even in the same place on the molecule) as shown in Fig. 4.51.

Oxygen and ozone attack can result in considerable degradation of natural rubber, especially in conjunction with sunlight. These effects can be combated by the addition of antioxidants and antiozonants. These additives interfere with the degradative processes and reduce the tendency to crack or perish.

Chlorination of natural rubber is used for chemical- and heat-resistant coatings, and epoxidised natural rubber (ENR) shows high strength and good oil resistance and low air permeability.

Table 4.57 shows a selection of properties of gum vulcanisates.

4.2.3.2 Styrene–Butadiene Rubber (SBR)

SBR is the other major rubber product used worldwide, although it must be said that the market is dominated by the pneumatic tyre, so that uses outside this area could show a rather different pattern. SBRs are copolymers of styrene and butadiene (Fig. 4.52), although the ratio of styrene to butadiene mers is usually about 1:6, so that the benzene ring does not have as much of an effect on the

Fig. 4.51. Schematic diagram of cross-linking of an elastomer using sulphur.

Table 4.57. Some properties of natural rubber gum vulcanisates

	Natural rubber	ENR-25	ENR-50	ENR-25 (black filled)	ENR-50 (black filled)
Modulus at 100% elongation (MPa)	0.93	0.89	0.90	2.93	4.10
Modulus at 300% elongation (MPa)	2.3	2.1	2.2	12.9	14.2
Tensile strength (MPa)	29.0	25.0	31.0	25.1	22.0
Elongation at break (%)	660	670	650	540	455
Tensile fatigue (kc to failure)					
0 to 100% extension	144	113	134	—	—
50 to 150% extension	1350	910	813	—	—
Crystallinity (400% elongation) (%)	11	11	10	—	—

Source: Brydson [93].
Note: ENR-25 = natural rubber with 25mol% epoxidation; ENR-50 = natural rubber with 50mol% epoxidation.

chains as is suggested by the formula. Divinyl benzene is occasionally used as an additional cross-linking agent.

The presence of unsaturation in SBRs makes their reactivity in terms of vulcanisation, chlorination and oxidation similar to NRs, although the mechanisms differ somewhat, owing to the absence of the activating methyl group adjacent to the double bond. On the other hand, owing to its synthetic nature, the molecular weight of SBR is controllable at manufacture, which is fortunate because the molecules do not break down on mastication.

As for natural rubbers, the properties of SBRs depend upon the vulcanisation process, and in particular the density and type of cross-links produced. Brydson [94] indicates that cross-link distributions are different between the two types of rubber, and Table 4.58 shows some properties of SBR vulcanisates.

Broadly the abrasion resistance, low T_g, and moderate ageing resistance of SBRs find them a number of engineering applications, although these are, as already mentioned, dominated by the tyre industry. Heat, weathering and oil resistance are properties which can be provided rather bet-

ter in other types of rubbers, and SBR does not quite match the elastomeric properties attained by the natural materials.

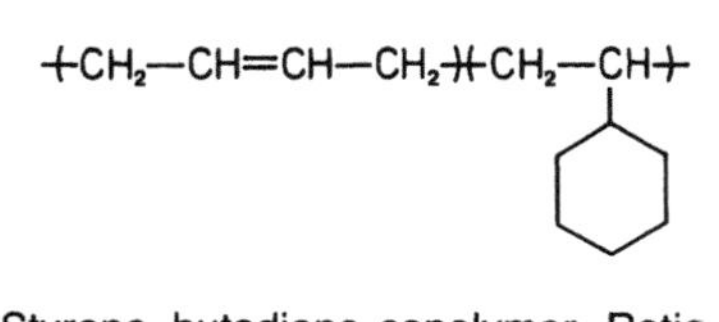

4.2.3.3 Butadiene Rubber (BR)

Butadiene rubber (one of the component copolymers of SBR) is the highest bulk product of the homopolymer synthetic rubbers, and finds most of its uses either in tyres or in the manufacture of high-impact polystyrene (Section 4.2.1.6).

It is made by the simple addition polymerisation of 1,3-butadiene as shown and three types of isomerism are possible in the repeating unit (Fig. 4.53), since the 1,2 and 3,4 polymers are identical. However, the 1,2 (or 3,4) structure contains an asymmetry and can therefore have tacticity. The aim is to produce rubbers which are as much as possible stereoregular (although this cannot be achieved with 100 per cent success), and thus the possible useful molecular structures are

cis-1,4

trans-1,4

isotactic 1,2-

syndiotactic 1,2

and the proportion of these present in the final product has a substantial effect on the melting and glass transition temperatures. The addition of vinyl also increases the glass transition temperature significantly (by about 1 degree per 1 per cent vinyl).

Resilience is very high at lower levels of deformation, but at higher levels (for engineering applications) heat

$$+CH_2-CH=CH-CH_2+CH_2-CH+$$

Fig. 4.52. Styrene–butadiene copolymer. Ratio of styrene to butadiene in commercial SBRs is about 1 : 6.

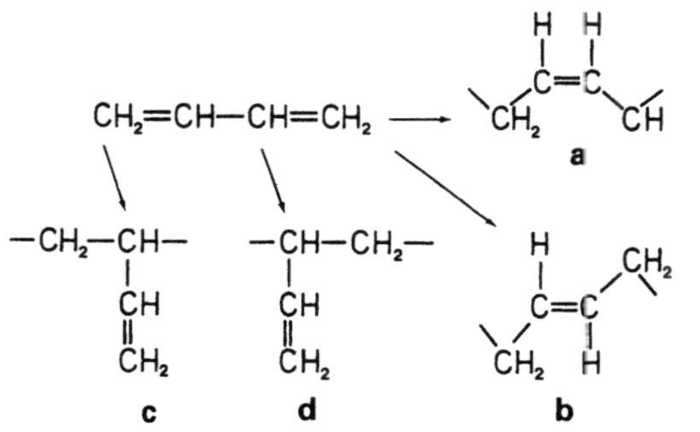

Fig. 4.53. The four isomers of butadiene: **a** *cis*-1,4-; **b** trans-1,4-; **c** 1,2-; **d** 3,4-.

Table 4.58. Some properties of styrene butadiene rubber vulcanisates
(CV, conventional vulcanising with 2% sulphur and 1% CBS accelerator; EV, 'efficient vulcanisation' using a variety of additives)

Property	CV	EV
Modulus at 300% elongation (MPa)	—	9 to 15
Tensile strength (MPa)	—	19 to 23
Tension fatigue (kc to failure)		
75% extension (unaged)	645	750 to 800
75% extension (aged)	192	400 to 445

Source: Brydson [93].

Table 4.59. A comparison of the properties of butadiene rubber (BR) with styrene butadiene rubber (SBR) and natural rubber (NR) all loaded with 50% high abrasion furnace black

Property	BR	SBR	NR
Tensile strength (MPa)	15.7	27.4	28.6
Modulus at 300% (MPa)	8.3	18.1	18.6
Elongation at break (%)	480	450	510
Tear strength (angle) (kN m^{-1})	39	49	108
Resilience, Lupke (%)			
at 23 °C	60	40	50
at 70 °C	70	55	60
Hardness, Shore	58	58	58
De Mattia cut initiation (kc)	> 500	> 500	100
Cut growth (kc)	10	20	100

Source: Brydson [93].

buildup is high, and this is why BRs are usually blended. Mechanical properties of BR are compared with SBR and NR in Table 4.59, and it may be noted that blending reduces tensile strength and modulus a little.

BRs find few applications alone and are usually blended with either NR or SBR.

4.2.3.4 Ethylene–Propylene Rubber (EPM and EPDM)

Ethylene–propylene rubbers are copolymers which, unlike the two homopolymers, have rubbery properties. The early binary copolymers (EPMs) have a certain amount of residual crystallinity because the copolymerisation is not entirely random and cross-linking cannot be achieved using sulphur alone, although sulphur is used as a co-agent. The terpolymers incorporate a third monomer (usually an unconjugated diene) to provide double bonds in the chains and hence aid cross-linking (EPDMs) using established methods for rubbers.

The saturation of the ethylene–propylene rubbers means that they have good resistance to oxygen and to polar solvents, but they can be halogenated. The degree of crystallinity (dependent upon the ethylene/propylene ratio) has an effect on the strength of the final material and also upon

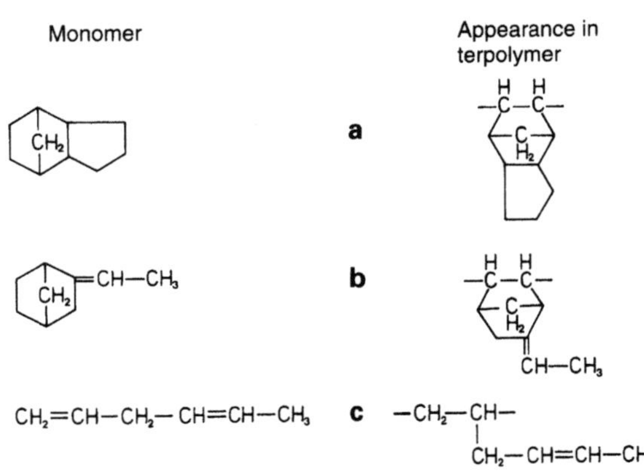

Fig. 4.54. The principal diene monomers used in EPDM terpolymers and the structure normally present in the terpolymer: **a** dicyclopentadiene, **b** ethylidene norbornene and **c** 1,4-hexadiene. (From Brydson [93].)

the glass transition temperature, and the rubberiness of the material is related to $T–T_g$.

Three principal diene monomers (Fig. 4.54) are used in commercial EPDMs, the key requirements being that the cure site double bond does not take part in the polymerisation reactions and that its scission does not cause a dramatic drop in molecular weight (i.e. it does not form part of the molecular backbone).

The vulcanisates are superior to the diene rubbers in their oxidation, ozone, chemical and heat ageing resistances, although protection against sunlight is normally required, and swelling can take place in hydrocarbons. Strengths of around 21 MPa (with carbon black reinforcement) are good (for a rubbery material). Continuous use up to about 125 °C is possible and resilience is as good as natural rubber, although less temperature-dependent above about 10 °C. As with NRs, resilience is limited at the low temperature end by T_g, which in turn depends on propylene content as described above.

The limited application of ethylene–propylene rubbers in tyre manufacture is due mainly to their increased cost and lack of tack. A major part of the non-tyre rubber sector is occupied by these materials, however.

Table 4.60. A summary of data on elastomers with particular reference to uses in offshore engineering

Resistance to	Natural rubber NR	Polychloroprene CR	Nitrile rubber NBR	Butyl rubber IIR	Synthetic isoprene IR	Fluorocarbon elastomers	Silicone rubbers
Abrasion	A	A	C to B	A to B	A	B	C
Low-temperature stiffening	A	C	C to B	B	A	C	C to B
Creep	A	B to C	C	—	—	—	—
Ozone	C to A[†]	A	C to B	A to B	A to B	A	B to A
Oil	C to B[††]	A to B	A	C	C	A	C to B
Fracture	A	B	C	B	A	C to B	C to B
Fatigue							
low strain	B	A	B	—	—	—	—
high strain	A	B	C	—	—	—	—

Source: Dutt [94] and Dexter [53].
Resistance: A, excellent; B, moderate; C, poor.
[†] Suitably protected with antiozonant additives.   [††] Depending on the dimensions of the components.

4.2.3.5 Other Elastomers

In the low-consumption list of rubbers are included butyl rubber (IIR), chloroprene rubber (CR), nitrile rubber (NBR), synthetic isoprene rubber (IR), the silicon rubbers and some others. The properties of some of these are listed in Table 4.60, which also includes some more general data on elastomers.

Apart from synthetic isoprene, which owes its limited success to its sterically more perfect natural counterpart and is only really manufactured for strategic reasons (in countries which do not have access to the natural material), the three most important elastomers mentioned in this section are butyl, chloroprene and nitrile rubbers.

Butyl rubbers (strictly isobutene–isoprene rubbers, or IIRs) occupy an important place in the speciality market, owing mainly to their low gas permeability and despite their low resilience, although the latter is occasionally an asset in damping applications. The gas permeability resistance is helpful in inner tubes, of course, but also in many chemical and process applications, and the low water absorption of these materials further enhances their potential application.

Chloroprene rubbers (polychloroprene or CR; also widely known by the trade name 'neoprene') are characterised by their combination of good oil, heat, flame and ozone resistance. Strength, abrasion resistance and weathering resistance are also useful, and applications include seals and cable sheathing.

Nitrile rubbers are copolymers of acrylonitrile and butadiene, and their principal feature is resistance to liquid hydrocarbon fuels. Their copolymeric nature, along with a number of other possible modifications and blends, means that these are a diverse group of rubbers, although all incorporate the major chemical feature on which their success relies. Although hydrocarbon resistance is not absolute (resistance in the presence of aromatics, sulphur-bearing additives or contaminants and oxygen-bearing liquids is more limited), NBRs are used widely where fuel, oil and grease are involved and, although much of this market is automotive, many other machinery applications exist.

The silicone rubbers are a much more complex group than any of those considered in this section because they are defined as any rubber which is made using organosilicon chemistry. There are five important chemical groups listed in Table 4.61, and three different production technologies:

RTV – room-temperature vulcanising silicone rubbers, which air cure and are used in adhesive applications.

RTV – high-temperature vulcanising silicone rubbers (solid materials in the uncured condition) using traditional rubber processing including curing.

LSR – liquid silicone rubbers which are heat-curable liquids.

Generally, the major strengths of these elastomers are a very wide usable range of temperatures (–100 to 250 °C), excellent ozone, oxygen and ultra-violet resistance, non-stick properties, low chemical reactivity, good electrical insulation and possible optical transparency. Their major limitations are low tensile strength, low resistance of some grades to hydrocarbons, relatively high cost and high permeability to gases, the last of which may of course also be an advantage in some applications.

Table 4.61. The ISO classification of silicone rubbers. VMQ are general purpose, PMQ and PVMQ are extreme low temperature grades and FVMQ are fuel, oil and solvent-resistant grades. MQ and PMQ are now rarely marketed

MQ	Silicone rubbers having only methyl substituent groups on the polymer chain (polydimethyl siloxanes)
VMQ	Silicone rubbers having both methyl and vinyl substituent groups on the polymer chain
PMQ	Silicone rubbers having both methyl and phenyl substituent groups on the polymer chain
PVMQ	Silicone rubbers having methyl, phenyl and vinyl substituent groups on the polymer chain
FVMQ	Silicone rubbers having fluoro, vinyl and methyl substituent groups on the polymer chain

Source: Brydson [93].

Silicone rubbers find uses in both electrical and mechanical applications, mostly for their insulating and sealing properties at relatively high temperatures.

4.3 Inorganic Materials

In bulk engineering use, inorganic materials are normally taken to comprise ceramics, glasses, stone, brick and, occasionally, cement. Only the first two of these are considered in this section, reserving a separate section for cement and concrete. In fact, 'inorganic materials' is a somewhat excessively broad categorisation to use for the materials considered here, since many are oxides or mixtures containing oxides. The term 'ceramic' is often used to group these materials, although some confusion exists as to whether glasses belong to this category or not.

The main areas of ceramic technology dealt with in this section are the developments in reinforcing materials for composites and those in the so-called 'engineering ceramics'. Glasses are considered for their applications both as optically transparent materials and also as reinforcements. Some reinforcing materials which are neither glasses nor ceramics are also considered.

The main features of glasses and ceramics are their extreme stability, their high strength and modulus and their brittle nature. In addition, the ceramics and some glasses show good high-temperature resistance.

4.3.1 Glasses

The glassy state of ceramic materials is one in which there is no abrupt change in physical property between the solid and liquid states [96]. For this reason, glasses cannot be regarded as having the long-range order associated with the solid crystalline state, and are treated in some ways as if they were highly viscous liquids and in others as if they were solids. In engineering, the latter is normally the case, but it is important to have an appreciation of the structure of glasses in order to understand the importance of composition on properties.

It is usual to consider glasses as semi-solidified silica where rapid cooling does not allow the glass to form into the ideal silica crystal structure. Of course, most glasses contain other oxides and other compounds, either as impu-

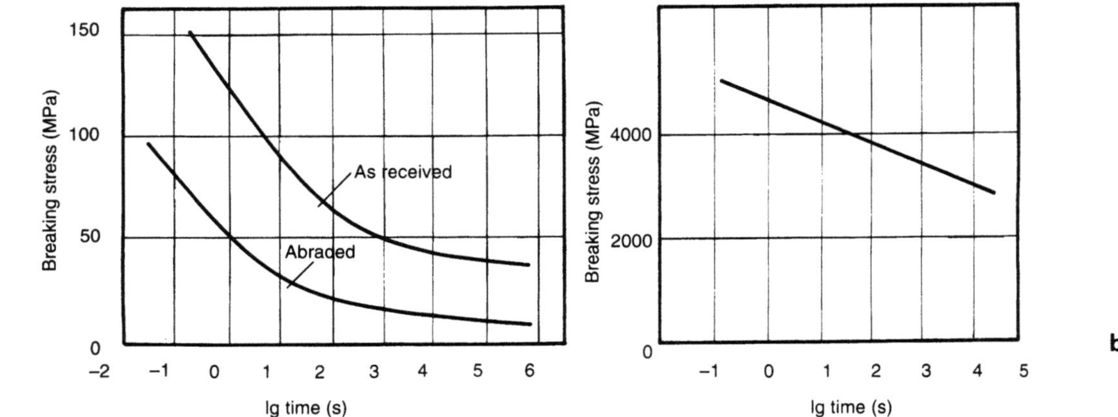

Fig. 4.55. Effect of defects on 'static fatigue' properties of glass: **a** soda–lime–silica rods and **b** 'flaw free' silica fibres. (From Holloway [95].)

rities or as deliberate additives. The additive oxides can be classified in terms of their M–O bond strength, the strongest being glass-formers and the weakest being referred to as modifiers. For example, sodium oxide is a very strong modifier and something of the range of possible oxide additives and their effects can be appreciated from Table 4.62.

The use of glass in engineering is dominated by either its optical or its corrosion resistant properties and it is used despite, rather than because of, its mechanical properties. For marine applications, the optical and corrosion-resistant properties of different types are not much of an issue, most glasses meeting these requirements, so that the remaining discussion centres around the mechanical properties for use as bulk solid and as reinforcement. The weathering resistance of relatively poor glasses in water-plus-sulphur-dioxide environments leads to a loss of only about 8 μm per year.

The three main types of bulk glasses are soda–lime, lead and borosilicate. Of these, the soda–lime glasses are the largest-volume glasses, particularly through their use in windows and containers. The borosilicate glasses are used for laboratory and chemical glassware and are the most resistant to weathering and chemical attack [96].

Whereas the theoretical strength of glass is very high, its low toughness means that this strength is difficult to achieve. In compression, glass can achieve strengths of around 10 GN/m^2 but, owing to the inevitable surface defects, the tensile values are much smaller. The compressive strength is of particular use in some subsea applications where spherical glass shapes under external pressure are a useful engineering structure. However, performance where a significant proportion of the stress is tensile needs also to be considered, and here safe working stresses are dominated by the material's toughness as described in Chapter 2.

Furthermore, stress-sensitive interactions between glass and water are possible in the atmosphere [96], so that allowable sustained loads are usually less than short-duration allowables, as illustrated in Fig. 4.55. Thus, although glass can be reliably considered to have a strength of at least 40 MPa when tested under laboratory conditions, it is unlikely that working stresses over 16 MPa can be used with confidence.

Table 4.62. Classification of the oxides in terms of their bond strengths and effect in glasses

Metal in oxide	Bond strength (kcal/mol)	Metal in oxide	Bond strength (kcal/mol)	Metal in oxide	Bond strength (kcal/mol)
Glass formers		Intermediates		Modifiers	
B	119	Ti	73	La	58
Si	106	Zn	72	Y	50
Ge	108	Pb	73	Sn	46
P	99	Al	60	Ga	45
V	99	Be	63	In	43
As	73	Zr	61	Pb	39
Sb	76	Cd	60	Mg	37
Zr	81			Ca	32
				Ba	33
				Sr	32
				Na	20
				K	13
				Cs	10

Source: Askeland [3].

The size and distribution of surface defects are not easily controlled, because the critical defect size resulting from the low fracture toughness of glass is so small. For glass, the value of G_c is about 0.001 to 0.01 kJ m^{-2}, compared with about 3 for cast iron and about 100 for mild steel. The work of deformation, G_c, is very like the parameter γ in the Griffith formulation and is related to fracture stress as follows:

$$\sigma_f = \sqrt{(EG_c/\pi a)}$$

The critical dependence of glass strength on flaw size can be demonstrated by carefully controlled experiments, e.g. Fig. 4.56.

Thus, although flaw-free glass can be produced, the maintenance of a flaw-free condition in service is practically impossible, and the best that can be done is to treat the glass to mitigate the problem. This is done in two ways, referred to as 'toughening' and 'laminating'. Glass can be reinforced with, for example, steel mesh, but this is not a particularly attractive or elegant solution. Laminating consists of making a laminated composite between the glass and a transparent plastic material (usually polyvinylbutyral) and toughening consists of a heat treatment (or occasionally a chemical treatment) whereby the glass surface is put into compression, inhibiting the initiation of fracture until the applied stress exceeds the residual surface stress. Because the thermal toughening process depends on cooling the surface more rapidly then the core, like residual welding stresses (Chapter 5) the magnitude of the effect depends on the coefficient of thermal expansion of the glass. Thus, for a plate of thickness 6 mm, soda–lime glass can be induced to have a compressive surface stress of about 100 MPa, but owing to its lower coefficient of thermal expansion the same thickness of borosilicate could only be induced to have a surface compressive stress of about 40 MPa. Nevertheless, thermal toughening of either of these glasses results in a substantial increase in their load-bearing capacity.

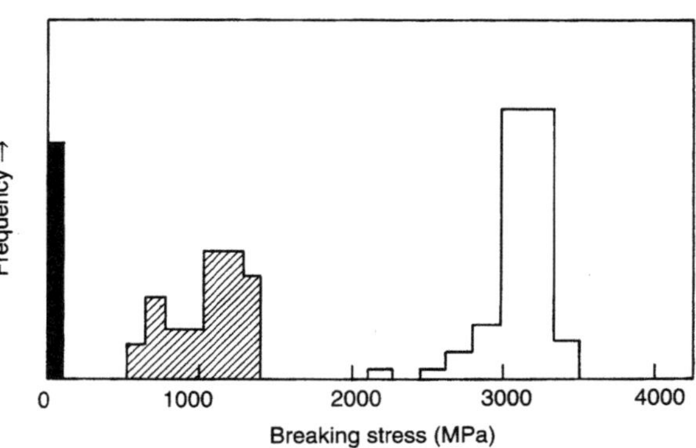

Fig. 4.56. Histograms of breaking strengths of 'off-the-shelf' glass rods (solid black), small rods free from abrasion cracks (shaded) and 'flaw-free' small rods (unshaded). (From Holloway [96].)

BS 6262 [97] gives design allowables for suspended glass areas subjected to wind loading and includes curves for annealed, laminated and toughened glass. Figure 4.57 shows the effect of toughening, which can be seen, for example, by comparing the required thickness of a 4 m^2 pane subjected 2 kN m^{-2} wind loading. Reed and Engelson [98] give a more quantitative description of the load-bearing properties of glass panels.

Another way of exploiting the relatively high modulus of glass is to impregnate it in resin as a reinforcing fibre, and glass is the commonest material used for reinforcing plastics materials in composites manufacture. Of course, because of their relative moduli, there is no advantage to be gained in reinforcing most metals with glass fibres.

The variable quantities in glass fibre use are glass composition, fibre diameter and fibre form. Diameters of glass fibres for reinforcement are usually in the range of a few to

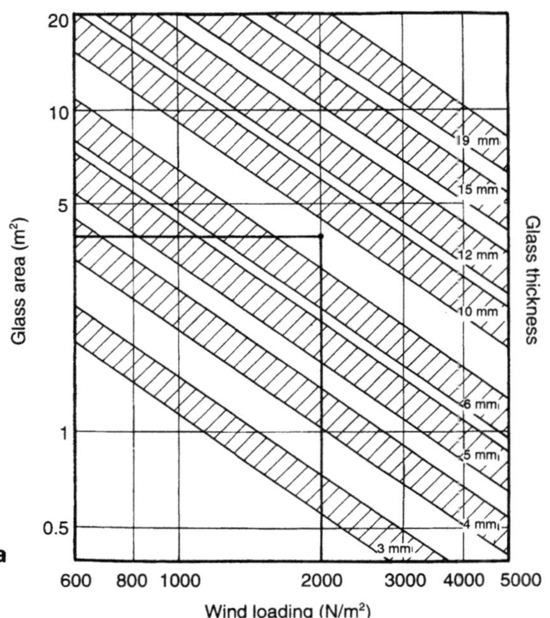

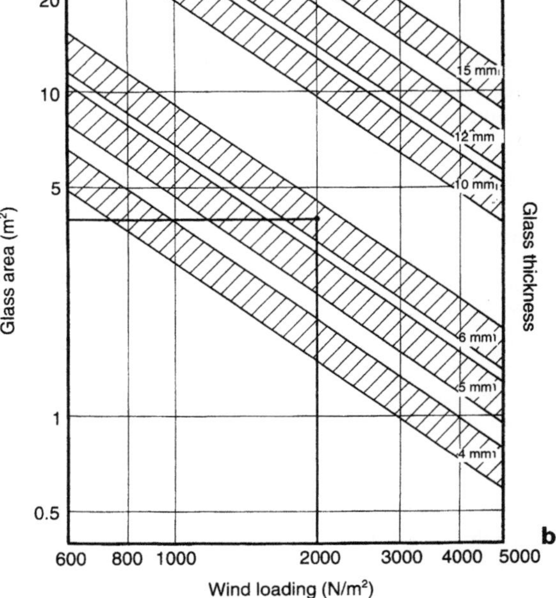

Fig. 4.57. Effect of toughening on the load-bearing capacity of glass panes: **a** annealed and **b** toughened glass illustrating required thickness for a 4-m^2 suspended area subject to a 2-kPa wind loading. (Adapted from BS 6262 [98].)

Table 4.63. Composition and properties of the main grades of reinforcing glass

	Grade of glass		
	C (chemical)	S (high-strength)	E (electrical)
Component			
Silicon oxide	64.6	64.2	54.3
Aluminium oxide	4.1	24.8	15.2
Ferrous oxide	—	0.21	—
Calcium oxide	13.2	0.01	17.2
Magnesium oxide	3.3	10.27	4.7
Sodium oxide	7.7	0.27	0.6
Potassium oxide	1.7	—	—
Boron oxide	4.7	0.01	8.0
Barium oxide	0.9	0.2	—
Miscellaneous	—	—	—
Property			
Specific gravity	2.62	2.48	2.58
Tensile strength (MPa)	3625	4590	3450
Tensile modulus (GPa)	72.5	86.0	72.5

Source: American Society for Metals [99].

20 μm. Although a wide range of glass fibre types are available, relatively few are in widespread use as reinforcements. The three main types are E-glass, S-glass and C-glass, of which the latter two are special-purpose, S-glass being a high-strength filament and C-glass being a chemical-resistant grade. The compositions and properties of the three grades are given in Table 4.63.

The quoted tensile strengths are usually measured on pristine single filaments or multifilament strands, and the influence of forming into yarns will be to reduce this somewhat. To resist the effect of inter-filament abrasion some form of size is usually applied to the fibres.

Individual filaments are combined into strands at manufacture and these strands, consisting of between 51 and 1224 filaments, constitute the basic building blocks of glass fibre reinforcing fabrics. There are a number of different ways of defining the amount of glass in a strand or yarn (a yarn being made from a number of strands twisted or plied together):

Tex: This is defined as the mass per unit length of glass in the strand in grams per kilometre.

Count: This is defined as one-hundredth of the total length (in kilometres) of glass in 1 pound (0.454 kg) of strand, for example a count of 130 represents a strand containing 13 km per pound. The relationship between Count and Tex is that Count × Tex = 4940.

Others. Two other, less common designations related to clothing fibres are denier (most conveniently given as 9 × Tex) and cotton yarn number (most conveniently given as 1/8.4 × Count)

Thus, a Tex strand count of 33 would correspond to a count of 150, or 200 denier or a no. 18 cotton yarn. Note that Tex and count are inversely proportional to each other, and Table 4.64 illustrates this with some basic strand designations.

The overall designation of a yarn also contains information on the glass type, twist and construction, and Table 4.65 illustrates the British and US designations with two examples.

Table 4.64. Basic reinforcing strand fibre diameters and count designations for a 200 filament basic strand

Filament diameter designation*		Strand count in Tex (g/100 m)	US strand count	Strand yards per pound
Metric (μm)	US			
3.5	B	5	B–900	90 000
4.5	C	9	C–600	60 000
5	D	11	D–450	45 000
6	DE	17	DE–300	30 000
7	E	22	E–225	22 500
9	G	33	G–150	15 000
10	H	39	H–125	12 500
13	K	66	K– 75	7 500

Source: Weeton et al. [100].
*For 200-filament basic strand.

Glass fibre reinforcement can be supplied in a variety of different forms, depending on their use, and some of these are described and illustrated in Chapter 2.

Glass flakes and particles can be used as fillers in composite materials. Flakes can produce a higher theoretical modulus than fibres in a composite, because they can be more tightly packed. Microspheres, whether hollow or solid, come close to providing the type of strengthening produced by precipitates in metals. It is difficult to measure mechanical properties of powder or flake materials, and anyway this is not of great relevance outside the composite, owing to the way in which load is transferred from the matrix to the reinforcement. One interesting application is hollow microspherical particles for reinforcement of buoyancy materials to great depths.

4.3.2 Crystalline Ceramics

Crystalline ceramics have traditionally been used in masonry, brick and pottery, and are often bonded with glasses to improve their structural qualities. However, modern manufacturing methods have allowed the development of a new breed of ceramics which have their own structural strength, and it is these 'engineering ceramics' which are dealt with briefly here. These ceramics are not always oxides and can involve a variety of different covalent compounds. Table 4.66 gives a brief indication of the properties of engineering ceramics against the more traditional types.

Most of the development in bulk engineering ceramics has been for the high-performance engine market, where the oxidation resistance and high temperature strength of the materials are unmatched. However, the other outstanding properties of ceramics, such as their extreme corrosion and wear resistance, coupled with a high Young's modulus, should keep them in consideration for other applications.

The principal mechanical engineering ceramics are oxides and low-density covalently bonded inorganics such as silicon nitride or the so-called 'sialons', which are phases in the silicon–aluminium–oxygen–nitrogen system (Jack [102]). Principal current applications are in cutting tools, turbine blades and other engine parts and as insulation. Some of the key properties of these ceramics and their microscopic make-up are discussed by Parrott [101] and by Lewis and Leng Ward [103].

Table 4.65. Examples of British and US designations for yarn yield, construction and twist for reinforcing

System	Single yarns						Plied yarns					
Tex	EC	9	33		40	Z	EC	9	33	×2	80	S
US	EC	G	150	1/0	1.0	Z	EC	G	150	1/2	2.0	S
Note	1, 2	3	4	5	6	7	1, 2	3	4	5	6	7

Source: Weeton et al. [100].

1. Glass type: E signifies that electrical grade glass is used.

2. Fibre type: C signifies that the yarn is made from continuous fibre. If staple fibre was used, S would be used to signify that type.

3. Filament diameter: the 9 and G are the corresponding Tex and US designations for 9 μm filament diameters.

4. Yarn count: the 33 [Tex] and 150 [Yardage] designate the yarn count.

5. Yarn constitutions: since the two systems differ somewhat, they are now explained in some detail:

Single yarn
Tex system: The convention in this system is to designate a single yarn as understood when no multiplier is indicated immediately after the count in Tex. The multiplier × 1 is not indicated and is understood by convention.
US system: the 1/0 designates a single yarn uncombined or plied with any others. The first digit signifies one strand. The second (the zero after the slash) indicates that the strand is unplied.

Plied yarns
Tex system: the designation × 2 signifies that the yarn is plied. Two single yarns twisted together form the yarn construction.
US system: the construction designation 1/2 signifies one strand plied with another yarn of the same type.

6. Magnitude of twist: the Tex and US twist equivalences are 40 and 1.0, respectively, for the single-yarn example in Tex, the 40 signifying forty turns per metre (TPM), which is the same as one turn per inch (TPI) or 1.0 in the US system. In the case of the single and plied yarns, the 80 and the 2.0 again indicate the magnitude of twist in turns per metre and turns per inch, respectively.

7. The twist is designated by the letter Z or S, and these letters have the normal significance for either a left-handed or right-handed twist, respectively.

As an example of the growing interest in marine applications of ceramics, Kyrtatos [104] has recently examined the potential uses of ceramic materials in the insulation and in the combustion chambers of large marine diesels. Broadly, this author concludes that the thermodynamic advantages of insulation are marginal, but that the possibility of improved heavy-fuel burning ability and the reduced wear resulting from the use of ceramics in combustion chambers could be quite advantageous.

Ceramic materials can also be produced in fibre form, for use in composites with resin, ceramic or metallic matrices. Ceramic fibres were originally manufactured for insulation, and indeed the naturally occurring ceramic fibre, asbestos, can be considered as one of these. However, continuous and discontinuous ceramic fibres based on oxides, nitrides and carbides have proven useful reinforcing materials. A large number of different compositions and forms of ceramic fibres are now available and these are discussed in detail by, for example, Johnson and Sowman [105].

It must be said that the main spur for the development of ceramic reinforcing fibre has been for high-temperature aerospace applications, but there are uses in the reinforcement of resins, more particularly for the continuous fibres. Whiskers, which are near-perfect single-crystal materials, are necessarily small and are of predominant benefit in the advanced metal- and ceramic-based composites.

Ceramic materials are widely used as fillers for resin-based composites, and some applications are noted in Table 4.67. Higher-grade ceramic particles are also used for reinforcing metal- and ceramic-matrix composites.

4.3.3 Other Reinforcing Materials

Apart from glasses and ceramics, there are a number of other fibres used for reinforcement. One of these is aramid

(see Section 4.2.1.5), and, although this is a polymeric material, some properties of these and other polymeric fibres will be discussed here for comparison. Apart from these, the other principal fibre reinforcing materials are carbon fibres. The key properties which distinguish these two are given in Table 4.68, with E-glass for comparison.

It should be noted that aramid is exceptional among the polymeric fibres, although some high strength, high-modulus polyethylene fibres are available. Table 4.69 shows a compilation of some of the high-performance polymeric fibres, and also shows some of the more prosaic textile fibres for thermoset and rubber reinforcement.

Carbon fibres can be seen to have developed from polymer fibres, because they are often manufactured from organic precursors. These precursors are converted to graphite-like substances having the strong covalent bonding of the graphene layer but the relatively weak bonding between planes. Because of its asymmetrical crystal structure, graphite has a very high theoretical modulus in the basal plane (1020 GPa) compared with diamond (20 GPa), but only a fraction of this modulus is achieved in fibres, even if steps are taken (usually by plastic deformation) to orient the graphene planes along the length of the fibre. The principal precursor materials for carbon fibres are rayon and isotropic pitches for low-modulus fibres, and polyacrylonitrile and liquid crystalline (or mesophase) pitches for the higher-modulus fibres. The effects of precursor and orientation on fibre properties are substantial, with tensile moduli as high as 725 GPa for VHM mesophase pitch precursor fibres to as little as 41 GPa for low-modulus rayon precursor fibres [107].

Among other reinforcing materials in general use are particulate carbon and steel, as fillers and reinforcement respectively for rubbers, and whiskers and boron fibres for application mostly in metal matrix composites.

Table 4.66. Summary of properties of 'conventional' and 'engineering' ceramics

Materials	Bond strength (MN m^{-2})		Tensile strength (MN m^{-2})		Compressive strength (MN m^{-2})	Toughness, K_{Ic} (MN m$^{-3/2}$)	Specific gravity	Elastic modulus (GN m^{-2})	Expansion coefficient (10^{-6} °C^{-1})	Thermal conductivity (W m^{-1} K^{-1})	Mohr hardness
	Room T	High T	Room T	High T	Room T						Room T
Glass (window)	50	—	30 (sheet) 112 (fibre)	—	600	1	2.2	65	9	1	6
Porcelain	100	—	60	—	500	2	2.4	81	3	2	6
Aluminium oxide (sapphire)	500	250 at 1200 °C	300	100 at 1200 °C	1200	4	3.9	450	9	20	9
Silicon nitride (hot-pressed)	700	500 at 1200 °C	350	250 at 1200 °C	2000	5	3.2	300	3	35	9+
Sialons	800	700 at 1300 °C	—	—	—	5	3	300	3	7	9+
Zirconia (tetragonal)	1000	300 at 1200 °C	—	—	—	25–30	6	170	10	3	9+

Source: Parrott et al. [101].

Table 4.67. Uses of ceramic and mineral materials as fillers for resin-based composites

Filler	Uses	Typical compatible resins
Alumina trihydrate	Fire-resistant filler	Polyesters
Carbon blacks	Electrical goods	Epoxies
Calcium carbonate		
Mineral	Tile, moulded goods	Most resins
Precipitated	Pipe, putty	Most resins
Clays	Flooring tile	Most resins
Feldspar	Plastisols	Thermoplastics
Mica	Sheet moulded goods	Most resins
Silica		
Diatomite	Films	Polyethylene
Novaculite	Electrical goods	Thermosets
Quartz flour	Moulded goods	Thermosets
Tripolite	Moulded goods	Thermosets
Wet process	Sheets, films	Thermoplastics
Vitreous	Electrical goods	Epoxies
Silicate glass		
Solid spheres	Moulded goods	Most resins
Hollow spheres	Moulded goods	Thermosets
Flakes	Electrical goods	Thermosets
Talc	Extruded and moulded goods	PVC, polyalkenes
Wood and sheet flour	Moulded goods	Most resins

Source: Weeton et al. [100].

4.4 Composite Materials

The use of composite materials in engineering requires more than the adoption by designers of a new set of properties. The opportunity to specify in detail materials requirements has slowly increased with increasing materials engineering knowledge. Particularly with polymers, this has reached quite a sophisticated level, where polymers can be almost 'made to order' within reason. The use of composites carries this a bit further, because the mixing of reinforcement and matrix allows the specification of properties in location and direction, with a consequent design responsibility to avoid provision where this is not necessary. It is at present rare for such 'clean' design to be adopted for composites use, but there is the option to accept the com-

Table 4.68. Comparison of the mechanical properties of the three principal continuous fibre reinforcing materials

Property	Kevlar 29	E-glass	Graphite
Tensile strength (MPa)	3620	2415	2760
Tensile modulus (GPa)	82.7	68.9	220
Specific gravity	1.44	2.52	1.74
Brittleness	Tough	Brittle	Brittle
Abrasiveness	No	Yes	Yes

Source: Weeton et al. [100].

posite materials as a bulk material with a given set of properties. The subject matter of this section is mostly concerned with this latter option, although those interested in developing the former should refer to Chapter 2.

It is worth while at this point to try to define exactly what is meant by a composite material. Broadly, a composite is a material made up of two or more components at the microscopic level. The definition of 'components' requires some clarification in that there is the implication that the components are separately manufactured, so that GRP (glass-reinforced plastic), where glass fibre strands or mattings are mixed with a resin, would be regarded as a composite material, whereas steel, which is essentially pure iron reinforced with iron carbide (see Section 4.1.2), would not normally be regarded as a composite. Likewise, the meaning of the term 'microscopic' needs to be appreciated. For example, in the bonded flexible pipes considered in Case Study 7.5, the cord-reinforced rubber layers can be regarded as being made of a composite material, whereas the entire pipe would not be regarded as being made of a composite material, though it is clearly of composite construction.

The primary means of classification of composites is normally in terms of the matrix, which can be polymeric, metallic or ceramic, and the form of the reinforcement, which may be fibrous or particulate, although the defining line here is not quite as clear as for the matrix. The discussion will therefore be divided between the types of matrices and, as will be seen, the impetus to introduce second components to each type of matrix is quite different. Resin-based composites will be discussed first, followed by metal matrix and then ceramic matrix composites, and this is roughly the order of their current or potential use in marine environments.

Table 4.69. Properties of high-performance and textile polymeric fibres

Fibre	Specific gravity	Tensile strength (MPa)	Tensile modulus (GPa)	Ultimate elongation (%)
Aramid – Kevlar 29	1.44	3620	83	4.4
Aramid – Kevlar 49	1.44	3620	124	2.9
Aramid – Kevlar 149	1.47	3400	186	2.0
HMPE – Spectra 900	0.97	2590	117	3.5
HMPE – Spectra 1000	0.97	3000	172	0.7
Polyamide	1.13	830	2.8	—
Polyester – Dacron Type 68	1.38	1120	4.1	14.5
Nylon – Du Pont 728	1.13	990	5.5	18.3

Source: Weeton et al. [100] and Pigliacampi [106].

4.4.1 Polymer-based Composites

Reinforced plastics are perhaps the earliest form of composite materials, leaving aside those natural materials which are of 'composite' construction. There are several reasons for reinforcing plastics, but the primary one is probably to improve modulus. The ways in which properties can be predicted from the fundamental properties of the components are dealt with in Chapter 2 and the purpose of this section is simply to summarise the range of types and properties available within the reinforced plastics. Reinforced plastics can be divided into those reinforced with particles (often called 'filled'), and those reinforced with fibres, either discontinuous or continuous.

It might be mentioned here that, for engineering application, the major use of filling is in elastomers, the major method of reinforcement of thermoplastics is with discontinuous fibres (sometimes referred to as 'glass-filled'), and the major use of continuous fibre reinforcement is for thermosetting plastics; the following discussion should be seen in this light.

4.4.1.1 *Filled Polymers*

Particulate-reinforced polymers are the simplest to manufacture of the reinforced plastics, often only requiring slight modification to the processes for the matrix materials. Fillers are occasionally used as reinforcement but are also used simply to increase bulk and hence cheapen a polymeric material.

Fillers are used on all three types of matrix materials – elastomers, thermoplastics and thermosets – although their use (principally as carbon black) in elastomers is almost an integral part of the material, and so has been discussed with the matrix materials in Section 4.2.3. Fillers, as distinct from short fibres, are not normally used to reinforce thermosets, mainly because they do not enhance toughness properties in what are already brittle materials. Thus their use in these materials (as talc, wood flour, clay and others) is principally for the purpose of extending the resin or for colouration or thermal properties.

The difficulty in applying continuous fibre reinforcement to thermoplastics, combined with the fact that these are inherently tough materials, means that there are some applications for filled thermoplastic composites (again as distinct from short fibre reinforcement). Table 4.70 summarises some of the available mechanical and other relevant property data on some specific types of filled thermoplastics, although it might be noted that the impetus is often not one of enhanced strength, but in improving other properties such as lubricating, electrical and flammability characteristics, wear resistance, cost effectiveness and processability. It might be noted that fillers usually reduce the toughness of thermoplastics.

4.4.1.2 *Reinforcement with Discontinuous Fibres*

This section deals with reinforcement with discontinuous fibres, the working definition of these being those fibres which produce a composite material with nominally isotropic properties. As seen elsewhere (Chapters 2 and 5), discontinuous fibres can be aligned or can be randomly oriented in two or three dimensions, and it is usually only the last of these which fulfil the above working definition.

Table 4.70. A selection of data on filled thermoplastics including elastomers

Matrix	Filler	Specific gravity	Tensile strength (MPa)	Flexural modulus (MPa)	Remarks/ Applications
PP	10% C	0.975	33	1720	Sheeting material
PP	20% C	1.04	29	1860	Outdoor leisure equipment
PP	40% T	1.24	35	4300	Housings
PP	30% M	1.13	32–43	4070–5030	Pool equipment
PTFE	15% G	2.12	9.5	1400	Low friction
PC	22% PTFE	1.33	45	1300	Lubricated
AC	22% PTFE	1.5	40	2100	Lubricated
PA	44% PTFE	1.43	38	2100	Lubricated
PI	15% G	1.51	45	3800	Lubricated
PI	15% Mo	1.59	41	3500	Lubricated
ABS	40% Al	1.54	23	4100	Electrically conductive

Source: Weeton et al. [100], Dexter [53], Brydson [93]. Abbreviations: Matrix type: ABS, acrylonitrile–butadiene–styrene; AC, acetal; NR, natural rubber; PA, nylon 66; PC, polycarbonate; PI, polyimide; PP, polypropylene; PTFE, polytetrafluoroethylene. Filler type: Al, aluminium flake; B, carbon black; C, calcium carbonate; G, graphite (lubricant); M, mica; Mo, molybdenum disulphide (lubricant); T, talc. (PTFE as above.)

The difficulty of producing continuous-fibre reinforcement in thermoplastics (see Chapter 5) means that, at present, the principal means of reinforcing these materials is by the incorporation of short fibres into the melt flow. Almost any thermoplastic can be reinforced in this way, but most engineering interest is centred around those polymer matrices which already show some useful properties. Thus, polyamides and thermoplastic polyesters are often reinforced, principally with glass but also with carbon and other fibre types.

Table 4.71 summarises some property data for fibre-reinforced thermoplastics. Mention should also be made here of the advanced matrix thermoplastic resins such as the polyimides, poly aromatic ketones and polycarbonates.

4.4.1.3 *Continuous Fibre Composites and Laminates*

To many, the materials in this section are the 'true' resin matrix composites in that the full benefits of the incorporation of a relatively high-modulus material can only be obtained by introducing a degree of continuity of this material in the final composite. Laminates are really an extension of this into two-dimensional reinforcement, because the reinforced (or reinforcing) sheets in the laminate can be seen as 'two-dimensional fibres'. These types of reinforced plastics are, of course, rather more complex to use in design than those already discussed, mainly due to the directionality of properties.

As pointed out above, the use of continuously reinforced thermosets awaits development of the manufacturing processes, which will allow good bonding between matrix and fibres in a production environment, and much work is

Table 4.71. Effect of reinforcement with discontinuous fibres (glass or carbon) on the properties of thermoplastics

Resin type	Tensile strength (MPa)	Flexural modulus (GPa)	Impact strength*, notched/unnotched (J/cm)	Heat-deflection temperature (°C)
Amorphous resins				
Acrylonitrile–styrene–butadiene (ABS)				
Base resin	55	2.8	1.9/>21	95
30% glass fibre	100	7.6	0.75/3.5	105
30% carbon fibre	130	12.4	0.59/2.4	105
Nylon				
Base resin	66	2.6	> 21/NB	125
30% glass fibre	148	7.9	0.64/3.7	140
30% carbon fibre	207	15.2	0.64/4.3	145
Polycarbonate				
Base resin	62	2.3	1.4/>21	130
30% glass fibre	128	8.3	2.0/9.34	150
30% carbon fibre	165	13.1	0.96/5.34	150
Polyetherimide				
Base resin	105	2.8	0.53/13	200
30% glass fibre	197	8.6	0.75/5.60	215
30% carbon fibre	234	17.2	0.75/6.67	215
Polyphenylene oxide (PPO)				
Base resin	66	2.5	1.4/>21	85–150
30% glass fibre	145	9.0	1.2/5.1	155
30% carbon fibre	159	11.7	0.53/3.0	155
Polysulphone				
Base resin	69	2.8	0.64/>21	170
30% glass fibre	124	8.3	0.96/7.5	185
30% carbon fibre	159	14.5	0.64/3.5	185
Styrene–maleic-anhydride (SMA)				
Base resin	54	3.2	0.2/NB	115
30% glass fibre	103	9.0	0.59/2.4	120
Thermoplastic polyurethane				
Base resin	14	0.1	>21/NB	32
30% glass fibre	57	1.3	5.1/15	170
Crystalline resins				
Acetal				
Base resin	61	2.8	0.69/>21	110
30% glass fibre	134	9.7	0.96/4.8	165
20% carbon fibre	81	9.3	0.53/1.6	160
Nylon 66				
Base resin	80	2.8	0.48/>21	77
30% glass fibre	179	9.0	1.5/11	255
30% carbon fibre	241	20.0	0.80/6.4	257
Polybutylene terephthalate (PBT)				
Base resin	59	2.3	0.48/>21	85
30% glass fibre	134	9.7	1.4/9.1	210
30% carbon fibre	152	15.9	0.64/3.5	210
Polyethylene terephthalate (PET)				
30% glass fibre	159	9.0	1.0/...	225
Polyphenylene sulphide (PPS)				
Base resin	74	4.1	0.16/1.9	140
30% glass fibre	138	11.0	0.75/4.5	260
30% carbon fibre	186	16.9	0.59/2.9	265

Source: Weeton et al. [100].
*NB = no break.

currently under way on thermoplastic composites based on resins such as polyether ether ketone (PEEK), polyphenylene sulphide (PPS) and polyimides [108]. Table 4.72 summarises some properties of candidate resins compared with competing thermosets.

In comparison with thermoplastic matrix composites, thermoset matrix composites are very mature. These materials have been categorised (see, for example [90]) in terms of strength and temperature resistance of the matrix resins. Polyester matrices are normally considered to give adequate performance for ambient temperature use, with phenolics being required at elevated temperatures and, for best temperature resistance, polyimides. Epoxy matrix composites are normally considered to give the highest-strength composites, and are suitable for use up to the medium temperature range. Table 4.73 shows collected data for thermoplastic composites with a variety of reinforcement and matrix materials.

Table 4.72. Comparison of two candidate thermoplastic matrices for continuous fibre reinforced composites with a more conventional epoxy

	Density (g/cm^3)	Melt temperature	Processing temperature (°C)	Tensile strength (MPa)	Tensile modulus (MPa)	Izod impact strength (J/m)	Hardness	Heat distortion temperature at 1800 kPa (°C)	Water absorption (%)	Creep modulus at 170 °C (kPa)	Continuous service temperature (°C)	Flammability limiting oxygen index	Solvent resistance
Epoxy	1.20	Decomposes	175	41	4500	10.7	M120	160	0.1 to 0.4	1400	204	<29	Fair
Polysulphone	1.24	Amorphous	315	70	2500	69.4	M69	174	0.3	950	204	31	Attacked
CM-X*	1.88	327 °C (620 °F)	340	38	3800	21.4	R-116	220	0.01	9150	288	60	Resistant

Source: American Society for Metals [90].
* A one-to-one, head-to-tail copolymer of hexafluoroisobutylene and vinylidene fluoride

Table 4.73. Properties of unidirectional composites

Property	Boron–epoxy	Boron–polyimide	S-glass–epoxy	High-modulus graphite–epoxy	High-modulus graphite–polyimide	High-strength graphite–epoxy	Aramid–epoxy	High-strength graphite–epoxy
Reinforcement content (vol %)	50	49	72	45	45	70	54	60
Density (g/cm^3)	2.02	1.99	2.13	1.55	1.55	1.61	1.36	1.58
Tensile strength (MPa)								
Longitudinal	1370	1040	1290	840	807	1500	1190	1520
Transverse	56	11	46	42	15	40	11	55
Tensile modulus (GPa)								
Longitudinal	201	221	61	190	216	145	84	110
Transverse	22	14	25	6.9	5.0	10	4.8	15
Compressive strength (MPa)								
Longitudinal	1600	1090	820	883	652	1700	290	1240
Transverse	123	63	162	197	70	246	65	248
Shear modulus (GPa)	5.38	7.65	12.0	6.2	4.48	6.9	2.83	4.96
Intralaminar shear strength (MPa)	63	26	45	61	22	68	28	69
Poisson's ratio:								
Major	0.17	0.16	0.23	0.10	0.25	0.28	0.32	0.25
Minor	0.02	0.02	0.09	—	0.02	0.01	0.02	0.034
Moisture coefficient (10^{-2} mm)								
Longitudinal	0.0762	0.0762	0.3556	0.0762	0.0762	0.1524	0.2032	0.1524
Transverse	4.267	4.267	3.251	3.277	3.277	3.277	3.835	3.277
Coefficient of thermal expansion (10^{-6}/°C)								
Longitudinal	6.1	4.9	3.8	-	0.0	0.02	−2.88	0.72
Transverse	30.4	28.4	16.7	33.3	25.4	22.5	56.3	29.5

Source: American Society for Metals [99].

Much has been written on the use of composites in marine environments; however, this is almost entirely confined to reinforced thermoplastics. The subject is discussed further in Case Study 7.3, but some of the major factors affecting composites use in marine applications are worth considering at this point.

The major structural marine application of composites has been in ships' hulls, particularly for mine countermeasure vessels (MCMVs). Most interest has centred around polyester matrices, owing to their weathering properties and despite their poorer fire performance compared with phenolic matrices, although these latter materials are used for internal partitioning in MCMVs [109]. Seamark [110] has discussed current and future applications of composites in offshore engineering, listing such current applications as fire protection panels, water piping systems, tanks and vessels, housings, lifeboats and buoys. The four candidate matrices were identified as polyester, vinyl ester, epoxy and phenolic, and discussion was confined to glass reinforcement, presumably for reasons of material cost. Table 4.74 summarises the critical parameters and rankings given by this author and, as can be seen, there is no obvious choice for all applications. The reasons for the use of polyester matrices in large construction can be seen as being dominated by cost and fabricability, despite their relatively poor engineering performance. The unmatched mechanical performance of epoxy matrices must be offset by their relatively poor fire resistance and high cost, whereas the converse is true for phenolic matrices. The potential for vinyl esters to supplant polyesters for large structures is clear if their higher cost can be compensated by the improved mechanical performance.

4.4.2 Metal Matrix Composites

The main impetus behind the development of metal matrix composites has been the aerospace industry, the object being to increase specific strength and/or specific stiffness. For this reason almost all development has been aimed at the two matrices in which these properties are already at a high value, namely aluminium and titanium. The choice of reinforcement then becomes limited to those of substantially higher modulus than the matrix materials, and the principal candidates are carbon, boron, silicon carbide and alumina, although carbon is not generally used to reinforce titanium [101].

As mentioned in the introduction to this section, it could almost be argued that steel is a metal matrix composite consisting of iron reinforced with iron carbide. Likewise, the mechanically alloyed nickel-based 'superalloys' used for aero-engine manufacture could be regarded as metal matrix composites consisting of the metallic material filled with a dispersant such as thoria or yttria. There is also a range of cutting tool materials fabricated in this way (for example, tungsten carbide in cobalt cermets). A natural extension of these ideas has led to a range of particle-reinforced metal matrix composites. These are relatively simple materials to manufacture (using powder metallurgy methods, casting or plasma co-spraying) and probably the most successful of these are the silicon-carbide-particle-reinforced aluminium alloys. Table 4.75 shows a selection of properties of such composites.

Short-fibre composites are the next easiest MMCs to fabricate, this typically involving squeeze casting techniques, although the process is relatively sophisticated, using preforms and the ability to reinforce selectively [111]. Again, the most widely available matrix is probably aluminium, this time reinforced with alumina, although silicon carbide (polycrystals and whiskers), zirconia and others are used. Magnesium and titanium alloys can also be treated in this way [90].

As with resin matrix composites, it is only with continuous-fibre reinforcement that MMCs reach their full potential. The most commonly encountered fibres are boron, silicon carbide, alumina, carbon and tungsten. A few examples are given below to illustrate the extent of property enhancement, which can be obtained by continuous-fibre reinforcement of metals.

Boron is widely used with aluminium and, for example, a 50 per cent by volume (50 vol%) loaded crossply laminate can achieve a modulus of 145 GPa in both in-plane directions [91]. Continuous silicon carbide reinforcement of titanium at about 40 vol% can result in a modulus of around 190 GPa with a UTS of 1500 MPa [91]. The extremely high performance of carbon-reinforced metals and the relatively high cost of the high modulus fibres has directed research towards space applications, and hence towards magnesium and aluminium as matrices. These composites are relatively sensitive to the fibre quality (because of the wide range of moduli of carbon fibres), but very useful increases in modulus can be obtained, particularly in magnesium alloys [90].

4.4.3 Ceramic Matrix Composites

The ceramic matrix composites, like the engineering ceramics, are primarily high-temperature materials, although they also find some applications in wear-resistant (e.g. cutting tools), extreme stability (e.g. body implants)

Table 4.74. Qualitative property selectors for thermoset resins for composites in offshore applications

Matrix resin	Relative cost*	Strength	Corrosion resistance	Fire performance
Polyester	1 to 1.3	2	2	1
Vinyl ester	1.8 to 2.2	3	3	1
Epoxy	>3	5	5	2
Phenolic	1.1 to 1.4	2	2	5

Source: Seamark [110].
*Cost relative to cheapest (polyester) material. Other ratings from 5 = most desirable value to 1 = least desirable value.

Table 4.75. Properties of some SiC-particulate-reinforced aluminium matrix composites

Composite	Modulus (GPa)	Yield stress (MPa)	UTS (MPa)	Elongation (%)
25% SiC 6061*	122.7	434	498	3.91
25% SiC 7090*	117	686	805	2.0
25% SiC 2124*	115.5	420	574	5.4
20% SiC 2014+	107.8	479	497	2.5
15% SiC 7075+	99.4	581	610	2.5

Source: Trumper [111].
*Fabricated by powder metallurgy then extruded, T6 heat treatment.
+Fabricated by casting route, extruded, T6 heat treatment.

and extreme reliability (e.g. aerospace) applications.

Unlike the other composites covered in this section, the reinforcement in CMCs is provided primarily to improve toughness, and for this reason it is not uncommon to find reinforcement and ceramic to be the same, or a similar, material. The effect of reinforcement on toughness is not simple, because it is the interfaces rather than the fibres which aid toughening. Table 4.76 illustrates this effect for silicon-carbide-whisker-reinforced alumina.

Multidirectional reinforcement gives further improvements in toughness, although strength and modulus can be less than in the monolithic material.

Carbon–carbon composites should also be mentioned here, as materials with quite exceptional sustained high-temperature properties.

Table 4.76. Effect of SiC whisker reinforcement on the fracture strength (modulus of rupture; see Chapter 2) and fracture toughness of alumina ceramic matrix composites

Whisker content (vol %)	Temperature (°C)	Fracture strength (MPa)	Fracture toughness (MPa $\sqrt{m}$)
0	22	—	4.5
	700	—	4.0
	1000	—	3.8
10	22	455 ± 55	7.1
	1000	320 ± 36	
20	22	655 ± 135	7.5 to 9.0
	700	535 ± 35	
	1000	570 ± 20	7.0 to 8.0
40	22	850 ± 130	6.0
	700	740 ± 61	
	1000	665 ± 88	6.2

Source: American Society for Metals [90].

4.5 Other Materials of Importance in Marine Environments

A number of material types and applications other than those already discussed are of importance in marine environments. The most notable of these is concrete, which finds wide structural use in coastal, marine and offshore applications. Timber, although of declining importance, still finds some marine applications and is therefore briefly discussed. Finally, materials used for coatings are discussed separately because of their importance and also in view of the fact that they belong to several of the groups already mentioned.

4.5.1 Cement and Concrete

As a material, concrete is a fairly complex, particulate-reinforced inorganic composite comprising principally a hardened cement paste (h.c.p.) with (usually) an irregularly shaped aggregate based on two or more size distributions and two or more materials. In addition, reinforcing bars or prestressing tendons may be used to counteract the inherent brittleness of concrete in areas where tensile stresses are likely to be encountered, as described in Chapter 2.

Unlike most other engineering materials, concrete is made by mixing all the dry ingredients together and casting the resulting wet mix between formers in an operation familiar enough not to require description. This means that concrete is not quite such a specifiable material as, for example, steel, which is purchased with an known set of finished properties.

The basic ingredients in a concrete mix are cement powder, water, sand and stones (coarse and fine aggregates), various additives and, of course, steel. With regard to the last of these it might be noted that it has been said [112] that concrete offshore platforms generally include as much steel as would the equivalent steel structure.

Cement powder is made by fusing a mixture of oxides, principally lime (CaO), alumina (Al_2O_3) and silica (SiO_2), with a little iron oxide (Fe_2O_3). A little gypsum ($CaSO_4$) is added to the clinker and the mixture is ground finely. It is important to control carefully the composition of the cement, as the range in which acceptable cementitious properties are obtained is relatively restricted, as shown in Fig. 4.58.

In cement chemistry, it is usual to abbreviate the oxide formulae to

Lime	C
Silica	S
Alumina	A
Iron oxide	F
Water	H

and this is especially helpful when dealing with the inter-oxide compounds involved, for example $CAS_2 \equiv CaO \cdot Al_2O_3 \cdot (SiO_2)_2$. As can be seen from the ternary (C—A—S) diagram (Fig. 4.58), most useful cement compositions lie in the subsystem C—C_5A_3—C_2S, and adding the fourth component identified above (iron oxide) allows most cement compositions to be contained within a tetrahedron bounded by the above three mixed oxides and a fourth, C_4AF, which, although not strictly a stoichiometric compound, is normally treated as such. The compounds present in most significant quantities are C_3S, C_2S, C_3A and C_4AF, and the following list [114] shows a typical Portland cement oxide analysis along with the calculated composition of the compounds identified above:

Oxide analysis	*(per cent)*	*Compound composition*	*(per cent)*
SiO_2	21.8	C_3S	45.7
Al_2O_3	4.9	C_2S	28.0
Fe_2O_3	2.4	C_3A	8.9
CaO	64.7	C_4AF	7.3
CaO (free)	2.0		
Na_2O	0.2		
K_2O	0.4		
MgO	1.1		
SO_3	2.6		
Ignition loss	1.4		

As can be seen, the main compounds present are C_3S and C_2S, and it is these which dominate the hydration behaviour and development of the final properties. In an ordinary Portland cement (OPC) the hydration reactions proceed first with the development of a gel layer on the individual cement grains (Fig. 4.59). These early reactions are con-

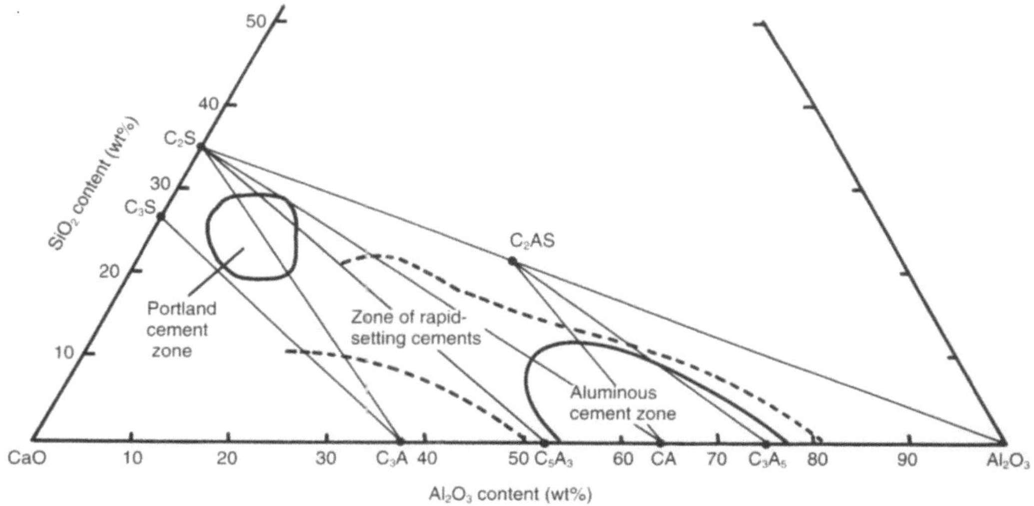

Fig. 4.58. Cementitious areas in the lime (CaO)/alumina (Al₂O₃)/silica (SiO₂) system. (From Lea [113].)

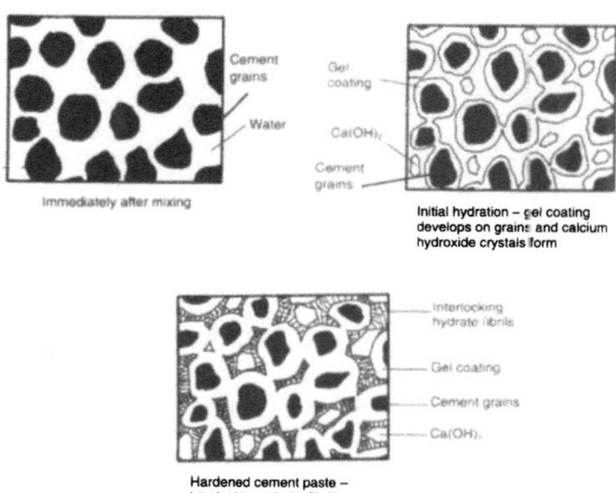

Fig. 4.59. Schematic view of the hydration of Portland cement.

trolled by gypsum (which impedes 'flash set'), and the gel consists mainly of hydrated calcium aluminates with the formation of portlandite, Ca(OH)₂, in the lime-saturated water between the grains. Some bridging takes place between the grains (setting), but little strength is developed until after about 3 to 5 hours when hardening, corresponding to the hydration of the calcium silicates, commences, and this process continues over several days as tubular fibrils of the calcium silicate hydrates grow and interconnect. Concrete hardening (as opposed to setting) is a relatively slow process and may take several weeks to become complete.

The main hydration reactions can be regarded in the following (simplified) manner:

Early hydration (C₃A and C₄AF)

Gypsum/C₃A – hydration of C₃A with the aid of gypsum:

$$C_3A + 3CaSO_4 + 32H_2O \rightarrow 3CaO . Al_2O_3 . 3CaSO_4 . 32H_2O$$

C₄AF/Ca(OH)₂ – hydration of C₄AF using calcium hydroxide:

$$C_4AF + 2Ca(OH)_2 + 10H_2O \rightarrow C_3AH_6 + C_3FH_6$$

C₃A – direct hydration of C₃A:

$$C_3A + 6H_2O \rightarrow C_3AH_6$$

Later hydration (C₃S and C₂S)

$$2C_3S + 6H \rightarrow C_3S_2H_3 + 3Ca(OH)_2$$

$$2C_2S + 4H \rightarrow C_3S_2H_3 + Ca(OH)_2$$

where the C₃S hydration is more rapid and produces more portlandite. Thus, although C₃A and C₄AF develop their full strength rapidly, their hydrates are much weaker than those of C₃S and C₂S, and so strength is dominated by the later reactions [115].

Five main types of cement are recognised (e.g. ASTM C150 [116]), and these are generally classified in terms of their compound composition, as shown in Table 4.77.

Ordinary portland cement (OPC) is usually regarded as having optimum properties for general use, whereas the modified cement develops its initial strength slightly more slowly than OPC, but reaches its maximum more rapidly.

Rapid-hardening cement has a higher C₃S content at the expense of C₂S. As seen earlier, this does not affect the final strength (since the hydrates are the same), but the approach to maximum strength is more rapid. Increases in rate of hardening can also be brought about by finer grinding, and the surface area of rapid-hardening cement is at least 325 m²/kg and can be up to 800 m²/kg. Both of the above factors (fineness and high C₃S content) tend to give a high rate of evolution of heat, which may be an important consideration in large structures. Low-heat cement uses a relatively small amount of C₃S, and was developed originally for dams where heat evolution within the structure could produce thermal stresses and hence cracks.

Sulphate-resisting cement is required where there is a danger of hydrates being attacked by sulphates present in the environment, such as magnesium sulphate in seawater or groundwater in gypsum-bearing soils. Both portlandite and C₃A can be attacked by sulphates, the latter producing ettringite (3CaO . Al₂O₃ . 3CaSO₄ . 32H₂O), with a large volume change which can lead to damage to the cement.

In addition to the above main types of cement, a number of other Portland cements are available, such as extra-

Table 4.77. Comparison of British and American Standards for cements and compound composition of ASTM types

Main types of Portland cement

British classification		American classification	
Description	BS	Description	ASTM
Ordinary Portland	12: 1978	Type I	C 150-84
Rapid-hardening Portland	12: 1978	Type III	C 150-84
Ultra-high-early-strength Portland	—	—	—
Low-heat Portland	1370:1979	Type IV	C 150-84
Modified cement	—	Type II	C 150-84
Sulphate-resisting Portland	4027:1980	Type V	C 150-84
Portland blast-furnace (slag cement)	146: Part 2: 1973	Type IS Type IS (MS)	C 595-83a
Low-heat Portland blast-furnace	4246: Part 2: 1974	—	—
White Portland	12: 1978	—	C 150-84
	4627:1970	Type IP	
Portland-pozzolana	6588: (Draft)	Type P	C 595-83a
	3892: Part 1: 1974	Type I(PM)	

Typical average values of compound composition of Portland cements of different types

Cement	Compound composition (%)							
	C_3S	C_2S	C_3A	C_4AF	$CaSO_4$	Free CaO	MgO	Loss on ignition
Type I	59	15	12	8	2.9	0.8	2.4	1.2
Type II	46	29	6 (8 max.)*	12	2.8	0.6	3.0	1.0
Type III	60	12	12 (15 max.)*	8	3.9	1.3	2.6	1.9
Type IV	30 (35 max.)*	46 (40 min.)*	5 (7 max.)*	13	2.9	0.3	2.7	1.0
Type V	43	36	4 (5 max.)*	12	2.7	0.4	1.6	1.0

Source: Neville and Brooks [117].
*Maximum or minimum values as specified by ASTM C 150–84.

rapid-hardening and ultra-high-early-strength versions, those using products from other manufacturing operations (e.g. blast furnace cement) and those with special colour properties (e.g. Portland–pozzolana).

As can be seen from Fig. 4.58, there is another useful area in the C–A–S system from the point of view of cementitious properties. This resides around the C–A line, with up to about 10 per cent silica, and these are called the aluminous, or high-alumina, cements. Commercial aluminous cements contain around 40 per cent alumina, 40 per cent lime, some ferrous and ferric oxides and up to 5 per cent silica and the main cementitious compounds are CA and C_5A_3 (or $C_{12}A_7$). The hydration reactions produce CAH_{10}, along with some C_2AH_8, and this is accompanied by a very rapid rate of strength development. Alumina gel $(Al_2O_3(aq))$ is also formed, and this eventually ages to gibbsite (AH_3). High-alumina cement was originally developed as a sulphate-resisting cement because portlandite is absent from the hydration products and also because of the relatively inert alumina gel. However, under certain conditions of humidity and temperature, the hydrates have been found to undergo a process known as 'conversion', with a resul-

tant loss of strength which depends upon the original water-to-cement ratio and the mix richness.

Aside from the composition of the cement, a number of other factors influence the final strength of concrete. These factors are mainly physical, including the fineness of the cement powder, the water-to-cement ratio used during mixing, and the amount, size and shape of aggregates. Other chemical factors include the composition of the aggregate(s), the quality of the mixing water, and any additives which are used to improve the mix.

Such admixtures can normally be categorised in terms of their effect on the mixture, i.e. accelerators (to accelerate hardening), retarders (to inhibit setting), water-reducers (plasticisers) and air-entraining agents, so that they can have an effect on the workability of the mixture and upon the hydration process. Such additives can be inorganic or organic compounds and some (e.g. lignosulphonates) can fulfil more than one function (e.g. as both plasticiser and retarder).

Apart from the hardened cement paste (h.c.p.), which can be regarded as the matrix, the other main constituent of hardened concrete is the aggregate, which can be regarded as reinforcement (in the sense which this is used for composites). The aggregate is usually based on two sizes, so that it consists of a loose structure of stones with the interstices filled with sand. Most of the remaining voidage is then taken up with the hardened cement paste. A wide range of minerals can be used as aggregates, and these are often siliceous but may contain lime or alumina. Thus, although the aggregate may be chemically similar to the cement, it does not contain the cementitious compounds and is therefore relatively inert compared with cement powder. The main important properties of aggregates are their porosity, shape, density and strength in as much as these affect the final properties of the concrete.

Because of its brittle nature, concrete cannot be used to resist appreciable tensile loads. The normal way of coping with this is by the use of steel reinforcement to carry the load in areas of tension, or by using prestressing tendons which induce a compressive stress in the concrete in the unloaded condition, thus enhancing its ability to resist external tensile loads.

In the general gamut of materials, concrete can be described as hard, brittle, durable (to weathering) and relatively strong. For the reasons described above, concretes are not normally compared in terms of their tensile properties, but rather compressive and flexural strengths are compared. Table 4.78 gives some general properties of concrete.

The extension of fracture mechanics to cover the (predominantly compressive) stress states existing in structural concrete is not widely used, although the Orowan criterion (covered in Chapter 2) can be used to determine failure in complex stress states. Other methods such as punching

Table 4.78.   Effect of mix variables on the properties of concrete

Proportions	Quantity of cement (kg/m³ of concrete)	Water/cement ratio	28-day strength (MPa)
1 : 1½ : 3	359	0.4	60.5
		0.5	47.0
1 : 2 : 4	289	0.4	60.5
		0.5	47.0
		0.6	36.0
1 : 3 : 6	197	0.6	36.0
		0.7	28.0

Source: Crane and Charles [21].

shear for impact assessment are also based on strength rather than toughness data. Fatigue in concrete as described in Chapter 2 may be dependent upon reinforcement or concrete behaviour, and most data reflect the effects of manufacturing variables rather than concrete chemistry.

Durability is a less quantitatively describable phenomenon for concrete, and is dominated by the corrosion of reinforcing bars (provided that alkali–aggregate reactions are avoided and C_3A levels are low). Apart from the permeability of the concrete, the degree of cover and the access of oxygen, reinforcement corrosion is also sensitive to local discontinuities in the condition or the quality of the concrete and this is a matter of quality control and degree of protection in service. Exposure tests of a few years' dura-

tion (Stillwell [118]), and the examination of structures that have been exposed to marine environments for several tens of years [119], would tend to indicate that reinforcement corrosion does take place in concrete in the splash zone, but tends to be confined to areas of poor quality or poor cover. Of the range of concretes examined, no detectable differences in these properties could be elucidated which were due to concrete type. There is also evidence that the quality of concrete construction has a considerable influence on durability [120], and studies have been carried out to determine the effect of cement content and cover on the durability of a number of existing concrete structures for the purpose of developing the UK guidance rules on the use of concrete in the marine environment [121]. Submerged concrete seems to be almost immune from deterioration over the time scales for which there is any field experience.

Concrete offshore structures in the North Sea have been constructed according to a number of different codes of practice, although Burdall and Sharp [122] have summarised some of the key parameters used in design. Most structures are made using OPC, and strength grades of 45 or 50 were commonly used. These (high) strengths are believed to be beneficial to durability [122]. Reinforcement and prestressing tendon cover depths are specified in the codes, and requirements vary, but about 45, 55 and 70 mm are probably appropriate for reinforcing bars at submerged, atmospheric and splash zone positions respectively. Burdall and Sharp suggest similar values for prestressing tendons, but recommendations in other codes go as high as 100 mm.

The susceptibility of concrete and its reinforcement to fatigue in large marine structures is a much less certain matter than is the case for steel jacket platforms. The main reason for this is that there appear to be no records of fatigue failure of either component in marine structures. Nevertheless, *S–N* data are available for the steel and the concrete, and one of the simplest approaches is to demonstrate that neither is susceptible by limiting the stresses in the concrete and the stress range in the steel. Some preliminary design curves for straight reinforcing bars embedded in concrete have been produced by Booth et al. [123], and the air and seawater curves are shown in Fig. 4.60.

The use of concrete in the marine environment is widely established and has a long history, particularly in coastal waters (see Chapter 1). The main modern uses are still in coastal structures such as harbours, breakwaters and lighthouses, but concrete is also widely used for offshore structures and submarine pipeline weight coatings as well as lining for (usually large-diameter) pipes carrying raw seawater. Its use in floating structures is also fairly extensive, from ships' hulls to large floating offshore structures. A typical example of a coastal structure is given by Tantry et al. [124] for an oil berth in Bombay consisting of a jetty 1.8 km long with a deepwater berth.

Concrete structures have been used widely in offshore oil exploitation, particularly in the Norwegian sector of the North Sea where a larger proportion of the structures are made from concrete than in the UK sector. The ability to construct such platforms requires deep, sheltered inshore sites, since concrete structures, although transportable, can only be moved by floating in their final attitude and then ballasting to the seabed. Most structures of this type operate in about 150 m of water and contain around 100 000 m³ of concrete. Many have oil storage capacity in cells at the seabed (e.g. Fig. 4.61).

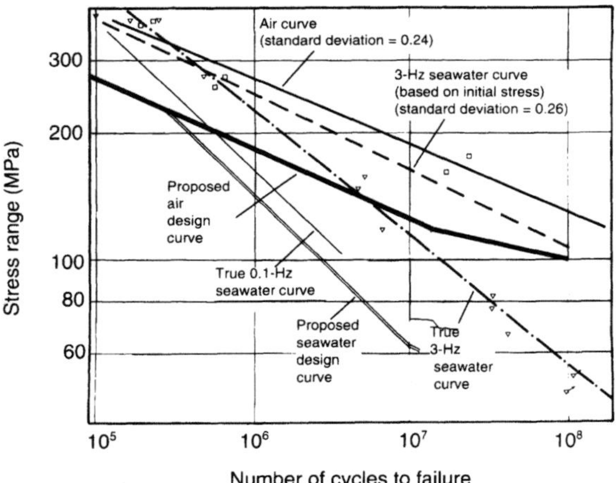

Fig. 4.60. *S–N* fatigue data for reinforced concrete beams in air and in seawater. (Adapted from Booth et al. [124].)

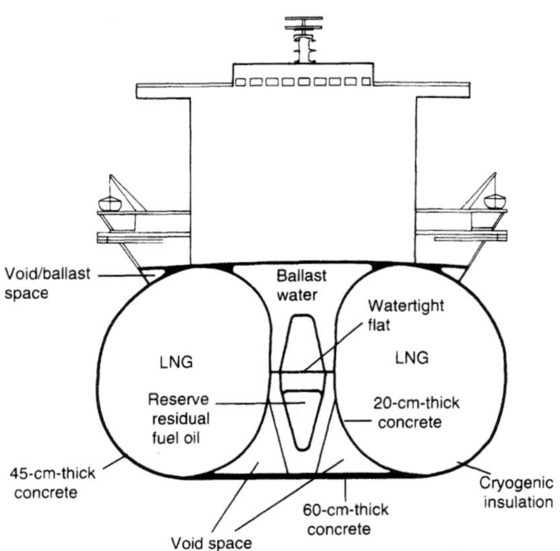

Fig. 4.62. Schematic view of a concrete-hulled LNG carrier. (From Stanford [128].)

The use of concrete on the seabed is typified by the concrete coatings applied to most submarine pipelines. Although the principal purpose of such coatings is to provide stability, the concrete also provides mechanical protection to such possible mishaps as interactions with fishing gear (e.g. Palmer [126]). This latter requirement has led to numerous studies of the impact resistance of such coatings, and the technology has been extended to other subsea protective structures such as those which might be used for remote hydrocarbon production wellheads. The ability to place concrete on the seabed might be useful in such applications and is, of course, essential for subsea repair work to both steel and concrete structures (see Chapter 5). For a number of reasons, chief among them being the tendency of underwater placed concrete to segregate, special mixtures have been developed for such applications (e.g. Yamaguche et al. [127]).

The use of cement for the hulls of boats goes back to about the mid-nineteenth century (see Chapter 1) and many modern vessels, particularly barges, are made from reinforced or, more often, prestressed concrete. The sophistication of such vessels varies from simple barge shapes [128] to more complex ones, for example for LNG carriers (Fig. 4.62).

The developments occasioned by concrete vessels and offshore platforms have led to a number of proposals for further floating uses of concrete, for example as platforms to carry a variety of engineering plant for operation offshore. Derrington [130] has identified a number of possible applications including petrochemical, desalination, offshore cold storage and mechanical workshops. Other floating concrete structures, such as tension leg platforms for hydrocarbon recovery, have also been suggested, as have applications of steel–concrete sandwich structures for ice walls to protect Arctic offshore structures [131].

4.5.2 Timber for Marine Use

Timber, despite its continuing use in building, has declined considerably as a marine engineering material. Part of the reason for this is the time and skill required to make large structures of timber, and also the shortage of good-quality timber. There is also a growing reluctance to use non-manufactured materials, and this is especially true of tropical hardwoods. For these reasons the treatment of this material is here quite brief.

As mentioned earlier, timber is a complex material with a highly anisotropic structure, and Fig. 4.63 gives some idea of this, along with the radial variability of the structure. Furthermore, superimposed on these two degrees of variability are local effects such as growing conditions. When all of these are considered along with the substantial influence of moisture content on properties (see Chapter 2).

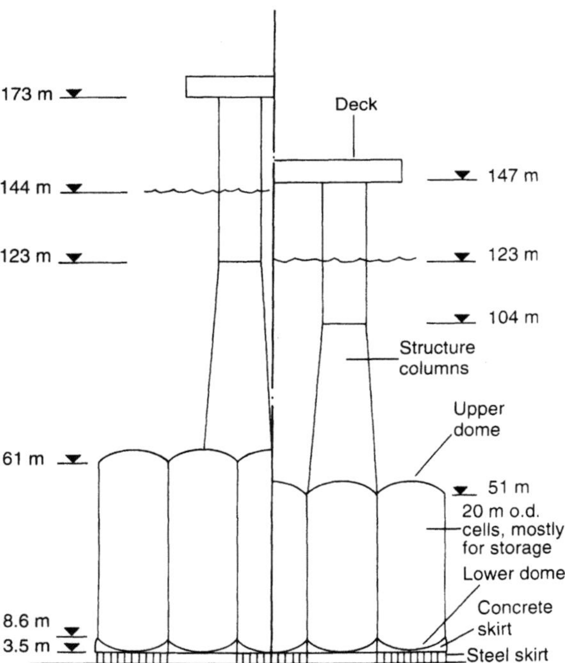

Fig. 4.61. Simplified vertical sections of two typical concrete offshore platforms. (From Moksnes [125].)

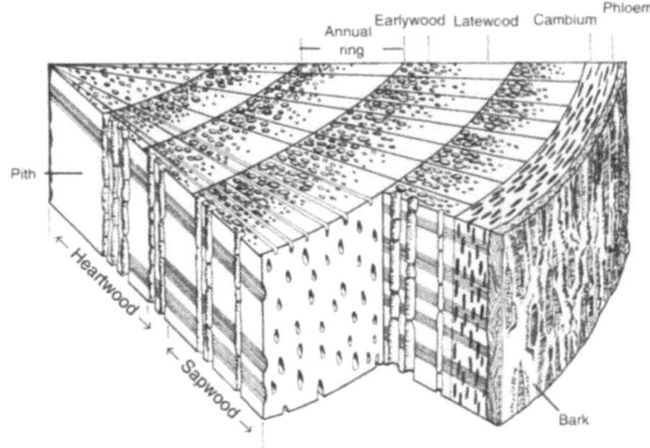

Fig. 4.63. The principal components of the bole of a typical hardwood. (From Mettem [132].)

it is clear that the specification of timber properties is not a simple matter of identification of the species of tree from which the material has been cut.

Lamination, where veneers are glued together in alternating directions, is a useful way of removing some of the anisotropy of properties. For example, a tropical hardwood with elastic modulus of 9800 MPa parallel to the grain and 220 MPa perpendicular to the grain can have this modified to 6000 MPa parallel and 5080 MPa perpendicular to the grain by the use of lamination.

Because of the variation in strength with moisture content, it is advisable when using timber in marine environments to take figures at maximum moisture content [25]. Table 4.79 shows some property values for a variety of timber types with moisture content.

'Marine grade' plywood [133] is manufactured using weather- and boil-proof adhesives (phenol and resorcinol formaldehyde resins), but of course the wood also requires to be resistant to decay. This means that either specific durable tropical hardwoods require to be used for the veneers or that some treatment needs to be applied to them. Timbers considered to be of sufficient natural durability include African mahogany, omu and utile.

Applications of timber in the marine environment include quays and groynes and small vessels from inshore fishing boats to sailing dinghies. Timber is also used widely for the construction of sea-cages for fish farming and some applications are found in construction yards, for example as skids for the launch of offshore structures.

Farmer [134] lists timbers resistant to marine borers (Table 4.80), and of these only eleven are recommended for marine applications such as piers, once strength has been taken into account. Other timbers require to be treated and this is usually done using high-pressure impregnation with creosote or copper–chromium–arsenic water-borne mixtures, for which reasonable permeability is, of course, required.

It is not normal to use these resistant timbers for boat construction owing to their density and their being difficult to cut, as well as the quantity required. Thus, a typical inshore fishing vessel might be constructed using an oak keel and ribs (because of its durability and likelihood of exposure to bilge water) and larch for planking and hull (owing to its strength, durability and availability in long lengths) [135].

Reinforced plastics have taken over from timber in ships which require to be non-magnetic (e.g. MCMVs; see Case Study 7.3), and marine plywood rather than timber is normally now used in the skins of other, smaller vessels.

4.5.3 Materials for Marine Coatings

Coatings are applied to structures for a variety of reasons, but by far the most important of these in the marine environ-

Table 4.79. Effect of moisture content on the mechanical properties of a range of timber types

Type	Moisture content (%)	Density (t/m³)	Static bending			Compression parallel to grain		Compression ⊥ to grain, proportional limit (MPa)	Shear parallel to grain, ultimate strength (MPa)
			Proportional limit (MPa)	Ultimate strength (MPa)	Flexural modulus (MPa)	Proportional limit (MPa)	Ultimate strength (MPa)		
Ash (white)	43	0.54	36.5	65.5	9.7	23.2	28.0	5.9	9.3
	12	0.58	61.4	100.7	11.6	38.5	50.2	10.4	13.2
Beech	54	0.56	29.6	59.3	9.5	17.6	24.5	4.6	8.9
	12	0.64	60.0	102.7	11.9	33.6	50.3	8.6	13.9
Elm	48	0.57	31.7	65.5	8.2	20.5	26.1	5.2	8.8
	12	0.63	55.2	102.0	10.6	32.4	48.6	10.5	13.2
Douglas fir	36	0.45	33.1	52.4	10.7	23.5	26.8	3.5	6.4
	12	0.48	55.8	80.7	13.2	44.5	51.2	6.3	7.9
Mahogany	58	0.45	42.2	63.7	8.9	—	31.3	4.9	9.0
	12	0.46	60.7	76.8	9.9	—	44.3	8.3	7.2
Oak	70	0.59	32.4	55.8	8.3	20.3	24.3	5.9	8.8
	12	0.67	54.5	95.8	11.2	30.0	48.5	9.7	13.0
Pine (yellow)	63	0.54	35.9	60.0	11.0	23.6	29.6	4.1	7.2
	12	0.58	64.1	101.4	13.7	42.4	58.2	8.2	10.3
Red cedar	37	0.31	22.1	35.2	6.3	17.0	19.0	2.3	4.9
	12	0.33	36.5	53.1	7.7	30.1	34.6	4.2	5.9

Source: Chalmers [25].

Table 4.80. Timbers generally recognised as being resistant (although not immune) to marine borers
(Those marked with an asterisk are generally believed to be the most suitable for marine applications)

African padauk	(*Pterocarpus soyauxii*)
Afrormosia	(*Pericopsis elata*)
Andaman padauk	(*Pterocarpus dalbergioidies*)
Basralocus	(*Dicorynia guianensis*)*
Belian	(*Eusideroxylon zwageri*)*
Brush box	(*Tristania conferta*)*
Ekki	(*Lophira alata*)*
Greenheart	(*Ocotea rodiaei*)*
Iroko	(*Chlorophora excelsa*)
Ironbark	(*Eucalyptus* spp.*)
Jarrah	(*Eucalyptus marginata*)
Kapur	(*Dryobalanops* spp.)
Manbarklak	(*Eschweilera subglandulosa*)*
Muninga	(*Pterocarpus angolensis*)
Okan	(*Cylicodiscus gabunensis*)*
Opepe	(*Nauclea diderrichii*)*
Pyinkado	(*Xylia dolabriformis*)*
Red louro	(*Ocotea rubra*)*
Southern blue gum	(*Eucalyptus globulus*)
Teak	(*Tectona grandis*)
Turpentine	(*Syncarpia laurifolia*)*

Source: Farmer [134].

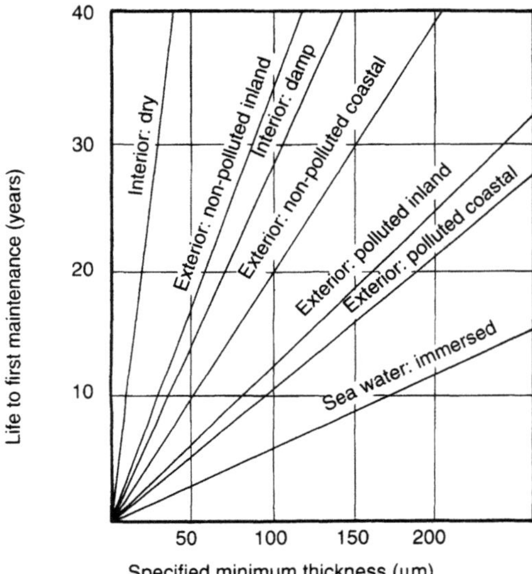

Fig. 4.64. Typical lives of zinc coatings in a variety of environments [136].

ment is for corrosion protection. In this section most space is devoted to these types of coatings, but those used for wear resistance and for fire protection are also briefly considered.

As mentioned in Chapter 3, corrosion protection systems can operate with sacrificial, barrier or inhibitive action, although often all three of these are operative simultaneously.

Among the sacrificial coatings are flame-sprayed and dipped metallic coatings such as zinc and aluminium metals, and design of such systems simply involves ensuring that a sufficient thickness is applied, to allow for the inevitable self-corrosion accelerated by the effect of any galvanic corrosion due to exposure of the substrate. Salama and Tetlow [47] indicate that an aluminium flame-sprayed coating of between 125 and 175 μm thickness was sufficient to provide long-term protection for the (submerged) mooring elements of a tension leg platform, while the British Standard code of practice for protection of steel against corrosion [136] suggests that zinc coatings will corrode at rates of between 1 and 15 μm per year in the atmosphere and that aluminium sprayed coatings 80 μm thick have lives in excess of 20 years in marine environments, also giving long life when totally immersed. The performance of zinc coatings is somewhat limited in polluted

conditions, however, and Fig. 4.64 shows the design curve for such coatings as a function of environment.

The selection of organic coatings for the protection of steel in marine environments is a rather more complex matter, and BS 5493 [136] gives guidance according to the type of service: non-polluted coastal, polluted coastal, splash zone and seawater immersion.

Morcillo [137] has provided life vs. film thickness curves for a range of coating systems in the marine atmosphere, and Table 4.81 shows his recommended minimum dry film thickness for each of the types of system considered.

Apart from the special case of pipelines mentioned below, it is relatively uncommon for coatings to be applied to offshore structures in the submerged zone, and the splash zone is generally recognised as being the most difficult to protect with working decks and helidecks requiring special consideration as regards abrasion resistance and anti-skid properties. Whitehouse [138] has surveyed UK practice for offshore structural protection, and a typical splash zone protection system might consist of blast cleaning followed by priming with zinc silicate or a zinc-rich epoxy primer with a coal tar epoxy, a high-build epoxy, a vinyl tar or a glass-flake-reinforced polyester top coat. The total dry film thickness for splash zone protection would be about 400 μm and, whereas a similar system might be used for

Table 4.81. Minimum dry film thickness in micrometres (primer–topcoat) for the protection of suitably cleaned steel exposed to marine environments

Required life (years)	Paint system					
	Oil	Phenolic	Alkyd	Vinyl	Chlorinated rubber	Epoxy
Short (2 to 5)	75 to 125	75 to 100	75 to 100	50 to 75	50 to 75	100 to 125
Medium (5 to 10)	125 to 200	100 to 175	100 to 125	75 to 100	75 to 100	125 to 150
Long (10 to 20)	—	175 to 280	125 to 150	—	100 to 280	150 to 200
Very long (> 20)	—	> 280	> 150	—	—	—

Source: Morcillo [137].

splash zone to underdeck and for topsides, the thicknesses then would be around 300 µm and 200 µm respectively. Aggregates such as corundum can be added to the top coat while it is still tacky to provide anti-skid properties for helidecks and walkways.

Although they are cathodically protected, it is generally accepted that subsea pipelines should receive an anti-corrosion coating as well, even if a concrete weight coating has been applied. Roche and Samaran [139] have reviewed experience on external pipeline coating performance, and Table 4.82 shows some pipeline coating types along with their relative performance.

Hot, high-pressure pipes such as some risers may be subject to unusually severe environments, and some consideration may be required to be given to a special coating. It is sometimes the case that heavy-duty epoxy mastics (3 to 4 mm thick), neoprene sheet wrapping or Monel sheathing are used to protect hot risers, but more conventional amine- or polyamide-cured coal-tar epoxy may be suitable in others [138].

Heavy-duty wear-resistant coatings are often provided by the deposition of a hard metal, or metal impregnated with hard particles, onto the surface. This is usually achieved by spraying (arc or plasma) or by weld overlaying. Hardfacing material selection is based on the three parameters: abrasion resistance, corrosion resistance and impact resistance; the interplay of these three factors is shown for some commercial hardfacing alloys in Table 4.83.

Fire resistance can be conferred upon surfaces by the application of coatings which supply a thermal barrier between the fire and the surface to be protected. The main methods of providing fire protection are by the use of inorganic or intumescent coatings, although other organic coatings are suitable for protecting polymers. The first two of these are described below [141].

Intumescent coatings are usually vinyl- or epoxy-based, with some inert filler or fibre to lend mechanical stability to the char which forms on exposure to flame. When heated, the organic materials undergo endothermic reactions and evolve gases which foam the char, decreasing its conductivity further.

Cementitious coatings use Portland cement and mineral mixtures, and significant heat absorption is provided by the evaporation of the bound water. Even after the water has evaporated, the coatings have some residual insulating properties and retain some structural integrity.

Table 4.82. External pipeline coatings for buried and subsea pipelines
(Data apply to the experience of one company only)

System	Application
Bituminous enamels	**Onshore/offshore**

Require to be protected with concrete weight coat if used offshore. These combined coatings show good service performance. A few instances of corrosion under disbonded coatings buried onshore without concrete protection have been reported. Limited performance with pipelines carrying hot fluids.

Cold-applied tapes   Onshore/offshore
Usually polyethylene with a coat of butyl or EVA adhesive. Rolled onto primed steel surface. For offshore field joints on concrete/bituminous enamel-coated pipelines, PVC tapes can be used. High-temperature formulations are available. Main performance limitation arises from the difficulty of quality control at the application stage.

**Heat-shrink tapes   Mostly offshore
and sleeves**
Generally used as sleeves for field joints in conjunction with polyethylene factory coatings. High-temperature versions available. Performance generally excellent, but the difficulty in achieving the high preheat temperatures necessary for application may lead to later loss of adhesion.

Polyethylene   Onshore/offshore
Usually factory applied by powder fusion or side extrusion over a copolymer adhesive coat. If concrete weight coating is required, care needs to be exercised over the shear strength of the bond between the concrete and the polyethylene. Performance appears to be excellent, although non-concrete-coated lines may suffer damage, for example from trawlers.

Fusion-bonded epoxy   Onshore/offshore
Because of the sophistication of the coating and its relative thinness, quality control is of the utmost importance. This includes checking thickness, checking for 'holidays' and checking cure. Performance is generally good even with hot pipelines, damage usually being confined to localised small blisters and a little disbonding around mechanical defects.

Polyamide   Mostly onshore
Good heat resistance and less brittle than epoxy. For good adherence and resistance to cathodic disbonding up to 100 °C, a liquid epoxy phenolic primer is necessary. Performance is generally good, provided coating does not receive excessive exposure to sunlight.

Source: Roche and Samaran [139].

Table 4.83. Abrasion, heat and corrosion resistance of some commercial hardfacing alloys

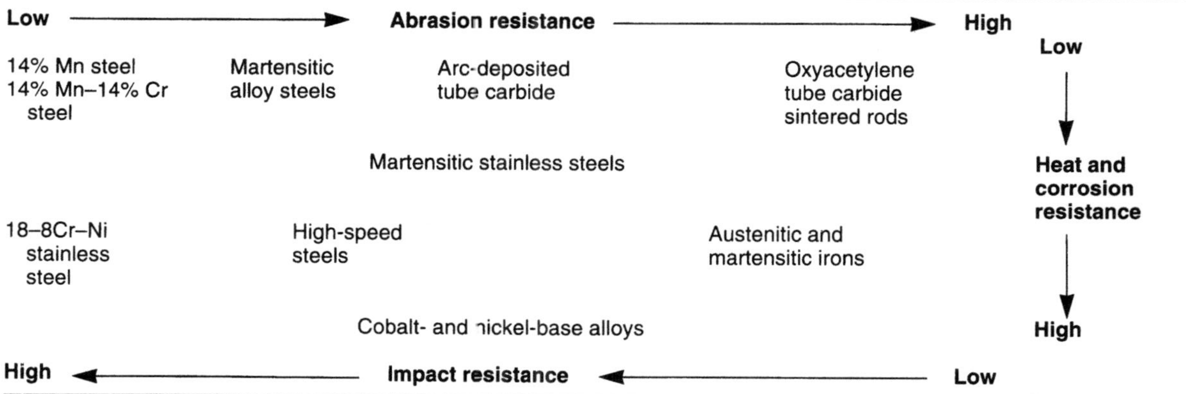

Source: Welding Institute [140].

A number of inorganic foams based on phosphate chemistry are available. These foams are relatively inert and have a low thermal conductivity and re-radiative outer surfaces.

References

1. American Society for Metals. Worldwide guide to equivalent irons and steels. ASM International, Metals Park, Ohio, 1987
2. American Society for Metals. Worldwide guide to equivalent non-ferrous metals and alloys. ASM International, Metals Park, Ohio, 1987
3. Askeland DR. The science and engineering of materials. Brooks/Cole, Monterey, 1984
4. Honeycombe RWK. Steels – microstructure and properties. Edward Arnold, London, 1981
5. Cohen M, Hansen SS. Microstructural control. In: Micro-alloyed Steels, Micon '78: Optimisation of processing and properties and service performance through microstructural control, ASTM STP 672. ASTM, Philadelphia, 1979, pp 34–52
6. Sage AM. Microalloyed steels for structural applications. Metals and Materials 1985; 5 (10): 584–588
7. Leslie WC. Solution and precipitation of aluminium nitride in relation to the structure of low carbon steels. Transactions of the American Society for Metals 1954; 46: 1470–1499
8. Platts GK, Vassiliou AD, Pickering FB. Developments in microalloyed high-strength low-alloy steels: an overview. Metallurgist and Materials Technologist 1984; 16 (9): 447–454
9. Baker R. Developments in steelmaking. Metallurgist and Materials Technologist 1984; 16 (4): 197–217
10. de Koenig AC. Developments in materials and welding technology for offshore structures. Metal Construction 1985; 17 (11): 727–734
11. UEG. Design of tubular joints for offshore structures. UEG Publication UR33, UEG/CIRIA, 1985
12. BS 4360. Weldable structural steels. BSI, London, 1986
13. Billingham J. Materials for offshore structures. Metals and Materials 1985; 1 (8): 472–478
14. BS 6235. Code of practice for fixed offshore structures. BSI, London, 1982 (now withdrawn)
15. Salama MM, Ellis N. Hutton TLP – A materials challenge. In: Proceedings, 4th Offshore Mechanics and Arctic Engineering Conference. ASME, 1985, pp 1–13
16. Offshore Supplies Office, Department of Energy. Lightweight materials for offshore structures, report no. WOL 161/87. OSO, London, 1987
17. Gurney TR. Fatigue of welded structures. Cambridge University Press, Cambridge, 1979
18. British Steel Corporation. Iron and steel specifications. BSC, London, 1989
19. Lieurade HP, Lecoq H. Potential use of high strength structural steel in offshore construction. In: Noordhoek C, de Back J (eds) Steel in marine structures. Elsevier, Amsterdam, 1987, pp 605–616
20. Suzuki M, Tsukada K, Watanabe I. Newly developed Arctic grade high Al/low N/micro Ti type offshore structural steel. In: Proceedings, Conference Offshore Welded Structures. Welding Institute, Cambridge, 1983, pp P16-1–P16-12
21. Crane FAA, Charles JA. Selection and use of engineering materials. Butterworths, London, 1984
22. Lloyd's rules and regulations for the classification of ships. Lloyd's Register of Shipping, London, 1987
23. Muckle W. Strength of ships' structures. Edward Arnold, London, 1967
24. Betts CV, Clayton BR. Ships and advanced marine vehicles. In: Morgan N (ed.) Marine technology reference book. Butterworths, London, 1990, pp 3/1–3/138
25. Chalmers DW. The properties and uses of marine structural materials. Marine Structures 1988; 1: 47–70
26. Masubuchi K. Analysis of welded structures. Pergamon, Oxford, 1980
27. Hadden PG. Materials in a modern warship. Metals and Materials 1988; 4 (7): 412–415
28. American Petroleum Institute. Specification for line pipe API-5L. API, Dallas, 1988
29. Det norske Veritas: Rules for submarine pipeline systems. DnV, Hovik, 1981
30. Cavaghan NJ, Hill ML, Lessells J. Production of line pipe in the British Steel Corporation. In: Proceedings, Conference on Steels for Linepipe and Pipeline Fittings. Metals Society, London, 1983, pp 192–200
31. Coolen A, Caron F, Leclerc J. Developments by Usinor to meet new pipeline requirements. In: Proceedings, Conference on Steels for Linepipe and Pipeline Fittings. Metals Society, London, 1983, pp 209–213
32. Sage AM. A review of the physical metallurgy of high strength low alloy line pipe and pipe fitting steels. In: Proceedings, Conference on Steels for Linepipe and Pipeline Fittings. Metals Society, London, 1983, pp 39–50
33. Nara Y, Kyogoku T, Yamura T, Takeuchi I. The production of line pipe in Japan. In: Proceedings, Conference on Steels for Linepipe and Pipeline Fittings. Metals Society, London, 1983, pp 201–208
34. Jones BL, Johnson DL. The metallurgical design of major pipelines. In: Proceedings, Conference on Steels for Linepipe and Pipeline Fittings. Metals Society, London, 1983, pp 14–21
35. Shiga C, Amano K, Hatomura T, Saito Y, Hirose K, Choji T. Ferrite – fine bainite steel line pipe of X70 and X80 grades for low temperature service. In: Proceedings, Conference on Steels for Linepipe and Pipeline Fittings. Metals Society, London, 1983, pp 127–135
36. Lander HN, Morrow JW, Coldren AP. Development of as-rolled X80 pipeline steels for 19 mm wall pipe. In: Proceedings, Conference on Steels for Linepipe and Pipeline Fittings. Metals Society, London, 1983, pp 136–146
37. Blondeau R, Cadiou L, Pont G. Development and production of pipe fitting steels. In Proceedings, Conference on Steels for Linepipe and Pipeline Fittings. Metals Society, London, 1983, pp 264–270
38. ASTM A707-74. Flanges, forged, carbon and alloy steels for low temperature service. ASTM, Philadelphia, 1974
39. Wada T, Diesburg DE, Boussel P, Lauprecht WE. High strength steels for pipeline fittings. In: Proceedings, Conference on Steels for Linepipe and Pipeline Fittings. Metals Society, London, 1983, pp 249–255
40. Ohba H, Susei S, Sakai T, Atsuta T, Ohkuma Y, Ohminami R, Tamura A. The development of casting leg nodes for a jack-up rig. Proceedings, 11th Offshore Technology Conference. Houston, paper no. OTC 3426, 1979, pp 601–610
41. Ohtake F, Sakamoto S, Tanaka T, Kai T, Nakazato T, Takigawa T. Static and fatigue strength of high tensile strength steel tubular joints for offshore structures. In: Proceedings, 10th Offshore Technology Conference. Houston, paper no. OTC 3254, 1978, pp 1747–1756
42. Ellis N, Salama MM, Beggs DV. Evaluation of structural steels castings for the Hutton tension leg platform. In: Proceedings, 15th Offshore Technology Conference. Houston, paper no. OTC 4450, 1983, pp 65–70
43. Wood AM, Gantke F, Webster S, Haneke M. Recent advances in the technology of cast nodes for use in offshore structures. In: Proceedings, 14th Offshore Technology Conference. Houston, paper no. OTC 4198, 1982, pp 359–369
44. Atkins M. Atlas of continuous cooling transformation diagrams for engineering steels. British Steel Corporation. Sheffield, 1977
45. American Petroleum Institute. Specification for drill pipe API-5D. API, Dallas, 1988
46. American Society for Metals. ASM metals reference book. ASM. Metals Park, Ohio, 1983

47. Salama MM, Tetlow JH. Selection and evaluation of high strength steel for Hutton TLP tension leg elements. In: Proceedings, 15th Offshore Technology Conference, Houston, paper no. OTC 4449, 1983, pp 57–64

48. Scott GA. Studbolting experience in the underwater environment of the North Sea. In: Proceedings, 17th Offshore Technology Conference, Houston, paper no. OTC 5050, 1985, pp 269–280

49. Chaplin CR, Potts AE. Wire rope in offshore applications, Marine Technology Directorate Report no. MTD 88/100. MTD, London, 1988

50. Breen AJ. Low alloy steels in refinery service – 1: materials selection. Metal Construction 1984; 16 (11): 671–677

51. Glendinning IS, Vincent JR. Low alloy steels in refinery service – 2: design against failure. Metal Construction 1985; 17 (1): 23–29

52. Cooper BC. Gearing trends in the Royal Navy. Transactions Institute of Marine Engineers 1986; 99: 43–53

53. Dexter SC. Handbook of oceanographic engineering materials. Robert E. Krieger, Malabar, Florida, 1985

54. Sedriks AJ. Corrosion of stainless steels. John Wiley, New York, 1979

55. Speich GR. The chromium–iron–nickel system. In: Metals handbook, vol. 8: Metallography, structures and phase diagrams. ASM, Metals Park, Ohio, 1973

56. Schneider H. A modified constitution diagram for stainless steals. Foundry Trade Journal 1960; 108: 562–571

57. Lula RA. Stainless steel. ASM, Metals Park, Ohio, 1986

58. Solomon HD and Devine TM. Duplex stainless steels – a tale of two phases. In: Lula RA (ed.) Duplex stainless steels. ASM, Metals Park, Ohio, 1983, pp 693–756

59. Brown BF. Stress corrosion cracking control measures. NBS Monograph 156, 1977

60. NACE. Standard materials requirements for sulphide stress cracking resistant materials for oilfield equipment. NACE-MR-01-75-90. NACE, Houston, 1990

61. Bernhardsson S, Oredsson J, Martenson C. The stress corrosion cracking resistance of duplex stainless steels in chloride environments. In: Lula RA (ed.) Duplex stainless steels. ASM, Metals Park, Ohio, 1983, pp 267–281

62. Rhodes PR, Welch GA, Abrego L. Stress corrosion susceptibility of duplex stainless steels in sour gas environments. In: Lula RA (ed.) Duplex stainless steels. ASM, Metals Park, Ohio, 1983, pp 757–803

63. Tynell M. Applicability range for a high strength duplex stainless steel in deep sour oil and gas wells. In: Lula RA (ed.) Duplex stainless steels. ASM, Metals Park, Ohio, 1983, pp 283–292

64. Whitehead T, Baloun CH. Evaluating the suitability of the NACE standard test TM-01-77 for testing 13% Cr martensitic stainless steel for SCC resistance. In: Haynes GS, Boaboian R (eds) Laboratory corrosion tests and standards. ASTM STP 866. ASTM, Philadelphia, 1985, pp 400–414

65. Shone EB. Stainless steels as replacement materials for copper alloys in seawater handling systems. Paper presented at the Institution of Marine Engineers, 5 April 1988. Institute of Marine Engineers, London, 1988

66. Jasner M. Application of an austenitic nitrogen alloyed 6% Mo stainless steel for desalination and waste water treatment. In: Proceedings, 12th International Symposium on Desalination and Water Re-use, vol. 4. Institute of Chemical Engineers, London, 1991, pp. 335–345

67. Walton CF, Opar TJ (eds). Iron castings handbook. Iron Castings Society, 1981

68. Angus HT. Cast iron – physical and engineering properties. Butterworths, London, 1976

69. Boyer HE (ed.). Atlas of fatigue curves. ASM, Metals Park, Ohio, 1986

70. Schumacher M (ed.). Seawater corrosion handbook. Noyes Data Corporation, Park Ridge, N. J., 1979

71. Ailor WH (ed.). Atmospheric corrosion. John Wiley, New York, 1982

72. Osgood CC. Fatigue design. Pergamon, Oxford

73. Ray MS. The technology and applications of engineering materials. Prentice-Hall, London, 1987

74. Hatch JE (ed.). Aluminium – properties and physical metallurgy. ASM, Metals Park, Ohio, 1984

75. BS CP1021. Code of practice for cathodic protection. BSI, London, 1973

76. Det norske Veritas. Cathodic protection evaluation. Technical Note TNA 703. DnV, Hovik, 1981

77. Polmear IJ. Light alloys. Edward Arnold, London, 1981

78. American Society for Metals. Metals handbook, vol. 2: Properties and selection – nonferrous alloys and pure metals. ASM, Metals Park, Ohio, 1979

79. West EG. The fire risk in aluminium alloy ships' structures. Metallurgist and Materials Technologist 1982; 14 (9): 395–399

80. Powell C. Development of cast copper–nickel–chromium alloy for naval service. Metallurgist and Materials Technologist 1990; 6 (12): 783–785

81. Carruthers DR. The use of 90/10 copper–nickel as a splash zone cladding. In: Proceedings, Conference on Copper Alloys in Marine Environments, Birmingham, 1985

82. Copper Development Association. The brasses – properties and applications. Publication no. TN24. Copper Development Association, Potters Bar, Middx., 1980

83. Copper Development Association. Aluminium bronze alloys corrosion resistance guide. Publication no. AB80. Copper Development Association, Potters Bar, Middx., 1981

84. Leidheiser H. The corrosion of copper, tin and their alloys. John Wiley, New York, 1971

85. Heaton WE, Edgeley J, Andrews EFC, Patient BA. Review of the factors leading to the use of titanium as a steam condenser tube material within the CEGB. In: Proceedings, Conference on Cooling with Seawater. Mechanical Engineering Publications, London, 1979, pp 79–90

86. Mills NJ. Plastics – microstructure, properties and applications. Edward Arnold, London, 1986

87. McCrum NG, Buckley CP, Bucknall CB. Principles of polymer engineering. Oxford University Press, Oxford, 1988

88. Brydson JA. Plastics materials. Butterworths, London, 1982

89. American Society for Metals. Engineered materials handbook, vol. 2: Engineering plastics. ASM International, Metals Park, Ohio, 1988

90. American Society for Metals. Engineered materials handbook, vol. 1: Composites. ASM International, Metals Park, Ohio, 1987

91. Det norske Veritas. Rules for certification of diving systems. DnV, Hovik, 1988

92. Scola DA. Polyimide resins. In: Engineered materials handbook, vol. 1: Composites. ASM International, Metals Park, Ohio, 1987, pp 78–79

93. Brydson JA. Rubbery materials. Elsevier, London, 1988

94. Dutt B. Specialised rubbers in offshore engineering. In: Stevenson A (ed.) Rubber in offshore engineering. Adam Hilger, Bristol, 1984, pp 208–230

95. Holloway DG. The physical properties of glass. Wykeham, London, 1973

96. Oliver DS. The use of glass in engineering. Oxford University Press, Oxford, 1975

97. BS 6262. British standard code of practice for glazing for buildings. BSI, London, 1982

98. Reed DA, Engelson F. Load capacities of glass panels. Engineering Structures 1986; 8: 64–69

99. American Society for Metals. ASM engineered materials reference book. ASM International, Metals Park, Ohio, 1989

100. Weeton JW, Peters DM, Thomas KL (eds). Engineers' guide to composite materials. ASM, Metals Park, Ohio, 1987

101. Parrott S. Engineering ceramics for aggressive environments. Metals and Materials 1990; 6 (4): 207–210

102. Jack KH. Sialons and related nitrogen ceramics. Journal of Materials Science 1976; 11: 1135–1158

103. Lewis MH, Leng-Ward G. Advanced engineering ceramics. Metals and Materials 1991; 7 (6): 356–361

104. Kyrtatos NP. The potential of ceramics and insulation in marine diesel engines. Paper presented at the Institute of Marine Engineers, London, 1990

105. Johnson DD, Sowman HG. Ceramic fibres. In: Engineered materials handbook, vol. 1: Composites. ASM International, Metals Park, Ohio, 1987, pp 60–65

106. Pigliacampi JJ. Organic fibres. In: Engineered materials handbook, vol. 1: Composites. ASM International, Metals Park, Ohio, 1987, pp 54–57

107. Diefendorf RJ. Carbon/graphite fibres. In: Engineered materials handbook, vol. 1: Composites. ASM International, Metals Park, Ohio, 1987, pp 49–53.

108. Harvey MT. Thermoplastic matrix processing. In: Engineered materials handbook, vol. 1: Composites. ASM International, Metals Park, Ohio, 1987, pp 544–553

109. Mableson AR, Osborn RJ, Nixon JA. Structural use of polymeric composites in ships and offshore. In: Proceedings, Conference on Polymers in Marine Environments. Institute of Marine Engineers, London, 1989, pp 79–86

110. Seamark MJ. Current and potential opportunities for fibre reinforced composites offshore. Paper presented at the Institute of Marine Engineers, London, 1990

111. Trumper RL. Metal matrix composites – applications and prospects. Metals and Materials 1987; 3 (11): 662–667

112. Conde JF. New materials for the marine and offshore industry. Transactions of the Institute of Marine Engineers 1985; 97: Paper 24

113. Lea FM. The chemistry of cement and concrete. Edward Arnold, London, 1970

114. Illston JM, Dinwoodie JM, Smith AA. Concrete, timber and metals. Van Nostrand Reinhold, New York, 1979

115. Bogue RH. Chemistry of portland cement. Van Nostrand Reinhold, London, 1955

116. ASTM C150 84. Specification for portland cement. ASTM, Philadelphia, 1984

117. Neville AM, Brooks JJ. Concrete technology. Longmans, Harlow, 1987

118. Stillwell JA. Exposure tests on reinforced concrete in seawater. Offshore Technology Report OTH87 246. HMSO, London, 1988

119. Taylor Woodrow Research Laboratory. Surveys of existing concrete marine structures. Offshore Technology Report OTH87 244. HMSO, London, 1991

120. Fookes PG, Simm JD, Barr JM. Marine concrete performance in different climatic environments. In: Proceedings, Conference on Concrete in Marine Environments. Concrete Society, London, 1986, pp 115–130

121. Lawrence CD. Permeability and protection of reinforced concrete. Cement and Concrete Association, London, 1986

122. Burdall AC, Sharp JV. Some aspects of revisions to the UK guidance notes for offshore concrete structures. In: Proceedings, Conference on Concrete in Marine Environments. Concrete Society, London, 1986, pp 37–48

123. Booth ED, Leeming MB, Paterson WS, Hodgkiess T. Fatigue of reinforced concrete in marine environments. In: Proceedings, Conference on Concrete in the Marine Environment. Concrete Society, London, 1986, pp 187–198

124. Tantry PV, Tandon MC, Alimchandani CR. Fourth oil berth at Butcher Island, Bombay. In: Proceedings, Conference on Concrete in the Marine Environment. Concrete Society, London, 1986, pp 299–310

125. Moksnes J. Condeep platforms for the North Sea – some aspects of concrete technology. In: Proceedings, 7th Offshore Technology Conference, Houston, paper no. OTC 2369, 1975, pp 339–350

126. Palmer AC. Concrete coatings for submarine pipelines. In: de la Mare RF (ed.) Advances in offshore oil and gas pipeline technology. Oyez, London, 1985

127. Yamaguchi M, Tsuchida T, Toyoizumi H. Development of high viscosity underwater concrete for maritime structures. In: Proceedings, Conference on Concrete in the Marine Environment. Concrete Society, London, 1986, pp 235–246

128. Stanford AE. LNG ship economics. In: Proceedings, Conference 'Concrete Afloat', Concrete Society and RINA. Thomas Telford, London, 1977, pp 83–96

129. Sare PN, Yee AA. Operational experience with prestressed concrete barges. In: Proceedings, Conference 'Concrete Afloat', Concrete Society and RINA. Thomas Telford, London, 1977, pp 71–82

130. Derrington JA. Prestressed concrete platforms for process plants. In: Proceedings, Conference 'Concrete Afloat', Concrete Society and RINA. Thomas Telford, London, 1977, pp 23–42

131. Nuijima K, Nojiri Y, Koseki K, Sasaki, M. Study on applicability of steel/concrete composite members to ice walls for Arctic offshore structures. In: Proceedings, Conference on Concrete in the Marine Environment. Concrete Society, London, 1986, pp 91–100

132. Mettem CJ. Structural timber design and technology. Longmans, Harlow, 1986

133. BS 1088 and BS 4079. Specifications for plywood for marine craft. BSI, London, 1966

134. Farmer RH. Handbook of hardwoods. HMSO, London, 1981

135. Desch HE, Dinwoodie JM. Timber – its structure, properties and utilisation. Macmillan, London, 1981

136. BS 5493. Code of practice for protective coating of iron and steel structures against corrosion. BSI, London, 1977

137. Morcillo, M. Minimum film thickness for protection of hot-rolled steel. In: Berger DM, Wint RF (eds) New concepts for coating protection of steel structures. ASTM STP 841. ASTM, Philadelphia, 1984, pp 95–112

138. Whitehouse NR. Survey of painting practice for protection of offshore structures. Paint Research Association, Teddington, 1983

139. Roche M, Samaran JP. Pipeline coating performance. Materials Performance 1987; 26 (11): 28–34

140. Welding Institute. Weld surfacing and hardfacing. Welding Institute, Cambridge, 1980

141. Marks PR. The fire endurance of glass reinforced epoxy pipes. In: Proceedings, Conference on Polymers in a Marine Environment. Institute of Marine Engineers, London, 1989, pp 69–78

5 Fabrication and Manufacture for Marine Technology

The demands of fabrication have dominated the marine technology industry almost from inception. Most of the major advances in shipbuilding technology have been related to fabrication, and nowhere (apart, possibly, from the case of boilers and pressure vessels) do the requirements of load-bearing capacity and fatigue resistance make welded fabrication such an attractive option. For this reason, most of this chapter is devoted to a description of how the particular structural and environmental considerations of offshore engineering and naval architecture have led to great advances in welding technology and also to a description of the properties of welded joints and the weldability of various alloys.

Fabrication and manufacturing processes applicable to polymers and composite materials are also described with particular reference to the limitations which these impose on the use of such materials in marine environments.

The manufacturing methods in use for marine concrete are also of considerable importance, and the advances in this field brought about by offshore engineering are also discussed.

5.1 Welding

The greater part by far of the metallic material used in marine construction is steel and this is due to the overwhelming cost and weight benefits brought about by the fabricability, availability, strength and rigidity of this material. This is despite the one obvious disadvantage of steel, namely that it corrodes in seawater.

5.1.1 Development of Welding Technology for Marine Applications

Since the first structural use of steel in about 1850, there has been a continued requirement to develop the highest possible strengths consistent with acceptable toughness, and nowhere has this been more difficult to achieve than in weld heat-affected zones (HAZs). The reason for this is that the weld thermal cycle is superimposed on any elaborate thermomechanical treatment which was used to achieve the properties of the plate, so that any such treatment must be developed with this limitation in mind. Furthermore, the weld metal itself can only be subjected to very simple treatments in large structures, and there is an attendant difficulty in ensuring that weld metals meet the strength and toughness values achieved in the plate.

In fact, brittle fracture in welded ships was quite a commonplace occurrence in the 1940s and these failures were still being analysed and correlated into the late 1950s (e.g. Williams [1], Williams and Ellinger [2]), but from the time that the American Bureau of Shipping issued specifications for steel grades and notch toughnesses for welded ships the incidence of brittle fractures reduced dramatically. However, such failures do still occur, a fairly recent example being that of a large integrated tug–barge in Port Jefferson Harbour, New York, in 1972 [3].

The loss of the drilling rig *Sea Gem* in 1965 lent further concern to the requirement for toughness in welded offshore constructions. The wreckage left on the seabed was described by one witness as presenting the aspect of 'a heap of broken glass', emphasising the very large number of brittle fractures which had taken place during the rig's collapse and sinking [4]. The prime cause of the collapse was put at the failure of the tie bars on which the barge was suspended, and, as well as the contributing factors of stress concentration at an abrupt change of section and low sea and air temperatures, the mean impact values of the material at service temperature were measured and found to be very low. A number of weld defects were also present, and the tie bars had not been heat-treated after welding. Modern offshore construction practice would not now allow a structure to go into service in such a condition, but the capsize of the rig *Alexander L Kielland* in 1980 served to highlight the importance of even non-structural welds on overall integrity [5]. A useful summary of the metallurgical aspects of this failure is contained in Easterling [6].

The critical considerations in the integrity of welded structures are: process and process control, weldability of the alloy in use and the properties of the welded joint, and these are now dealt with in turn.

5.1.2 Welding Processes and Process Parameters

The energy required to produce a weld can come from a variety of different sources, including electric resistance heating, friction and electron or laser beams. However, the majority of welding processes in use employ an electric arc as the power source.

It is not the place of this text to describe in detail the welding processes, and the reader is referred to the American Welding Society's *Welding Handbook* [7] for a detailed description of the processes referred to in the following. As for application, the manual and semi-automatic arc processes (MMA, MIG and FCAW) are widely used in yard fabrication, whereas TIG is used in more difficult situations requiring deep penetration or exceptional control of the weld bead. The submerged arc and electroslag processes are used for simple geometries and high deposition rates, and the remaining processes are only used under exceptional circumstances.

It is useful to have an appreciation of the importance of energy and heat flow in welding, as this helps to define some of the important process parameters and hence the development of weld procedures. The first consideration in welding is as to whether sufficient power density can be applied to achieve melting but not so much as to vaporise the metal. Figure 5.1 shows this range of power density, and clearly the arc processes are those which achieve the melting requirement with heat loss from the weld area being dominated by conduction in the plate. Depending upon the particular geometry of the process, the heat source in welding can be approximated as a point, a line or a plane (Fig. 5.2). Examples of the first two might be arc processes where incomplete and complete penetration are achieved respectively, and the third might be represented by a flash-butt, resistance or friction weld. Most analytic work has centred around the arc processes and, in particular, the heat flow in the plate resulting from the application of a moving point- or line-source. Solutions to the Fourier equation for the 'pseudo-stationary state' (using a moving origin progressing along the x-axis) are available (e.g. Easterling [7] and Masubuchi [3]) but these are rarely used directly, although the effect of process parameters on the solution gives useful information for the development of weld procedures.

Firstly, it is helpful to consider the effect of process parameters on the size of the resulting weld bead. If an electric

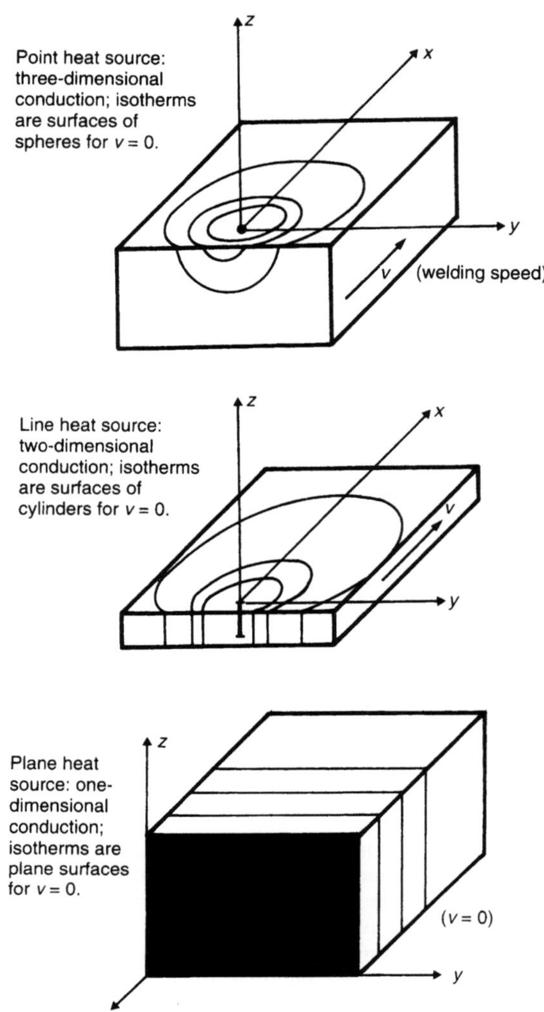

Fig. 5.2. Dimensionality of welding heat sources and resulting solutions for conduction in the plate. (Adapted from Lancaster [9].)

arc dissipating a power EI (where E is the arc voltage and I is the arc current) progresses along the weld groove at a speed v, the total volume of weld metal (that which has been melted, and consisting of some material from the parent plate and some from the filler wire) produced will depend upon the total amount of heat dissipated in the weld run and some characteristic of the metal itself (including latent heat and heat required to raise the metal to its melting temperature), which can be expressed as Q, the heat required to melt a unit volume of metal. Thus, it can be seen that the cross-sectional area of the weld bead A_w is proportional to the arc power divided by the welding speed, and this latter collection of terms is normally referred to as the 'arc energy', 'energy input' or 'heat input', none of which terms suggests its physical meaning of the amount of energy delivered per unit length of weld:

$$H = EI/v = KA_w$$

The important point is that the cross-sectional area of the weld bead is increased by increasing the arc energy, but arc

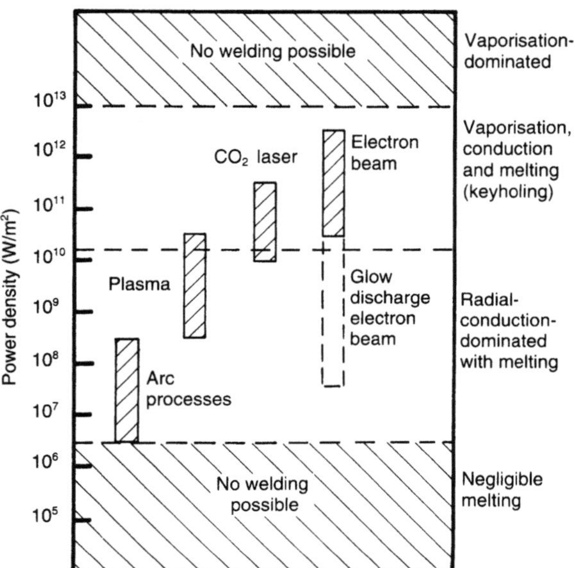

Fig. 5.1. Range of power densities for the various classes of welding processes. (From Lancaster [8].)

energy has an effect on other thermal features of the weld through its appearance in the solutions to the conduction equation.

As mentioned above, the energy used in melting metal is finally lost from the plates, mostly by conduction into the plate material, so that any point on the plate experiences a thermal cycle which is partly dependent upon its distance from the centre of the weld bead and partly upon the welding conditions (Fig. 5.3). The important features of this thermal cycle are shown in Fig. 5.3, and these are the peak temperature, time above any critical temperature (for example that at which overageing might occur in an aluminium alloy), and the cooling rate. Any part of the plate which is metallurgically altered by the welding thermal cycle (for example annealed, overaged or quenched) is regarded as constituting the heat-affected zone (HAZ), and so there is generally a band of metallurgically altered material around the weld metal in a typical weld cross-section.

The pseudo-stationary-state solutions can be used to give a qualitative indication of the effects of process parameters on weld thermal cycles, and the most important of these are as follows:

1. Increased arc energy results in a higher peak temperature at a given distance from the weld centre-line (i.e. a wider heat-affected zone), but also results in a lower cooling rate for a given peak temperature.

2. The application of pre-heat (increasing initial temperature of the plate) has a similar effect to increasing energy input.

The arc processes dominate the offshore construction and shipbuilding industries, with the manual metal arc (MMA) or shielded metal arc process slowly giving way to those processes which are more automatible and/or have a higher production rate. The reason for this can easily be seen if it is considered that a typical main leg node tubular fabrication for an offshore structure would involve a 6-m-long, 3.5-m-diameter tube of wall thickness 100 mm. The seam weld itself, with a 60 ° included angle, requires about 300 kg of weld metal. It can be seen that at a typical MMA deposition rate of 2.3 kg/h such a weld would involve a substantial manpower input. Fortunately, other processes such as submerged-arc welding, can achieve much higher deposition rates (perhaps 20 kg/h or more, using multiple electrodes [10]) and are semi-automatic in that the welding head is not manually guided and travels on a boom. However, the same node will also typically have 16 one-piece internal ring stiffeners 30 to 80 mm thick, and will

have at least six stub/brace connections from 1 m diameter and 30 mm wall thickness to 2.5 m diameter and 80 mm wall thickness (Keeler [11]). These welds would all require to be made with full penetration and the geometry of the connections, as well as the fact that a number of the connections need to be made in the yard, makes the application of automatic welding rather difficult. For this reason, any processes which can provide improved deposition rates are of interest to offshore constructors, and an example of this is given by Keeler [11] with regard to 'Innershield' (actually flux-cored arc) welding of offshore structures.

The shipbuilding industry has also in the past been alert to this problem and has, for example, adopted variations of the MMA process such as 'firecracker' welding and gravity welding for improved operability, as well as the electroslag and electrogas processes for improved deposition rate. There is, however, a tendency to be more traditional in choice of steels for ships, mainly because designs are stiffness-dominated rather than strength-dominated (see Chapter 2), so that there is less need for those steels which have higher strengths and are more difficult to weld.

Pipelines also present weld process problems. In smaller diameters these are produced seamless, but larger diameters require seam welding, again with the seam weld often being produced by the submerged-arc process. In this latter case, production rate is again a consideration, and it is important to achieve reasonable rates without a loss of property or an unacceptable defect rate in the weld metal. The butt welding associated with the laying of subsea pipelines presents a different problem. To achieve a reasonable lay rate the pipeline must be welded at a number of stations on the deck of a laybarge (Fig. 5.4), and a typical pipeline would require about 85 such girth welds per kilometre of length. Lay rates vary with weather and operational factors but are usually between 1 and 3 km per day,

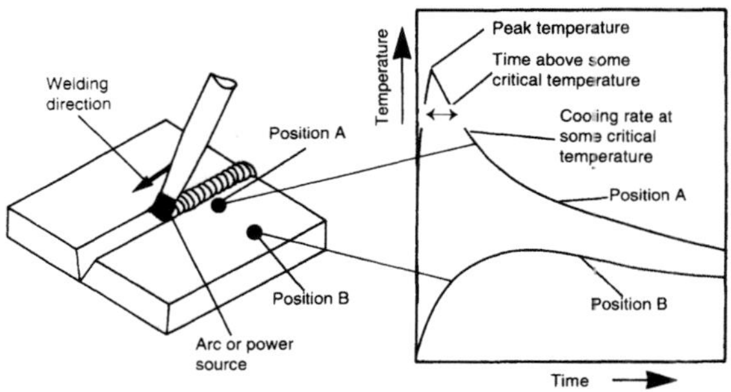

Fig. 5.3. The welding thermal cycle as experienced by a point in the plate away from the weld centre-line.

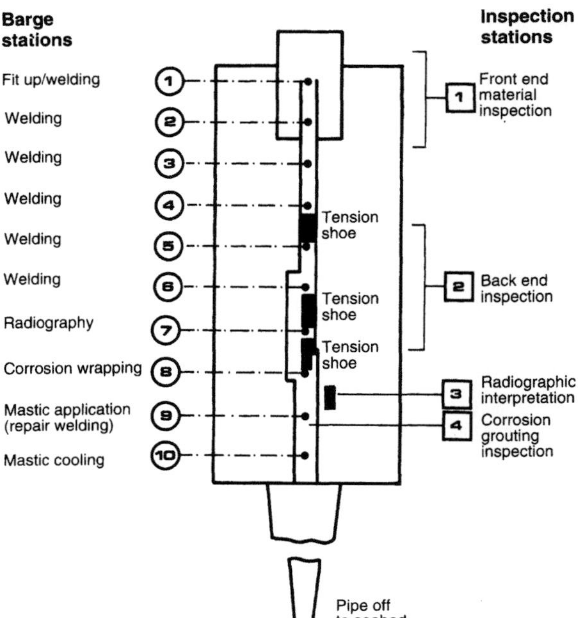

Fig. 5.4. Typical arrangement of welding and inspection stations on a pipe laybarge for the installation of subsea pipelines. (From Still and Rae [12].)

and this includes pre-heating, post-weld heat treatment, inspection and the making good of the coating system in the weld area. Also, anodes and buckle arrestors need to be fitted where appropriate. The traditional welding method for pipelines is the so-called 'stovepipe' method which employs the MMA process in the vertical down direction and was developed in the USA in the late 1920s (Letchford [13]). Modern pipeline welding may still employ the MMA process, typically with cellulosic electrodes, but variants of the MIG process, used in an automatic or semi-automatic mode, are also found (Culbertson and Gumm [14]).

The 'S-type' lay procedure (Fig. 5.5) is only appropriate up to about 800 m of water depth. For deeper pipelines a steeper configuration is required on the laying vessel (Fig. 5.5), called the 'J-lay technique'. This reduces the opportunity for the laybarge approach, which uses multiple welding stations, and so rapid welding processes are required to achieve acceptable lay rates. In principle, this can be done either by using a process which simultaneously attaches all parts of the butt (such as flash-butt welding [15] or friction welding [16]) or by using a very fast welding process such as electron beam [17]. Proposals have been put forward for all of these approaches, and there have also been suggestions of explosive welding combined with bottom tow as a pipeline construction technique [18].

The necessity of carrying out welding operations underwater is a problem which occasionally arises. Such cases include underwater damage to ships, offshore structures and pipelines, of which the latter two are probably the most important in view of the vast expense of bringing such elements to the surface or shore. Furthermore, underwater welding may be required at some stages of pipeline construction, such as tie-in or recommencement of laying after the line has been allowed to sink to the seabed.

Underwater welding can be carried out either dry or wet, according to whether or not the ambient seawater has been excluded from the area of the weld. In most common applications, wet welding is carried out using the MMA process and dry welding using the MMA and/or TIG (tungsten inert gas, or gas tungsten arc) processes. However, a few proposals have been made for wet and dry applications using other processes such as friction, stud, MIG, firecracker and submerged-arc processes. Table 5.1 summarises some of

Table 5.1. Summary of literature on underwater welding

	Process	Reference
Wet	MMA	24, 25
	MIG/GMA/FCAW	20
	TIG	
	Submerged-arc	23
	Firecracker	19
	Stud	21
	Explosive	22
Dry	TIG	26, 29, 30, 31, 33
	MMA	31, 33
	MIG/GMA/FCAW	31, 32

the literature reports of underwater welding processes and their application.

Wet welding is normally considered to be inferior to dry welding, especially for steels. The reasons for this are obvious in that, although a bubble is formed in the arc region, arc stability is disrupted, weld metal and HAZ cooling rates can be high, the hydrogen potential of the process is high and the welder's view of the arc is impeded. This means that wet MMA is only likely to be used for non-critical attachments to non-critical structures, although ways are still being developed of overcoming the difficulties. For example, Masubuchi [24] has demonstrated that 'fairly efficient' wet welds can be produced in a quenched and tempered steel, provided that these were relatively thin. More recently, Gooch [25] has carried out a systematic study of wet MMA welding of steels of carbon equivalents (see Section 5.1.7.1) from 0.3 to 0.49 per cent with a thickness range of 25 to 100 mm. It was found that the main factor affecting final joint performance was the quality of the weld metal, with particular regard to the incidence of defects which, in turn, depends on factors affecting arc stability. However, it is very likely that once such factors are overcome, hydrogen control will become critical. Some automated and mechanised methods have been reported for underwater wet welding (e.g. Asnis and Savich [26]) and these are useful in eliminating some of the defects of shape associated with welder skill. In wet welding, it is also important to consider the safety aspect of the electric field produced around the electrode. Diesen [27] has presented methods of calculating the risk distance for an MMA electrode submerged in water, and suggests ways of minimising the danger to the diver.

Dry, or hyperbaric, welding on the other hand, is usually the preferred option for underwater welding. The method involves the exclusion of water, usually by the erection of an 'underwater habitat' around the weld zone and the application of gas at ambient (i.e. hyperbaric) pressure (Fig. 5.6). The diver–welders then enter this habitat to carry out cutting, aligning, preheating and reinstatement operations (e.g. Haux [28]). Other methods are available which include more local dry areas, cofferdams (for work close to the surface) and even habitats which are held at one atmosphere, although this last is much more difficult to achieve than a hyperbaric habitat. These alternatives are illustrated schematically in Fig. 5.7, although with the exception of cofferdams they are not widely used. Apart from the work conditions, the humid atmosphere and the condition of the workpieces, which may have been underwater prior to welding, weld procedures for cofferdams and one-

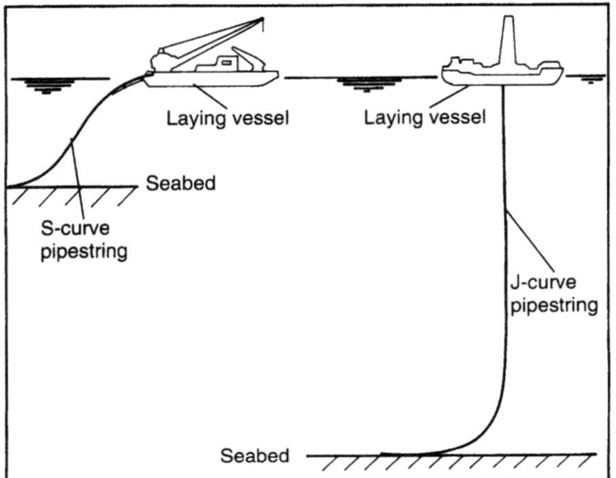

Fig. 5.5. Comparison of J- and S-lay methods for subsea pipeline installation.

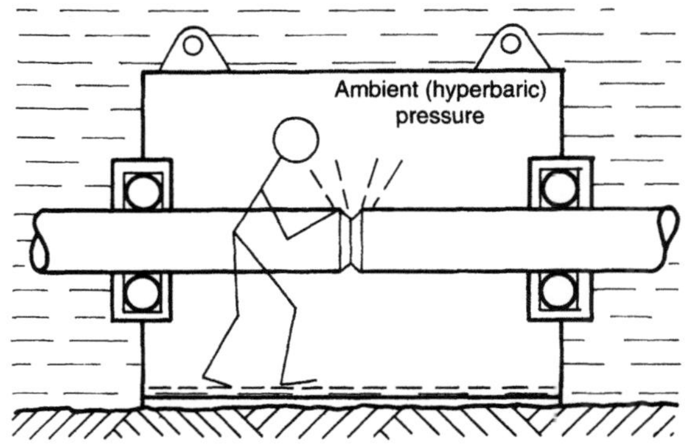

Fig. 5.6. An underwater habitat for subsea hyperbaric welding. (From Haux [28].)

atmosphere habitats are largely the same as those which would be used at the surface. Although simpler to achieve, the local-dry-spot approach has found only a few applications for subsea welding. This is probably because visibility can be poor for the diver–welder, and also because it is desirable to minimise diver exposure when carrying out what is a fairly dextrous task even on the surface. The operational difficulty of the hyperbaric welding operation has led to developments where the cutting, preparation and welding tasks can be carried out in a mechanised or automated fashion using orbital (usually TIG) heads which can be controlled and/or monitored from the surface (e.g. Scott Lyons and Middleton [29], van der Torre and Sipkes [30]). Diver intervention is, however, still necessary in the setting up of such systems, and divers are usually present at the work site while the welding is being carried out.

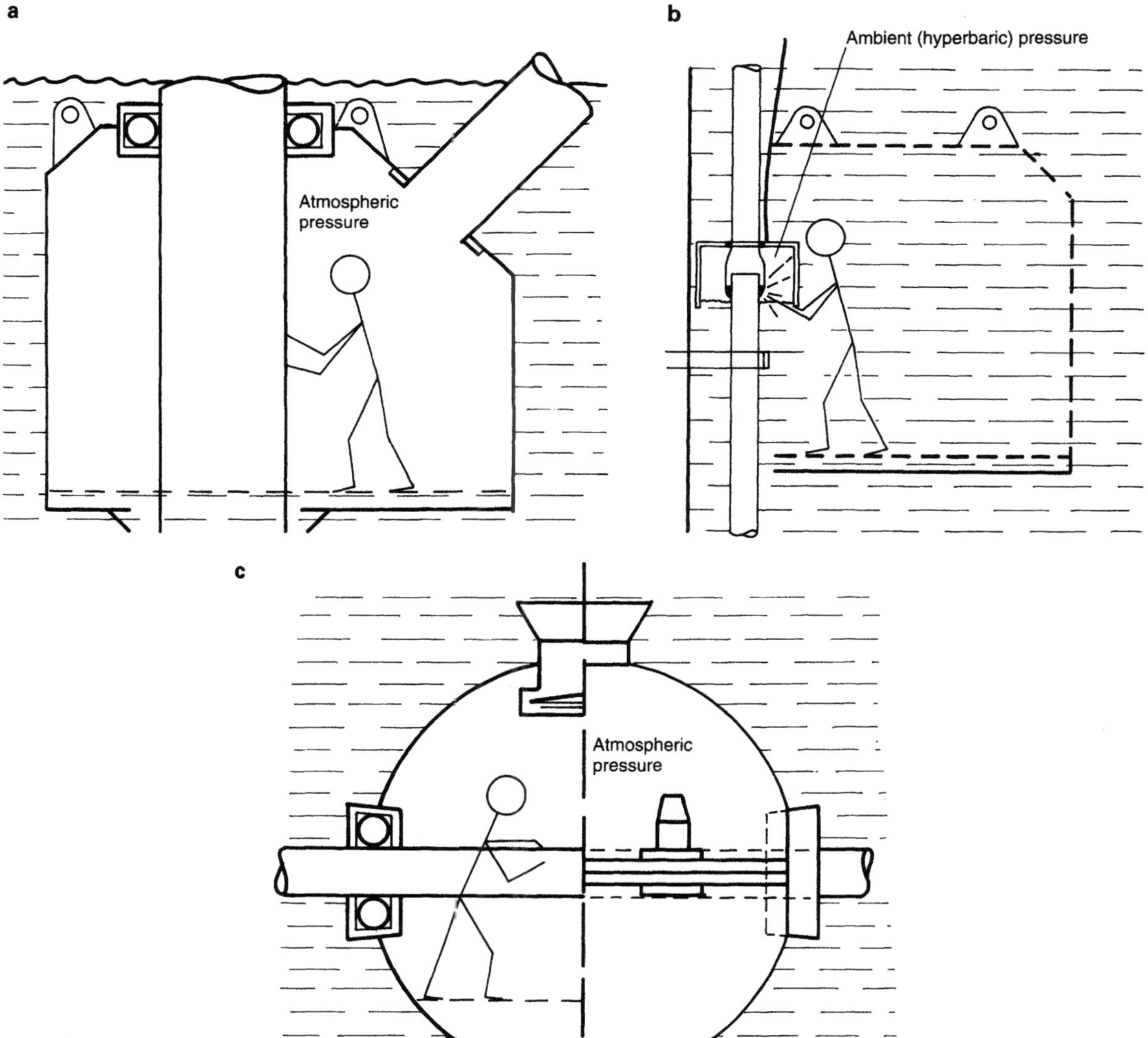

Fig. 5.7. Alternative methods for dry subsea welding. **a** Cofferdam. **b** Portable dry spot. **c** One-atmosphere habitat. (Adapted from Haux [28].)

The existence of a hyperbaric atmosphere during welding has some additional difficulties. First, although carefully scrubbed by a rapid recirculating system, the habitat atmosphere is not suitable for breathing, especially during TIG welding, where the local ventilation system which would be required to capture the fume close to its source would disrupt the arc stability through its effect on the shielding gas (Worral and Gibson [31]). This means that when trying to carry out such a manual operation divers must still wear breathing apparatus while inside the habitat, with the concomitant difficulties this brings. Second, electric arcs are sensitive to ambient pressure, and stability problems can arise at pressures as low as 10 bar (e.g. Hamasaki [32]), although at this pressure these can usually be overcome by appropriate choice of process or process parameters. Nevertheless, hyperbaric welds in very deep water may well not be possible. TIG welds using argon shielding have been carried out at up to 80 bar in the laboratory, whereas definite difficulties can arise with the consumable arc processes (Allum [33]). Associated with such effects on the arc, bead shape and penetration characteristics can be markedly different from those at ambient pressure, and it is usually necessary to develop weld procedures which are specific to the pressure at the worksite. A number of published accounts of subsea hyperbaric welding operations exist (e.g. DeJong et al. [34] and Gaudiano [35]), some including details of weld procedures [36].

An interesting application of underwater welding which has been considered (Mellon [37], Lynch and Pilia [38]) on a few recent occasions is hot tapping. This process has been widely studied for onshore pipelines as the method whereby pipe connections can be made without depressurisation, the welding being carried out on a 'live' pipeline. Although there are purely mechanical methods for hot tapping, there are some advantages in the welded methods. If a welded hot tap is to be carried out, it normally involves the application of a split 'T' (Fig. 5.8). The important considerations are that blowthrough due to excessive penetration is avoided, and also that the internal fluid does not cause excessive chilling of the heat-affected zone of the weldment (Phelps [40]). Since hoop stresses are larger than longitudinal stresses in a pressurised pipe, it is desirable to avoid welding directly onto the pipe in the longitudinal

direction, and methods such as the use of backing strips and undercutting of the longitudinal weld are employed. Penetration of the circumferential welds cannot, however, be avoided, and substantial efforts have gone into establishing the procedures for these welds to minimise the risk [39, 40].

A number of the less common welding processes find uses in subsea and offshore engineering for site work and also for the manufacture of components. For example, the tool joints of drilling tubulars for oil wells are often attached by inertial welding (a variant of friction welding), stud welding has found some applications for attachment of, for example, pad-eyes for lifting operations, and flash-butt welding is often used for the manufacture of anchor and mooring chains. Explosive bonding has also seen some proposed and actual uses, for example as a method of repairing pipelines underwater (Chadwick [22]) and in the production of transitions between aluminium superstructures and steel hulls for warships (Trethewey [41]). Explosive welding is particularly suited to this latter application because of the difficulty in fusion welding two alloys of such different thermal and metallurgical properties. Bonding using pressure is also important in the production of lined material such as pipes for high-pressure applications. Loose linings are normally considered to be unacceptable for this type of service for two reasons. First, the development of an interface pressure between liner and main pressure containment may lead to implosion of the liner (Yoshida et al. [42]). Such a pressure could be developed by puncturing of the liner through pitting, stress corrosion cracking or other means, followed by rapid depressurisation of the bore. Secondly, if puncture does take place this could produce unacceptable galvanic corrosion (for the reasons outlined in Chapter 3) if fluid were to come between the lining and the cladding. Such problems can be minimised by bonding the cladding to the surface or by the employment of interfacial compressive pressure. Bonding methods include explosive cladding and the use of forge welding during drawing of the tube. Of course, special weld procedures are required when welding clad pipes together. Welded cladding has also been used on the exterior of offshore structures in the splash zone and also for large ships' hulls (Carruthers [43]). The antifouling properties of copper–nickel alloys often make these attractive for such purposes and, to avoid the purchase of clad plate, various techniques have developed for the attachment of the cladding to the substrate by welding. This can be achieved by the use of butt welds between cladding plates which penetrate to the substrate, and care must be taken to avoid excessive dilution which could compromise the corrosion resistance of the weld metal and introduce galvanic effects (Sandor [44]).

5.1.3 Residual Stress and Distortion

When designing for welding, it is important to remember that the substantial local temperature gradients and freezing contractions involved can give rise to distortion or to residual stresses where these distortions are opposed either by the bulk of the weldment or by the application of fixturing during welding. As a general rule, the control of distortion is a problem of relatively slender fabrications, and a number of preventive and curative measures are possible [3, 45]. In heavier sections, the physical forces are usually manifested as residual stress. Residual stress can lead to a

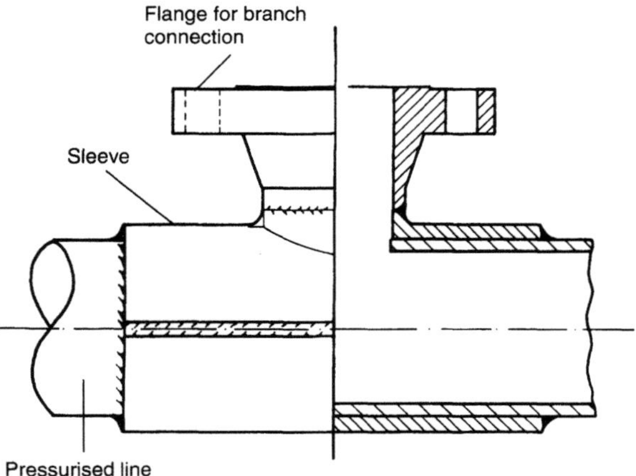

Fig. 5.8. Split 'T' method for hot-tap welding. (From Osborn [39].)

number of problems with strength, fracture resistance, fatigue resistance and weldability, and it is important to have an estimate of the likely extent and possible alleviation of the effects of residual stress.

The likely degree of distortion in a welded fabrication can often be roughly estimated by making a few assumptions about restraint conditions. As an example of how such an estimate might be made, the longitudinal distortion of a single weld run (Gray and Spence [46]) can be expressed as the radius of curvature resulting in a long body of cross-sectional area A and second moment of area I, to which a longitudinal weld has been applied at a distance z from the neutral axis (Fig. 5.9):

$$1/R = \epsilon_1 z A / I$$

where ϵ_1, the longitudinal contraction strain, is given by:

$$\epsilon_1 = 0.35 \frac{H_{net}\alpha}{C_v A}$$

where H_{net}, the net energy input, is the energy input defined earlier multiplied by an 'arc efficiency', α is the coefficient of thermal expansion and C_v is the volumetric thermal capacity. Arc efficiencies vary from about 0.99 for the submerged arc process to as low as 0.3 for the TIG process operating in a.c. mode. A typical value for MMA welding is about 0.7.

The degree of simplification required, the likelihood of unknown contributions (such as pre-existing residual stresses resulting from plate manufacture) and the computational difficulty combine to make distortion analysis an unattractive proposition for any but the simplest geometries or the most critical applications. However, analysis of the final distorted shape is possible in cases of complex geometries and an extensive literature, reviewed by Masubuchi [3], exists, some of it semi-empirical. One area in which distortion is of considerable interest is in shipbuilding and, for example, the US Navy has standards of allowable distortion in steel and aluminium fabrications. The reason for

such standards is to avoid the danger of buckling during service, but it is important that these are not so stringent that fabricators will have difficulty in meeting them. One important example of a distortion problem which occurs in ship structures is that of a panel with longitudinal and transverse stiffeners (Fig. 5.10). This problem is discussed in detail by Masubuchi [3] and the results of several analyses are compared with the US Navy specifications.

Distortion can to a certain degree be corrected by thermal or mechanical means, but it is rather better to attempt to avoid it by careful joint design, welding sequence and choice of process parameters. The application of restraints will, as indicated above, lead to an increased incidence of residual stress.

Residual stress distributions can also be calculated by considering the thermomechanical behaviour of the weldment. Again, calculation is usually only feasible in the simpler cases. One simple, well-known such case is that for a butt weld where the transverse and longitudinal residual stress distributions are familiar; see e.g. Masubuchi [3].

The effect of applying restraint can be appreciated by considering the example of the effect of applying a transverse restraint. If this is done, the transverse residual stress pattern is modified as shown in Fig. 5.11.

A particular residual stress problem of importance in offshore and marine engineering is the tubular connection. The residual stresses in pipe butt welds have been measured by, for example, Burdekin [47], with the result shown in Fig. 5.12.

More recently, residual stress distributions have been investigated for more complex tubular geometries (e.g. Porter-Goff et al. [48]) such as 'Y' and 'I' configurations,

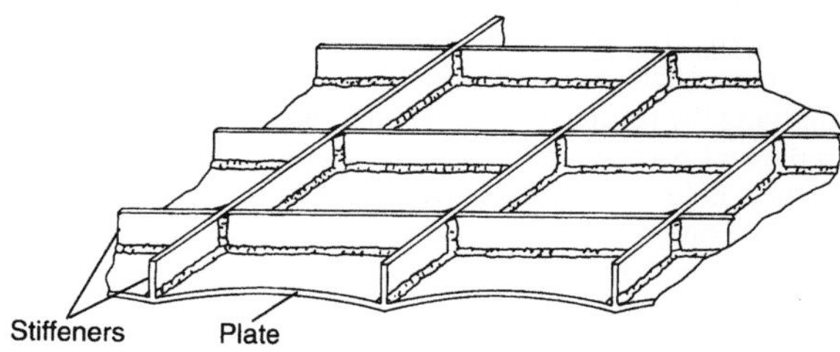

Fig. 5.10. Panel structure with longitudinal and transverse stiffening such as might be used in the construction of ships' hulls. (From Masubuchi [3].)

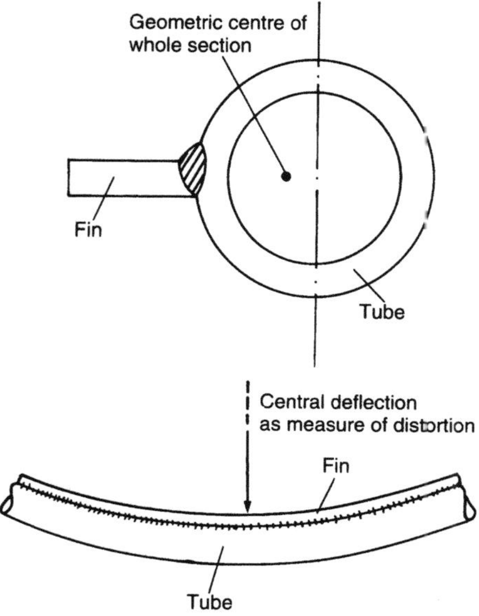

Fig. 5.9. Bowing distortion produced when welding a fin to one side of a hollow cylinder. (From Gray and Spence [46].)

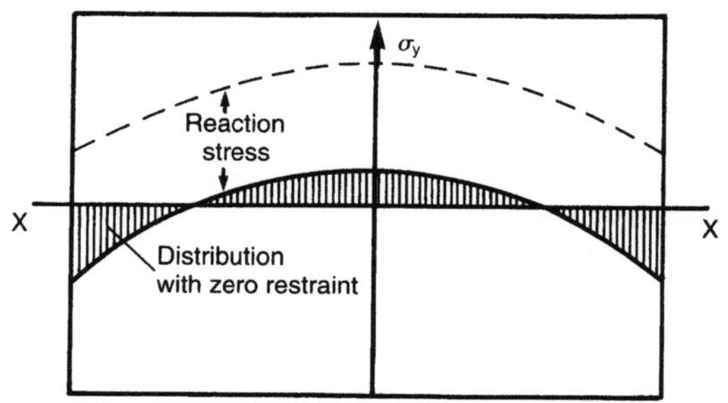

Fig. 5.11. Effect of restraint on the transverse residual stress distribution in a simple butt weld. (From Masubuchi [3].)

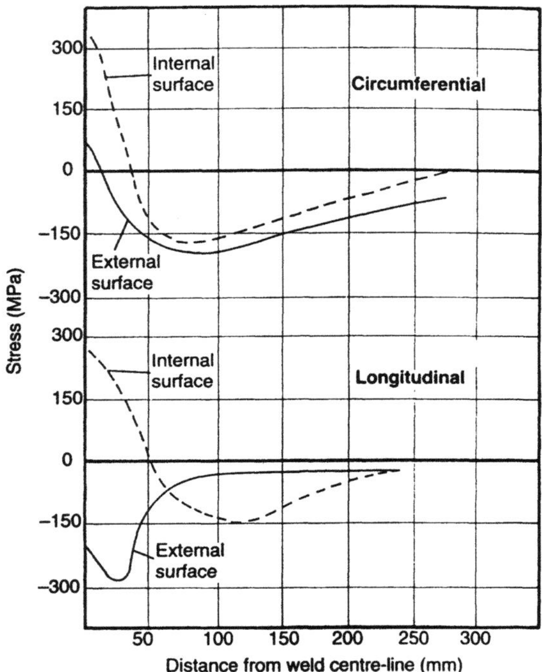

Fig. 5.12. Residual stress patterns in a girth-welded low-carbon steel pipe of length 76 mm, diameter 76 mm and wall thickness 11 mm. (From Burdekin [47].)

but little has been done on multibranched connections. For these cases it is usually assumed that residual stresses of the order of yield will occur, so that some plastic deformation will have taken place.

The reduction of welding residual stress can be carried out thermally, thermomechanically or mechanically. The thermomechanical method involves relieving the stress by applying the reverse phenomenon to that which generated it in the first place. Careful analysis should be able to produce a stress field which almost cancels the one due to welding, but the complexity of this process even for simple geometries means that it is not often applied, especially where simpler methods are feasible.

Purely mechanical stress relief is also possible, on the principle that raising material above the yield point does not result in further elastic strain being locked into the material. Thus, a load is applied in the direction of maximum tensile residual stress 'washing out' the peaks in the distribution. This method is of limited applicability in offshore engineering or shipbuilding, because substantial cold work may be necessary and the application of a suitable deforming load could be rather difficult.

Thermal stress relief is simple in concept and can be relatively easily applied. The principle is that the yield stress of most materials reduces with increasing temperature, so that raising the material to a temperature at which the yield stress is relatively low and at which creep may take place can remove most of the residual stress distribution, as is the case in purely mechanical stress relief. The temperature at which this is normally carried out for structural steels is around 650 °C. Obviously, it is important to remain below the critical temperature (around 715 °C), as the transformation to austenite could cause much more severe problems. Temperature and time both affect the degree of stress relief

experienced, and Fig. 5.13 gives an indication of the relative effects which might be expected by local treatment of the weld metal.

In structural steels post-weld heat treatment (PWHT) can produce other benefits, for example in tempering the heat-affected zone and allowing the diffusion of hydrogen from the weld area. In the fabrication of very large structures, the application of PWHT can be very difficult without introducing thermal gradients elsewhere and perhaps aggravating the problem. This has been overcome, for example, in the case of offshore structures by the use of very large furnaces in which an entire node can be heat-treated [49]. The subsequent butt welds for attachment to the nodal pieces do not normally present the same degree of problem, since there is generally less welding and these connections are of less critical structural significance. Alternatively, large fabrications can be locally heat-treated if there is no other practicable method (Saunders [50]), although here it is important to minimise temperature gradients, because in steel a temperature change of 100 °C can produce a change in length of about 0.14 per cent, which is close to the yield strain. In heat-treatable low-alloy steels PWHT is often necessary for other reasons than stress relief, and this aspect is dealt with under 'weldability'. One important aspect of stress relief which applies to stainless steels is that the temperature of 650 °C is quite inappropriate because of the danger of sensitisation or formation of sigma phase (see Section 5.1.2.3). Stress relief can of course be very important for stainless steels from the point of view of stress corrosion cracking and, if this is a danger, it is necessary to stress-relieve at a temperature of about 1000 °C. This does not produce the same problems as in mild steels, since no transformation takes place, but the implications for duplex stainless steels are more complex. Some aluminium alloys are sensitive to heat treatment at stress-relieving temperature, and the decision to heat-treat a weldment made from a heat-treatable aluminium alloy will normally be taken in the context of its age-hardening properties.

It is sometimes necessary to reduce residual stresses in fabrications subject to fatigue loading. The argument as to the effects of residual stress on fatigue is not simple; it is compounded by the fact that the only way to produce a control for any experiment on the measurement of the effect is to remove residual stress entirely from a sample welded joint and that this cannot be guaranteed.

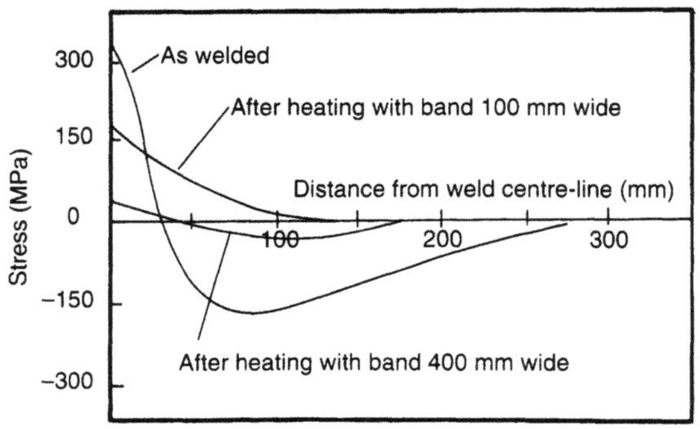

Fig. 5.13. The effect of post-weld heat treatment on the circumferential residual stress pattern on the internal surface of a girth-welded low-carbon steel pipe of length 76 mm, diameter 76 mm and wall thickness 11 mm. (From Burdekin [47].)

Comparison between PWHT joints and those to which no stress relief has been applied certainly suggests that there is an effect, but this is difficult to quantify because the degree of stress relief brought about by PWHT is itself difficult to quantify [51].

5.1.4 Properties of Welded Joints

Almost every weld made for offshore use is carried out to full penetration, so that the strength of the welded connection is primarily controlled by the properties achieved in the heat-affected zone and in the weld metal, and this is the subject of Section 5.1.5.

One aspect of the properties of welded joints for offshore use dominates all others, namely their behaviour under fatigue loading. The subject of fatigue design of tubular joints is covered in detail in Case Study 7.2, and the intention in this section is simply to describe the effects of the weld design on fatigue.

Apart from the geometry of the connection and the wall thicknesses involved, both of which affect the stress concentration factors and hence the hot-spot stresses, but are a matter of global design, some methods can be used as part of the fabrication process to modify the fatigue properties of the joint. The so-called weld improvement techniques consist of mechanical treatments which would intuitively be expected to enhance fatigue life. These are essentially shot-peening and dressing of the weld using TIG remelting or grinding. The effectiveness of these has been demonstrated, and Fig. 5.14 shows one example. However, there is generally insufficient evidence to allow design codes to recommend an increase in fatigue life assessment on the basis that such techniques have been used.

For PWHT there is some evidence that the removal of residual stress can increase fatigue life. However, the stress ratio (trough value of stress/peak value of stress) is an important factor when assessing the likely effect of residual stress, and little effect on stress range and hence on fatigue life is to be expected with residual stresses of yield point magnitude if the stress ratio is varied between zero and −1 [51]. It is probably safer to assume the most conservative case and to apply PWHT to welded tubular joints, especially in large wall thicknesses where fracture considerations dominate the requirement for stress relief.

5.1.5 Weld Procedures and Qualification

All welds of structural importance need to be produced with adequate control of quality. This is normally achieved through the adoption of a weld procedure and qualification of that procedure. Weld procedures include details of all process parameters which are to be used during welding, including the choice of process type, weld preparation, electrode, process parameters, the use of pre-heat and PWHT.

The two principal standards of importance in developing and qualifying weld procedures are the American Welding Society's Structural Welding Code (AWS D1.1) [52] and British Standards, particularly BS 5135 [53], BS 4360 [54] and BS 4871 [55]. Broadly, a procedure needs first to be developed; this is a matter of design, often coupled with experience of similar constructions, although many guides are available [53, 10] which are helpful even when working from scratch. Following this, the procedure needs to be qualified, which consists of carrying out a series of test welds using a qualified welder and subjecting these welds to a series of destructive and non-destructive tests, specified in the appropriate code. After the weld procedure has been qualified, all welders working on the particular job need to be qualified for that procedure, again requiring that they carry out test welds and that these be examined both destructively and non-destructively. Finally, production can commence with the specified level of non-destructive testing of the welds, although sometimes requalification of the procedure is required after a certain length of weld has been carried out.

Table 5.2 illustrates the type of information which is required to properly define a manual weld.

5.1.6 Arc Blow

In certain cases, it may be important to account for magnetic effects during welding. Since the arc is a current-carrying conductor, it can be influenced by local magnetic fields. This influence takes the form of an arc deflection (commonly called 'arc blow'), which can be quite variable and unpredictable and can therefore produce weld defects. One way in which these arise is through interaction between the arc and the earth return for the welding current through the workpiece if this is asymmetrically disposed.

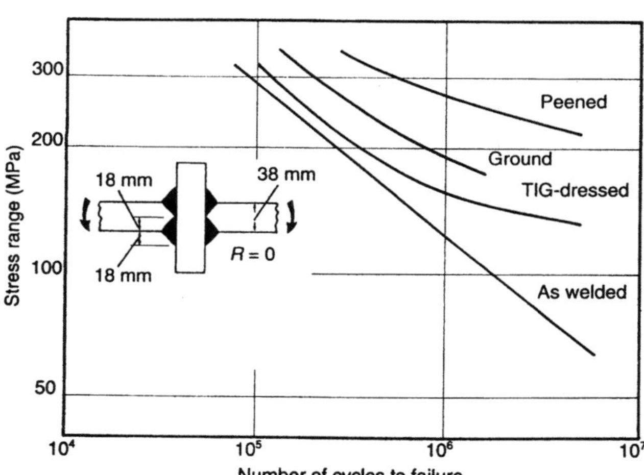

Fig. 5.14. Effect of weld improvement techniques on the fatigue life of moment-loaded weldments. (From [51].)

Table 5.2. Illustration of the type of information which might be included in a weld procedure specification

---

Procedure specification
Base metal specification ⎱ Used for procedure test
Base metal thickness ⎰

Welding specification
Welding process
Consumables: trade name, specification (e.g. BS or AWS)
Bare wire, coated electrode, flux, shielding gas and flow
   rate, as appropriate

Preparation and finishing
Bevel preparation method, e.g. flame cut and dress
Treatment to back side, e.g. backing bar
Weld finish, e.g. shot peen

Heat treatment
Pre-heat temperature
Pre-heat method, e.g. torch
Interpass temperature
Post-weld heat treatment temperature
Post-heating method, e.g. furnace
Maximum cooling rate

Procedure test preparation details
Sketch of bevel details including any dimensions and
   showing weld layer details with passes numbered

Welding technique
Welding position, e.g. overhead (or AWS or BS designation)
Direction of travel, e.g. vertical upwards
Process parameters:

Pass no.	Filler diameter	Current	Polarity	Travel speed	Energy input
Root	—	—	—	—	—
2	—	—	—	—	—
•					
•					
•					
Backing	—	—	—	—	—

Residual magnetism in ferromagnetic material may also produce problems. The elimination of arc blow can be achieved by altering welding sequences, or even occasionally using the welding cable to produce a cancelling field. In hyperbaric welding, where otherwise minor nuisances can become important, arc blow can be a serious operational problem (Waller and Gaudin [56]).

5.1.7 Weldability

This is a qualitative term which defines the ease with which an alloy can be welded. As a general rule, all alloys can be welded and the difficulty is to define the precautions which need to be taken and the degree of compromise in final properties of the joint which must be accepted.

The following discussion is aimed mostly at the weldability of carbon–manganese steels, about which most is known and from which most large fabrications are made. However, it is also important to have an appreciation of the welding implications of other material choices, so these are considered also, albeit in lesser detail. The reader is referred to the AWS *Welding Handbook* [57] for a detailed discussion of weldability of various alloys.

5.1.7.1 Weldability of Carbon–Manganese Steels

Most materials which are welded for marine technology applications are steels, often of the 'structural' variety. Such steels are, by necessity, of good weldability. Most structural steels are of the carbon–manganese type as defined in Chapter 4, that is to say that they have a ferrite–pearlite microstructure. When such steels are welded a heat-affected zone is formed from material which is raised above the critical temperature without melting. It is usual to identify three distinct metallurgical zones in C–Mn steel weldments: weld metal, heat-affected zone (HAZ) and parent plate (Fig. 5.15).

Weldability considerations affect all three of these areas, and they will therefore be considered separately.

The main difficulties which can arise in weld metal for structural use are possible internal defects such as fisheye (a type of porosity, Keeler [11]) or slag entrainment and the failure to develop sufficient strength and toughness. The internal defect problem is normally dealt with through non-destructive examination and the adoption of rigorous control of welding parameters through a weld procedure. Electrode manufacturers and research institutions have been working for many years to develop acceptable and reproducible properties in weld metals, and electrodes are supplied in a package on which is printed an optimum set of welding parameters to be used in order that the specified properties can be reached. Because there are no opportunities for thermal or mechanical treatments, the properties of weld metals usually have to be developed in the as-welded or as-welded-and-stress-relieved condition. It is normally relatively straightforward to develop the required tensile properties in weld metal, and most recent attention has been focused on toughness. The usual requirement is that COD values (see Chapter 2) should reach a minimum of 0.25 mm (Taylor and Evans [58]), and Fig. 5.16 shows some data from a recent compendium [59] carried out over a period of 3 or 4 years from a variety of offshore fabricators. As can be seen, there is a high degree of variability in weld metal toughness, only part of which is attributable to the variability in the weld procedures.

Weld metal is metallurgically unlike the parent plate or its heat-affected zone in that its composition is different, emphasising the fact that properties are to be developed under a specific thermal history not experienced by the parent plate. A typical weld metal will solidify in an epitaxial

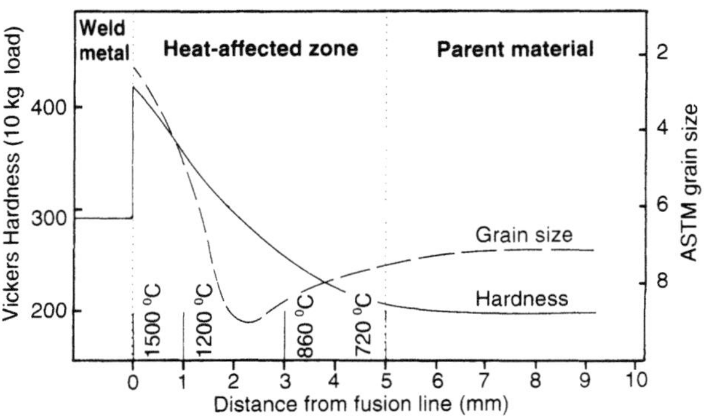

Fig. 5.15. Effect of welding on the metallurgy and properties of carbon–manganese steels. (From Easterling [6].)

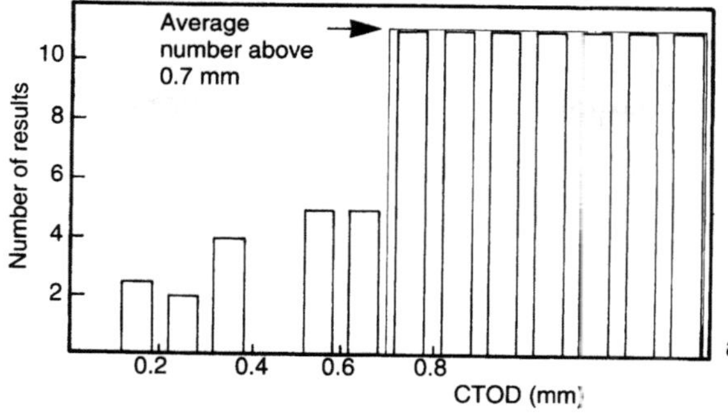

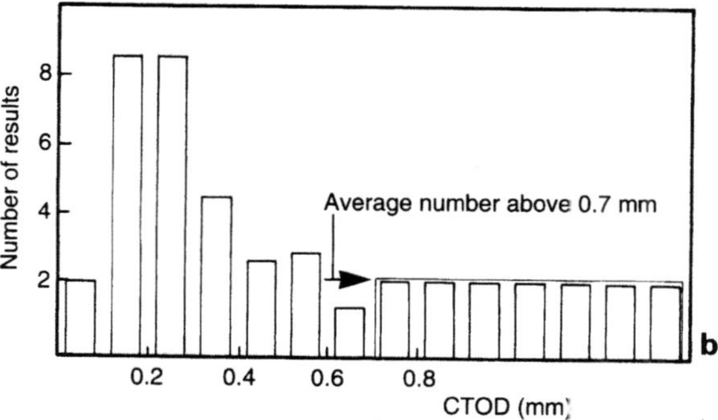

Fig. 5.16. Sample of fracture toughness data for structural steel weld metal gathered from offshore fabricators [59]. **a** Stress-relieved. **b** As welded.

manner by losing heat through the HAZ, and crystals will therefore grow from the solid walls toward the weld centre-line. In the case of steel, there will be a condition during cooling where relatively large columnar grains of austenite are present, and after these have transformed the original grains can still be seen outlined with ferrite [6, 9].

The transformation product is a complex one and has been the subject of intense study as far as characterisation is concerned (e.g. Dolby [60]). For the purposes of this discussion, it is sufficient to consider this material to be of a very fine ferritic nature and sometimes containing austenitic or martensitic components.

The requirements for achieving adequate weld metal toughness in offshore welded fabrications continue to be a matter for research and a number of recommendations as to consumable selection and procedural requirements can be found (e.g. Dawson and Judson [61] and Judson and McKeown [62]).

Weld metal composition is dependent upon the electrode wire composition, but also quite heavily upon the transfer processes. Weld pool chemistry is an important part of electrode development and the formulations of coatings, wires and gases, and the recommendations as to procedure, all have to be considered. In submerged-arc welding, the weld pool chemistry aspect is more important (due to the nature of the process) and is more within the control of the fabricator than the specialist electrode manufacturer. As a result, guidance on the development of optimum toughness in submerged-arc welds is more complex (e.g Bailey and Pargeter [63]).

The way in which weld metal solidifies can give rise to the segregation of impurities to the centre-line. This is not normally a problem in ferritic steels, but the very high heat inputs used for submerged arc welding require the application of some care to avoid such solidification cracking (or hot cracking). Bailey and Jones [64] have developed an equation which relates solidification cracking susceptibility to weld metal composition:

$$U_{cs} = 230 \times \%C + 190 \times \%S + 75 \times \%P + 45 \times \%Nb - 12.3 \times \%Si - 5.4 \times \%Mn - 1$$

where values of greater than about 20 (depending upon geometry and bead shape) represent a danger of cracking in submerged-arc welds. Hrivnak [65] has also suggested a formula for hot cracking susceptibility in carbon–manganese steel weld metals.

As stronger plate materials become adopted for pipeline and structural applications, there is a continuing pressure on electrode manufacturers to develop weld metals which can provide strengths to match the parent plates without compromising on toughness. The problems of developing strength and toughness in the parent plate belong only partly in the realm of weldability. However, as is shown below under the discussion of the heat-affected zone (p. 172), it is important to develop strength without an unacceptable increase in alloy or carbon content. Since these are important methods of strengthening steel, it is obvious that more subtle means need to be adopted for weldable applications.

Increasing the strength of steels while retaining their ferrite–pearlite structures (i.e. without heat treatment) requires either the use of cold work or the refinement of the grains (see Chapter 4). Cold work is not normally considered an acceptable alternative for structural or pressure containment applications because of the loss of toughness which arises, whereas a reduction in grain size not only increases strength but also increases toughness. For modern structural steels it is therefore necessary to refine grains as much as possible while retaining the chemistry within the limits normally associated with good weldability (i.e. a carbon equivalent of less than about 0.55). This is usually done using advanced thermomechanical treatments at the plate-rolling stage. Such treatments use combinations of accelerated cooling (but not quenching) and dynamic recrystallisation. The basis of such methods is that, by hot working in a certain temperature range in the austenitic state, simultaneous plastic deformation and recrystallisation can be made to take place. If very fine dispersions of thermally stable precipitates (such as Al_2O_3 or NbC) are present, these precipitates can limit the size to which the new austenite grains will grow. Subsequent transformation to ferrite and pearlite will therefore result in an even finer ferritic grain size. Accelerated cooling from hot rolling can help to enhance this latter effect. Steels to which small alloying additions have been made for this purpose are normally referred to as 'microalloyed' and an addition of, for example, 0.1%Nb is sufficient to achieve this effect.

One further advance in structural steel which has been prompted by the offshore industry is the reduction in sulphur content required to provide through-thickness ductility. The measurement of this property has been referred to in Chapter 2, and the means by which sulphur can be reduced has been referred to in Chapter 4.

Normally, sulphur in steels does not present any particular difficulties, because it is bound by manganese, which is present (for other purposes) in quantities more than sufficient to react all sulphur to manganese sulphide. When formed shortly after solidification, these sulphides are spheroidal, and during rolling are flattened to a pancake-like shape. The spherical inclusions are typically of diameter 10 to 20 microns and their distribution is such that they have an insignificant effect on mechanical properties. However, after rolling, their planar diameter may become as much as 50 μm; if the plate is stressed in the through-thickness direction then such discontinuities become much more significant and can result in a substantial drop in ductility. In some cases this can give rise to cracking, particularly in welds where a significant through-thickness stress can develop. The bulk welding of thick structural plates for offshore structures provides the conditions required for the occurrence of this phenomenon (Farrar and Dolby [66]). Thus lamellar tearing can result when a plate is subjected to through-thickness welding stresses, whereby cracking between the flattened sulphides (and other inclusions) and hence local delamination can occur in the plate. Further application of static or fatigue loads can lead, in extreme cases, to pullout of the welded connection.

Many structural steel grades can now be purchased with enhanced through-thickness properties, and these are usually achieved by the reduction of sulphur content as well as some inclusion shape control (e.g. Arrowsmith and Shenton [67]). Figure 5.17 illustrates some of the effects of inclusion control on through-thickness properties [68].

Chapter 4 charts the development of high-strength structural and pipeline steels for the offshore and shipbuilding industries. Briefly, the early plates in the 1960s were of high strength (Grade 50) and were defined as being resistant to brittle fracture on the basis of Charpy impact values. Compositions were restricted to ensure good weldability, and it was recognised that the steel should maintain its properties after hot forming and after subassembly stress relief at around 580 to 620 °C (Arrowsmith and Shenton [67]). It became apparent, for example, after examination of the semi-submersible drilling rig *Sea Quest* after 1 year's service, when considerable lamellar tearing was found, that further developments in the specification of steel quality for offshore structures were required.

The area of a steel weldment which has received the most attention from metallurgists is the heat-affected zone. Because the steel is locally heat-treated in an only partially controlled way, it is important that its properties are not excessively sensitive to small changes in heat treatment. The two most important considerations in ferritic alloy heat-affected zones are the possibility of unacceptable toughness and the possibility of cold cracking (or hydrogen cracking).

Hydrogen cracking is a phenomenon that has been known for many years and has been the major factor to be considered in the development of steels for weldable applications. It is best known in ferritic steels, where the presence of hydrogen dissolved in the heat-affected zone can cause brittleness and possibly local fracture under residual welding stresses (hence the term 'cold cracking'). Hydrogen cracks normally are fairly small, but they severely compromise the bulk properties of an already sensitive area of the weldment and can propagate under fatigue stresses. The main factors which influence the likelihood of hydrogen cracking are the susceptibility of the plate material to embrittlement, the heat-affected zone cooling rate, the amount of hydrogen available to the weld area and the degree of restraint applied to the plates which are being welded.

Bailey in 1970 [69] suggested a way in which these factors could be quantified into a rationale for the development of welding procedures for carbon–manganese steels. This was modified slightly and adopted as a British Standard [54]. Although some aspects require further consideration, Bailey's methodology gives a useful appreciation of the factors governing carbon–manganese steel weldability.

The effect of parent plate susceptibility is defined essentially by the steel's hardenability (as described in Chapter 4), because steels of higher hardenability will be more likely to form hard transformation products in the heat-affected zone than those of lower hardenability. The hardenability in welding is usually quantified using the carbon equivalent; this concept dates back to 1940, although in a slightly different form (Cottrell [70]). The formula seeks to quantify parent plate weldability with a single index related to its composition and is usually expressed as follows:

$$CE = C + \frac{Mn}{6} + \frac{Cu + Ni}{15} + \frac{Cr + Mo + V}{5}$$

The heat-affected zone cooling rate is quantified partly by the thermal severity of the weld, defined as its combined thickness (Fig. 5.18), which is a measure of the size of the conduction path available to cool the weld. As described in Section 5.1.1, the other main factors controlling HAZ cooling rate are the energy input and any pre-heat which is applied to the weld.

The restraint is only qualitatively defined and is combined with the hydrogen level of the process into a critical hardness or carbon-equivalent scale (Table 5.3). The carbon-equivalent scale is a means by which the maximum tolerable hardness in the heat-affected zone can be represented for later use in the empirical nomograms. For example, a 'normal structural fit' fillet weld carried out using baked hydrogen-controlled electrodes would be able to tolerate a hardness of 400DPN in the HAZ without danger of cracking and is given a carbon-equivalent scale of 'C'.

Using the energy input, the carbon equivalent, the carbon-equivalent scale and the combined thickness as inputs, it is possible to use nomograms (e.g. [53]) con-

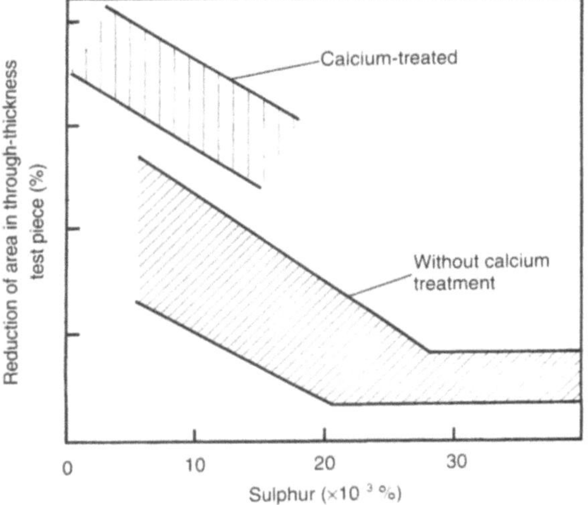

Fig. 5.17. Effect of inclusion shape control on through-thickness properties of structural steel Grade StE 355. (From Baumgardt et al. [68].)

Table 5.3. Effect of restraint and process hydrogen level on the critical hardness in carbon–manganese steel heat-affected zones

Weld hydrogen level	Examples of corresponding processes and consumables	Critical hardness value and CE scale									
		Bead-on-plate		Butts				Fillets			
		Continuous surface	Abutting surfaces	Normal	Misaligned	Partial penetration, high restraint	Full penetration, patch	Normal structural fit	Machined fit	High restraint	Patch
High	Manual metal-arc electrodes (non-hydrogen-controlled coating) Submerged-arc process Flux-cored-wire processes	A 350	A 350*	A 350	A 350	220*	300*	A 350	A 350	300*	300*
Medium	Basic manual metal-arc electrodes dried up to 250 °C Submerged-arc process Flux-cored-wire processes	C 400	B 375*	B 375	B 375	250*	A 350*	B 375	C 400	A 350	A 350*
Low	Basic manual metal-arc electrodes dried at 350 °C and above Submerged-arc and flux-cored-wire processes Gas-shielded metal-arc with solid dirty wires	D 450	C 400*	C 400	C 400	300*	B 375*	C 400	D 450*	B 375*	B 375*
Very low	Basic manual metal-arc electrodes dried at 450 °C and above Submerged-arc process Gas-shielded metal-arc with solid clean wires	500	D 450*	D 450	D 450	A 350*	C 400*	D 450*	500*	C 400*	C 400*

Source: Gooch [71].
*Tentative values.

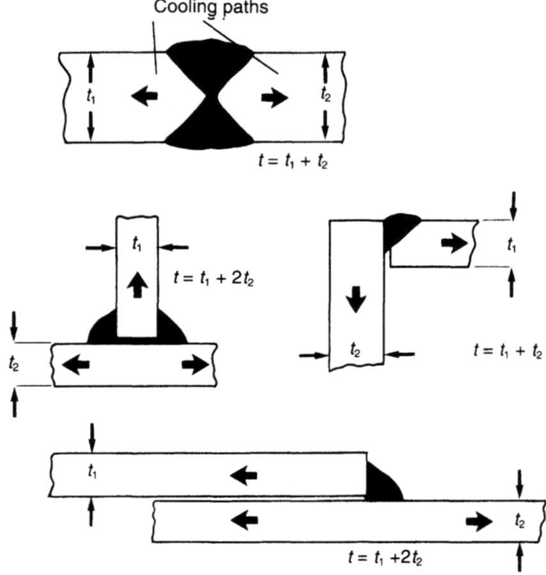

Fig. 5.18. Illustration of the quantity combined thickness for assessing the cooling severity offered by a welded plate. (From Bailey [69].)

structed from empirical data to assess the amount (if any) of pre-heat required to prevent hydrogen cracking.

Carbon-equivalent control is now firmly built into materials standards for welded constructions such as structures and pressure vessels, so that hydrogen cracking is now a relatively isolated phenomenon. However, with the development of new types of structural steels using novel strengthening techniques there is a continued need to update experience on hydrogen cracking limits.

Recently Cottrell [70] has reviewed weldability equivalents for steels, and Hrivnak [65] has reviewed the various formulae in use. Carbon equivalent is still widely used, although Japanese welding engineers have found the low-alloy steel cracking parameter more useful for a wider range of steels [6]:

$$P_{cm} = \%C + \%Si/30 + \%Mn/20 + \%Cu/20 + \%Ni/60 + \%Cr/20 + \%Mo/15 + \%V/10 + 5 \times \%B$$

McKeown et al. [72] have identified some weldability problems which may occur in low-sulphur steels. The issue is an interesting one, in that 'cleaner' steels have low interstitial (oxygen and sulphur) contents, and this in turn can affect hydrogen levels during the steelmaking process. As a result, it was found that low-sulphur steels were initially often associated with an increased risk of hydrogen cracking, and this was eventually traced to a higher HAZ hardenability in these materials (Dolby [60]). Dolby has estimated the increased carbon equivalent, when going from a sulphur level of 0.03 to 0.005%, to be about 0.03%CE, so that this effect is only critical for marginal procedures.

Heat-affected zone hardness is also normally a consideration if there is a requirement to avoid SSC (see Chapter 3) in service, although other mechanisms of hydrogen damage in wet, sour environments (such as blistering and HIC) are less critically affected by the steel hardness, susceptibility being dictated more by inclusion type, content and distribution. This latter phenomenon is intricately tied up with plate manufacture, sulphur content and other factors outside the

realm of weldability, and is discussed in more detail in Chapter 4. For SSC the compositional requirements of NACE [73] are simple enough to follow, although some of these can give problems to electrode manufacturers for high-strength applications. Formulae which allow an estimation of likely heat-affected zone hardness are available (e.g. Hrivnak [65]), so that a procedure development can include this to avoid later problems of performance in SSC or HIC tests. Lorentz and Duren [74] have estimated the relationship between HAZ hardness and composition for linepipe steels. Aside from the recommendations of NACE, a number of authors have attempted to correlate hardness with SSC cracking susceptibility, and some of this work is reviewed by Terasaki et al. [75].

Heat-affected zone toughness is rather more difficult to quantify than HAZ hardness. As with weld metal, a recent survey of welded connections carried out for offshore fabrications showed a wide variation in measured toughness properties within and between procedures, and an indication of this is given in Fig. 5.19. Indeed, it is not even certain whether the HAZ or weld metal is the critical area with regard to overall weldment performance. Of course, this situation is not helped by the difficulty of locating the fracture wholly within the HAZ during a toughness test, but probably says more about the fracture process (in testing the line of least resistance rather than testing an average property) than about the quality control of welded joints. This points to the need for a reliability approach to structural integrity, a matter which is dealt with later in Chapter 6.

Such difficulties cannot, however, be allowed to hamper the search for steels which produce intrinsically tougher HAZs. Apart from carbon-equivalent or other compositional parameters, a number of other useful indicators of

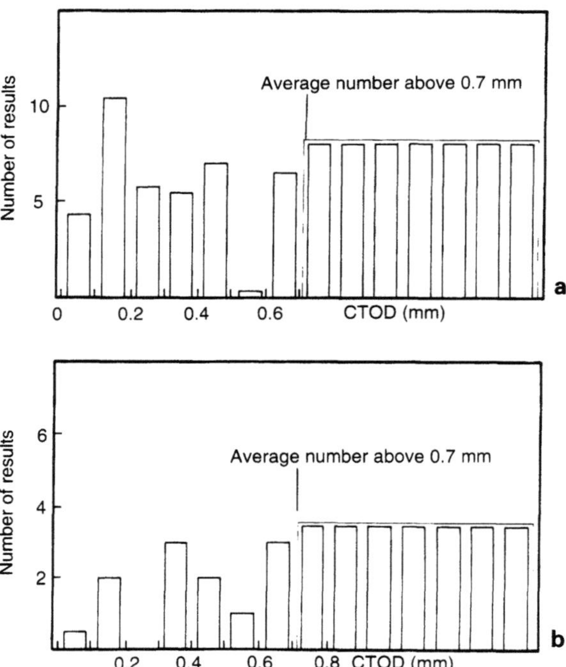

Fig. 5.19. Sample of fracture toughness data for structural steel HAZ material gathered from offshore fabricators [59]. **a** As welded. **b** Stress-relieved.

HAZ toughness have been identified over recent years. Perhaps the most significant of these concerns the effect of stable nitrides on the transformation behaviour of steels in the HAZ. Dolby [60] has traced the development (mostly in Japan) of titanium-treated steels, showing the distinct effect of titanium additions on the heat-affected zone toughness of structural steel.

In summary, the welding of carbon manganese steel can be satisfactorally carried out with a careful control of welding conditions and consumable selection. Care should always be taken, however, that the procedural development is applicable to the steel being welded. For example, according to Gooch [71] some pressure vessel steels and other special steels may be more or less hardenable than their carbon equivalents would suggest, and additional precautions may be required.

5.1.7.2 Weldability of Low-alloy Steels

Higher-strength steels with carbon equivalents greater than about 0.55 or alloy contents (excluding manganese) greater than about 1 per cent are generally considered to belong in a different weldability category than the structural steels considered above. The reason for this is that some type of global temperature control is likely to be necessary to ensure freedom from difficulties of toughness and/or cracking in the heat-affected zone. Also, the limited amount of processing which can be carried out in the weld metal and HAZ defines the achievable properties in these areas.

The difficulties encountered in the welding of high-hardenability steels are a function of composition, but indicators such as carbon equivalence no longer apply. Not only must the importance of a wider range of alloying elements at higher composition be considered, but also a wider range of hardenability. Generally, three strategies are available:

1. To control the temperature changes in the HAZ by preheat or post-heat.
2. To use isothermal or near-isothermal transformation to avoid martensite formation by holding above the M_s temperature (see Figs 4.13 and 4.14).
3. To allow hardening in the HAZ and limit the welding stresses by the use of an austenitic weld metal.

A fourth strategy involving total heat re-treatment will be outlined below. Clearly strategies 2 and 3 result in a weld that has properties rather different from those of the original steel, and a decision is then obviously required as to whether the high-strength steel was needed in the first place. Strategy 1 is feasible only under certain conditions, and an assessment of these conditions can be made knowing the composition of the parent material and the degree of restraint to be applied [71].

First, the steel can be classified according to a scheme which takes into account the quantity and type of alloying elements. One such classification (Gooch [71]) gives a designation 'K', 'L' or 'M' to a low-alloy steel, according to whether its susceptibility is low, medium or high respectively for a given HAZ hardness or carbon content. An assessment of the qualitative strategy can then be made using the hardness and pre-heat diagram (Fig. 5.20) and the carbon content of the steel.

Two examples of low-alloy steels in which there is considerable interest from the point of view of marine applications, described in Chapter 4, are the so-called petroleum

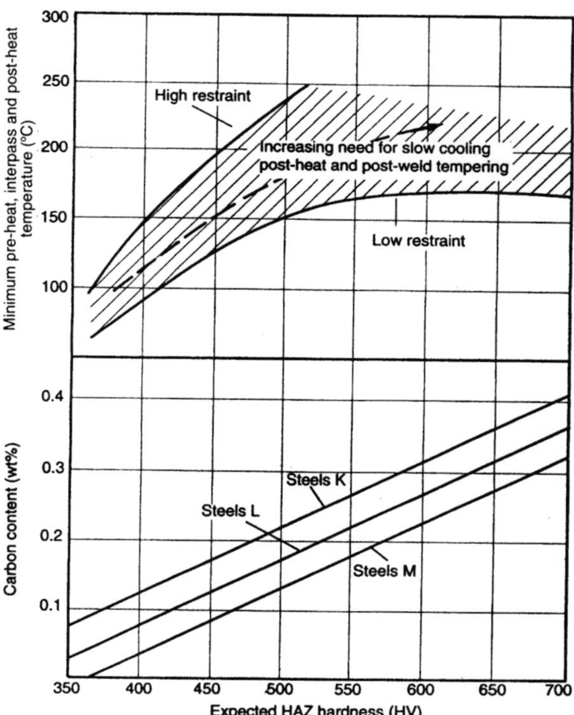

Fig. 5.20. Diagram for selecting minimum pre-heat, interpass and post-heat temperatures for low-alloy steels with fully hardened heat-affected zones. (From Gooch [71].)

grades, usually of compositions 4140, 4340 or 4145, and the so-called refinery grades, usually based on Cr–Mo compositions. Some examples of each are given in Table 5.3, and it can be seen that the first group are distinct from the second because their carbon content (in the range 0.25 to 0.45) means that they usually have to be welded in a softened condition to prevent hydrogen cracking [71].

As can be seen from Table 5.4 and Fig. 5.20, the carbon content and generic type of the petroleum grade steels put them in a region where slow cooling needs to be considered. In such cases, loss of strength can be expected in the HAZ and in the weld metal, and therefore the connection cannot develop the properties for which the parent plate was chosen. This can be overcome to an extent by using the fourth strategy referred to above. This consists of overtempering or annealing the steel prior to welding and then heat-treating the entire fabrication after welding.

The welding of the refinery-grade steels is more viable because of their relatively low carbon contents. Cr–Mo

Table 5.4. Typical compositions of a selection of important low alloy steels. These are *not* specifications and are merely intended to give an idea of the major alloying additions which contribute to strength and hardenability

Alloy	%C	%Mn	%Cr	%Ni	%Mo
AISI 4140	0.4	0.9	1.0	—	0.2
AISI 4340	0.4	0.7	0.8	1.8	0.25
1¼Cr–½Mo	0.15	0.5	1.3	—	0.5
2¼Cr–1Mo	0.13	0.45	2.3	—	1
9Cr–1Mo	0.13	0.45	9	—	1

steels, however, are still air-hardening, and pre-heat and post-weld tempering are almost always necessary, even for low-restraint joints. Turnell and Marsh [76] have carried out a comprehensive study of these steels from the fabrication point of view for compositions from C–1/2Mo up to 9Cr–Mo. MMA electrodes which match the Cr–Mo steels in composition are available, and baking a short time prior to use is necessary to reduce the danger of hydrogen cracking. The C–Mo steels can normally be treated in the same way as carbon–manganese steels with a carbon equivalent of 0.5. For the other steels, the importance of pre-heating below M_f is illustrated by Fig. 5.21. With pre-heat above M_s it is possible for austenite to be retained until the end of PWHT, where it may then transform to martensite without any tempering operation being applied. With pre-heat below M_f, martensite is formed immediately after welding and is tempered in the PWHT phase. Using the diagram (Fig. 5.20) and recognising the 'L' classification of these steels, it can be seen that a pre-heat of 200 to 225 °C is required for high-restraint conditions, and this is in general agreement with pressure vessel codes. On the other hand, codes were found to differ considerably [76] in their recommendations for PWHT, and it is often the final required properties which dictate the PWHT temperature.

A number of low-alloy steels are susceptible to cracking when welds are reheated to the temperature range 500 to 650 °C. This phenomenon is known as 'reheat cracking', and since the susceptibility range contains the temperature often used for PWHT to achieve stress relief, some care has to be taken to avoid this problem. The exact mechanism of reheat cracking is not known in detail, but it is probable that it differs for different types of low-alloy steels. The steels which have been most widely studied from the point of view of reheat cracking are the Cr–Mo–V and

Ni–Cr–Mo–V steels for creep-resistant applications, but some care also needs to be taken with the 2¼ Cr–Mo steel described above.

5.1.7.3 Weldability of Stainless Steels

As mentioned in Chapter 4, stainless steels can be classified into a number of groups, depending upon the metallurgical condition of their matrix. In marine technology, the two most important groups are the austenitic and duplex stainless steels, and these are also the most widely studied. Weldability problems can occur in austenitic stainless steel weld metal or in the heat-affected zone.

As seen earlier, the HAZ is subjected to a thermal cycle which can include heating in the range 500 to 800 °C. In some forms of stainless steel, this can result in the precipitation of intermetallic compounds which can have an effect on strength or corrosion resistance. The main concerns are the formation of σ-phase and the precipitation of carbides. Sigma-phase is an iron–chromium intermetallic compound which will form in some austenitic stainless steels. Other intermetallics are possible in austenitic stainless steels, and the general problem is that under prolonged exposure between 500 and 900 °C there is the possibility that these embrittling phases can form. For kinetic reasons, sigma rarely forms in the simpler 18/8-based austenitic stainlesses, but some alloying elements, notably molybdenum, promote its formation.

Because of the affinity shown for carbon by chromium, temperatures at which carbon is mobile in austenite will allow the formation of chromium rich carbides. As with tempering of low-alloy steels, this is a diffusion-controlled phenomenon which commences at a significant rate in austenite at around 500 °C. An analysis of the kinetics of the phenomenon was first carried out by Stawstrom and Hillert [77], who identified a chromium-depleted zone next to grain boundaries caused by the different rates at which carbon and chromium reach the reaction site; these authors were able to produce a modified time–temperature–precipitation curve which explained the loss in corrosion resistance in an 18/8 steel. This phenomenon of sensitisation, or 'weld decay', is now well known and is avoided either by using low-carbon stainless steels or by stabilising the steels with additions of titanium or niobium; these getter, i.e. scavenge, the carbon into carbides which are stable to temperatures in excess of 1000 °C.

Both sigma and related intermetallics can also reduce corrosion resistance by virtue of the fact that the mobility of transition elements is low at their formation temperature, again giving rise to depleted zones; however, the gradients in chromium are not usually as steep, first because it is not competing with carbon and second because the chromium content of the intermetallics is sometimes lower than in the carbides.

Aside from the depleted zone effect, precipitation of either intermetallics or carbides can have a further influence on austenitic stainless steels. One of the reasons why austenite has good corrosion resistance is its homogeneity, and the introduction of a second phase disrupts this homogeneity and hence reduces corrosion resistance, as indicated in Chapter 3. Where precipitation cannot be avoided, the damage caused by intermetallic and carbide formation can be eliminated by redissolving them at about 1050 °C and cooling and this is also the recommended temperature for stress relief of stainless steel weldments to minimise any

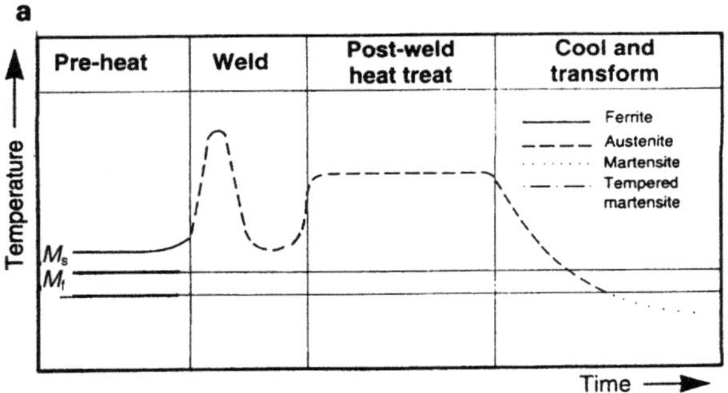

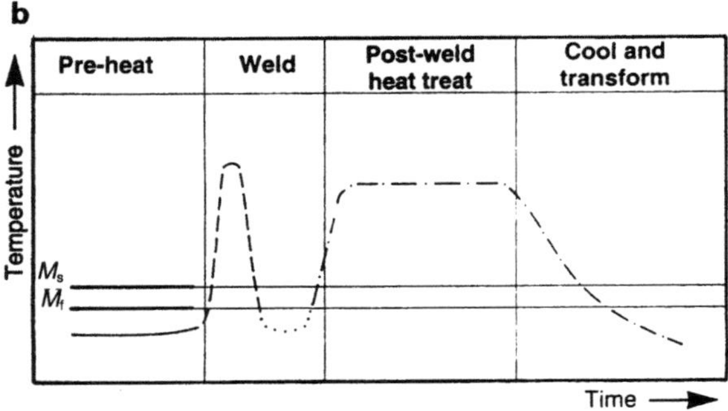

Fig. 5.21. Transformations occurring during welding and PWHT of Cr–Mo low-alloy steels. **a** Pre-heat above M_s. **b** Pre-heat below M_f. (From Turnell and Marsh [76].)

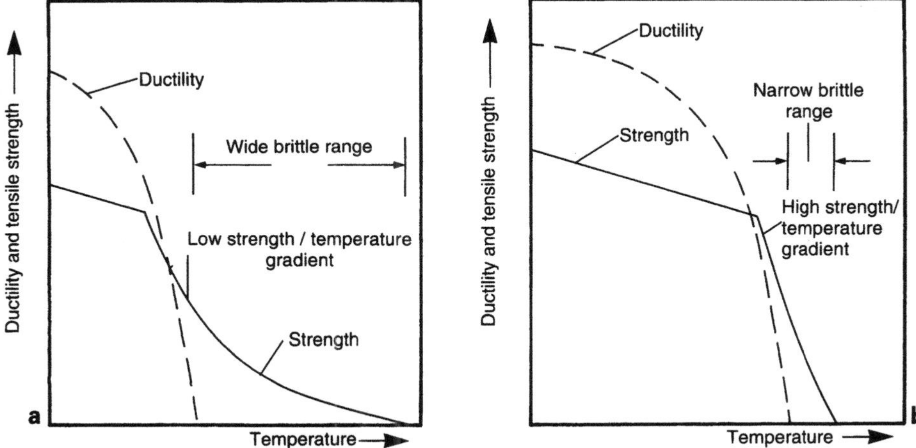

Fig. 5.22. Influence of mechanical properties close to the temperature of complete solidification on the susceptibility to hot cracking during welding. **a** Crack-sensitive. **b** Crack-resistant. (From Lancaster [9].)

danger of stress corrosion. Subsequent rapid cooling is normally recommended for unstabilised or non-low-carbon grades and lower temperature treatments (< 480 °C) are usually sufficient for all grades if it is only necessary to remove peak residual stresses [78].

The main weldability problem in stainless steel weld metal is the control of composition and hence constitution. It is well known that austenitic material can suffer from solidification (or hot) cracking, where segregation towards the weld centre-line produces a plane of weakness which can fail in the final stages of solidification of the weld metal. The magnitude of the freezing range is an important factor in hot-cracking susceptibility, as illustrated schematically in Fig. 5.22.

The susceptibility of austenitic weld metal to hot cracking can be reduced drastically by introducing ferrite formers to the weld metal, thereby producing a duplex ferrite–austenite structure soon after freezing and hence a greater number of grain boundaries at which tramp elements (segregating impurities) can be distributed. It is generally accepted that a final ferrite content of about 4 to 8

per cent [77] is sufficient to eliminate fissuring in the weld metal, and this is achieved by balancing the ferrite-forming power of the weld metal against its austenite forming power. Two pseudo-phase diagrams have been developed to allow this, namely the Schaeffler and DeLong constitution diagrams (Figs 5.23 and 5.24).

The DeLong diagram is a slightly later adaptation designed to account for the effect of nitrogen on ferrite-forming power. This diagram also addresses the practical aspects of measuring ferrite content without sectioning the weld, and the ferrite number used on the diagram is often used by electrode manufacturers to indicate the ferrite content of as-deposited weld metal from the electrode. The use of the diagrams is quite simple and consists of calculating the chromium and nickel equivalents of the parent plate and candidate electrodes. The dilution of the weld (proportion of the weld metal which consists of parent plate material) is then estimated, and the constitution of the weld metal can be assessed by graphical proportion (e.g. [79]).

As indicated in Chapter 4, duplex stainless steels find particular uses in pipelines and associated applications,

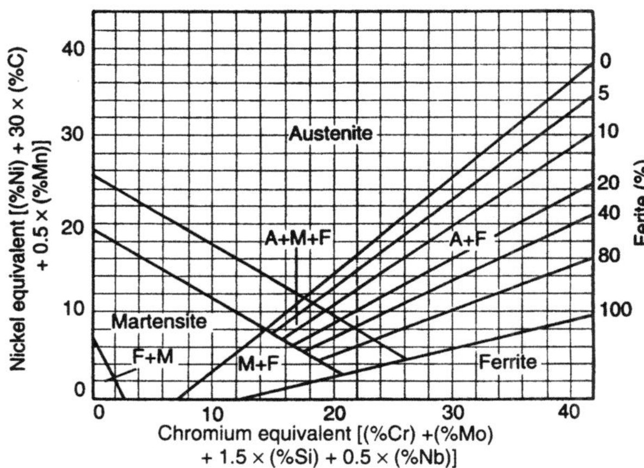

Fig. 5.23. Schaeffler constitution diagram for the control of weld metal composition in weldments involving stainless steels [79].

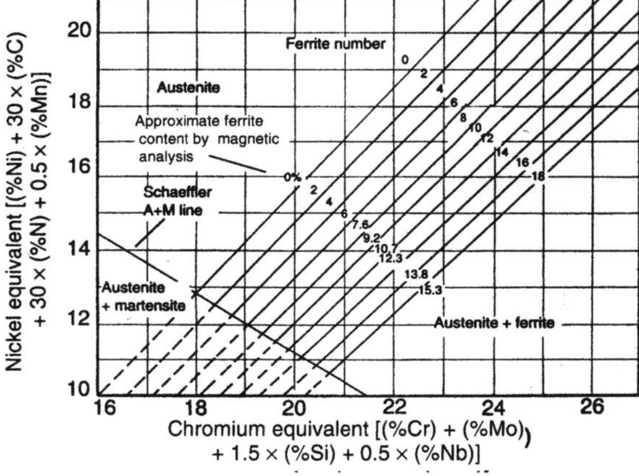

Fig. 5.24. The DeLong constitution diagram for the control of weld metal composition in weldments involving stainless steels [79].

because their mixed ferrite–austenite microstructure combines the advantages of ferrite (strength) with those of austenite (corrosion resistance) [80]. The welding of duplex stainless steels does require some care, however, because of the increased metallurgical complexity of these alloys over the austenitic stainless steels.

In the weld metal, the considerations are broadly similar to austenitic steel, with the need to control ferrite number and for the same reasons. However, there is an additional requirement to match strength and corrosion resistance with the base metal. Kotecki [81] has extended the ferrite number system indicated above to cover duplex alloys, and has found that the 'EFN' provides a useful parameter for predicting the mechanical properties of duplex weld metals.

Gooch [25] has indicated the main areas which need to be considered in heat-affected zones for duplex stainless steels. First, the heating of duplex stainless steels results in the formation of delta ferrite, in which there can be substantial grain growth. The subsequent reversion to a duplex structure means that the phase morphology and distribution can be considerably different from that of the parent plate, which is a uniform dispersion of ferrite in austenite. The way in which this dispersion is modified during welding depends upon the original steel composition, and also upon the welding parameters (as well, of course, as the distance from the fusion line). A degree of solution treatment (see Chapter 4) will also take place in the heat-affected zone (again dependent upon peak temperature), with subsequent reprecipitation on cooling or on reheating by other welding passes. Because of the transformation behaviour of duplex alloys and partitioning of solubilities, the pattern of precipitation is different than in austenitic stainless steels, but sensitisation may nevertheless occur.

Although there are some circumstances under which post-weld heat treatment is required for homogenisation, solution treatment and precipitation (in P-H grades), duplex stainless steels are generally used in the as-welded condition [25]. Some more specific applications of the technology of welding duplex stainless steel pipelines have been discussed by, for example, Lefebre and Dufrane [82] and Doyen and Demuzare [83].

For ferritic stainless steels, the main problems are sensitisation, sigma formation and excessive grain growth in the HAZ. A phenomenon known as '475 °C embrittlement' can also occur after exposure in the temperature range 400 to 565 °C. This problem is more severe for higher chromium contents and is associated with the precipitation of a very fine chromium rich phase [77].

Martensitic steels clearly suffer from the same difficulties as the hardenable low-alloy steels, but to a greater extent. Pre-heating and post-heating are generally required; often the final HAZ properties are different from the parent material, requiring a full weldment heat treatment to restore them [77]. Filler metals may be of the same composition as the parent plate or may be of austenitic composition, in which case the weld metal will, of course, be of inferior strength to the plate.

The welding of precipitation-hardening stainless steels depends upon the matrix type. The martensitic grades are not generally susceptible to hot cracking, but austenitic grades can suffer from hot cracking in the heat-affected zone. In general, it is advisable to carry out a full re-heat-treatment of the weldment, and if this is not practicable then they should be welded in the solution-annealed condition and aged after welding [57].

5.1.7.4 Weldability of Aluminium Alloys

The weldability of aluminium alloys depends critically on whether or not the alloy in question is of heat-treatable grade. Because the ageing process requires to be fairly precisely controlled (see Chapter 4) there is the potential for a substantial loss of strength in the heat-affected zone. As a rough guide, the properties in the HAZ of a heat-treatable aluminium alloy approximate to the T4 condition [84]. Some improvement in this situation can be brought about by welding in the solution-treated condition and ageing after welding, but, as with the precipitation-hardening stainless steels above, the only way to develop full strength in the HAZ is to re-heat-treat the entire weldment.

Non-heat-treatment alloys which have been strengthened by cold-working will lose most of this strengthening in the heat-affected zone owing to recovery, so the strength there will be substantially the same as in the annealed condition [84]. The great difficulty with this latter effect is, of course, that strengthening cannot be recovered by re-heat-treatment of the weldment.

The solid-solution-strengthened aluminium alloys are the most resistant to HAZ softening, and any loss of properties is only associated with the region of grain growth close to the fusion line. On the other hand, these alloys are among the weaker of the aluminium alloys.

All of the non-heat-treatable aluminium alloys (i.e. those belonging to series 1xxx, 3xxx, 5xxx and 8xxx) are considered to be readily weldable, and alloys which have a relatively high degree of solid-solution (as opposed to cold-work) strengthening will suffer from less HAZ softening. Such alloys can be easily identified as those which have the highest strengths in the annealed condition. It is worth noting that a number of the alloys which are of particular use in marine applications fall into this category.

In general, the medium-strength heat-treatable alloys are more easily welded than the higher-strength ones. Broadly this means that most of the 2xxx series (Al–Cu and

Table 5.5. Filler metals for commonly welded aluminium alloys to meet specific requirements

Base metal	Recommended filler metal for				
	High strength	Good ductility	Colour match after anodising	Saltwater corrosion resistance	Least cracking tendency
1100	4043	1100	1100	1100	4043
2219	2319	2319	2319	2319	2319
3003	4043	1100	1100	1100	4043
5052	5356	5654	5356	5554	5356
5083	5183	5356	5183	5183	5356
5086	5356	5356	5356	5356	5356
5454	5356	5554	5554	5554	5356
5456	5556	5356	5556	5556	5356
6061	5356	5356	5654	4043	4043
6063	5356	5356	5356	4043	4043
7005	5556	5356	5356	5356	5356
7039	5556	5356	5356	5356	5356

Source: Kotecki [81].

Al–Cu–Mg) and some of the 7xxx series (Al–Mg–Zn–Cu) are of limited weldability [84]. Coincidentally, this limitation also corresponds with a number of the alloys which are unsuitable for marine use for other reasons.

Filler metals for aluminium alloys are usually of the 1xxx, 2xxx, 3xxx, 4xxx or 5xxx types and, as with the base metals, some of these are heat-treatable if the application demands this and circumstances allow. Table 5.5 shows a rough guide to filler metal selection for aluminium alloy fillers and, as can be seen, the selection is quite heavily dependent upon the application [57].

5.1.7.5 Weldability of Other Alloys

The weldability of alloys other than those considered above is beyond the scope of this text, and the reader is referred to [57] for detailed weldability considerations. A few broad remarks can, however, be made here.

Cast irons are generally considered to be difficult to weld because of the severe embrittlement caused in the HAZ. However, since cast irons are anyway relatively brittle materials, this can often be tolerated. The commonest way of approaching the welding problem in cast irons is to minimise the thermal stresses by the use of a relatively ductile weld metal, usually a nickel-based alloy.

Many of the softer copper alloys are strengthened by cold work, and welding has the effect of locally annealing these and hence softening the HAZ. In the more complex alloys, there is sometimes a danger of hot cracking, owing to preferential liquation or segregation and wide freezing range. Nickel alloys are generally readily weldable, but are known to be susceptible to contamination of plates or consumables, particularly by substances which can liberate sulphur.

Titanium and its alloys can all be embrittled by oxygen, nitrogen and hydrogen, and should be protected from these gases until cool. This may involve the use of argon-purged chambers, which can render welding rather expensive and unwieldy. Apart from this, most alloys can be welded readily, but some care is required over the metallurgy, as titanium shows an allotropic transformation in the solid state.

Welding of dissimilar combinations requires at least that the precautions required for the two plate materials be taken into consideration. For example, the welding of low-alloy steel to austenitic stainless steel would require that the effects in the HAZ of the low-alloy steel should be accounted for, but also that the weld metal was of acceptable composition to avoid cracking. There may be other difficulties in dissimilar combinations and, for example, the joining of aluminium alloys to mild steels is extremely difficult because of the relative melting points and thermal conductivities. At least one author [41] has suggested that explosive bonding is appropriate for such incompatible alloys.

The foregoing considerations have been applied only to wrought alloys. As has been indicated in Chapter 4, many engineering alloys are available in both wrought and cast forms. It may seem like a misnomer to consider the weldability of cast materials, but this may be necessary for a variety of reasons, not the least of which are repair and joining of cast-to-cast or cast-to-wrought subassemblies.

In general, the considerations of weldability which apply to wrought alloys apply also to their cast counterparts, but a few differences can arise. These are usually specific to the alloy system in question, and a specialist text should be consulted (e.g. [7, 57, 85]).

5.2 Manufacture with Polymers and Composites

This section deals briefly with the basic plastics manufacturing processes and joining methods. This is extended to the manufacture of composite materials which employ a resin matrix, the object being to indicate how the design philosophy needs to be altered to accommodate a change from, for example, metals to composites.

The discussion deals only with manufacture at the product formation stage and not the actual production of the material, except in those cases where the two processes coincide.

5.2.1 Plastics Manufacturing

The basic plastics processing methods are extrusion, injection moulding, thermoforming, calendering, rotational moulding, compression moulding and transfer moulding [85]. As might be expected, all of these processes involve flow, unlike metals production, where forming and casting take a less important place than welding for final fabrication.

The basic activities involved in plastics processing depend upon the type of polymer involved. Thermoplastics are heated to melt, shaped, and then cooled to solidify to shape. Thermosets require to be mixed and in some cases heated to a state at which curing can commence. This usually involves a shaping operation, because there is no later opportunity for this once curing has taken place. Rubbers are generally mixed and then shaped, the mixing process possibly including mastication. The rubber is then usually cured to form cross-links. Figure 5.25 summarises the basic plastics manufacturing processes.

The extrusion process is used for products of constant cross-section, and is essentially the same as a metallic extrusion process where the polymer is forced through a die of the required cross-section. The process is suitable only for thermoplastics and quite complex shapes can be produced directly from the extruder. Quite often extrudates are fed directly into a further process such as blow moulding.

Injection moulding is one of the commonest plastics processing methods. It is very similar to pressure die casting and there are many variants of detail, but essentially the polymer is forced into the die in liquid form and then allowed to cool. The die can have any shape provided the moulded piece can be removed afterwards, and some design modifications may be necessary to accommodate this latter requirement. This process was originally developed for thermoplastics, but a variant known as reaction injection moulding allows the polymerisation reaction to take place in the mould.

Compression moulding is the commonest process for manufacture of thermosets. The preform is usually partially polymerised, and the application of heat and pressure in the mould performs the final cure. Transfer moulding is essentially the same process except that the pre-heating and pressurisation take place in a separate chamber before injection into the mould. The improved viscosity and mixing in this

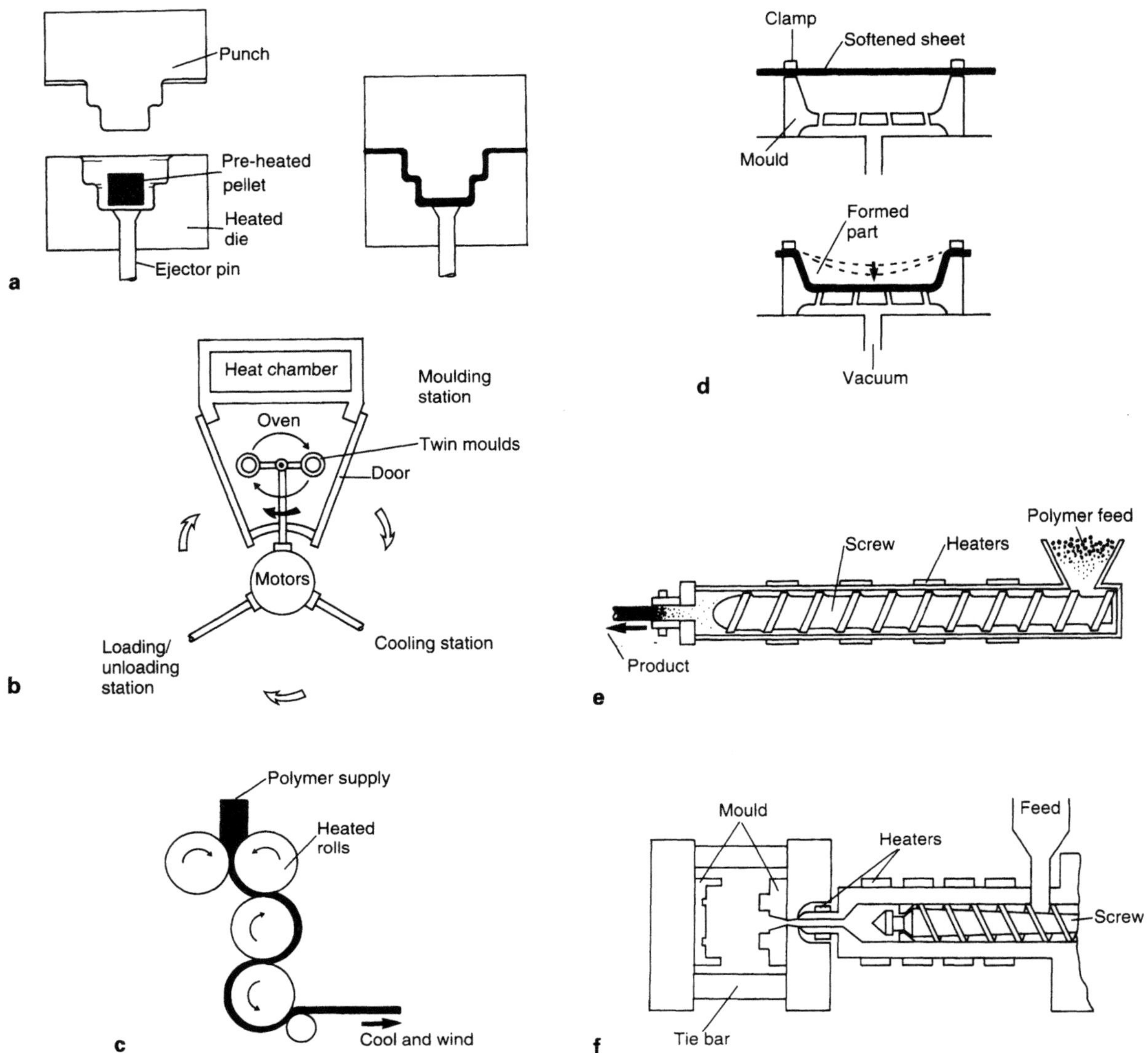

Fig. 5.25. Summary of some of the basic plastics manufacturing processes. **a** Compression moulding. **b** Rotational moulding. **c** Calendering. **d** Vacuum forming. **e** Extrusion (direct). **f** Injection moulding. (Adapted from various sources including [85], [86] and [87]).

latter case usually mean that more intricate shapes can be produced.

Rotational moulding and blow moulding are used to form hollow objects where a die could not otherwise be removed. In blow moulding, pressure is used against a die and in rotational moulding two halves of the die are clamped together and the die rotated about two axes, rather like a three-dimensional version of centrifugal casting. Again, these processes are applicable only to thermoplastics.

Thermoforming is again applicable only to thermoplastics and is normally applied to sheet material. This is done by heating the sheet and applying a pressure against a former. This can be done either by use of positive pressure or by a vacuum on the former side.

Calendering is a film-forming process, again applicable only to thermoplastics. It is directly related to metal rolling and paper manufacture.

5.2.2 Processing of Reinforced Plastics

Reinforcement in plastics materials can be of a particulate or a fibrous nature, and as far as processing is concerned short-fibre reinforcement with random orientation can be regarded in the same way as particulate reinforcement; however, where continuous fibres, mats, fabrics or prepregs are used this is an important factor governing processing. Figure 5.26 summarises the processing options for reinforced plastics with regard to reinforcement disposition.

The processing of the resin in a reinforced plastic usually follows one of the routes outlined in Section 5.2.1, and for particulate and chopped short-strand reinforced plastics nothing much is different from the unreinforced form. If there is a large element of flow during processing it is possible that fibres which should be randomly oriented may become oriented in the direction of flow, thus affecting properties.

Composite type	Suitable forming method
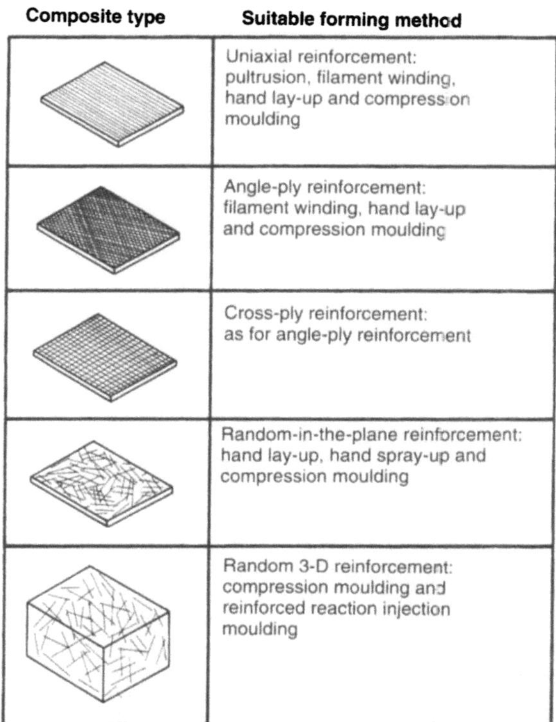	Uniaxial reinforcement: pultrusion, filament winding, hand lay-up and compression moulding
	Angle-ply reinforcement: filament winding, hand lay-up and compression moulding
	Cross-ply reinforcement: as for angle-ply reinforcement
	Random-in-the-plane reinforcement: hand lay-up, hand spray-up and compression moulding
	Random 3-D reinforcement: compression moulding and reinforced reaction injection moulding

Fig. 5.26. Summary of the options for processing reinforced plastics depending upon disposition of reinforcement. (Adapted from McCrum et al. [86].)

For continuous-fibre-reinforced plastics, the reinforcement form and final required geometry have an important bearing on the manufacturing process. However, as with the unreinforced materials, the primary variable is matrix type.

Until fairly recently, reinforcement of thermoplastics with long or continuous fibres was uncommon, largely because of the difficulties of bonding the matrix to the reinforcement and achieving the necessary intimacy of contact to do this in what is essentially a rather viscous medium. At present, a limited number of continuous-fibre-reinforced thermoplastics is available and most of these are supplied as preimpregnated forms ('prepregs') where the fibres are completely wetted and impregnated by the resin, the impregnation being carried out from solution, from melt or indeed from monomers or prepolymers depending upon the properties of the particular thermoplastic. Some resins are in fact pseudothermoplastic in that the polymerisation chemistry continues during processing until the required matrix formulation is achieved. Postimpregnated forms simply have the resin (in the form of powder, fibre or film) and fibre closely intermixed.

Manufacturing from prepregs depends upon whether the prepreg is in the form of preconsolidated sheet or a tape or tow. For sheet, forming processes are commonest, using hydroforming, press forming, roll forming or moulding into matched dies. Figure 5.27 illustrates the diaphragm forming method.

For tape and tow reinforcement, filament winding, tape laying, braiding, pultrusion, autoclave and vacuum bag moulding and diaphragm forming can all be used, the first two of these being shown in Fig. 5.28.

Postimpregnated thermoplastic continuous-fibre compos-

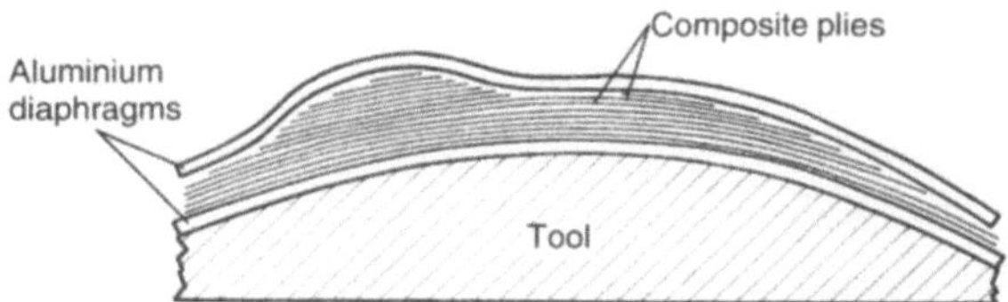

Fig. 5.27. Diaphragm forming method for manufacturing of continuous fibre reinforced thermoplastics from prepregs. (From Harvey [88].)

ites have some important additional processing considerations, these being mostly concerned with creating the correct conditions for matrix-fibre bonding. This usually means higher pressures and temperatures for longer times, and there are also severe limitations upon flow.

In contrast, the manufacture of continuous-fibre and fabric-reinforced thermosets is a far more traditional practice than for thermoplastics [89] and is arguably of more relevance to the subject of this text, since these materials have found much more marine application than others discussed in this section.

As can be seen from Fig. 5.26, there are five established techniques for the manufacture of products from reinforced thermosets and these are discussed briefly below.

Hand lay-up is the traditional technique for fibre-reinforced thermosets, and simply involves hand placement of the reinforcement, usually as matting, followed by application of resin with a brush. Despite its labour intensive nature, this method is still widely used (see Case Study 7.3) owing to its versatility. Hand spray-up involves spraying of reinforcement onto the material to be reinforced and application of resin, again by brush. Clearly this can only work for random in-the-plane reinforcement (Fig. 5.29).

Pultrusion and filament winding are suitable for the production of specific shapes, pultrusion being most suitable for sections which might otherwise be produced by extrusion, with fibres being oriented along the axis of the prism (Fig. 5.30). Curing takes places after the polymer

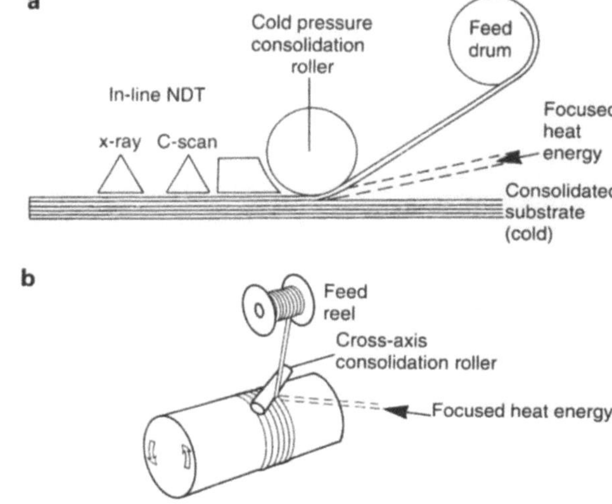

Fig. 5.28. Some methods of manufacturing continuous-fibre-reinforced thermoplastics from tape or tow. **a** Automated tape-laying process. **b** Thermoplastic filament winding with continuous consolidation. (From Harvey [88.])

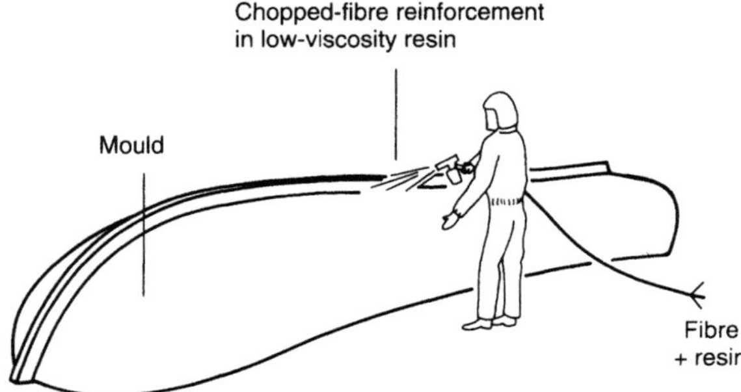

Fig. 5.29. Hand spray-up for random in-plane fibre-reinforced thermosets. (From Edwards and Endean [87].)

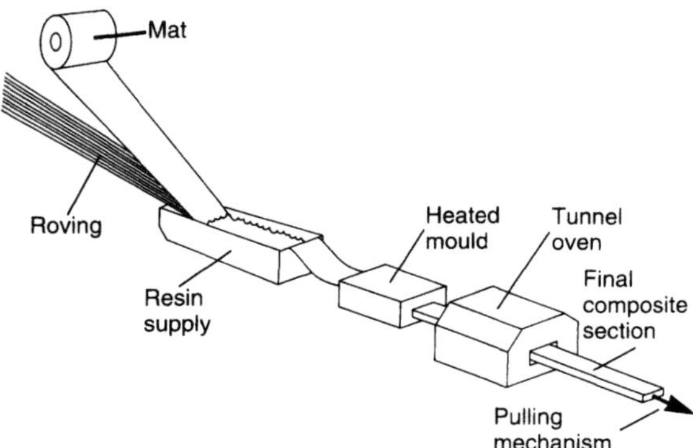

Fig. 5.30. The pultrusion process for the manufacture of composites with fibres oriented along the axis of the prism. (From Crawford [85].)

precursor and reinforcement have been formed. Filament winding is basically suitable for cylindrical objects, the reinforcement being wound onto a mandrel after impregnation with the polymer precursor. This geometry of reinforcement clearly gives good resistance to hoop and axial stresses, although, of course, far more complex objects and filament geometries can be produced.

Compression moulding is an exact analogue to the process for the unreinforced material. However, the sheet needs to be manufactured separately from reinforcement and a thickened resin (not yet fully cured). The sheet moulding compound (SMC) is then moulded in the way already described in Section 5.2.1.

5.2.3 Joining of Polymers and Composites

Like the manufacturing processes described above, the joining processes which can be applied to polymers and composites are critically dependent upon matrix type. Choice of this is dominated by the following facts:

Thermoplastics are generally good hot-melt adhesives.

Thermosets are generally good two-pack curing adhesives.

Elastomers generally possess the property of tack, which allows them to stick to each other (but not generally to other surfaces) in their unvulcanised state. Subsequent curing further consolidates the rubber-to-rubber bond, but also allows bonding to other materials.

Joining is not a common method of fabrication for plastic or composite products. The two main reasons for this are (a) that the materials are sufficiently mouldable to be able to be produced to near-net shape and (b) that the use of fasteners is not governed by the limiting strength which can be obtained in the fastening materials, as can be the case in, for example, steel-bolted joints.

As mentioned above, thermoplastics are good hot-melt adhesives, so that the main method of joining for these materials is by the use of welding. Because the thermal conductivities and melting temperatures are so different from metals, the sources and means of application of heat are also totally different. The three main heating methods used are electromagnetic, friction and hot-plate, and these are applicable to most thermoplastics with only minor modifications (such as pre-drying for polycarbonates [90]). In hot-plate (or heat) welding, the parts are melted on the surfaces of a heated platen and the platen is removed in order that the melted surfaces can be brought into contact and held in place until set. Induction, or electromagnetic, welding uses an insert made from a thermoplastic impregnated with micrometre-sized particles of metal. These particles are excited by a high-frequency a.c. source which melts the thermoplastic, thus providing bonding. Spin and vibrational welding both use friction to melt the plastic surface, and ultrasonic energy can also be directed to the weld site by use of a horn and shaped workpiece. This latter method is rather more limited in its applicability because of the requirement to focus the standing waves.

Thermoplastics can also be solvent-bonded to themselves or to other plastics which are soluble in the same solvent, provided of course that compatibility of both components and the solvent can be established. Adhesive bonding is also possible, subject again to compatibility and bond strength criteria.

In joining reinforced thermoplastics it is important to consider the possible reorientation of the reinforcement owing to flow in the bond area, as this may have implications for bond strength.

The joining of thermosets is normally carried out using adhesives or using fasteners. In the latter case, inserts (as would be necessary for thermoplastics) are not usually required, but great care needs to be taken with stress concentration, owing to the inherently brittle nature of the resins.

Adhesive joint design is essentially the same for thermoset composites as it is for the neat resin, except that in the latter case the strength requirements for the bond are likely to be less stringent. Adhesive bonds can be reinforced with fasteners, which help to resist the peel and cleavage loads to which adhesives are susceptible. Unlike welded joints, adhesive joints must be designed to be loaded in shear as far as is possible, and joint designs such as might be used for brazing or soldering are normally used to ensure this [91].

5.3 Manufacturing Processes Involving Cement and Concrete

Concrete, as seen in Chapter 4, ranks with steel in its broad range of applicability in marine environments, ranging in use from breakwaters to ships. As well as its durability, much of its popularity is owed to ease of use, even for very large structures. This section describes only briefly the manufacture of structures involving concrete, not because it is unimportant but rather because, once the structure has been designed and the mix chosen, the actual materials technology involved in manufacture is relatively straight-forward.

5.3.1 The Casting of Concrete

Perhaps the most important factor in the construction of concrete structures for the marine environment is the control of quality, from the raw materials through the mixing to the maintenance of specified cover and good compaction. Although this is covered in general codes of practice for concrete structures, there are also specific codes (e.g. [92]) which cover concrete construction for marine applications.

Several accounts exist of the concreting procedures for the construction of concrete offshore platforms (e.g. [93]), but essentially these consist of variants of the process of slip-forming, where shuttering is moved up the structure as the walls are cast (Fig. 5.31).

One interesting example of large-scale, relatively inaccessible concrete placement is provided by the seawater transfer system described by Hulshizer and Kodal [94]. The tunnel lining contained 92 000 m^3 of concrete and required round-the-clock pumping; the concrete had to be placeable for up to three hours, had to achieve enough early strength to allow stripping and jumping of the formwork, and of course had to develop the durability required of reinforced concrete in seawater. Consolidation was achieved by the use of form vibrators and internal vibrators inserted through windows in the formwork as the concrete advanced.

For large concrete ships, Tosswill [95] has suggested that the best means of construction is to use prestressing rather than reinforcement. Furthermore, he indicates that prefabricated, prestressed units should be produced under factory conditions and then post-tensioned one to another in both transverse and longitudinal directions at the construction site.

5.3.2 The Use of Cement and Concrete in Repair

Cement and concrete can be used not only to repair concrete structures but also to repair steel structures.

Concrete repair may become necessary during construction (due, for example to poor placement), after impact damage or after deterioration by weathering or reinforcement corrosion. The repair procedure (if reinstatement of concrete is to be carried out) consists essentially of a 'cut out and patch' approach. This must begin with careful removal of crumbling or otherwise suspect concrete and cleaning of any exposed rebars. If repairs are underwater, special consideration has to be taken in the selection of grouting materials with regard to their ease of application, bond with the substrate and resistance to cracking, dimensional stability, resistance to the environment and protection of the steel. Coote et al. [96] have examined a number of repair systems with regard to these criteria and have found that best protection is afforded by the use of inhibitive (rather than barrier) coatings on exposed steel, followed by reinstatement using OPC mortar, preferably modified with SBR. Acceptable short-term (about 2 years) protection was afforded to spalled or deteriorated concrete by the use of epoxy or SBR–glass-fibre repair systems.

It occasionally becomes necessary to repair or strengthen the steel tubulars of offshore structures, owing for example to fatigue or collision damage or simply to increased code requirements. One method of doing this is by using a grouted repair scheme [97]. These schemes can be divided into the following five types: grouted connection, grouted clamp, stressed grouted connection, stressed grouted clamp and grout-filled tubular. The advantage of such systems is that they are relatively easy to install and permit a reasonable tolerance on the steel tubular constructions to which they fit. The main disadvantage is that their effect on strength and fatigue resistance requires to be evaluated fully before their adoption. There now exists a wide body of data on these aspects, and this is summarised in a design manual for this type of repair [97].

Some further aspects of these and other repair methods are dealt with in Chapter 6.

References

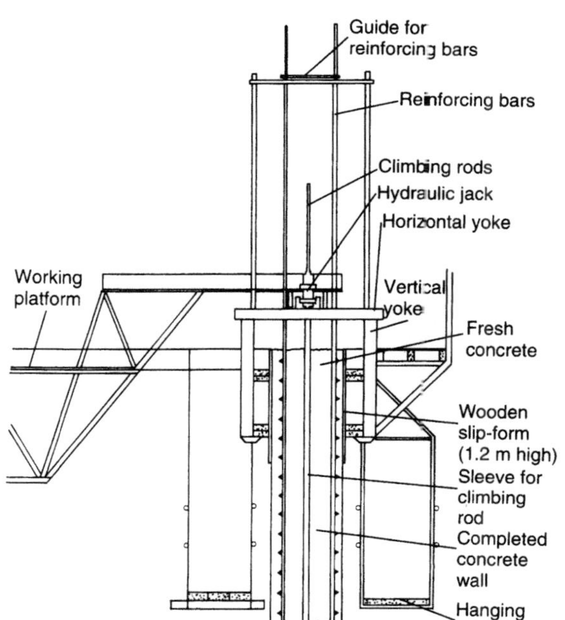

Fig. 5.31. Slip-forming method for casting large concrete structures. (From Moksnes [93]).

1. Williams ML. Correlation of metallurgical properties and service performance of steel plates from fractured ships. Welding Journal Research Supplement 1958; 37: 445s–454s.
2. Williams ML, Ellinger GA. Investigation into structural failures of welded ships. Welding Journal Research Supplement 1953; 32: 498s–527s
3. Masubuchi K. Analysis of welded structures. Pergamon, Oxford, 1980
4. Cmnd. 3409. Report of the Enquiry into the causes of the accident to the drilling rig 'Sea Gem'. HMSO, London, 1967
5. Almar-Naess A, Aagensen PJ, Lian B, Moan T, Simonsen T. Investigation of the Alexander L Kielland failure – metallurgical and fracture analysis. Proceedings, 14th Offshore

Technology Conference, Houston, paper no. OTC 4236, 1982, pp 79–84

6. Easterling K. Introduction to the physical metallurgy of welding. Butterworths, London, 1983

7. American Welding Society. Welding handbook. vol. 2: Welding processes. AWS, 8th edn, Miami, 1991

8. Lancaster JF (ed.). The physics of welding. Pergamon, Oxford, 1984

9. Lancaster JF. Metallurgy of welding. Allen & Unwin, London, 1980

10. Lincoln Electric Company. The procedure handbook of arc welding. Lincoln Electric Company, Cleveland, 1973

11. Keeler T. Innershield welding – development and applications. Metal Construction 1981; 13: 667–673

12. Still JR and Rae G. Laybarge inspection of submarine pipelines. Metal Construction 1984; 16 (5): 268–275

13. Letchford PW. An overview of pipeline welding – systems available and quality assurance. In: Recent developments in pipeline welding practice. Welding Institute, Cambridge, 1979, pp 2–5

14. Culbertson RP, Gumm WG. Construction procedures and quality control in offshore pipeline construction. In: Recent developments in pipeline welding practice. Welding Institute, Cambridge, 1979, pp 25–30

15. Muesch H, Langer J, Dueren CF, Luegger H. Flash butt welding for large diameter pipes. Proceedings, 13th Offshore Technology Conference, Houston, paper no. OTC 4103, 1981, pp 327–335

16. Nicholas ED, Lilly RH. Radial friction welding. In: Recent developments in pipeline welding practice. Welding Institute, Cambridge, 1979, pp 50–56

17. deSivry B, Sudreau BG, Anselme OR, Bonnet C. A new welding concept for pipelines – electron beam welding. Proceedings, 13th Offshore Technology Conference, Houston, paper no. OTC 4101, 1981, pp 299–310

18. Redshaw PR, Stalker AW. Explosive welding combines with bottom tow for new subsea pipeline construction technique. Proceedings, 11th Offshore Technology Conference, Houston, paper no. OTC 3523, 1979, pp 1431–1438

19. Christenson BC, Evans RM, Caudy DW, Balogh SP, Tierney JM. Underwater firecracker welding for attaching pipe-connection flanges. Proceedings, 11th Offshore Technology Conference, Houston, paper no. OTC 3468, 1979, pp 967–972

20. Hamasaki M, Sakakibara J, Watanabe M. MIG welding underwater. Welding Design and Fabrication June 1976: 78–80

21. Hamasaki M, Yoshikawa S, Takaoka H. Underwater stud welding aids reclamation of corroded pipes. Metal Construction 1983; 15: 622–623

22. Chadwick MD. Pipe-to-pipe welding by explosives. In: Explosive welding. Welding Institute, Cambridge, 1976, pp 31–39

23. Hasui A, Suga Y, Kishi S, Teranishi M. On underwater submerged arc welding. Transactions of the Japan Welding Society 1980; 11 (1): 9–15, 16–20

24. Masubuchi K, Meloney MB. Underwater welding of low carbon and high strength (HY-80) steels. Proceedings, 6th Offshore Technology Conference, Houston, paper no. OTC 1951, 1974, pp 177–190

25. Gooch TG. Properties of underwater welds. Metal Construction 1983; 15 (3): 164–167; (4): 206–215

26. Asnis AE, Savich IM. The new method of mechanised underwater welding. Proceedings, International Conference on Underwater Welding, IIW, Trondheim, 1983. Pergamon, Oxford, 1983, pp 311–318

27. Diesen A. Shock hazard in the hyperbaric environment. Proceedings, International Conference on Underwater Technology, April 1980. Pergamon, Oxford, 1980, pp 301–314

28. Haux, G. Subsea manned engineering. Ballière Tindall, London, 1982

29. Scott Lyons R, Middleton TB. Orbital TIG system simplifies underwater welding. Metal Construction 1984; 16 (10): 627–631

30. van der Torre D, Sipkes MP. Remote controlled underwater welding in the dry. Metal Construction 1982; 14 (11): 588–592

31. Worrall DA, Gibson DE. Sampling and analysis of welding fumes and gases produced under experimental hyperbaric conditions. Proceedings, International Conference on Underwater Welding, IIW, Trondheim, 1983. Pergamon, Oxford, 1983, pp 283–292

32. Hamasaki M, Sakakibara J, Arata Y. Underwater MIG welding – high pressure chamber experiments. Metal Construction 1976; 8 (3): 108–112

33. Allum CJ. Effect of pressure on arcs. Proceedings, International Conference on Underwater Welding, IIW, Trondheim, 1983. Pergamon, Oxford, 1983, pp 171–178

34. DeJong GJ, Feller MD, Clementson DR. Thistle 'A' platform pipeline connections. Proceedings, 10th Offshore Technology Conference, Houston, paper no. OTC 3347, 1978, pp 2557–2570

35. Gaudiano AV. A summary of 26 underwater welding habitat jobs. Proceedings, 7th Offshore Technology Conference, Houston, paper no. OTC 2302 1975, pp 559–569

36. Stevenson AW, Sleveland A. Damaged brace on offshore platform replaced using hyperbaric welding. Metal Construction 1983; 15 (12): 720–723

37. Mellon B. Applications of subsea hot tapping to extend the life of offshore pipeline systems. Conference on New Technologies to extend the Life of North Sea Facilities. Oyez, London, 1985, pp 25–26

38. Lynch RP, Pilia FJ. Pipeline hot-tap welding under 110 ft of sea water. Welding Journal 1969; 48 (3): 183–190

39. Osborn CW. Problems associated with welding onto pipelines under pressure. In: Recent developments in pipeline welding practice. Welding Institute, Cambridge, 1979, pp 8–10

40. Phelps B. British Gas studies of welding on pressurised gas transmission pipelines. In: Recent developments in pipeline welding practice. Welding Institute, Cambridge, 1979, 36–39

41. Trethewey KR. The use of explosion-bonded transition joints for joining aluminium superstructures to steel hulls. Paper presented to Institute of Marine Engineers. November 1988. Marine Management Holdings, London, 1988

42. Yoshida T, Shigetomo M, Atsuta T, Itoga K. Corrosion problems of pipeline and a solution. Proceedings, 12th Offshore Technology Conference, Houston, paper no. OTC 3891, 1980, pp 361–370

43. Carruthers R. The use of 90/10 copper–nickel as a splash zone cladding. Seminar on Copper Alloys in Marine Environments. Copper Development Association, Birmingham, 1985

44. Sandor LW. Joining techniques for copper–nickel sheathing of offshore structures and ships' hulls. Seminar on Copper Alloys in Marine Environments. Copper Development Association, Birmingham, 1985

45. Welding Institute. Control of distortion in welded fabrications. Welding Institute, Cambridge, 1982

46. Gray TGF, Spence J. Rational welding design. Butterworths, London, 1982

47. Burdekin FM. Local stress relief of circumferential welds in cylinders. British Welding Journal 1963; 10 (9): 483–490

48. Porter-Goff RFD, Free JA, Tsiagbe WY. Residual stresses in Y-nodes and PWHT joints. Department of Energy Offshore Technology Report OTH 89 315. HMSO, London, 1989

49. Elliott DS, Brown JM. Welding considerations in the construction of steel platforms in the North Sea. Welding and Metal Fabrication 1978; 46: 141–153

50. Saunders GG. Influence of welding and postweld heat treatment on the heat affected zone fracture toughness of carbon–manganese and low alloy steels. Proceedings, Conference on Heat Treatment of the Joint Committee of the Iron and Steel Institute. Institute of Metals and Institute of Metallurgists, London, 1971, pp 18–28

51. UEG. Design of tubular joints for offshore structures. UEG Publication UR33. UEG/CIRIA, London, 1985

52. American Welding Society. Structural welding code, AWS D1.1-84. AWS, Miami, 1984

53. BS 5135. Specification for the metal arc welding of carbon- and carbon–manganese-steels. BSI, London, 1974

54. BS 4360. Specification for weldable structural steels. BSI, London, 1979

55. BS 4871. Approval testing of welding procedures/Approval testing of welders working to welding procedures: Fusion welding of steel. BSI, London, 1981

56. Waller DN, Gaudin JP. Investigation of magnetism and its elimination in submarine pipelines. Proceedings, International Conference on Underwater Welding, IIW, Trondheim, 1983. Pergamon, Oxford, 1983, pp 247–254

57. American Welding Society. Welding handbook, vol. 1: Welding technology, 8th edn (Connor LP (ed.)). AWS, Miami, 1987

58. Taylor DS, Evans GM. Development of MMA electrodes for offshore fabrication. Metal Construction 1983; 15 (7): 438–443

59. Department of Energy. Compendium of fracture toughness data on weldments. Offshore Technology Information Report, OTI 88 534. HMSO, London, 1988

60. Dolby RE. Advances in welding metallurgy of steel. Metals Technology 1983; 10: 349–362

61. Dawson GW, Judson P. Procedural guidelines for the achievement of tough welded joints in structural steels for offshore applications. In: Cotton HC (ed.) Proceedings, 2nd International Conference on Offshore Welded Structures. Welding Institute, Cambridge, 1983, pp 2-1–2-13

62. Judson P, McKeown D. Advances in the control of weld metal toughness. In: Cotton HC (ed.), Proceedings, 2nd International Conference on Offshore Welded Structures. Welding Institute, Cambridge, 1983, pp 3-1–3-14

63. Bailey N, Pargeter RJ. The influence of flux type on the strength and toughness of submerged-arc weld metal. Welding Institute, Cambridge, 1988

64. Bailey N, Jones SB. Solidification cracking of ferritic steels during submerged arc welding. Welding Institute, Cambridge, 1977

65. Hrivnak I. The mutual relationship and interdependence of developments in steel metallurgy and welding technology. Welding in the World 1978; 16 (7/8): 130–151

66. Farrar JCM, Dolby RE. Lamellar tearing in welded steel fabrication. Welding Institute, Cambridge, 1972

67. Arrowsmith JM, Shenton DC. Steel plate for offshore structures. Metal Construction 1976; 8: 396–400

68. Baumgardt H, de Boer H, Musgen B. High strength steels for offshore technology. Metal Construction 1984; 16: 15–19

69. Bailey N. Welding carbon–manganese steels. Metal Construction 1970; 2 (7): 442–447

70. Cottrell CLM. Hardness equivalent may lead to a more critical measure of weldability. Metal Construction 1984; 16 (12): 740–744

71. Gooch TG. Welding steels without hydrogen cracking. Welding Institute, Cambridge, 1990

72. McKeown D, Judson P, Apps RL, Pumphrey WI. The weldability of low sulphur steels. Metal Construction 1983; 15 (11): 667–673

73. NACE MR-01-75. Sulphide stress cracking resistant metallic materials for oil field equipment. NACE, Houston, 1984

74. Lorentz K, Duren C. Evaluation of large diameter pipe steel weldability by means of the carbon equivalent. Proceedings, Conference on Steels for Linepipe and Pipeline Fittings. Metals Society, London, 1983, pp 322–332

75. Terasaki F, Ohtani Y, Ikeda A, Nakanishi M. Steel plates for pressure vessels in sour environment applications. Proceedings, International Conference on Fusion Welded Steel Pressure Vessels. Mechanical Engineering Publications, London, 1984, pp 45–64

76. Turnell AJ, Marsh PG. Low alloy steels in oil refinery service – 3b Effects of fabrication – welding. Metal Construction 1985; 17 (5): 293–296

77. Stawstrom C, Hillert M. An improved depleted zone theory of intergranular corrosion of 18–8 stainless steel. Journal of the Iron and Steel Institute 1969; 207 (1): 77–85

78. Lula RA. Stainless steel. ASM, Metals Park, Ohio, 1986

79. American Welding Society. Welding handbook, vol. 4: Metals and their weldability, 7th edn (Kearns WH (ed.)). AWS, Miami, 1982

80. DeBold TA, Martin JW, Tverberg JC. Duplex stainless offers strength and corrosion resistance. In: Lula RA (ed.) Proceedings, Conference on Duplex Stainless Steels. ASM, Metals Park, Ohio, 1983, pp 169–189

81. Kotecki DJ. Duplex alloy weld ferrite content: Measurement and mechanical property effects. In: Lula RA (ed.) Proceedings, Conference on Duplex Stainless Steels. ASM, Metals Park, Ohio, 1983, pp 415–430

82. Lefebre G, Dufrane J-J. Production and welding of Zeron 100 super duplex. In: Denys R (ed.) Proceedings, Pipeline Technology Conference, Oostende, October 1990. K.UIU-Antwerpen and RV-Gent, 1990, pp 11.7–11.16

83. Doyen R, Demuzere R. Welding of ferritic–austenitic pipes. In: Denys R (ed.) Proceedings, Pipeline Technology Conference, Oostende, October 1990. K.UIU-Antwerpen and RV-Gent, 1990, pp 11.17–11.22

84. Polmear IJ. Light alloys – metallurgy of the light metals. Edward Arnold, London, 1981

85. Crawford RJ. Plastics engineering. Pergamon, Oxford, 1987

86. McCrum NG, Buckley CP, Bucknall CB. Principles of polymer engineering. Oxford University Press, Oxford, 1988n

87. Edwards L, Endean M (eds). Manufacturing with materials. Butterworths and Open University Press, London and Milton Keynes, 1990

88. Harvey MT. Thermoplastic matrix processing. In: Engineered materials handbook, vol. 1: Composites. ASM International, Metals Park, Ohio, 1987, pp 544–553

89. Johnson CF. Compression moulding. In: Engineered materials handbook, vol. 1: Composites. ASM International, Metals Park, Ohio, 1987, pp 559–563

90. American Society for Metals. Engineered materials handbook, vol. 2: Engineering plastics. ASM International, Metals Park, Ohio, 1988

91. Williams J, Scardino W. Adhesives selection. In: Engineered materials handbook, vol. 1: Composites. ASM International, Metals Park, Ohio, 1988, pp 683–688

92. Det norske Veritas. Rules for design, construction and inspection of fixed offshore structures. DnV, Hovik, 1974

93. Moksnes J. Condeep platforms for the North Sea – some aspects of concrete technology. Proceedings, 7th Offshore Technology Conference, Houston, paper no. OTC 2369, 1975, pp 339–350

94. Hulshizer AJ, Kodal A. Design considerations, construction and performance of an extensive seawater conveyance system. Proceedings, Conference on Concrete in Marine Environment ACI SP-109. American Concrete Institute, Detroit, 1988, pp 677–704

95. Tosswill SIG. Prestressed concrete in ship construction. Concrete afloat. Thomas Telford, London, 1977, pp 147–158

96. Coote AT, McKenzie SG, Treadaway KWJ. Repairs to reinforced concrete in marine structures. Proceedings, Conference Concrete in the Marine Environment. Concrete Society, London, 1986, pp 333–348

97. Harwood RG, Shuttleworth EP. Grouted and mechanical strengthening and repair of tubular steel offshore structures. Offshore technology report OTH 88 283. HMSO, London, 1988

6 Inspection, Testing and Reliability

In marine technology, perhaps more than in any other area, a knowledge of the condition of a structure or component is essential to the extent that life may be lost as a result of component or structural failure. For instance, it has been said that a saturation diver is every bit as remote from medical help as is an astronaut, although this is now less believable with the advent of hyperbaric medicine. The potentially disastrous consequences of offshore failures, whatever their cause, have been felt more than once in UK waters alone, and there is continuing growth in awareness of the need to look carefully at all aspects of reliability of components and structures for the marine environment. More prosaically, the cost of *in situ* maintenance offshore or underwater can be very high, and the cost of bringing a mobile unit to shore for unscheduled maintenance may be equally punishing.

All this means that the related disciplines of inspection, testing and reliability are of paramount importance in the marine environment, a fact that is recognised by legislatory systems throughout the world. It is not the purpose of this chapter to describe such legislation, but rather to illustrate the ways in which quality and integrity can be ensured by appropriate design, fabrication, installation, operation, inspection and testing.

After a brief discussion of destructive testing, the methods of non-destructive testing are outlined, including an assessment of the reliability of these. This is followed by a description of monitoring techniques and also methods of underwater intervention, inspection and repair. The chapter concludes with some discussion of mechanical and structural reliability techniques.

6.1 Destructive Testing

Since most of the methods of destructive testing have already been dealt with (Chapters 2 and 5), only a brief description of their applicability is included here.

Destructive testing is normally carried out to determine material quality and is not, with the exception of weld procedure qualification, common for components or structures. Thus, aside from the exception noted above, destructive testing is normally the responsibility of the primary (material) supplier, and the designer ensures material quality by use of an appropriate specification (whether a standard, an internal company specification, or a mixture of both).

Material destructive tests are normally mechanical, involving tensile testing (for strength and ductility) and, sometimes, some type of toughness test; this is usually an impact test, because few materials standards carry a requirement for fracture toughness testing. Aside from these materials tests, some structural components, for example pipelines, are subject to destructive testing (e.g. drop weight tear test) on a batch basis.

Because of the metallurgically disruptive nature of welding, weld procedures for critical (e.g. structural or pressure-containing) components are subject to destructive testing. This normally consists of producing a test weld as described in Chapter 5 and subjecting this to a variety of destructive mechanical tests. Similarly, welders usually also require to be qualified by welding to a procedure and having the weld subjected to destructive examination.

Proof testing of various types, although not usually destructive, might be considered under the above heading. Two examples of such testing are for lifting gear and for pressure vessels. The purpose of proof testing is to take a structure to a specified proportion above design load and to take the absence of failure as evidence of fitness for service. This method may seem obvious, but in fact it can lead to difficulties. A chastening example of this is provided by the Comet Mk I airliner failures described by Polak [1].

6.2 Non-destructive Testing

Most of the non-destructive test methods are aimed at discovering defects of manufacture in a material or a structure or at finding defects which develop during service. Thus, the types of defects sought depend upon the material, its method of manufacture and fabrication, and any expected deterioration mechanism which can occur in service.

The types of manufacturing defects to be expected in composites, concrete and metals are quite different from each other, as are those to be expected in metals which are cast to final shape as opposed to those which are welded. Also, the service defects which might develop in concrete (e.g. spalling, reinforcement corrosion) are unlike those which might develop in steel (wall thickness loss by corrosion, fatigue crack propagation). Table 6.1 summarises the types of defects which might be expected in the main classes of structural materials after manufacture and after a service period.

Table 6.1. Possible sources of defects for three types of structural material

Material	Defect sources
Structural steel	*Primary processing.* Defects of various types may arise in the steel mill, although these are largely eliminated by quality control. Variations in mechanical properties are inevitable, but these are normally specified.
	Secondary processing. Mostly from welding – porosity, cracking, slag inclusions, defects of shape (e.g. undercut, distortion), metallurgical changes and residual stress. Machining (usually grinding) may give rise to some defects, and casting defects should be considered if the secondary process involves casting.
	Service. Fatigue, corrosion (including pitting and SCC), wear (fluid and/or solid). Direct mechanical damage (e.g. impact).
Concrete	*Primary processing.* Not applicable.
	Secondary processing. Defects can arise during casting – lack of soundness (usually local), incorrect mix (water/cement ratio, aggregate content, aggregate type, inappropriate water and/or additive chemistry), poor construction joints and poor curing conditions. Most of these give rise to later service damage.
	Service. Fatigue, reinforcement corrosion, cracking and spalling due to water penetration, internal reactions and weathering. Direct mechanical damage (e.g. impact).
Composites	*Primary processing.* Not applicable.
	Secondary processing: Defects of fabrication – incorrect internal bonding, poor wetting of reinforcement, incorrect placement of reinforcement, poor curing conditions and/or poor chemical control of resin.
	Service. Fatigue, weathering, creep, chemical attack, water damage, overloading and impact.

Defects in materials can also usefully be classified in terms of the scale at which they occur, from the sub-microscopic such as vacancies, impurities and dislocations through the microscopic such as grain boundaries, inclusions, segregation, porosity, or incorrect or unwanted microscopic constituents such as martensite in steels or carbides in stainless steels. Macroscopic defects can include cracks, large porosity, cavities and defects of shape and surface (for example profile or corrosion pits).

The process of non-destructive testing (NDT) is designed to find unacceptable defects without the necessity to remove samples or otherwise alter any part of the structure. The question as to what constitutes an acceptable defect has to be decided on a case-by-case basis, and the design should include an assessment of this, either by fracture mechanics or by strength considerations as outlined in Chapter 2.

The problem of NDT is to find a probe which will search the material both externally and internally and will discover, with an acceptable degree of reliability, those defects which are considered unacceptable. This is rarely achievable with a single method, and is never achievable with 100 per cent reliability. The traditional probes used are light, sound, ionising radiation and electrical and magnetic fluxes. The ways in which these probes interact with solids are sensitive to the presence of certain types of defects, and this can be used in a qualitative or quantitative way to yield the required information. Some more modern methods use, for example, laser light, heat flux and neutron beams as probes.

The question of whether or not a defect is of significance can be understood in terms of the critical defect size which the structure or component can stand under service conditions, coupled with an appropriate factor of safety. This can be found by a fracture mechanics analysis or by a strength calculation, but is sometimes specified by a code for particular applications. For example, the DnV Code for pipelines [2] specifies defect acceptance levels for visual and radiographic inspection of submarine pipelines, and these are illustrated in Table 6.2.

For marine applications most concern is felt over welded construction, because weld defects can be relatively easily introduced owing to the inherently manual nature of much construction welding and also because such defects can be propagated under fatigue loading. For this reason, much effort has gone into the development of non-destructive testing (NDT) methods for welded steel construction, and the following brief description of the main NDT methods gives an indication of their use and application for marine and offshore engineering. The list is not intended to be exhaustive, but rather concentrates on the more important methods; nor are the descriptions very detailed, the reader being referred instead to recent texts on the matter (e.g. Halmshaw [3] and Hull and John [4]). The discussion is separated into those techniques which are only suitable for surface-breaking defects, those involving electrical and/or electrical methods, those using ultrasonics and those using radiography. Table 6.3 summarises these methods and their capabilities.

6.2.1 Methods for Surface-breaking Defects

From the point of view of NDT, defects can be conveniently classified as those which appear on an external surface (surface-breaking defects) and those which are enclosed within the material (body defects). Whereas the methods for body defects can often be used for surface defects, this section will be used to describe those methods which cannot detect body defects.

In large engineering structures, defects of any significance are usually large enough to be visible to the unaided eye and the function of the surface methods is to render these visible, or enhance their visibility. The main reason why this is necessary is that whereas the depth and length dimensions of significant defects may be quite large, their width may be quite small. This is especially true of cracks or other narrow defects; also, of course, for a given depth and width, cracks are more significant than wide defects (Fig. 6.1). Despite this, one of the most important

Table 6.2. Acceptance levels for defects in pipeline welds – visual and radiographic inspection
(This is not a specification; source material should be consulted for details)

	Acceptance limits
Internal defects	

Porosity	Scattered porosity is to be a maximum of 3 per cent by projected area. Largest pore dimension $t/4$, maximum 4 mm.
	    Cluster porosity is not to exceed an area of 12 mm in diameter in any continuous 300 mm of weld length. Maximum dimension of any individual pore is not to exceed $t/8$, maximum 2 mm.
	    Porosity on line is not to penetrate weld surface, largest pore dimension $t/8$, maximum 2 mm.

Slag inclusions	Isolated slag: length $\leq t/2$, width $\leq t/4$, maximum 4 mm. Slag lines: length $\leq 2t$, maximum 50 mm, width ≤ 2 mm. For 'wagon tracks' width of each parallel slag line is not to exceed 1.5 mm.
Lack of fusion or incomplete penetration	Length $\leq 2t$, maximum 50 mm.
Cracks	Not acceptable.
Surface defects	
Misalignment of adjoining pipe ends	$\leq 0.15t$, maximum 3 mm.
Dents	Depth: maximum 6 mm, length: maximum o.d./2.
Cold-formed gouges, grooves, notches and arc burns	Not acceptable. May be removed by grinding.
External and internal weld reinforcement	For $t \leq 12.5$ mm: maximum 3 and 2 mm respectively. For $t > 12.5$ mm: maximum 4 and 3 mm respectively.
External and internal concavity	External concavity is not acceptable. Internal concavity is acceptable provided that the density of the radiographic image of the concavity does not exceed that of the adjacent base metal.

Undercut	Depth $\leq t/10$, maximum 0.8 mm.
	For girth welds the length of an undercut in any continuous 300 mm of weld length is not to be more than:
	maximum 50 mm for depth $\leq t/10$, maximum 0.8 mm
	maximum 100 mm for depth $\leq t/20$, maximum 0.4 mm
	    For longitudinal or spiral welds the length is to be no more than a maximum of one-fifth of the above limits for girth welds.
	    For depth ≤ 0.3 mm, undercut may be accepted regardless of length provided its shape and notch effect are not considered detrimental.

Lack of fusion or incomplete penetration	Length $\leq t$.
Hollow bead	Length $\leq t/2$, maximum 12 mm.
Burn-through	Maximum dimension 6 mm of any single defect, total length maximum 12 mm in any 300 mm weld length.
Cracks	Not acceptable.

Source: Det norske Veritas [2].

Table 6.3. Capabilities of the various NDE methods in routine use

Method	Materials	Defect types	Minimum detectable and measurable defect size
Liquid penetrant	Most	Clean, surface breaking	8 mm min., depth not measurable
Magnetic particle	Must be ferro-magnetic	Clean, surface breaking or just sub-surface	0.25 mm min.; depth not measurable
Ultrasonics	Most	Mostly internal but can be used for surface	5 mm long × 2 mm deep at surface 2.5 mm long × estimate of depth sub-surface
Eddy current	Metals and coating thickness measurement	Mostly surface but some sub-surface	Approx. 8 mm × 8 mm, but detectable depth varies
Radiography	Most	Mostly internal but will show surface defects	Varies with orientation relative to beam

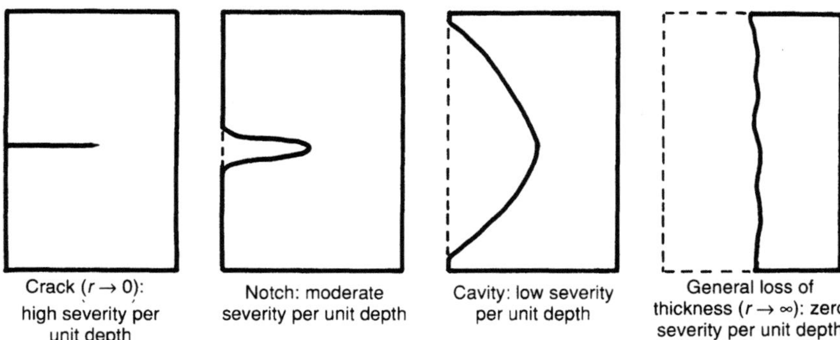

Fig. 6.1.   Effect of defect width (or tip radius of curvature) on defect severity.

NDT methods for surface-breaking defects remains simple visual inspection.

Because the most dangerous defects are very narrow, the search for surface-breaking defects is not always simply a case of surface cleaning followed by close visual inspection, although this would always precede any attempt at using an NDT method.

The two most widely used methods of enhancing surface defects so that they may be discovered by the naked eye are magnetic particle inspection (MPI) and dye penetrant (DP) inspection.

Magnetic particle inspection, like DP inspection, requires a clean surface on which to operate. In marine applications this often necessitates the use of surface blasting to expose underlying metal free of oxide, fouling or other surface contamination. The method then simply consists of producing a magnetic field which is guided through the metal, and provided that the lines of flux cross the surface defect then application of a 'magnetic ink' (usually a suspension of iron filings) will result in these aligning themselves so as to bridge any surface discontinuities, hence highlighting any cracks. Several methods are used to further enhance contrast. These include the use of white background paint which shows the particles as black on a white ground, and the use of various dyes on the particles which fluoresce, usually under ultra-violet light. The magnetic field can come from a permanent magnet or may be induced by the use of prods with a d.c. voltage, as shown in Fig. 6.2.

The obvious limitations of the MPI method are that the flux lines need to be aligned correctly within the solid, and of course the material must be magnetic. The former can be overcome by testing in two mutually perpendicular directions, provided that the geometry of the structure to be tested permits this. Spurious indications can be produced by shallow laps or machining marks on the surface, and some operator experience is necessary to ensure that excessive reject levels do not occur without compromising the effectiveness of significant defect detection.

DP inspection is very similar to MPI, except that the method of enhancement is slightly different. Here, a penetrating liquid is applied to the cleaned surface and left for a period to be drawn into any surface defects. Excess penetrant is then wiped from the surface and a powder 'developer' is applied. The penetrant is drawn out of the surface traps into the developer, staining it and indicating the presence of defects. This method will not, of course, work on porous or absorbent surfaces such as concrete.

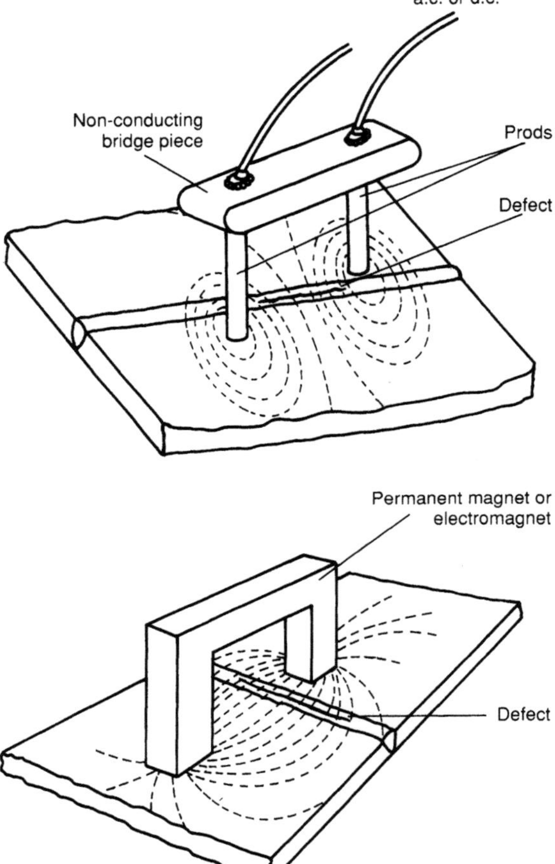

Fig. 6.2.   Methods of application of magnetic flux in MPI.

6.2.2   Methods Involving Electrical and/or Magnetic Measurements

A quantitative extension of the MPI method involves the measurement of electrical or magnetic characteristics of the component to be tested. Such methods can be used to yield information about surface coatings, as well as revealing the presence of surface or body defects.

Perhaps the most important of these is eddy current testing, where eddy currents are induced in the test material from a coil carrying alternating current placed in proximity to the surface. The eddy currents produce a back e.m.f. in the coil, which changes its impedance. The coil impedance is measured and any changes as the surface is scanned can be used to make inferences as to the condition of the structure. Because the resultant current is dependent on a wide range of different factors such as electrical conductivity and magnetic permeability of the specimen material, as well as process parameters of the set such as probe-to-work stand-off distance, a.c. frequency and coil size, the important variables (from the point of view of NDT) such as specimen size, metallurgical condition of the specimen and the presence of flaws can be relatively difficult to isolate. A number of applications to specimens of cylindrical geometry especially useful for production testing are described by Halmshaw [3].

For the more versatile approach needed for crack detection, a ball-nosed probe containing a very small coil at relatively high frequency is used. These probes are most useful for the detection of surface cracks, owing to the fact that the penetration, δ, of eddy currents is related to the frequency f (the so-called 'skin effect'):

$$\delta = (\pi f \mu \sigma)^{-1/2}$$

where μ and σ are the permeability and conductivity of the test material respectively.

As the probe is moved across the surface, any crack will distort the eddy current pattern, allowing detection. Quantification of the depth of cracks is possible in principle, but this can be quite difficult in practice, owing to effects such as bridging of crack faces with conducting material. Nevertheless, some success is claimed in quantifying crack depth with such instruments, and Halmshaw [3] shows a crack depth calibration curve for a twin-frequency eddy current instrument.

Non-destructive testing can also be achieved by applying a pair of contacts to the surface of a specimen and measuring the potential drop between the two points (Fig. 6.3). Of course, the potential drop is likely to be very small, especially in metals, and may indeed be masked by contact resistances at the probes. It is commoner to use a.c. in such cases where the skin effect will confine the current to the surface and increase sensitivity. It is also normal to use a probe point between the two contacts, so that internal calibration can be achieved for the effects of relative permeability, frequency and conductivity on the a.c. resistance of the test material. Taking two contacts relatively far apart, so that the field is relatively uniform, an a.c. voltmeter can be used to measure the p.d. between two probes in the uniform region. If the distance between the probe points is D, then the potential drops measured in intact material, V_1, and in defective material, V_2, can be related through the different path lengths, D and $D + h$, where h is the crack depth [3]:

$$h = [(V_2/V_1) - 1]\, D/2$$

The sensitivity and accuracy of such a method can be quite good (down to a few mm and ± 1 mm) on flat surfaces and where cracks are sufficiently long.

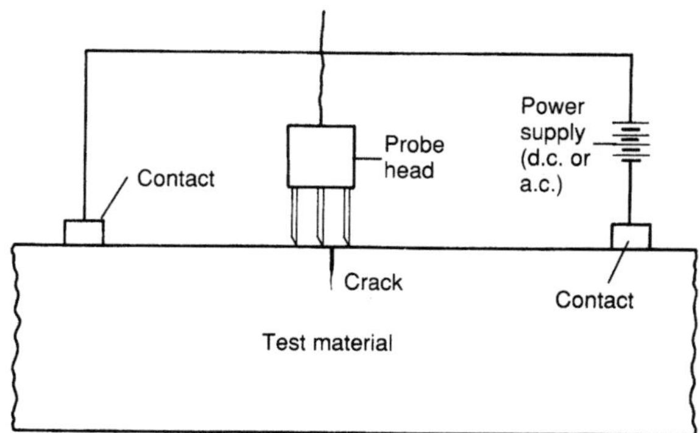

Fig. 6.3. Principle of the potential drop method for measuring defect depth. (From Halmshaw [3].)

6.2.3 Methods Involving Ultrasonics

Very-high-frequency sound is a useful method of probing materials for defects, and has been used in a qualitative and a quantitative way for many years. To understand why ultrasonics are useful in NDT, a knowledge of a few of the properties of such waves when they are being propagated in solids and other media is useful.

The frequencies of ultrasound used in non-destructive testing are of the order of a few MHz, and these waves are propagated in solids as elastic waves. The surface is usually excited by a piezoelectric crystal through some coupling medium, and the crystal is itself excited by an oscillating electric potential. In many ultrasonic techniques a short pulse is used at a frequency close to the resonant frequency of the crystal, and the transducer acts as both transmitter and receiver.

The cut of the piezoelectric crystal, the backing material in which it is mounted, its orientation within the mounting and the design of the shoe all contribute to the characteristics of the probe. In general, compression and shear wave probes are available, capable of injecting pulses of longitudinal and transverse excitation into the solid through an appropriate couplant. Stress waves can be propagated in solids as one of two different types of bulk waves, namely 'compression waves' where the particle vibrations are along the direction of propagation and 'shear waves' where the direction of particle vibration is normal to the direction of wave propagation. A variety of different types of surface waves (of which Rayleigh waves are probably the best known) are also possible, but these are of less importance in ultrasonic NDT.

The velocities of compression waves, V_c, and shear waves, V_s, are given by

$$V_c = \left[\frac{E(1 - \nu)}{d(1 + \nu)(1 - 2\nu)} \right]^{1/2}$$

and

$$V_s = (G/d)^{1/2}$$

where E = Young's modulus
$G = E/2\,(1 + \nu)$ is the shear modulus
γ = Poisson's ratio
d = the material density

Another useful property of the material in which the sound is propagating is its specific acoustic impedance:

$$Z = Vd$$

When a wave is propagated through a material from a transducer it is subject to beam spread, reflection, refraction and attenuation, and all of these have some effect on the analysis of the reflected signal in ultrasonic NDT.

Beam spread commences at the probe and is a diffraction effect resulting from the edges of the circular aperture, diameter D, from which the ultrasonic beam of wavelength λ is emitted from the probe. For the purposes of this application, it is sufficient to use a simplified diffraction analysis which yields a near field length of

$$N = D^2/4\lambda$$

which defines the depth beyond which useful measurements can be made, and a beam spread angle:

$$\sin(\theta/2) = 1.2\ \lambda/D$$

although the intensity has dropped to one-tenth that at the centre-line by:

$$\sin(\theta'/2) = 1.08\lambda/D$$

Thus, beam spread limits the effective probe diameter, and near zone limits the useful depth for pulse-echo applications in ultrasonic NDT.

The amount of reflection, upon which the pulse-echo technique depends, is determined by the specific acoustic impedance of the two media, which are separated by the reflecting boundary. When an ultrasonic beam meets a plane boundary between two media A and B, some of the incident energy, E_i, is reflected and this reflected energy, E_r, is given by

$$E_r = [(Z_A - Z_B)/(Z_A + Z_B)]^2\ E_i$$

for the case where the second medium is of thickness t (in the direction of propagation), much greater than the wavelength of the incident beam. Otherwise, the ratio between incident and reflected energies is given by:

$$\frac{E_r}{E_i} = \frac{\left(Z_A/Z_B - Z_B/Z_A\right)^2}{4\cot^2(2\pi t) + \left(Z_A/Z_B + Z_B/Z_A\right)^2}$$

One implication of this in flaw detection is that very narrow cracks may be transparent to ultrasonic NDT pulses, but Halmshaw [3] estimates that no practical problem should arise (in steels at least) for openings of about 1 μm and over.

When waves are obliquely incident on surfaces the phenomenon of refraction takes place, embodied in Snell's law:

$$\frac{\sin(i)}{\sin(r)} = \frac{V_A}{V_B}$$

One major implication of this law is that there is a critical incident angle, i, at which the refracted angle, r, reaches 90° and the incident beam cannot enter the second medium, and that this angle is dependent upon the ratio of the ultrasonic velocities in the two media. For a water–steel interface this critical angle is 15°, and this represents a limit on the launch angle for ultrasonic energy from a probe into a steel specimen using a water-based couplant.

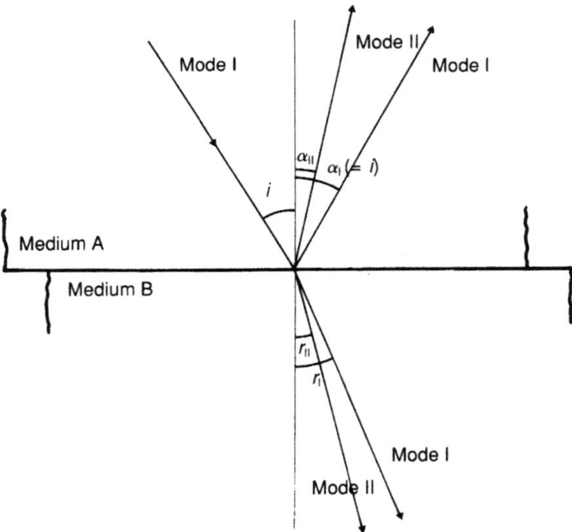

Fig. 6.4. Combinations of refraction, reflection and mode conversion at a boundary. (Adapted from Halmshaw [3].)

Owing to the different velocities, the critical angles for compression and shear waves are different, and furthermore it is possible for conversion from shear to compression mode and vice versa to take place at a boundary. For the general case of a wave of mode I (compression or shear) incident at a boundary between media A and B (Fig. 6.4), the relationships between the incident (mode I), reflected (modes I and II) and refracted (modes I and II) beams are:

$$\frac{\sin(i)}{V_{I,A}} = \frac{\sin(\alpha_I)}{V_{I,A}} = \frac{\sin(\alpha_{II})}{V_{II,A}} = \frac{\sin(r_I)}{V_{I,B}} = \frac{\sin(r_{II})}{V_{II,B}}$$

provided, of course, that the angles are appropriate and hence the beams exist. The amplitude of each wave can also be calculated, but this is a rather more complex matter.

As well as the above effects, quantitative NDT using ultrasonics requires some estimate of the degree to which the beam is attenuated by the test material, in order that appropriate amounts of energy can be used and to allow an assessment of the losses due to effects other than scattering, viscous friction, elastic hysteresis and heat conduction. The foregoing attenuation effects can be lumped together in an attenuation coefficient and the effect described by an equation of the following form:

$$I = I_0\exp(-\mu x)$$

This relates the intensity at depth x into the material to the intensity at the surface, I_0. The quantity μ is known as the 'absorption coefficient' or 'attenuation coefficient', and it varies with frequency and material. Attenuation losses are usually very small in ultrasonic NDT, when compared with the losses in returned energy associated with significant flaws.

By far the commonest application of ultrasonics in NDT is the pulse-echo method, where a short ultrasonic pulse is propagated through the test material. The transmitter probe is also used as a receiver, and an oscilloscope is triggered as the pulse leaves the transmitter, thus allowing an estimate of the time for any echoes to return to the transmitter–receiver. As can be seen from Fig. 6.5, not only does the

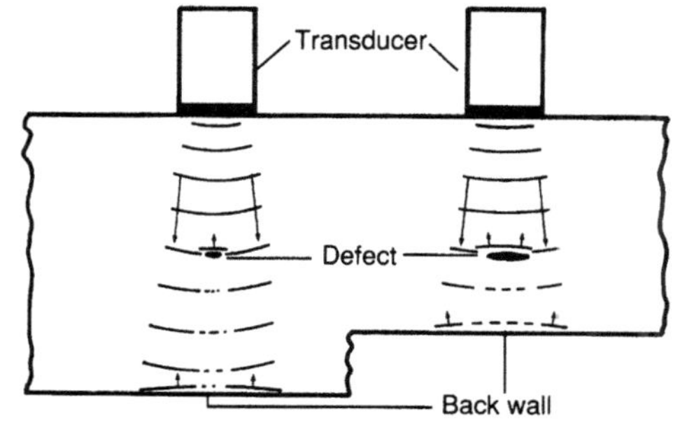

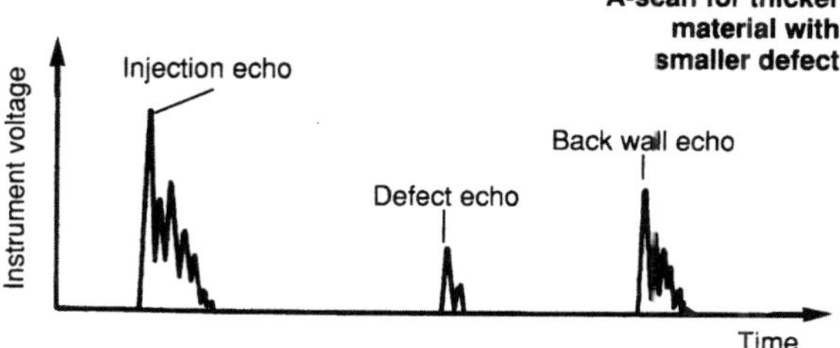

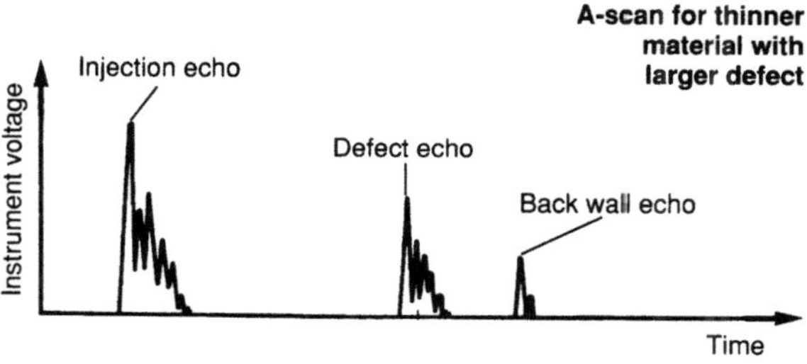

Fig. 6.5. Wall thickness and defect size effects in A-scan ultrasonic NDT.

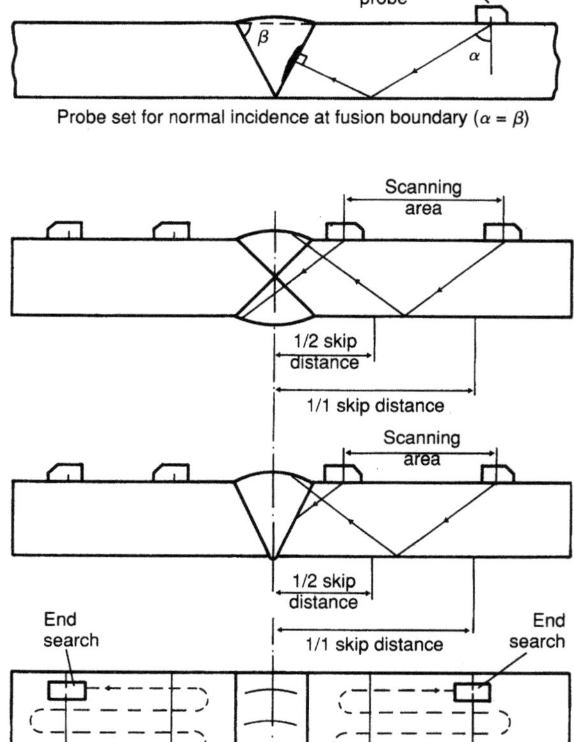

Fig. 6.6. Probe geometry and search strategy for the ultrasonic testing of V- and X-groove butt welds [5].

method give an estimate of flaw size and location, but it also gives an estimate of wall thickness, which can be very useful for corrosion monitoring as well as quality control.

Ultrasonic NDT is used widely in a semi-quantitative application to search for flaws in, for example, welded constructions for offshore use. Often, the presence and location of a flaw can be estimated with sufficient accuracy by qualitative examination of an oscilloscope screen. Nevertheless, specific search methods are required for specific defects and geometries, and, for example, DnV [5] recommends the search methods shown in Fig. 6.6 for lack of fusion defects in butt welds. Occasionally it is desirable to have a semi-quantitative estimate of flaw size and position, and methods exist to do this. DnV [5] describes a method where first the beam profile (to a specific attenuation, say 20 dB) is plotted using a calibrated flaw and then this information, coupled with probe movement to a 20-dB drop in flaw echo, is used to locate and size the defect.

'Acoustic emission' is the term used for stress waves which are generated as a result of events occurring inside the material. For example, plastic deformation and cracking processes generate acoustic emission, and a transducer placed on the surface of a structure should in principle be able to monitor such acoustic emission. Much work has been done on determining the signatures of acoustic emission (in either the time or the frequency domain) generated

by specific events, but one limitation is that the signals are often complex and not exactly reproducible, so that it can be difficult to separate noise associated with unwanted events (such as, for example, the cracking of surface oxide layers) from that emanating from the event to be monitored. For 'clean' signals, it is also possible to locate flaws by using transducer arrays.

6.2.4 Radiographic Methods

The relative absorption by different materials of high-energy electromagnetic radiation is a well-known method of body inspection and is typified by the medical X-ray. Both X-rays (wavelengths from 0.5 to 0.0004 nm) and γ-rays (wavelengths from 0.01 to 0.0005 nm) can be used for non-destructive testing, and it is often a question of convenience which to use. Of more interest than the wavelength is the energy of the radiation, because this in large part determines its penetrating power. The photon energy of electromagnetic radiation is given by

$$E = hf$$

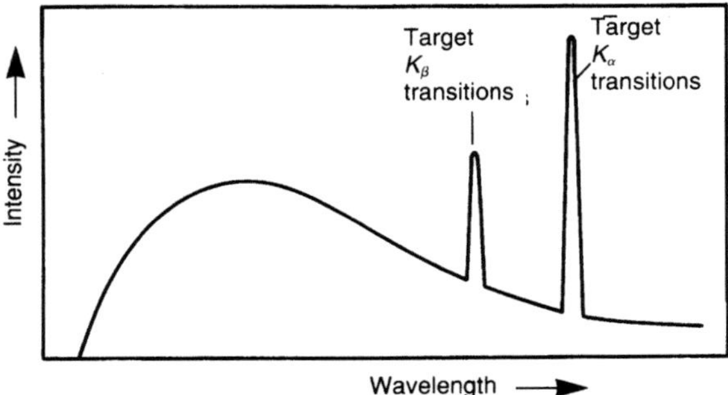

Fig. 6.7. X-radiation spectrum including white radiation and target characteristic frequencies. (From Hull and John [4].)

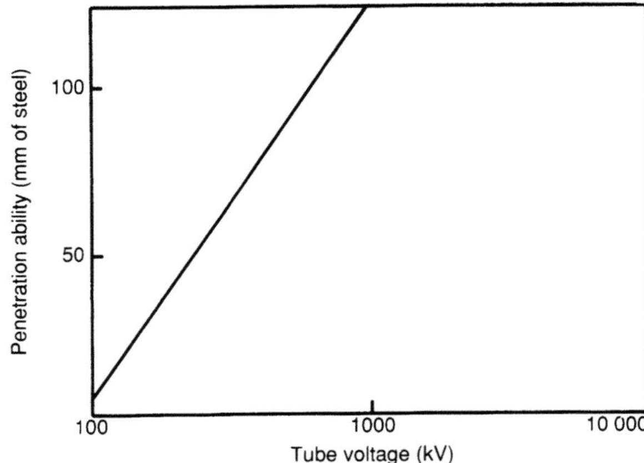

where f is the frequency and h is Planck's constant. The softest X-rays of wavelength about 10 nm have very little penetrating power, whereas hard X-rays of wavelength about 10^{-4} nm can penetrate several hundred millimetres of steel.

X-rays are generated by accelerating electrons towards a target material under very high voltages (usually 50 kV to 1 MV); the emitted X-rays are partly characteristic of the target material (owing to electronic transitions in the target) and partly 'white' radiation due to collisions between electrons and target atoms. One of the commonest target materials is tungsten, and this produces characteristic spikes in the X-ray spectrum (Fig. 6.7). These spikes are associated with the relaxation of atomically bound electrons in the target material after excitement to higher energy levels by the bombarding electrons. The spectrum is affected by tube voltage, but it is normal to consider an average penetrating power which is related to tube voltage, owing principally to the decrease in short wavelength limit with voltage. The effect of tube voltage on short wavelength limit and hence on penetrating power is illustrated in Fig. 6.8.

For the higher energies (short wavelength limit > 1 MeV) normal accelerating methods using thermionic cathodes are no longer possible, and the so-called 'high-energy sources' such as linear accelerators are required to provide a sufficient energy of bombarding electrons. This generally means that radiographic equipment becomes more expensive for thicker sections of materials, particularly metals.

γ-ray sources are radioactive materials which are continually producing radiation as part of the process of decay. Because of this, the intensity of emitted radiation decreases with time and this is described by:

$$I_t = I_o \exp(-kt)$$

where I_o and I_t are emitted intensities at the initial condition and at time, t, respectively. The constant k is related to the radioactive half-life of the material as follows:

$$k = \ln(2)/t_{1/2}$$

Each source material has its specific half-life (usually a few tens of days to a few tens of years) and photon energy, normally in the range 0.1 to 1.5 MeV.

Although X-ray and γ-ray absorption in materials is a complex phenomenon, much quantitative NDT can be done by assuming simple absorption according to the relationship:

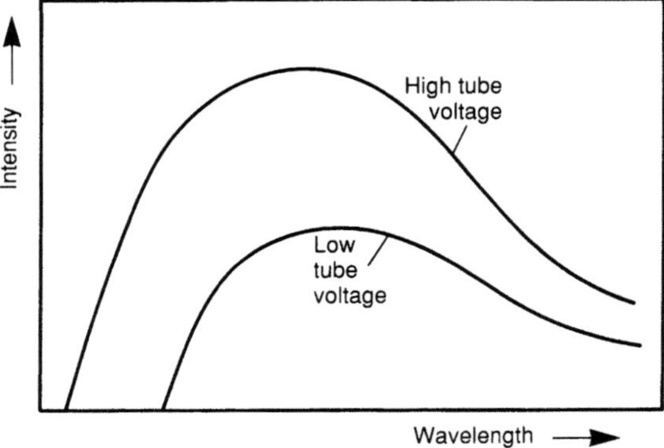

Fig. 6.8. Effect of tube voltage on X-radiation intensity, frequency distribution and penetrating power. (Adapted from Hull and John [4].)

$$I = I_o \exp(-\mu x)$$

where I_o is the incident intensity and I is the intensity at depth x into the material. The absorption coefficient, μ, is a function of radiation energy and also of material type, and this relationship can be quite complex.

Furthermore, the absorption process is not entirely permanent, in that some of the absorbed energy is re-emitted as scattered radiation. This means that the total radiation falling on a plate will be made up partly of transmitted beam (which has travelled in a straight path through the material and is hence image-forming) and partly of multiply scattered radiation (which is not image-forming). The ratio of total intensity recorded at the image $(I_d + I_s)$ (direct plus scattered radiation) to total direct intensity received at the image, I_d, is known as the 'build-up factor', and this can be measured by comparing broad- and narrow-beam experiments, the latter of which uses a collimator after the specimen to remove most of the scattered radiation.

Radiographic images are almost always recorded on film, and the behaviour of the film can be described by the relationship between photographic density, D, and exposure, E, defined as the product of radiation intensity and time. This relationship is normally described by the gradient of the curve of D against lg E, known as the 'film contrast' (Halmshaw [3]):

$$G_D = dD/d(\lg E)$$

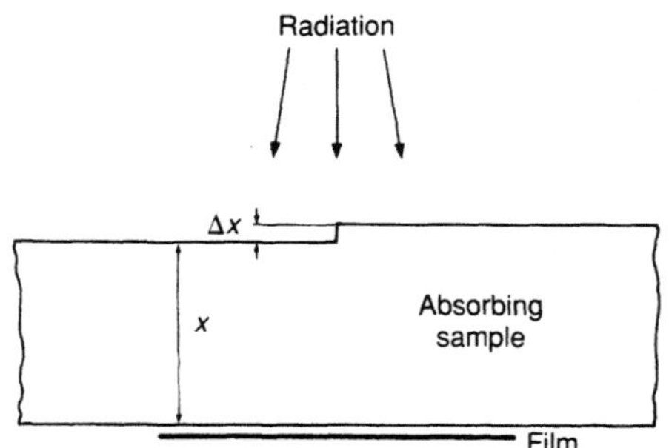

Fig. 6.9. Contrast sensitivity of a radiography procedure. (From Halmshaw [3].)

and this quantity is usually a function of photographic density. Considering a plate of stepped thickness as illustrated in Fig. 6.9, the step increases in thickness, Δx, will result in reduction in film density, ΔD [3]

$$\Delta x = \frac{2.3\,\Delta D\left(1 + I_s/I_d\right)}{\mu G_D}$$

so that the sensitivity of the proposed technique (including radiation conditions, material, film type and exposure) can be evaluated. The sensitivity can be expressed as the minimum value of step size, Δx_{min}, which can be detected for the smallest discernible density difference on the film ΔD_{min}:

$$\Delta x_{min}/x = \frac{2.3\Delta D_{min}\left(1 + I_s/I_d\right)}{\mu G_D x}$$

In practice, a combination of radiographic technique charts and the use of penetrameters is used to obtain suitable exposure conditions and sensitivity for the detection of

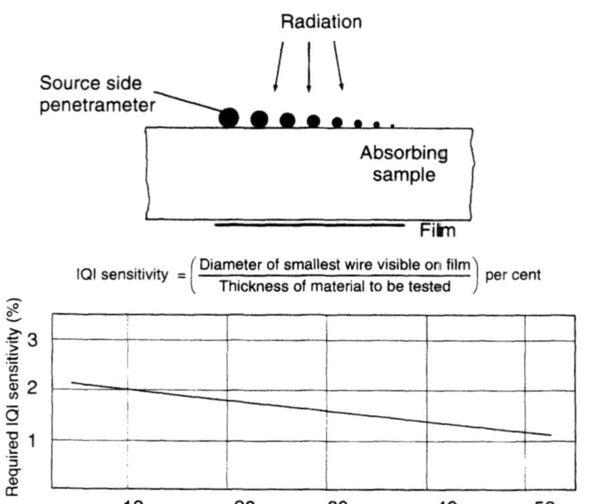

Fig. 6.10. Sensitivity required by DnV and method of measurement. (Adapted from Det norske Veritas [2].)

body defects. The method is illustrated in Fig. 6.10, which shows the required image quality indicator sensitivities for a range of material thickness suggested by DnV [2]. Such quality would be ensured beforehand by a knowledge of the source and film parameters indicated above.

6.2.5 Special Methods for Concrete and Composites

The science of NDT for other structural materials such as concrete and composites is not as advanced as it is for steel and other metallics. For composites, the main reason is that these materials are still finding novel applications and fabrication techniques where the types and the significance of defects are yet to be fully quantified. For concrete, the nature of the material means that design is conservative with regard to defects, so that those of significance are usually found with relatively simple techniques.

Apart from the corrosion of reinforcement (dealt with in Section 6.3.1), the NDT of concrete centres around the detection of fairly large surface features (such as spalling), and can usually be done visually. A number of instrumented methods do exist however, and these are discussed briefly below.

The rebound hammer measures the elastic properties of the surface concrete by measuring the degree of rebound of a known mass with known kinetic energy from the surface. In practice, the degree of energy absorption is quite difficult to relate to strength.

X-ray methods can locate voided or poorly compacted concrete, although implementation is rather inconvenient. Backscattered gamma-radiography can be related to concrete density up to 150 mm deep, but only gives a point reading and is therefore of limited use as a routine technique.

The 'covermeter' detects reinforcing bars or prestressing tendons by distortion of an applied magnetic field, and can therefore be used to give a measure of depth of reinforcement cover.

Finally, ultrasonic pulse velocity methods can be used to find voids and assess strength (Fig. 6.11).

The main defects of manufacture likely in composite materials are voids, foreign objects and fibre misalignment, bunching or absence. Those which may arise in service include matrix cracking, fibre–matrix debonding, delamination between plies and fibre breakage. The principal NDT methods are stiffness measurement, radiography, ultrasonics, acoustic emission and thermography [6].

The stiffness of a composite material is known from a combination of experience and laminate theory (see Chapter 2) so that, in principle, measurements of stiffness can be used to assess the state of a composite. Although a method like the rebound hammer could be used, stiffness is at present mostly used in the laboratory to measure damage accumulation in, for example, fatigue tests, although the nature of the damage is difficult to elucidate.

The usefulness of radiography in composites NDT depends somewhat upon the reinforcing material, graphite and aramid fibres being generally undetectable. Low-energy X-ray sources are usually required (15 to 25 kV), and even then some fibres and matrices are very nearly transparent. However, image-enhancing penetrants can be used to help detect surface-breaking defects.

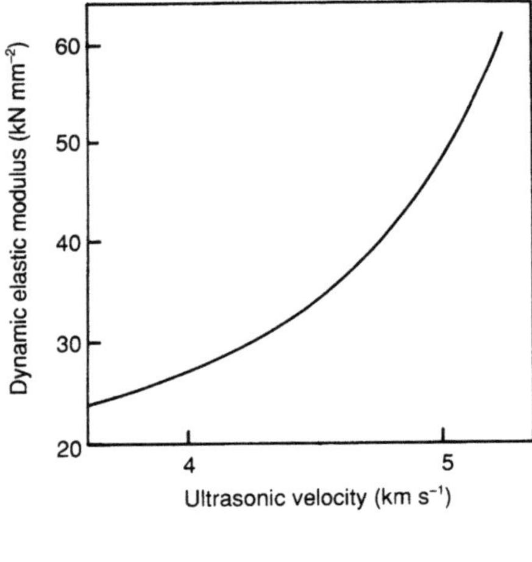
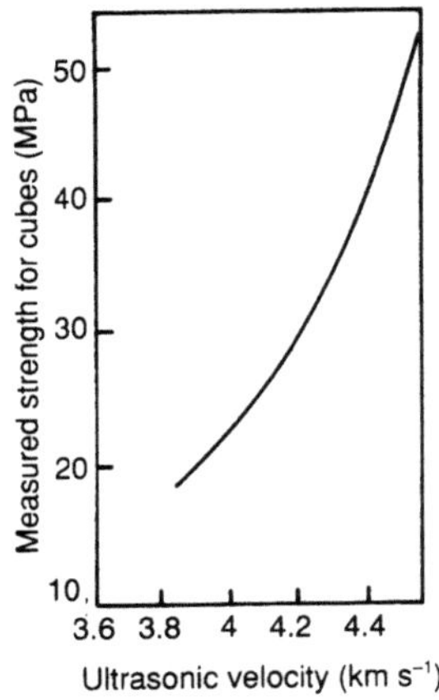

Fig. 6.11. Sensitivity of ultrasonic velocity in concrete to dynamic elastic modulus and strength. (From Halmshaw [3].)

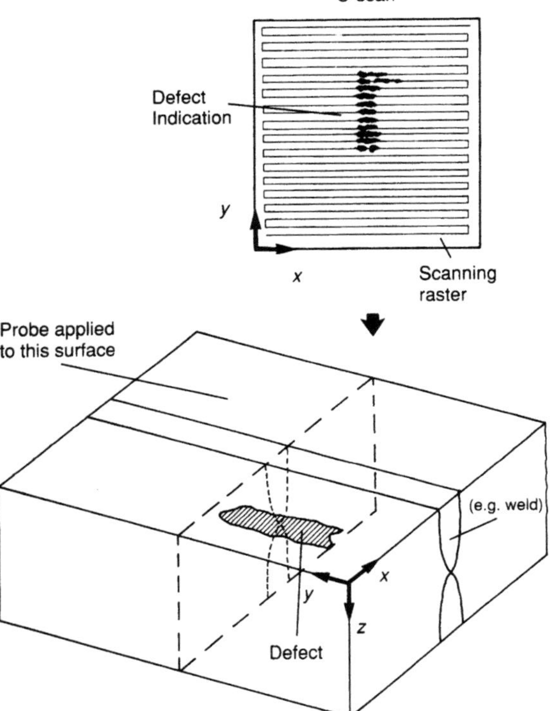

Fig. 6.12. C-scan display for ultrasonic NDT.

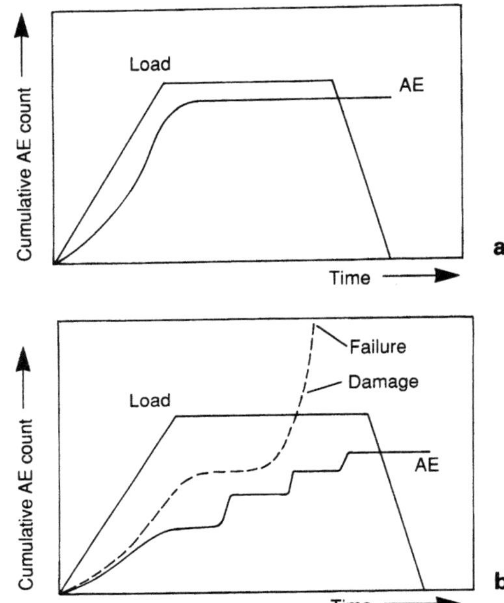

Fig. 6.13. Acoustic emission from FRP rocket cases under loading for a sound material and one of poor quality. a Sound material. b. Defective material (From Halmshaw [3].)

Ultrasonics are one of the most widely used NDT methods for composites. Although pulse-echo methods are helpful in locating discrete defects, the C-scan is widely used for factory quality control because of its ability to convey two-dimensional information (Fig. 6.12).

Acoustic emission (AE) is a powerful technique for damage assessment in composites, because many of the damage mechanisms are brittle and therefore produce strong AE signals. However, unless induced, for example during a proof test (Fig. 6.13), acoustic emission is really a monitoring technique apparent only while damage is taking place.

Most successful thermographic methods for composites consist of detecting differences in thermal conductivity caused by the existence of defects. Because of this, voids are probably most easily detected, and a typical arrangement for producing a thermographic image in a composite sheet is shown in Fig. 6.14.

6.2.6 Acceptable Defect Levels and Reliability of Techniques

There is little point in carrying out a programme of non-destructive testing unless clear acceptance levels are available and the method is sufficiently reliable to find unacceptable defects. This argument can be extended to planning of test programmes during service life, and this aspect is dealt with later.

All manufactured products will contain defects of some type, and it is essential that time and money are not wasted

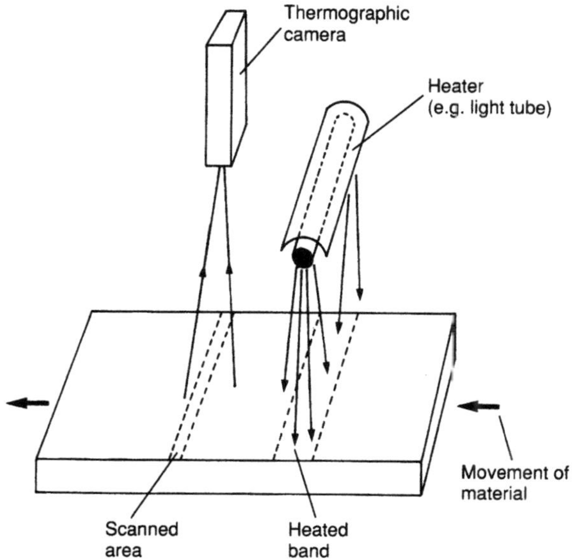

Fig. 6.14. Schematic of one method of thermographic NDT. (From Halmshaw [3].)

in searching for and repairing defects which do not significantly impair the integrity of the structure or component under investigation.

The reliability of inspection methods is a matter of some concern, since ultimately most techniques are subjective to some degree. A number of trials are reported (e.g. Nicholas [7]) for specific industries (particularly nuclear), and these normally take the form of round-robin tests followed by a comparison of success rates using different methods. Packman et al. [8] suggest four different statistical methods for presenting such data, each with its own strengths and weaknesses. Figure 6.15 shows one example of lower-bound probability data for detection of fatigue cracks in an aluminium alloy by liquid penetrant inspection, but the authors [8] warn against direct use of these data without careful analysis of the techniques used.

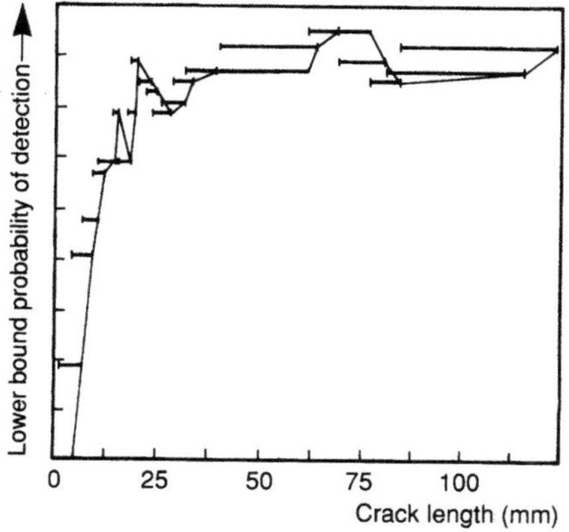

Fig. 6.15. Sample data illustrating the reliability of detection of fatigue cracks in an aluminium alloy using the dye penetrant method. Data plotted for lower bound probability of detection using the equal sample size method. (From Packman et al. [8].)

The foregoing applies only to whether or not a given technique will detect a flaw when one is present. In practice, there are four possible inspection results [8]:

Flaw detected when one is present;

Flaw undetected when one is present;

Flaw indicated when none is present;

No flaw detected when none is present.

Of course, only two of these are correct results, and the possible wrong decisions are that a component is rejected when in acceptable condition, or that a component is accepted when in an unacceptable condition. It can be argued that the latter may be dangerous, whereas the former merely leads to a financial penalty, but nevertheless it is often necessary to consider both by combining probabilities.

The question of reliability in underwater NDT is dealt with later in Section 6.4.2, but there are other major considerations in offshore and marine engineering, particularly in the pre-service inspection of welded constructions such as offshore structures and pipelines.

One example of the importance of inspection reliability assessment is given by Still and Rae [9], who have examined the quality control of submarine pipelines with regard to inspection of girth welds on the laybarge. They found, for example, that the practice of interpreting radiographs before these were dry led to an underestimate of defect frequency, and also that specific types of defects were more difficult to find on wet radiographs. Other conclusions were that lay rates could be substantially improved by the use of engineering critical assessment (see Case Study 7.1) instead of the prescribed standards for defect tolerance.

The above example illustrates a more general requirement that proper quality control involves an integrated approach, which includes an assessment of defect tolerance, fitness for purpose, service history, reliability and the programme of inspection.

Birchon [10] gives a method of quantifying these using the 'LEO' technique, where the 'L' is related to defect tolerance, the 'E' to service and inspection and the 'O' to inspection performance. One of the Birchon's examples relates to 76-mm-diameter cylinder head studs on marine diesel engines which were prone to corrosion fatigue failure. The tolerance parameter, L, was the maximum fatigue crack depth which could be withstood by the bolt in normal service, and this was found to be 63.5 mm from examination of a number of fracture surfaces. The derivation of E is somewhat more subjective, but Birchon suggests a set of rules based on the toughness of the material, the frequency of inspection and other complicating factors such as aggressive environments and degree of experience of defect type. Each of these complications results in a subtraction from the ideal value of E ($= 1$).

In the case in question, E is built up from the following subtracting factors:

Fracture toughness greater than 120 MN m$^{-3/2}$ in steel	-0.1
Infrequent inspection relative to loading frequency	-0.2
Previous experience (extensive)	-0
Duplex failure mode	-0.1
Multiple initiation	-0.1
Value of E	$1 - 0.5 = 0.5$

A similar set of factors was identified for the parameter O:

Prior experience	–0
Difficult access	–0.2
No available standard	–0.1
Poor surface finish	–0.1
Multiple echoes from stud geometry	–0.1
Value of O	$1 - 0.5 = 0.5$

Thus, the product LEO was found to be $63.5 \times 0.5 \times 0.5 = 16$ mm as the defect size at which the studs should be removed from service. Separate experiments showed that defects of a depth greater than 9.5 mm could be reliably detected in the studs using ultrasonic NDT, so that inspection *in situ* was recommended using ultrasonics as a method of rejecting studs at routine inspection prior to failure.

A more quantitative approach is possible using NDT reliability data and quantitative crack growth prediction methods to replace the factors E and O. This is dealt with in detail in Section 6.5, but an example of how the O-factor might be approached is given by Forli and Pettersen [11], who have made attempts to quantify NDT reliability. They address both the aspects of detection and characterisation, the latter of which includes sizing. Although the conclusions were limited, it would appear that radiographic interpretations were rather more accurate and consistent than ultrasonic ones, especially in the area of defect sizing. This latter problem has been addressed by Fortunko and Schramm [12], with particular reference to large-diameter (girth) welds for the detection of two-dimensional defects which might escape radiographic inspection due to their orientation (Fig. 6.16).

These authors recognised the quantitative difficulties in conventional ultrasonic NDT and proposed a modification using electromagnetic acoustic transducers operating at a relatively long wavelength.

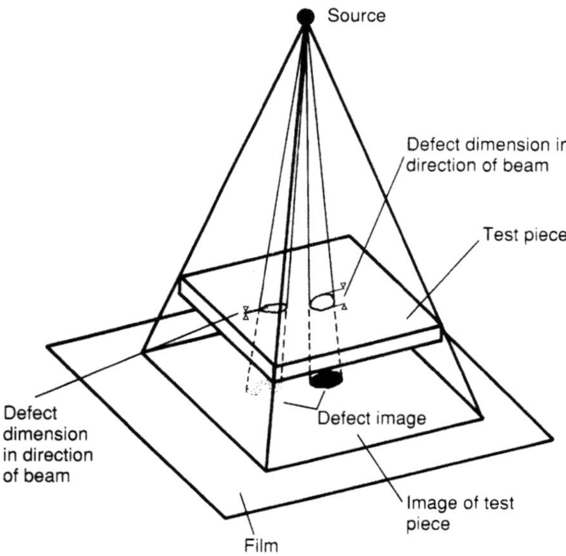

Fig. 6.16. Schematic diagram illustrating the detectability of flaws of different orientation using radiography. (Adapted from Hull and John [4].)

6.3 Monitoring

In its strictest sense, 'monitoring' means the continuous examination of the condition of a structure or component, although the term is sometimes used for examination which is only semi-continuous. Monitoring methods (usually in the semi-continuous sense) are most often used for corrosion, although there are some structural monitoring methods as well.

6.3.1 Corrosion Monitoring

The philosophy of corrosion monitoring originated in the process industry where corrosivities were often not well known, and it was necessary to have a method other than design corrosion allowance for ensuring reactor vessel integrity. Nowadays, corrosion monitoring is practised as a backup for systems where corrosion is expected to proceed at a slow rate or where inhibition is used. The most important area where corrosion monitoring is applied in marine engineering is in the process, production and transport streams of hydrocarbon production systems, as well as more generally in seawater handling systems.

Corrosion monitoring techniques either are usually aimed at measuring corrosion by electrochemical methods or are variants of the non-destructive test methods. As seen in Chapter 3, corrosion need not take the form of general loss of wall thickness, but a whole range of localised phenomena from pits to cracks are possible, and the choice of monitoring technique should be made in the light of a knowledge of the likely morphology. In addition, corrosion mechanisms involving the evolution of hydrogen can also be monitored by detecting the amount of hydrogen evolved on the surface of a test membrane (Yamakawa et al. [13]). The electrochemical methods vary from simple potential monitoring to polarisation resistance measurements made on probes of metal similar to the plant wall material. The commonest NDT methods used are ultrasonics, radiography and eddy current methods, although acoustic emission can be of particular use for stress corrosion cracking and damage involving cavitation. Finally, a range of simple analytical methods can be used, from corrosion coupons (which are pieces of plant material immersed in the process stream and removed periodically for weight loss measurement) to analysis of the process stream for dissolved metal ion, residual inhibitor concentration, pH, or oxygen concentration. A summary of the various techniques and applications is given in Table 6.4.

One example of a corrosion monitoring application has been published concerning a ship's seawater cooling system, including the condenser [14]. The copper-based alloys from which these components are normally made show good resistance to flowing clean seawater, but some difficulties may be encountered with polluted water, especially if this is stagnant. Such problems were normally encountered during prolonged fitting-out periods, and it was found that the use of polarisation monitoring probes helped to ensure an appropriate inhibition programme, which was otherwise difficult to predict owing to the variable conditions, thus avoiding, for a relatively small outlay, the danger of requiring a complete replacement of heat-exchanger tubes and seawater system components. However, such strategies are not always successful and Cameron and Coker [15] discuss an attempt to carry out

Table 6.4. Summary of some of the corrosion monitoring methods and their applications

Type	Method	Application
Electrochemical	*Linear polarisation resistance.* Corrosion rate is measured by the electrochemical polarisation resistance method with 2 or 3 electrode probes.	For most engineering metals and alloys in process fluids of suitable conductivity.
Electrochemical	*Potential monitoring.* Potential change of monitored metal or alloy (preferably plant) with respect to a reference electrode.	To indicate the corrosion state of the plant, e.g. active, passive, pitting, stress-corrosion-cracking as characterised by a specific potential region. Directly measures plant behaviour.
NDT	*Ultrasonics.* Thickness of metal and presence of cracks, pits etc. by changes in response to ultrasonic waves.	Generally used as an inspection tool for metal thickness.
NDT	*Radiography.* Flaws and cracks by penetration of radiation and detection on film.	Not widely used in corrosion monitoring.
NDT	*Eddy current.* Uses an electromagnetic probe to scan surface.	Detects surface defects such as cracks and pits.
NDT	*Acoustic emission.* (a) Leaks, collapse of cavitation bubbles, vibration level in equipment. (b) Cracks by detection of the sound emitted during their propagation.	(a) Leak detection and the possibility of cavitation erosion, fretting and corrosion fatigue. (b) Stress corrosion cracking and fatigue cracking in vessels and lines. Not strictly a monitoring tool and currently still a new technique.
Analytical	*Corrosion coupon testing.* Average corrosion rate over a known exposure period by weight loss or weight gain.	Very satisfactory when corrosion is a steady rate. Useful in hazardous areas where electrical devices prohibited. Moderately cheap corrosion monitoring. Indicates corrosion type.
Analytical	*Fluid analysis.* (a) Hydrogen probe used to measure hydrogen gas liberated by corrosion. (b) Concentration of the corroded metal ions or concentration of inhibitor. (c) pH of process stream. (d) Oxygen concentration in process stream.	(a) Typically in petrochemical industry and involving mild steel corrosion in sulphide. (b) Can be used to identify specifically corroding equipment. (c) To monitor change of pH, such as in effluent streams where acidic conditions can cause severe corrosion. (d) Control of oxygen level, usually through oxygen getters to mitigate corrosion.

similar inhibitor optimisation using field electrochemical monitoring of corrosion rate in oil wells, but found difficulties in making reliable long-term measurements using linear polarisation resistance (LPR).

Because of their critical nature and difficulty of maintenance, much effort has been expended in the corrosion monitoring of subsea pipelines. In general, the 'belt and braces' approach of coating and cathodic protection means that little concern is felt over the external surface; in any case, access to the metallic surface for monitoring would represent an unacceptable breach of the integrity of the pipeline. Although hydrocarbons are dried and inhibited prior to transport and the corrosion risk is very low, a number of investigators have proposed methods of internal cor-

rosion monitoring. Britton [16] has discussed the current methods used for corrosion monitoring of subsea pipelines and offshore topsides, and lists coupons, electrical resistance probes, LPR probes, coupon-deposit analysis, analysis for iron in produced fluids, ultrasonics and 'intelligent vehicles'. These latter are adaptations of the 'pig', a vehicle which travels along the inside of a pipeline, usually driven by a differential pressure, the adaptation being that this device is capable of making measurements of pipe wall thickness. De Raad [17] has compared two different types of such inspection pigs which use ultrasonics and magnetic flux respectively, and lists the relative merits of each. Such vehicles are in a stage of active development, and at least one pipeline operator [18] has concluded that they provide an economic and effective alternative to periodic pressure testing for pipeline condition monitoring.

Although it is not strictly a method of corrosion monitoring, the monitoring of cathodic protection (CP) systems is of some importance for submerged structures. As opposed to CP inspection (discussed in Section 6.4.2), CP monitoring involves the installation of permanent sensors (usually half-cells for potential measurement) with facilities for data transmission to the surface. The principal problem would appear to be in the telemetry aspect, since current methods are prone to operational or weather damage, although it may be possible to overcome this with through-water transmission methods.

6.3.2   Structural Monitoring

The monitoring of structural integrity is a wide subject, which can include structural inspection on a batch basis. Since this latter is dealt with in Section 6.4.2, this section is confined to structural monitoring on a continuous basis. For offshore and marine structures, the main purpose of structural monitoring is to detect fatigue cracking at its early stages, and this is a difficult matter because it requires some knowledge of where such cracks are likely to occur (because of the extent of typical structures).

Since offshore structures contain a high degree of redundancy, there is scope for monitoring methods which might detect the redistribution of load resulting from a minor member failure, or even the failure event itself.

The various monitoring methods possible for offshore structural monitoring have been listed by the Marine Technology Directorate [19]:

Vibration monitoring.

Acoustic emission monitoring.

Monitoring of internal pressures in steel tubulars.

Witness devices.

Monitoring of performance in comparison with design predictions.

Monitoring for member flooding.

Crack propagation monitoring.

Vibration analysis for offshore structural integrity monitoring is well documented (e.g. Begg et al. [20]). Briefly, the spectral response of the structure to mechanical excitation will be a function of the condition of the structure, so that the loss of a member or a change in compliance can be detected in this spectrum. The sensitivity of the technique is related to the design efficiency of the structure itself, more efficient structures showing larger changes of stiffness when a member is lost or damaged.

Usually several transducers are used, and these can be above or below the water line. The most widely used application employs ambient excitation from wave action, although it is possible to use local or global excitation at a selected range of frequencies using electromagnetic or hydraulic impactors or vibrators.

Acoustic emission, as indicated in Section 6.2.3, is the monitoring equivalent of ultrasonic NDT, using transducers to detect internally produced stress waves rather than detecting echoes of pulses produced at the transducer. The main interest in acoustic emission for offshore structural monitoring has centred around its ability to detect cracking, and it is the propagation of the crack rather than its

presence which generates AE. The major difficulties with AE monitoring lie in the question as to whether a detectable level of signal is given out during fatigue crack propagation, and whether this can be separated from other emission sources. Nevertheless, such a method can form a useful part of a multi-sensor monitoring programme.

If structural members were pre-pressurised at fabrication and sealed, it would be possible to use internal pressure as a monitoring technique to detect through-wall cracking. In principle, such a method is not difficult, but would need to be envisaged at the design process.

A number of possible witness devices have been proposed over the years as having potential for use as detectors of surface cracking. The use of brittle materials bonded to the surface provides the possibility of continuity monitoring for surface crack detection, and this is extended in the idea of optical fibre devices (e.g. Hockenhull et al. [21]), which can be specially treated and, in conjunction with the use of an appropriate adhesive, can be induced to fracture at predetermined levels of surface strain.

Strain gauges represent another type of surface device which can in principle be used to gauge structural performance in comparison with that expected from the design of the structure or with that measured at installation. Other possible performance indicators might be inclinometers for tilt and settlement measurements.

The monitoring of selected members for flooding can give an indication of through-wall penetration in much the same way as pressure monitoring mentioned just above.

Crack growth monitoring can be useful for the stage between a defect discovery and repair. The present course is normally to monitor known defects carefully at planned inspections until remedial action can be taken, but the installation of a monitoring device at such known defect locations is attractive in terms of saved diver time and the added security of continuous knowledge of the condition of the structure. The most suitable method for doing this could be chosen from the monitoring methods described above or even from some of the NDT methods, such as ultrasonic probes fixed to the defective member.

6.4   Underwater Intervention, Inspection and Repair

For all fixed marine structures, and many mobile ones, it is essential to be able to carry out inspections in the underwater environment. The ease and feasibility of this depend upon an understanding of the limitations of intervention and the operability of the inspection methods underwater. If defects are found then the additional problems of underwater repair need to be addressed. This section examines in turn ways of achieving underwater intervention, the application of inspection methods underwater, and finally a brief discussion of suitable underwater repair methods.

The need for underwater inspection is coincident with the need to avoid failure, and for offshore structures and ships this need is usually embodied in a legislative requirement to demonstrate that a structure or ship is fit for its purpose. Such legislative requirements usually include the systematic inspection of the structure at predetermined intervals.

A typical requirement for fixed offshore structures would indicate that the following should be carried out every five years:

Visual examination of the entire structure above and below the water level.

Detailed visual examination of selected welds and of recorded repairs.

Additional close inspection and provision of further information required during survey.

Seabed-level inspection including scour survey.

Survey and inspection of cathodic protection system.

This would normally be met by an annual programme whose summation amounted to the statutory requirement every five years. A typical annual programme might include the following:

Complete general visual survey of structure and risers.

Selection of CP readings covering about 10 per cent of anodes.

Close visual survey of 10 per cent of nodal joints, including all critical nodes every year; this may lead to NDT of some areas as necessary.

Preparation of a scour map.

Visual survey of pipelines and riser connections to 20 m from platform perimeter.

Of course, such inspection is pointless without some previous knowledge of what is being sought. As an example, one of the certifying authorities (DnV [22]) have listed 25 'significant areas of inspection' along with the possible defects which may arise, the possible reason for such defects and the possible consequences for structural performance. One example of such a significant area is one where a repair has previously been carried out. DnV list possible defects which may be found there as: failure of repair material, rebar corrosion (in concrete structures only), cracking and weld failure (in steel structures only). Possible reasons listed are: poor workmanship, unsuitable materials or repair procedure, shrinkage (concrete) and unforeseen stress concentrations. Finally, the envisaged consequences are: accelerated deterioration through corrosion, reduction in load-bearing capacity and leakage.

Other 'significant' areas include (not in order of importance):

General inspection of structure.

Areas of inferior construction.

Areas of high stress.

Areas of cyclic stress variation.

Welded joints with steel thickness greater than 50 mm.

Cut-outs and penetration.

Bolted connections.

Structural members in compression.

Structure foundation including piles and any anti-scour system.

Corrosion protection system.

Areas for potential measurement.

Areas for material thickness measurement.

Areas with signs of corrosion.

Construction joints.

Splash zone.

Areas of mechanical damage (abrasion or collision).

Embedded steel parts in concrete.

In the main, it is normally expected that an inspector is looking principally for cracks, corrosion and signs of direct mechanical damage. A recent survey [23] has examined 61 incidences of repair to offshore structures from 15 operators. The main findings were that dents and cracks were the commonest types of defect, and that the commonest causes of defects requiring repair were collision and fatigue. The type of repair most often used was welding (air or underwater), followed by friction clamps. One case study reported in this review concerned the routine inspection of a horizontal tubular steel conductor guide frame at 2.3 m above LAT. The inspection had been carried out four years after installation and the frame was found to have severe cracks at its connections to the main structure. After attempting local repair and finding the cracks to continue to grow, the frame was eventually replaced by one of different design, once it had been recognised that an unforeseen vertical fatigue loading was acting on the frame.

6.4.1 Underwater Intervention

Like maintenance, repair and installation, underwater inspection requires the choice of an underwater work system, but inspection differs in that the type of work is light but requires a high degree of mobility, unlike, for example, an underwater weld as described in Chapter 5.

Underwater work systems can be classified as to whether they are manned or unmanned; hyperbaric or monobaric; and tethered or untethered. For example, a submarine is a manned, monobaric, untethered work system, whereas a diving bell with gas supplied from the surface is a manned, hyperbaric, tethered work system. At this point it is worth making the distinction between a person who operates underwater at elevated (or ambient) pressure, a diver, and one who works underwater at normal atmospheric pressure.

Each type of work system has its own limitations and advantages.

Manned systems have the advantage that the judgement and, to a large extent, the direct feedback of having a human being at the worksite cannot be simulated. The principal disadvantage is that there is always a risk to human life and health.

Hyperbaric systems have the advantage that pressure differences between worker and work are small and the human being can have direct access to the worksite, making full use of their versatility and dexterity. The principal disadvantage is that there are physiological limitations to endurance in time and depth.

Tethered systems allow for increased endurance in that life support and power can be 'piped' to the worksite. The principal disadvantages are the risk of entanglement and the reduced mobility that this brings, along with the limitation of fluid loading on very long umbilicals.

For the purposes of engineering (as opposed to sport), underwater intervention is usually brought about using either; manned hyperbaric systems which are always tethered; manned monobaric systems which may be tethered, or free-swimming; or unmanned systems which are, of course, always hyperbaric and may be free-swimming but are most often tethered. These will be discussed briefly below.

6.4.1.1 Diver Work Systems

Physiologically speaking, diving can be carried out in three modes, depending upon the breathing gas used and the type of decompression practice employed. These, in turn, depend upon the depth of the dive and the time required at the worksite and are known, respectively, as: air diving, mixed gas bounce diving and saturation diving.

As the pressure of the inhaled gas (i.e. depth) increases, a number of physiological factors become important, and these limit the type of work that can be carried out using a specific type of equipment.

Inert-gas narcosis occurs owing to the breathing of high-pressure inert gas. The effect is variable, in that different pressures may be required to produce an effect for different individuals, but also the effect may vary for the same individual from time to time. The phenomenon has been called 'rapture of the deep', as it may produce a dangerous lack of concentration and even mis-orientation. The effect is worst for nitrogen, and much higher partial pressures are required to produce narcosis with helium.

Oxygen becomes toxic at excessive pressures, so that there is a limitation in the use of normoxic air. For bottled supplies it is of course important to ensure that a mixture is breathable at depth and also at atmospheric pressure.

As depth and time at depth increase, the amount of diluent gas (i.e. not oxygen) dissolved in the blood and tissue increases, thus incurring a decompression debt. This means that a diver needs to be brought to atmospheric pressure in a series of stages to allow dissolved gas to evolve at a moderate rate avoiding decompression sickness (or bends). In general, the more severe the exposure, in depth and time, the longer the decompression period required. At some depths, the decompression time can far outweigh the allowed bottom time, so that it eventually becomes economic (of time) to use saturation diving. In this case, divers are allowed to saturate (with respect to dissolved gas) at bottom pressure, and are kept at this pressure in a chamber on the deck of the dive support platform. Transfer to and from the work site is also done under pressure by locking and unlocking the submersible bell onto the deck chamber. Typical systems for doing this are illustrated in Haux [24].

In practice, diving can be carried out using compressed air at depths of up to 50 m, and decompression is usually carried out in a chamber on the surface. Transfer is usually effected using an open cage to help divers through the splash zone, and the time from the worksite to the surface chamber for decompression has to be strictly controlled to avoid the occurrence of bends. Light exposures do not generally require decompression, and these are detailed in appropriate tables.

For deeper dives, it is necessary to use mixed gas, usually a mixture of helium and oxygen. Because of the rapidly increasing decompression penalty with depth (30 min at 75 m requires 2 h decompression and 30 min at 137 m requires 12 h decompression), so-called 'bounce diving' is relatively uncommon, it being much more economic to use saturation, keeping the divers at bottom pressure for the duration of the job, bringing them to a surface compression chamber between shifts and decompressing after the work is completed.

The absolute physiological limit for diving is unknown, although divers have been experimentally exposed to depths of 700 m and more, and some work has been done using hydrogen rather than helium as a diluent. Commercial diving can certainly be carried out to 350 m, but there has always been some uncertainty as to the long-term consequences of exposure to high pressures (even under appropriate conditions of decompression). This, along with the expense of diving, has meant that the drive continues to replace human beings at the underwater worksite as far as is possible.

Figure 6.17 summarises the different types of commercial diving systems in a schematic form.

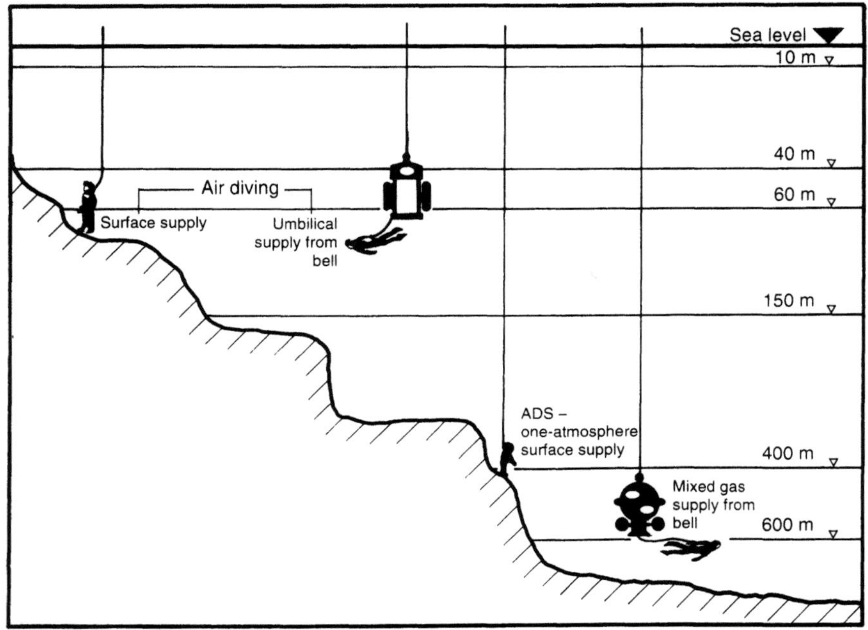

Fig. 6.17. The deployment and range of some commercial diving systems. (Adapted from Haux [24].)

6.4.1.2 One-atmosphere Systems

One-atmosphere systems are invariably manned, because this is the only reason for making them monobaric. Such systems are all variants of submarines but are, of course, much smaller and usually have some type of manipulating capacity along with improved visibility. One-atmosphere systems can be generally categorised as:

Observation–manipulator bells.

Atmospheric 'diving' suits.

Tethered one- or two-man submersibles.

Figure 6.18 illustrates these.

Atmospheric diving suits (ADSs) are so called because they are recognisably human-shaped. The first such suit was developed by Lethbridge in 1715, and led to successful dives to 20 m. The major difficulty in the design of such suits was in the provision of articulated joints which can operate at large pressure differences. Neufeld and Kuhnke solved this problem to a large extent in 1913, allowing dives to 125 m. Modern-day ADSs are either fully or semi-anthropomorphic (in the second case articulated legs are replaced by thrusters), and limitations (about 620 m) are principally due to the pressure hull, although this promises to be overcome using CFRP hulls [25].

Tethered one-man submersibles are an extension of the ADS principle and are for bottom work. Thrusters replace the ADS articulated legs, and manipulators replace the articulated arms; the pilot is in a prone rather than an upright position, and the depth limitations are similar to ADSs.

Figure 6.19 illustrates the three morphologies.

Observation–manipulation bells are simply a one-atmospheric extension of the diving bell designed to be lowered to the worksite. The first important systems of this type were the bathyspheres of the 1930s, which were used to go as deep as 1000 m for scientific observation. The modern extensions of this concept usually have some sort of manipulation capability, and typical depth limitations are 1500 m with manipulation and 5500 m without manipulation.

6.4.1.3 Remotely Operated Vehicles

Unmanned systems (ROVs) are a rapidly and continually developing area of underwater technology, and are clearly

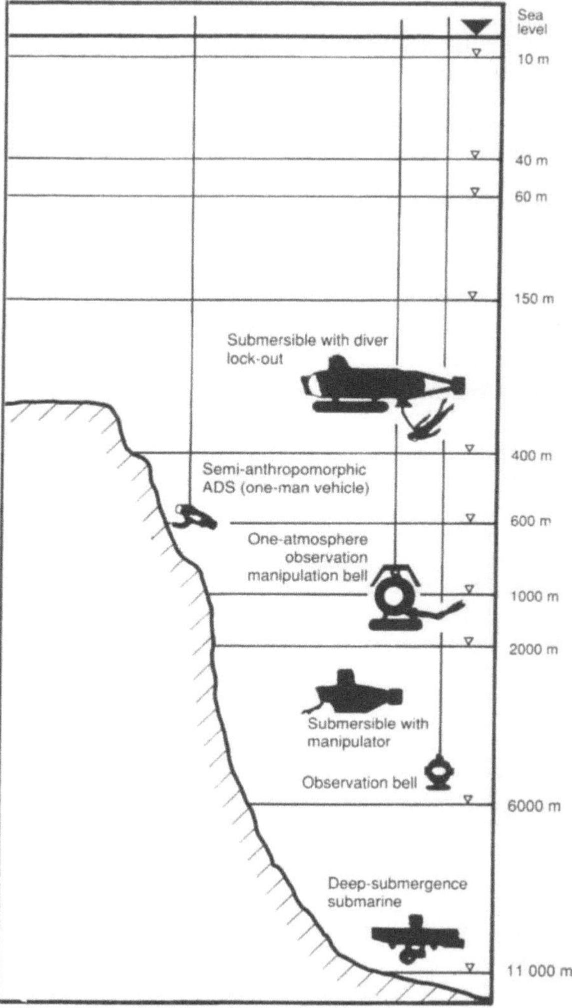

Fig. 6.18. One-atmosphere subsea intervention systems and their range of applicability. (Adapted from Haux [24].)

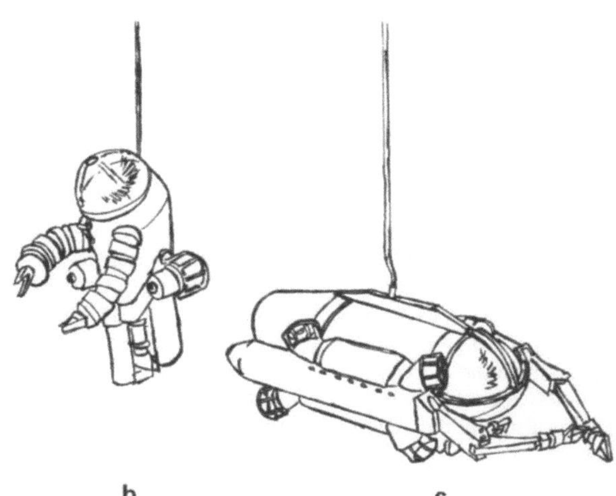

Fig. 6.19. The atmospheric diving suit (ADS) and its variants for one-man, one-atmosphere tethered submersibles. **a** Fully anthropomorphic. **b** Semi-anthropomorphic. **c** Non-anthropomorphic. (From Melegari [25].)

a

b

c

the preferred way of carrying out underwater work, provided that there is sufficient sentience and dexterity to perform the task in hand.

Almost all ROVs are tethered in some way or other, and are often launched from a submersible 'garage'. They are operated by a pilot on the surface, mainly using visual feedback through CCTV cameras, although some developments are being made in force feedback for manipulators.

A convenient way of classifying unmanned subsea work systems is as follows:

Observation–documentation systems.

Observation–manipulation systems.

Towed systems.

Seabed systems.

In the first category are the 'eyeball' ROVs, which are virtually supplanting divers for routine inspection tasks. CCTV is the basis of the usefulness of such vehicles, and they can be regarded as remote-controlled TV cameras. These ROVs are highly mobile, and tend to be of simple hydrodynamic shapes such as spheres.

The observational/manipulative submersibles vary in sophistication, from small, limited-work vehicles with single simple-function (i.e. extend–open–rotate) manipulators to full-work vehicles with two manipulators and instrumentation packages. CCTV and lighting are normal additional functions.

Towed systems are more often used in surveying than in inspection, and a typical application is in seismic surveying. Functions are often very simple.

Seabed systems can be used for trenching and burial operations. Some of the specialist pipeline follower vehicles can also be regarded as belonging in this category.

Figure 6.20 illustrates some of the possible types of ROVs.

6.4.1.4 Underwater Operability and Dexterity

From the point of view of control, it is instructive to examine the different interfaces which come between the human and the task to be performed. These barriers to effective working are illustrated schematically in Fig. 6.21 for the general cases of diving, manned vehicles and ROVs [26].

In the (simplest) case of the diver, sensory information from the work enters the diver's 'input interface' (e.g. work appears through the diver's eyes). This information is perceived, knowledge is gained, decisions are taken and commands are sent to muscles, all within the human brain. The output is transferred through the 'output interface' (i.e. the force from a hand) to the work; as a result there occurs, for example, a change in appearance and the cycle continues.

In the case of the manned vehicle (such as an ADS), the input interface is still relatively direct and similar to a normal diving suit. However, in this case, the output from the diver's muscles is an action by the pilot on the machine (e.g. operating the manipulator) and the machine then operates on the work.

In the case of a remotely operated vehicle, the input interface is also modified. Here, the information from the work is captured by sensors on the vehicle, and this is converted to some type of information display (e.g. CCTV), which is

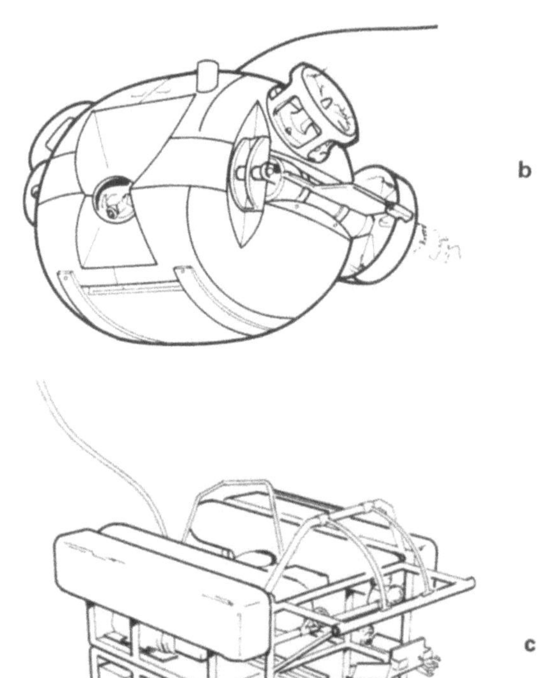

Fig. 6.20. Three of the wide range of remotely operated vehicles (ROVs). **a** Pipeline inspection ROV. **b** 'Flying eyeball' (observation only) ROV. **c** Observation manipulation ROV. (After manufacturers' and operators' literature.)

then captured by the pilot, the remainder of the control loop being similar to the case of the manned vehicle.

It can therefore be seen that the ability of an ROV to perform a piece of work depends upon how well the input and output interfaces can and need to be simulated, and for this reason many purely observational tasks can readily be taken by ROVs.

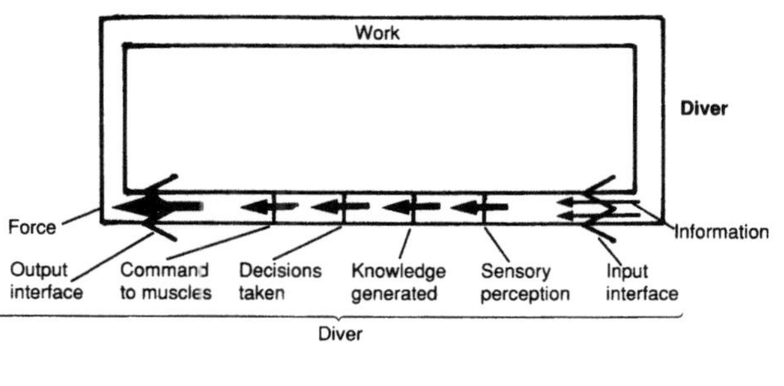

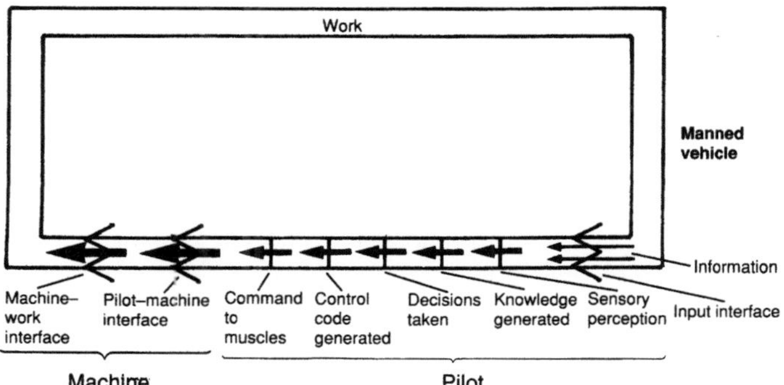

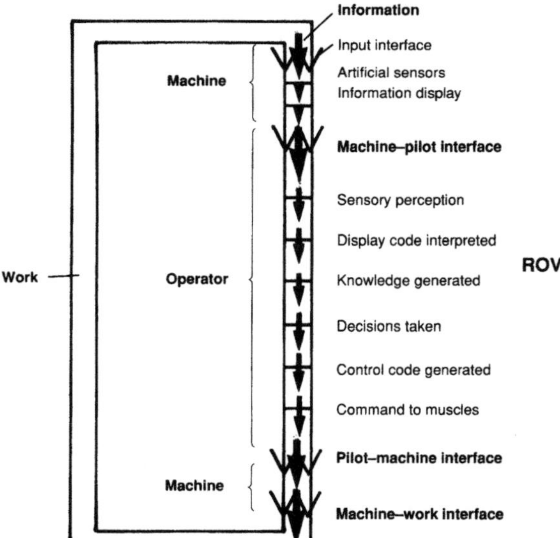

Fig. 6.21. Schematic view of the barriers to effective working in the subsea environment. (From Balch [26].)

6.4.2 Underwater Inspection Methods

The principal difference between underwater and land-based inspection is that the former is made difficult by the presence of water, particularly from the point of view of human access to the work site, but also because of the effect which such an environment has on the operation of inspection techniques.

As mentioned above, underwater inspection is most commonly applied on a routine basis to fixed, offshore structures, and most countries have a statutory requirement for such inspection. As an example of the type of inspection that is required, the certifying authority DnV [22] recognises three levels of in-service underwater inspection for offshore structures:

Type I. General visual inspection to detect any obvious damage. This type of inspection does not require any prior cleaning.

Type II. Close visual inspection to detect less obvious damage. This normally requires prior cleaning of the area to be inspected.

Type III. More detailed inspection, usually involving NDT methods to detect incipient damage, again requiring prior cleaning.

From this and the typical five-year programme described briefly above, two facts are immediately obvious: much of the underwater inspection required is simply of a visual nature, not requiring prior cleaning (and can therefore be carried out using ROVs), and any detailed inspection requires prior cleaning, but this is only carried out if there is a specific reason for doing so.

Since so much underwater inspection is of a visual nature, it is worth considering a few aspects of underwater visibility. The two most important factors to consider are refraction and absorption.

Refraction, as indicated in Section 6.2.3, is described by Snell's law, which in the case of light is conveniently written as follows:

$$\mu_2 \sin \phi = \mu_1 \sin \theta$$

where μ_1 and μ_2 are the refractive indices of the two media at the boundary, and the angles are those of incidence and refraction respectively.

Because of the relative refractive indices of air, water and glass, a diver looking through a face mask will have a reduced field of vision, and objects will appear to be closer (Fig. 6.22). This has some consequences for dexterity and may be important for quantitative photographic work.

The absorption of light by water may be many times that by air and the following consequences arise:

Daylight will only penetrate to a certain depth.

Artificial lights will only illuminate objects at a certain distance.

Some wavelengths are absorbed more than others, so that colour distortion is encountered.

Scattering of light results in less contrast and blurring of edges, as in objects viewed through a misty atmosphere.

Such effects need to be borne in mind for planning of underwater work. In particular, some seasons of the year are associated with poor underwater visibility due to algal blooms, and disturbance of the bottom by currents, tides or the presence of divers may make visibility difficult.

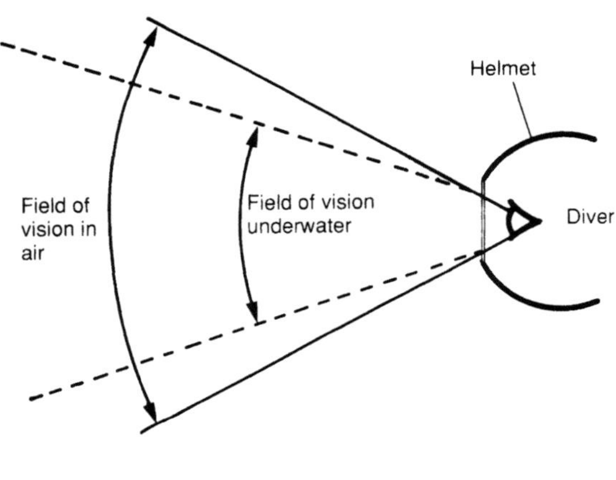

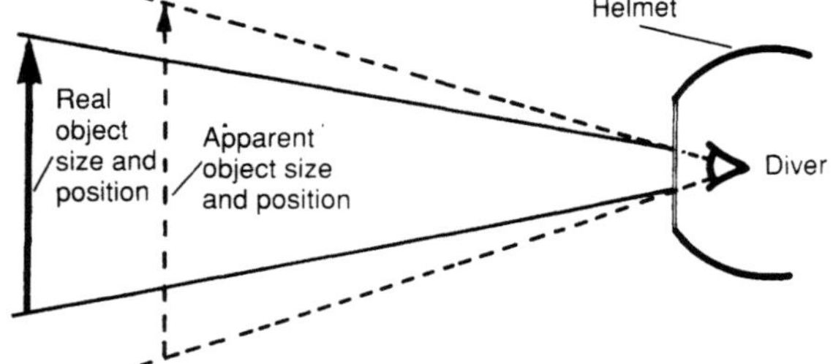

Fig. 6.22. Effects of refraction on diver perception underwater.

Much recent work has been done on the enhancement of field of vision and depth perception using appropriately angled cameras. For example, the use of photogrammetry for accurate underwater surveying using still photographs has been proposed on a number of occasions (e.g. Turner [27]), and it is recognised that the use of a number of CCTV cameras on ROVs enhances pilot orientation and perception.

As mentioned above (p. 190), more detailed inspection requires cleaning to remove any adherent scale or fouling which might otherwise obscure possible defects. Methods of cleaning vary, and are largely dependent upon the type of deposit to be removed, although it is generally recognised that hard, calcareous growths are the most difficult to remove. Methods used include wire brushes, scrapers and chisels, which may be hand- or power-operated. Water jets with and without entrained solid particles are also widely used. Great care has to be taken, especially with the more aggressive cleaning methods, that the underlying structural material is not damaged and possible defects masked. Typical cleaning rates for the removal of fouling with water jets are in the region of 10 to 15 m²/h, and rotary brushes or cutters remove hard fouling at about 30 m²/h [19], so that the cleaning of large areas such as ships' hulls and offshore structures is a time-consuming and (especially if carried out underwater) costly task.

The NDT methods available for use underwater are largely modified from the above-water methods, although some are not applicable. Sisman [28] lists MPI, ultrasonics and radiography as the principal underwater NDT methods, although radiography is rather more difficult to use and is relatively uncommon, except, for example, for repair welds where water has been displaced.

Probably the dominant NDT technique used underwater is MPI, where modified magnetisation equipment is used with fluorescent inks which are sensitive to ultra-violet light (which is provided from a source carried by the diver).

Ultrasonic methods are again modified, but there is no need for couplant, because water is such a good transmitter. Ultrasonics are most commonly used for thickness measurement, but the use of shear waves for defect sizing and time-of-flight methods has also been documented [19].

As mentioned above, most underwater radiography is carried out in dry habitats, the in-water applications being confined to pipe-like geometries. Real-time radiography, where the image is shown on a fluorescing screen, and source and screen can be moved around, offers an attractive possible future application.

Eddy current testing and a.c. potential drop have also found some underwater application.

The major concern over underwater NDT is its reliability. In other words, given the presence of a defect, it is important to know how likely it is that this will be found during a routine inspection. Furthermore, given the expense of underwater repair, it is also very important to minimise the number of spurious indications and also the number of repairs of defects, which are, in fact, benign. A number of studies of NDT reliability have been carried out for underwater applications. One of these [29] compares results from four separate trials using MPI, ultrasonics and eddy current testing. The conclusions were quite complex, but it would appear that, among other things, only an MPI apparatus gave as good performance underwater as above water. Some of the other sets were equally good at indicating the presence of defects, albeit under the relatively ideal condi-

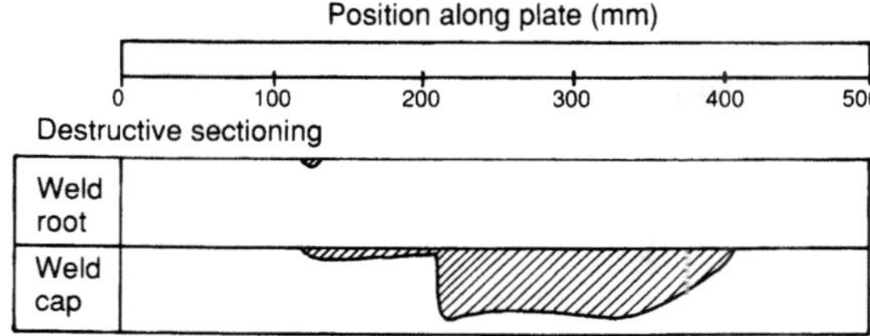

Fig. 6.23. Comparison of defect measurements made with above-water and underwater test methods under controlled conditions. Destructive sectioning (top) gives 'dead reckoning' of defect sizes in sample and tests 1 to 7 are made with different MPI sets (From Department of Energy [29].)

tions of clear, shallow water and no fouling, but their measurement of length and position were less good. A typical set of results for a specific defect is shown in Fig. 6.23.

Underwater inspection of concrete is largely confined to the visual, and is particularly aimed at the detection of areas of poor surface quality. The main problem here is one of classification of damage and the assessment of its significance. For instance, surface spalling may be associated with poor construction, reinforcement corrosion, alkali aggregate reaction or all or none of these. The significance of surface cracks is also difficult to assess, especially where this may lead to water penetration. Some further details of this are given in [30].

Steel within concrete may become susceptible to corrosion, although for reasons outlined in Chapter 3 this is more of a problem in non-submerged areas. The main method of assessing the degree of passivation of steel in concrete is by the use of electrochemical potential measurements [31], usually using a copper–copper-sulphate standard electrode with a porous plug to achieve electrochemical contact between the copper sulphate and the water within the concrete. The main problem in extending this underwater is that the porous nature of the reference electrode would quickly lead to its contamination, and to date the method appears not to have been adopted for underwater use.

6.4.3 Underwater Repair and Maintenance Methods

The function of repair and maintenance is to restore a structure to its initial state once a defect has been discovered. In practice, a decision usually has to be made both as to when (or whether) a repair needs to be effected, and also by which method. It is reasonable to say that repair never restores a structure exactly to its initial state, and if the main failure causes are considered then the choice of method is usually taken from a relatively small number of alternatives.

For steel structures, the main likely defects are fatigue cracking, collision damage (including that due to dropped objects) and corrosion. The main repair methods available are welding and clamping. The type of repair required obviously depends upon the degree of damage, but, as mentioned above, repair is most likely to be required to eliminate cracking or to strengthen dented or buckled members. Strengthening of intact members is also occasionally necessary, owing, for example, to a change in service conditions or an alteration in code requirements.

For fatigue cracks, there is some interest in determining whether a local repair around the crack is adequate, or whether it is necessary to repair the entire joint. Tubby [32] has investigated a number of different local repair methods, and has compared the fatigue performance of the repaired joints with that of a 'new' joint. The local repair methods investigated were as follows:

Grinding away of partially penetrating cracks with no reinstatement of steel.

Grinding away of cracks with reinstatement of steel using manual metal-arc welding.

As above but with weld metal ground flush with the original surface.

Drilling of through-thickness holes at crack tips followed by cold expansion of the holes to produce a compressive residual stress around the hole circumference.

These methods are illustrated in Fig. 6.24.

The broad finding of this investigation was that the weld reinstatement methods were the most effective, restoring fatigue life to close to (or above) the unrepaired value. Grinding without reinstatement of the steel was also found to be effective, but the drilling of holes at crack tips was not found to be a useful means of delaying crack growth.

More extensive welded repair is also possible with the removal of parts and replacement using new sections of pipe (see e.g. Stevenson and Sleveland [33]). Such repairs are logistically difficult and expensive in diver time, and also require jigging of the repair area and the exclusion of water by means of either a cofferdam or a hyperbaric chamber (Chapter 5). As a result, a number of alternative mechanical means have been investigated for the repair of punctures and dents in tubular members [34]. These methods usually contain a combination of direct mechanical and/or grouted connections, and some are described in Chapter 5.

Briefly, grouted and mechanical underwater repair systems can be divided into the following types:

Grouted connections.

Grouted clamps.

Mechanical connections.

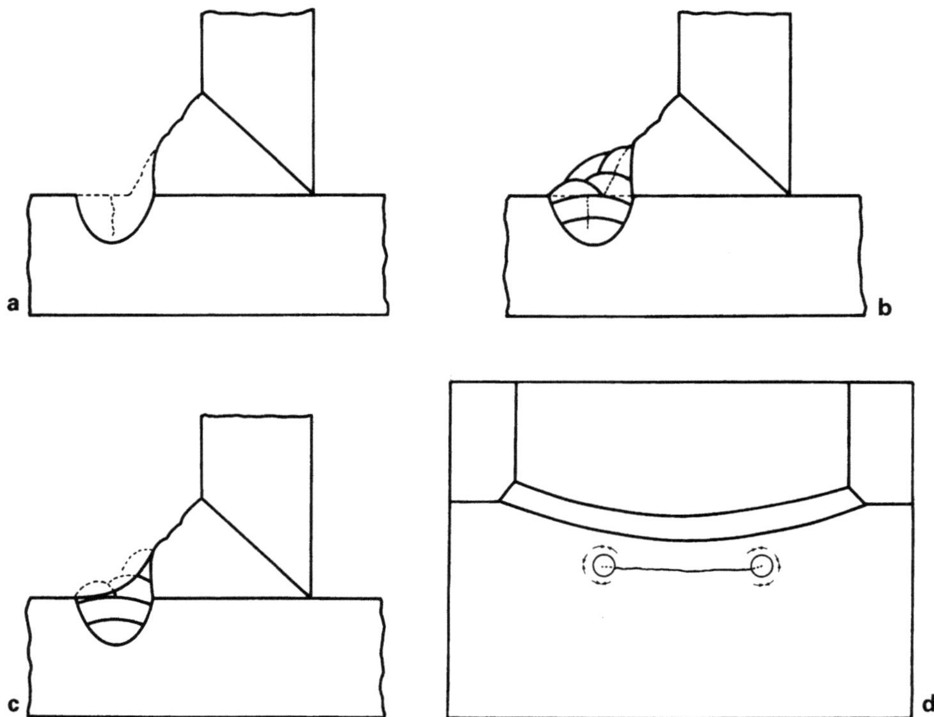

Fig. 6.24. Illustration of candidate methods for fatigue crack repair. **a** Grind away partially penetrating crack with no reinstatement of steel. **b** Grind out crack and reinstate using MMA welding. **c** As b with weld metal dressing. **d** Drill through-thickness holes at ends of crack. (After Tubby [32].)

Mechanical clamps.

Stressed grouted connections.

Stressed grouted clamps.

Grout-filled tubulars.

Each repair should be regarded as unique, however. Such methods can be used for alterations of the global strength of the structure achieved, for example, by the introduction of new members which are connected to the existing structure. They can also be used for local repair or strengthening of existing parts of the structure.

The installation of the purely mechanical sleeves and clamps is perhaps not complex (at least in principle), but there are a number of factors to be considered in the grouting procedure [34]; the major problem is the reliable detection of whether a volume has been filled with grout, since NDE methods for doing this are somewhat lacking [34]. Many data exist on the performance of grouted connections and [34] provides a summary of these.

For concrete structures, the main repair methods involve the placement of new concrete or a substitute material such as epoxy resin. Major damage, such as that caused by collision or explosion, requires careful engineering to ensure that the pattern of stresses extant before the damage occurred is restored as closely as possible, and that the previous factors of safety are preserved for all anticipated loadings [35]. The methods for dealing with prestressed concrete depend upon whether any of the prestressing cables have been broken and, if so, whether these are to be rejoined. In any case, some preparation work is necessary before reinstatement. This consists basically of removing all severely damaged concrete and reinforcing steel and sealing and injecting any smaller cracks with epoxy resin. After this, the following possible methods of proceeding have been suggested [35]:

1. Where no prestressing tendons have been broken, all exposed tendons should have ducts removed and the strands cleared of grout. New ducts should be fitted and left ungrouted. The reinstatement should then proceed as indicated in Fig. 6.25.

2. In the case where prestressing tendons are broken and are to be rejoined, they can be either (a) overlapped, anchored and stressed using conventional anchorages or (b) curved outside the wall, joined together and stressed using a centre-stressing anchorage. Otherwise the repair proceeds as for the first method.

3. Where prestressing tendons are not to be rejoined, any exposed tendons are to be cut at the edges of the repair. Thereafter, two possible approaches can be taken. The first of these is to use high-strength reinforcing bars to carry the tensile stresses at ultimate condition which would have been supplied by the lost tendons. This is achieved by ensuring that repair concrete remains in compression at all loads and by jacking before casting of the concrete. The second method uses a steel plate splint which will carry tensile loads. Unlike the reinforcing bar solution above, the splint is attached by an epoxy adhesive layer to the rest of the structure.

4. For smaller-scale repair, where only concrete needs to be reinstated, a range of cementitious grouts and epoxy resins capable of being placed underwater are available. Such placement depends on the nature of the repair and the properties of the filler material. The strength of a

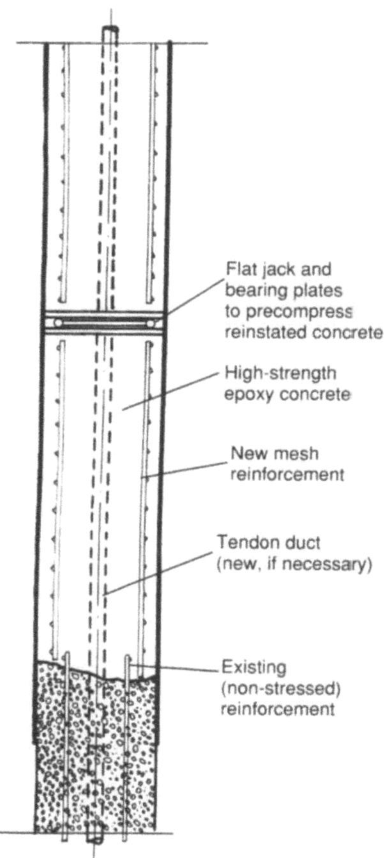

Fig. 6.25. Possible method of repair to concrete structure when no prestressing tendons have been broken. (From [35].)

range of such materials in simulated repairs has been assessed by Perry and Holmyard [36], with the conclusion that cementitious grouts appeared to behave more consistently in repairs than epoxy resins; however, it was concluded that a large proportion of the original strength can be regained by injecting cracks underwater with material of either type. The problems of placement and durability remain the most difficult.

6.5 Mechanical Reliability

In order to plan an optimum inspection strategy (as opposed to one which simply satisfies regulations) it is important to have some conception of structural reliability with particular regard to fatigue. The discipline of mechanical reliability is based around the load–strength interaction, where 'load' and 'strength' are interpreted in their widest sense. In the case of, say, a structure which is made from a material of known thickness and strength and is subject to a static load of known value, the load–strength problem is simply one of ensuring that strength exceeds load, as shown in Fig. 6.26.

In a more realistic situation, both load and strength will be distributed quantities and there will be some interaction of the two distributions, which represents the probability

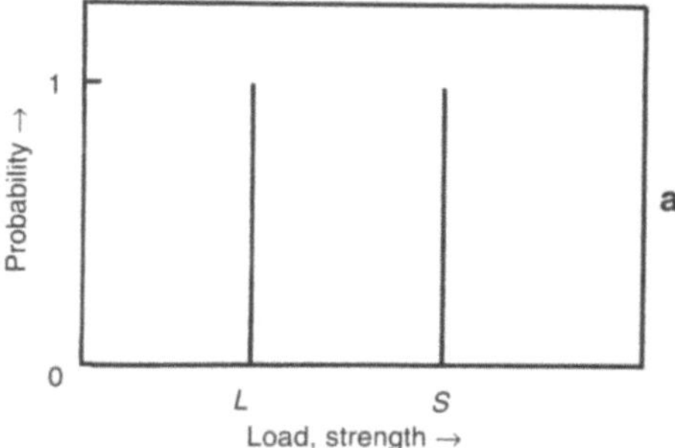

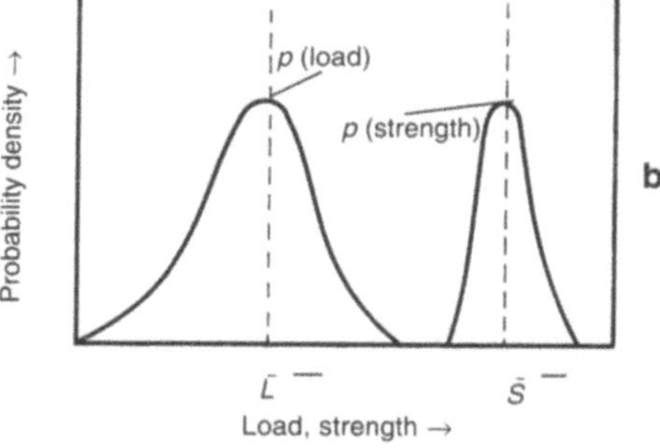

Fig. 6.26. Schematic representation of the load–strength relationship for **a** non-distributed, **b** distributed variables. (From O'Conner [37].)

that load will exceed strength and hence some probability of failure, even if the structure is 'properly' designed (Fig. 6.26). The distributions of load and strength need not be static in time. For example, corrosion may decrease the wall thickness of a structure so that the strength decreases with time. Also, actions such as inspection and consequent repair may alter a distribution, say of defect sizes, truncating at the inspection threshold with obvious implications for the strength distribution. These contributions are illustrated in a general way in Fig. 6.27.

In order to develop a methodology some very general indicators of structural reliability can be defined. For example, the safety margin can be defined as a measure of the separation of two normal load and strength distributions:

$$\text{Safety margin} = \frac{S_m - L_m}{\sqrt{\left(\sigma_S^2 - \sigma_L^2\right)}}$$

where S_m and σ_S are the mean and standard deviation of the strength distribution and L_m and σ_L are the mean and standard deviation of the load distribution.

The loading roughness is a quantifier of the relative spreads of the load and strength distributions given by (e.g. O'Conner [37]):

$$\text{Loading roughness} = \frac{\sigma_L}{\sqrt{\left(\sigma_S^2 - \sigma_L^2\right)}}$$

The most reliable situations are those where loading roughness is low and safety margin is high, and the least reliable situations are where safety margin is low and loading roughness is high. The effect can be quantified by calculating the load–strength interference (the probability Pr that strength exceeds load), which for a single load application is given by [37]

$$\Pr(S > L) = \int_0^\infty f_S(S)\left\{\int_0^S f_L(L)\,dL\right\}^n dS$$

and for multiple-load applications by

$$\Pr(S > L) = \int_0^\infty f_S(S)\left\{\int_0^S f_L(L)\,dL\right\}^n dS$$

where $f_S(S)$ and $f_L(L)$ are the probability density functions (p.d.f.s) for strength and load respectively, and n is the number of applications of load.

The above relationships illustrate that, for a single-load application, reliability ($R = \Pr(S > L)$) is a function only of safety margin, whereas for multiple-load applications reliability is dependent upon both safety margin and loading roughness. The conventional way to express mechanical reliability is as a plot of the logarithm of failure probability per load application against safety margin, and this is illustrated in Fig. 6.28.

If the failure probability per load application (p) is small (as it usually is) then the reliability can be found from:

$$R = (1 - p)^n \approx 1 - np$$

As can be seen from Fig. 6.28, there is a value of safety margin (dependent upon loading roughness) at which the failure probability becomes very low and the design is intrinsically reliable. Because of the shape of the curve, the region immediately below this in safety margin shows a very high sensitivity of reliability to both loading roughness and safety margin.

Until now the terms 'load' and 'strength' have been considered in a broad sense. To be more specific, these quantities could be expressed as stress (load) and, for example, ultimate tensile strength (strength). More commonly, probabilistic methods are applied to fracture mechanics, where

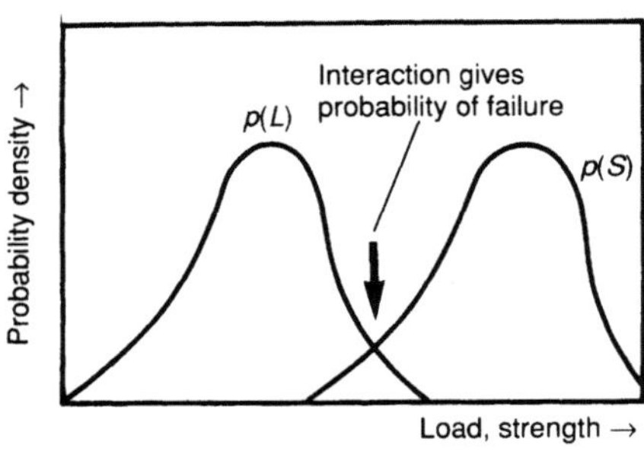

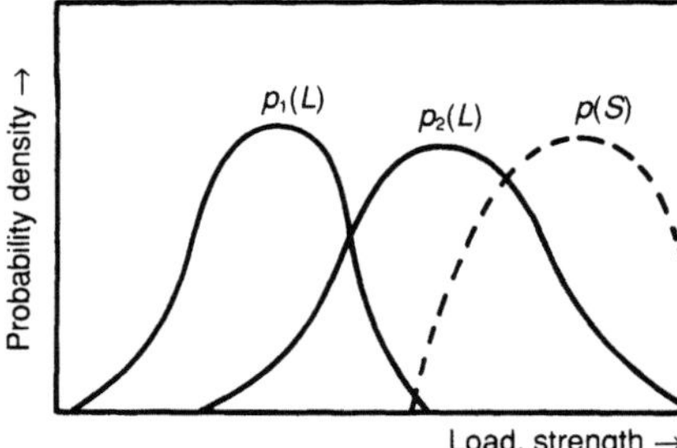

Fig. 6.27. The load–strength interaction and the effect of changing one or other of the distributions.

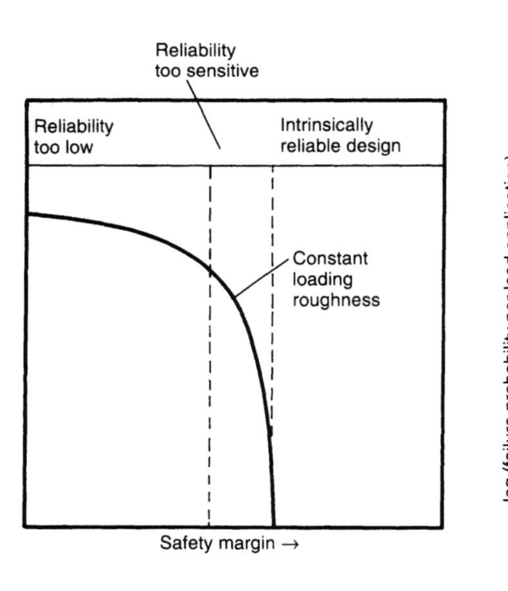

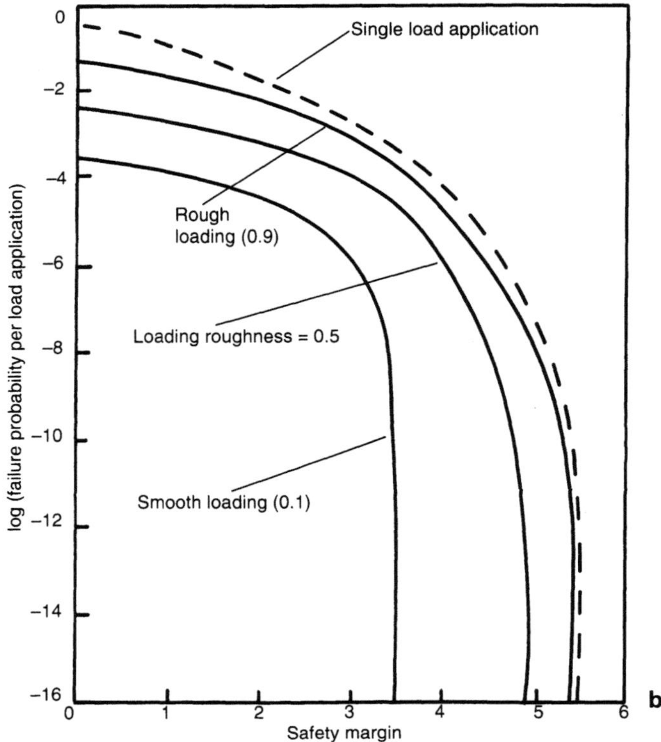

Fig. 6.28. Failure probability vs. safety margin. **a** Schematic relationship showing characteristic regions. **b** Relationship where both load and strength are normally distributed showing effect of loading roughness. (From O'Conner [37].)

the strength could be regarded as K_{Ic} and load as K_1 but general approaches allowing elastic–plastic fracture mechanics methods are also possible. Failure can occur by overload, where an abnormally high load is applied, or by strength degradation under service loading, where strength might be reduced by fatigue, wear or corrosion.

In the simplest analysis, both wear and corrosion can be regarded as phenomena which reduce wall thickness with time, and hence result in a strength distribution which changes with time, approaching that of the load distribution and hence increased probability of failure. If the corrosion or wear rate is known, this effect can be quantified and the time at which the region of excessive sensitivity is entered can be assessed. For fatigue, a similar approach can be taken where the load and strength distributions converge with number of load applications.

However, an alternative approach can be taken to fatigue using the probabilistic fracture mechanics method. Here, an initial distribution of crack sizes approaches a distribution of allowable crack sizes (owing to a distribution of toughness), and the interaction of these distributions gives an assessment of failure probability.

6.6 Structural Reliability and Maintenance Strategies

Although the compliance with legislation remains a priority in offshore structural engineering, there is growing interest in maximising efficiency in the planning of maintenance and repair of marine structures and components. The basis of this approach is usually probabilistic fracture mechanics, and the formalism is laid out briefly below.

A given component or structure will contain a distribution of defect sizes, which can be represented by a probability density function, say $f(a)$, where a is the defect size. If material toughness is distributed and/or loading is distributed then the maximum allowable defect size, a_{max}, will also be distributed, say $p(a_{max})$. The distributions are illustrated schematically in Fig. 6.29, and probability of failure is then given by the interaction of the two distributions (Mudge and Williams [38]):

$$\Pr\left(a > a_{max}\right) = \int_{a=0}^{\infty} \int_{a_{max}=0}^{a} f(a)\, p\left(a_{max}\right) \mathrm{d}a_{max}\, \mathrm{d}a$$

Thus, to determine structural reliability using probabilistic fracture mechanics, it is necessary to establish the distributions for both defect size and material toughness. For fatigue, and for the effects of inspection and repair, appropriate operations need to be carried out on the defect size distribution. For corrosion and wear, it may also be necessary to operate upon the maximum allowable defect size distribution (retaining the original material toughness distribution).

Rogerson and Wong [39] have found for offshore structures that a three-parameter Weibull distribution is appropriate for the description of defect sizes. This distribution can be expressed analytically as:

$$f(t) = \frac{\beta}{\alpha} \left[\frac{t - \gamma}{\alpha}\right]^{\beta - 1} \exp\left[-\left[\frac{t - \gamma}{\alpha}\right]^{\beta}\right]$$

where α, β and γ are the parameters of the distribution. The cumulative distribution can be written:

$$F(t) = \int_{-\infty}^{t} f(t)\, \mathrm{d}t = 1 - \exp\left[-\left[\frac{t - \gamma}{\alpha}\right]^{\beta}\right]$$

If a component or structure is subjected to inspection, the resulting repair or rejection will produce a modification of the defect size distribution and hence the reliability. Thomas [40] has given an example of how this might apply to components (see Fig. 6.30).

The effect of pre-service inspection can be seen between the p.d.f.s **a** and **b** where the inspection threshold results in a very low probability of defects above the threshold but also some rejection of components containing non-significant defects. The effect of putting the component into

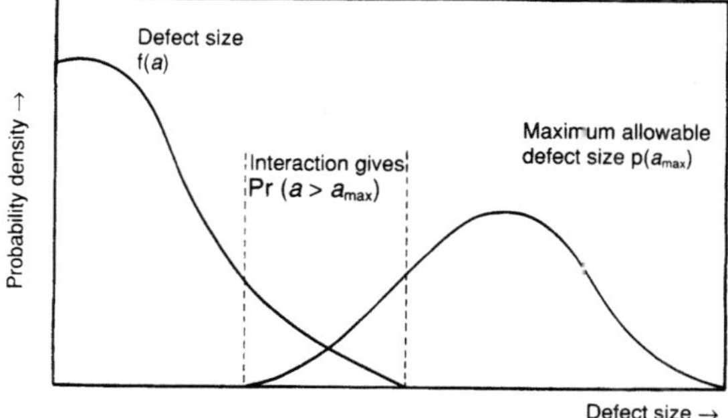

Fig. 6.29. Probability of failure when a p.d.f. of defect sizes interacts with a p.d.f. of maximum allowable defect size (due to distributed toughness).

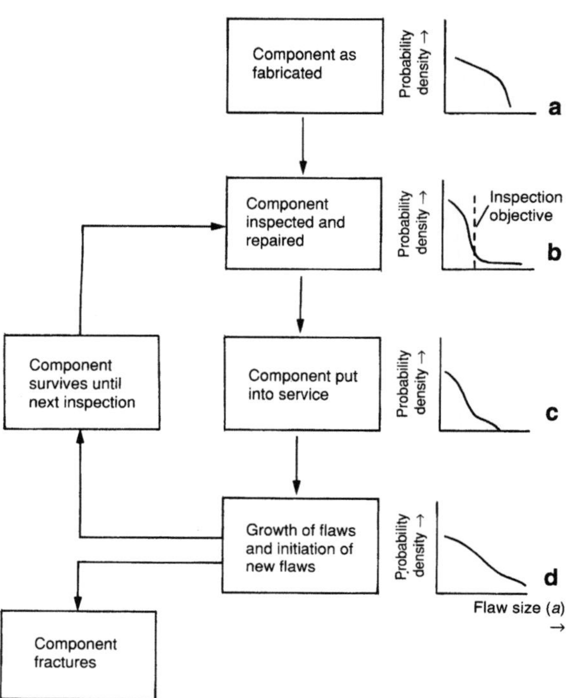

Fig. 6.30. Component reliability model. (From Thomas [40].)

service in, for example, a fatigue environment is for defects to grow and some new defects to be initiated. Depending upon the frequency of inspection, the component may survive to the next inspection, resulting in a restoration of the original size distribution (or something close to it). The question then arises as to a suitable rejection level to set, since excessive sensitivity will result in excessive repair. Thomas presents some results of a Monte Carlo simulation of the effect of inspection sensitivity (measured as percentage of a reference signal) on risk and found an optimum balance between excessive numbers of repairs of benign defects and excessive numbers of failures due to undetected defects, and this optimisation curve is shown in Fig. 6.31.

The effect of applying an inspection threshold (with 100 per cent inspection reliability) on the defect size distribution can be calculated (in the simplest analysis) by truncating the distribution at the inspection threshold, a_t:

$$f'(a) = f(a) \, K \qquad \text{for } 0 \le a \le a_t$$

$$f'(a) = 0 \qquad \text{for } a < 0 \text{ and } a > a_t$$

where K is a normalising factor given by:

$$K = \left[\int_0^{a_t} f(a) \, da \right]^{-1}$$

and $f(a)$ and $f'(a)$ are the defect size distributions before and after truncation.

More generally, Tang [41] has proposed that the modified defect size distribution can be given by

$$f'(a) = \frac{n(a) \, f(a)}{\Pr(n)}$$

where $n(a)$ is the p.d.f. of non-detection of a defect of size a and $\Pr(n)$ is the probability of non-detection of all defect sizes. The process of arriving at the modified distribution is illustrated in Fig. 6.32.

Rogerson and Wong [39] have reported defect size distributions for offshore structures in terms of the three-parameter Weibull distribution along with data from a num-

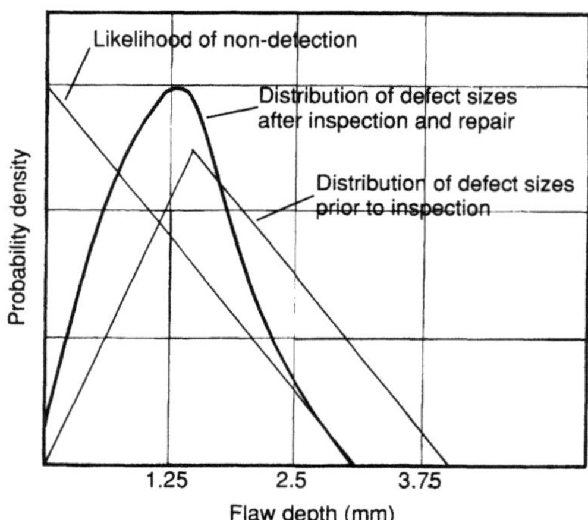

Fig. 6.32. Determination of the defect size distribution after NDT and repair. (From Tang [41].)

ber of investigators. However, Wong and Rogerson [42] have indicated that toughness distributions in offshore steel weldments are difficult to evaluate, owing to the expense involved in carrying out large numbers of COD tests. Using Charpy impact data, they suggest that a normal distribution is usually adequate but that this becomes bimodal in the transition region (Fig. 6.33). On the other hand, Mudge and Williams [38] suggest that CTOD values for MMA welds and HAZs in offshore structural steel can be described by a two-parameter Weibull distribution.

In either case, the allowable defect size distribution can be obtained from the toughness distribution by a simple transformation of coordinates. This is most easily illustrated with respect to CTOD measurements, where an allowable defect size can be obtained from CTOD (incorporating a factor of safety) by the following relationships, [43]:

$$a_{max} = \frac{\delta_c E \sigma_y}{2\pi\sigma^2}; \qquad \sigma/\sigma_y \le 0.5$$

$$\text{and } a_{max} = \frac{\delta_c E}{2\pi(\sigma - 0.25\sigma_y)}; \; \sigma/\sigma_y \le 0.5$$

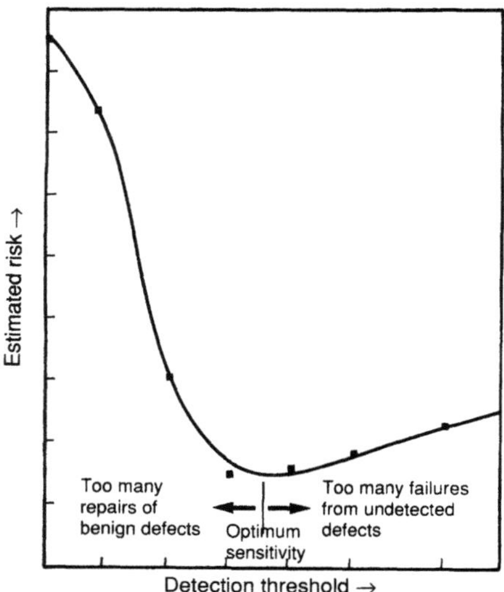

Fig. 6.31. Optimisation of inspection strategy to set detection threshold for minimum risk. (From Thomas [40].)

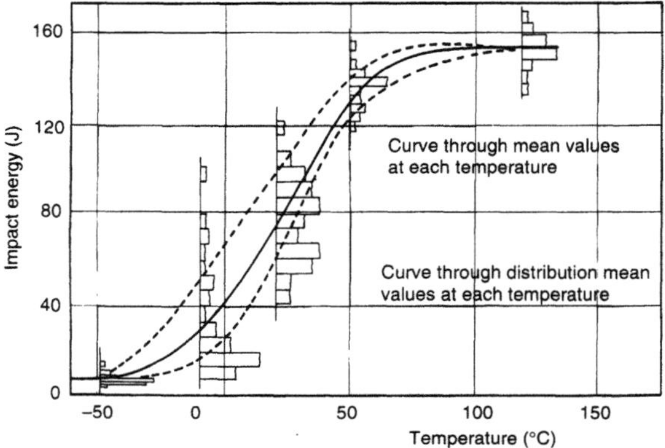

Fig. 6.33. Distribution of Charpy impact results for C–Mn steel weld metal. (From Wong and Rogerson [42].)

where σ and σ_y are the applied (design plus residual) and yield stress respectively. Thus, for a two-parameter Weibull distribution of CTOD:

$$d(\delta_c) = \frac{b}{e^b}\,(\delta_c)^{b-1}\,\exp\left\{-(\delta_c/e)^a\right\}$$

giving, for the allowable defect size distribution (if $\sigma > \sigma_y$ and assuming both to be non-distributed):

$$p(a_{max}) = k\frac{b}{e^b}\,(ka_{max})^{b-1}\,\exp\left\{-(ka_{max}/e)^b\right\}$$

where $k = 2\pi\,(\sigma - 0.25\sigma_y)/E$ and is required as a normalisation factor for the p.d.f., $p(a_{max})$.

Having evaluated $p(a_{max})$ and $f(a)$, it is possible to determine failure probabilities, using the interaction function given above. Mudge and Williams have also indicated how normally distributed sizing errors can be taken into account in such an analysis.

Larsen et al. [44] have applied a much more generalised (limit state) reliability analysis to the problem of pipelines under the influence of scour-induced freespans in conditions where there are major uncertainties in the load as in the other possible distributed variables outlined above. There is a major difficulty, however, in defining functions for all the distributed quantities.

Wolfram [45] has suggested a means of optimising inspection strategy based on the cost consequence of various failure types and on computing the risk cost profile with time for a range of inspection intervals. His main conclusion was that sampling of the state of the structure should be the aim of periodic inspection rather than an attempt to inspect the entire structure, and that this should be done using a properly designed sampling procedure. This approach is echoed by MTD [19], who suggest that the aim of inspection should be to ensure a level of confidence in the condition of each component of a structure commensurate with the consequences of failure of that component. While recognising the difficulties in obtaining p.d.f.s for all the distributed variables (such as load, strength and inspection reliability), these authors indicate that the extra time and cost in developing such strategies should be recoverable on increased efficiency of inspection. Three basic approaches are discussed: the fail-safe approach, which ensures that the inspection interval is such that the minimum detectable defect cannot grow to critical (or unacceptable) size before the next inspection; the ranking tree approach, which allocates an inspection priority to components in terms of their likelihood of failure and also upon the consequence of failure; and the approach of design for minimal inspection, which trades off the high capital cost of conservative design against the consequent reduced inspection costs.

On the above considerations, an overall strategy can be identified which makes decisions as to where and when to inspect based on the following hierarchy:

What is the consequence of component failure?

What is the likelihood of component failure?

What are the costs and reliability of inspection?

The flow diagram shown in Fig. 6.34 summarises the method.

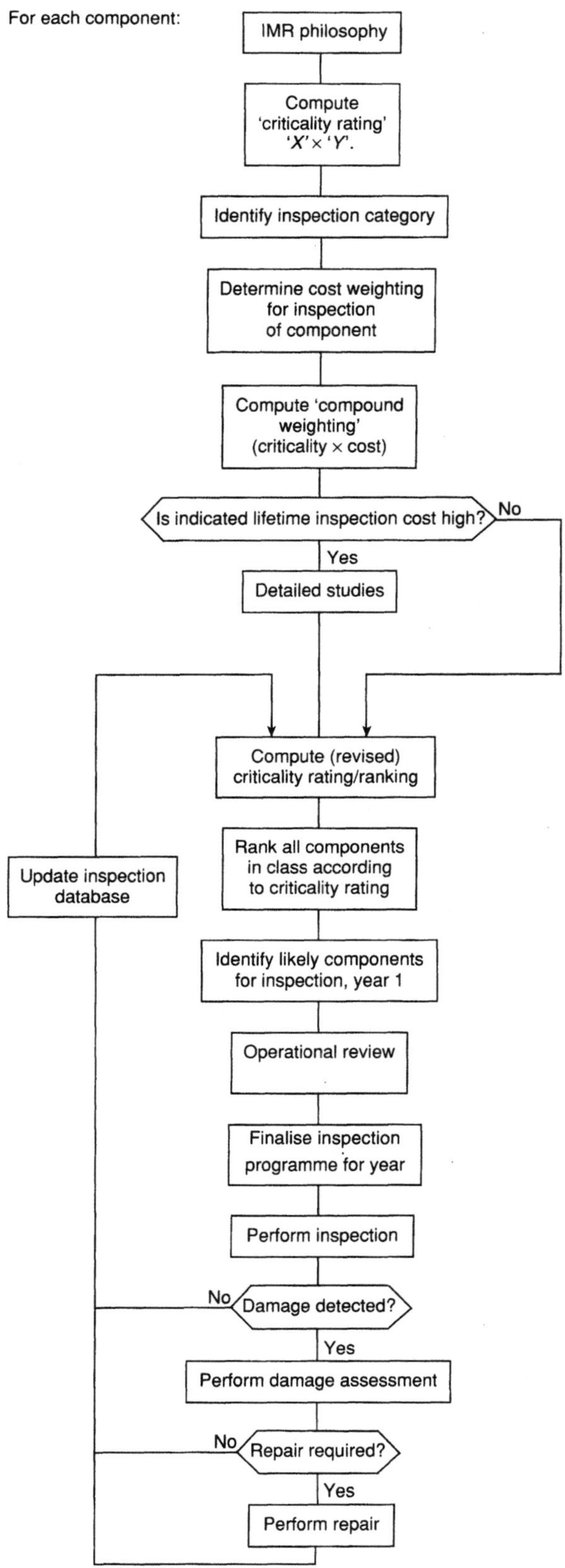

Fig. 6.34. Flow diagram for formulation of an optimised annual inspection/repair programme for offshore structures [19].

At commencement, the criticality ratings of individual components are quantified. These ratings are based on the consequence and likelihood of failure respectively, and combined will allow an inspection category (priority) to be assigned. By considering the difficulty of inspection, a cost can be assigned to the inspection of each component, and this is combined with its criticality (i.e. frequency of inspection) to obtain a compound weighting. At this stage, it is possible to calculate the likely lifetime inspection costs; if these are excessive, a more detailed study of the compound weightings should be carried out, which may result in re-ranking or even a decision to strengthen parts of the structure to reduce future inspection costs. Once an acceptable set of inspection frequencies has been decided, the inspection loop can be entered, which starts with the criticality rating of the components to be inspected. This leads to a decision on the components to be inspected in the following season. At this point it will be necessary to carry out an operational review which will, in addition to the above, identify any components with mandatory inspection requirements, plan the provision of dive support vessels to cover mandatory inspection requirements, and any other subsea operations required (such as repairs) and inspection of those components with high ranking. A contingency for the inspection of other accessible components is also advisable. The final inspection programme for that year can then be carried out, and, depending upon whether or not damage is detected and whether or not repair is required, the condition record is updated (revised criticality ranking) and the cycle repeated. In updating the criticality ranking, it is necessary to take cognisance of inspection reliability, so that the criticality ranking should be based upon the minimum size of defect that is certain to be discovered under the circumstances (e.g. depth, accessibility) of the particular component, rather than on the minimum size of defect which can be discovered by the technique itself.

References

1. Polak P. Designing for Strength. Macmillan, London, 1982
2. Det norske Veritas. Rules for submarine pipeline systems. DnV, Hovik, 1981
3. Halmshaw R. Non-destructive testing. Edward Arnold, London, 1987
4. Hull B, John V. Non-destructive testing. Macmillan, London, 1988
5. Det norske Veritas. Ultrasonic inspection of weld connections, Classification Note no. 7. DnV, Hovik, 1980
6. Henneke EG. Destructive and nondestructive tests. In: Engineered materials handbook, vol. 1: Composites, ASM International, Metals Park, Ohio, 1987, pp 774–778
7. Nicholas RW (ed.). Advances in non-destructive examination for structural integrity. Applied Science Publishers, London, 1982
8. Packman PF et al. Reliability of flaw detection by non-destructive inspection. Metals handbook, vol. 11: Nondestructive inspection and quality control. ASM, Metals Park, Ohio, 1984, pp 414–424
9. Still JR, Rae G. Laybarge inspection of submarine pipelines. Metal Construction 1984; 16 (5): 268–275
10. Birchon D. Non-destructive testing. Design Council, London, 1975
11. Forli O, Pettersen B. Reliability of ultrasonic and radiographic testing. Proceedings, Conference on Fitness for Purpose Validation of Welded Constructions. Welding Institute, Cambridge, 1982, pp 27-1–27-9
12. Fortunko CM, Schramm RE. Nondestructive evaluation of large diameter girth welds using electromagnetic–acoustic transducers. Proceedings, Conference on Fitness for Purpose Validation of Welded Construction. Welding Institute, Cambridge, 1982, pp 20-1–20-8
13. Yamakawa K, Tsubakino H, Yoshizawa S. A new electrochemical hydrogen probe. In: Moran GC, Labine P (eds.) Corrosion monitoring in industrial plants using nondestructive testing and electrochemical methods, ASTM STP 908. ASTM, Philadelphia, 1986, pp 221–236
14. Department of Industry. Industrial corrosion monitoring. HMSO, London, 1978
15. Cameron GR, Coker LG. Oil production corrosion inhibitor optimisation by laboratory and application of electrochemical techniques. In: Moran GC, Labine P. (eds) Corrosion monitoring in industrial plants using nondestructive testing and electrochemical methods, ASTM STP 908. ASTM, Philadelphia, 1986, pp 251–265
16. Britton CF. Internal corrosion monitoring of subsea pipelines. Pipes and Pipelines International 1986; Nov.–Dec.: 12–18
17. de Raad JA. Comparison between ultrasonic and magnetic flux pigs for pipeline inspection. Pipes and Pipelines International 1987; Jan.–Feb.: 7–15
18. Shannon RWE, Knott RN. On-line inspection – development and operating experience. Proceedings, 17th Offshore Technology Conference, Houston, paper no. OTC 4923, 1985, pp 259–272
19. Marine Technology Directorate. Underwater inspection of steel offshore installations – implementation of a new approach, MTD Publication 89/104. MTD, London, 1989
20. Begg RD, Mackenzie AC, Dodds CJ, Loland O. Structural integrity monitoring using digital processing of vibration signals. Proceedings, 8th Offshore Technology Conference, Houston, paper no. OTC 2549, 1976
21. Hockenhull BS, Billingham J, Christodoulou G, Hale KF. Optical fibre witness devices for monitoring the integrity of offshore structures. In: Atteraas L (ed.) Underwater technology – offshore petroleum. Pergamon, Oxford, 1980, pp 341–350
22. Det norske Veritas. Rules for design, construction and inspection of fixed offshore structures. DnV, Hovik, 1974
23. UEG Offshore Research. Repairs to N Sea offshore structures – a review, Report no. UR 21. UEG/CIRIA, London, 1982
24. Haux G. Subsea manned engineering. Ballière Tindall, London, 1982
25. Melegari JE. The evolving role and expanding array of submersibles and underwater vehicles. Proceedings, Conference Subtech '83. Society for Underwater Technology, London, 1984
26. Balch J. Manned vehicles – philosophy. Proceedings, Conference Subtech '83. Society for Underwater Technology, London, 1984
27. Turner J. Underwater photogrammetry – of what value to the offshore industry? Journal of the Society for Underwater Technology 1983; Autumn: 7–13
28. Sisman D (ed.). The professional diver's handbook. Submex, London, 1982
29. Department of Energy. The effectiveness of underwater nondestructive testing, Offshore Technology Report, OTH 84 203. HMSO, London, 1984
30. McAlpine Sea Services. Classification and identification of typical blemishes visible on the surface of concrete underwater, Offshore Technology Report OTH 84 206. HMSO, London, 1985
31. Carney RFA, Lawrence PF, Wilkins NJM. Detection of corrosion in submerged reinforced concrete, Offshore Technology Report OTH 87 239. HMSO, London, 1990
32. Tubby PJ. Fatigue performance of repaired tubular joints, Offshore Technology Report OTH 89 307. HMSO, London, 1989
33. Stevenson AW, Sleveland A. Damaged brace on offshore platform replaced using hyperbaric welding. Metal Construction

1983; 15 (12): 720–723

34. Harwood RG, Shuttleworth EP. Grouted and mechanical strengthening and repair of tubular steel offshore structures, Offshore Technology Report OTH 88 283. HMSO, London, 1988

35. Wimpey Laboratories. Repair of major damage to the prestressed concrete towers of offshore structures, Offshore Technology Report OTH 87 250. HMSO, London, 1988

36. Perry SH, Holmyard JM. Assessment of materials for repair of damaged concrete underwater, Offshore Technology Report OTH 90 318. HMSO, London, 1990

37. O'Connor PDT. Practical reliability engineering. John Wiley, Chichester, 1991

38. Mudge PJ, Williams S. Performance of ultrasonic nondestructive methods for defect assessment in relation to fracture mechanics requirements. Proceedings, Conference on Fitness for Purpose Validation of Welded Constructions. Welding Institute, Cambridge, 1982, pp 44-1–44-11

39. Rogerson JH, Wong WK. Weld defect distributions in offshore structures and their influence on structural reliability. Proceedings, 14th Offshore Technology Conference, Houston, paper no. OTC 4237, 1982, pp 95–107

40. Thomas JM. Quantitative risk analysis of offshore welded structures. Proceedings, Conference Welded Offshore Structures. Welding Institute, Cambridge, 1983, pp 1-1–1-10

41. Tang WH. Probabilistic updating of flaw information. Journal of Testing and Evaluation 1977; 1 (6): 459–467

42. Wong WK, Rogerson JH. A probabilistic estimate of the relative value of factors which control the failure by fracture of offshore structures. Proceedings, Conference Welded Offshore Structures. Welding Institute, Cambridge, 1983, pp 9-1–9-7

43. BS PD6493. Guidance on some methods for the derivation of acceptance levels for defects in fusion welded joints. BSI, London, 1980

44. Larsen EN, Skjong R, Madsen HO. Assessment of pipeline reliability under the existence of scour-induced free spans. Proceedings, 18th Offshore Technology Conference, Houston, paper no. OTC 5343, 1986, pp 475–482

45. Wolfram J. A techno-economic approach to underwater inspection and maintenance strategies. Journal of the Society for Underwater Technology, 1983; summer: 27–32

7 Case Studies and Applications

The purpose of this final chapter is to highlight some of the considerations given in the foregoing ones by drawing detailed examples and case histories from the literature. The examples are not intended to illustrate exhaustively all of the material considered in the previous chapters, but rather have been chosen so that the designer can gain some appreciation as to the process of materials selection and its importance in design and manufacturing considerations.

Reference to relevant codes and standards has been made as far as is possible to illustrate how useful these can be in the design and specification process, and all examples are of course of relevance to marine technology.

7.1 Fracture Mechanics Applied to Pipelines

The methods of fracture mechanics when applied to pipelines provide an example of the usefulness of this science in a system where the loading and geometry are relatively simple and where the material properties are typical of those used for high-stress applications.

Since pipelines are essentially pressure vessels, and because of the nature of the fluids they are designed to transport, it is necessary to ensure that the possibility of failure is minimised, and such situations often benefit from the use of engineering critical assessment based on fracture considerations.

The main application of fracture mechanics in pipelines is in fitness-for-purpose evaluations in order to set defect acceptance criteria, although conservative acceptance levels based on 'good-workmanship' criteria are available in many standards, notably API 1104 [1] and BS 4515 [2]. Jones et al. [3] have summarised the workmanship criteria given for girth welds in a number of standards, and this is reproduced in Table 7.1.

The general approach to defect assessment in pipelines is usually centred around the welds, and slightly different methods are required depending upon whether circumferential (girth) welds are under consideration or longitudinal (seam) welds. In general, girth weld defects can be considered to propagate under longitudinal stresses and seam weld defects under circumferential (hoop) stresses. The toughness of pipeline steels is such that linear elastic fracture mechanics is inadequate, and the possibility must be considered that failure due to a given defect may be by plastic collapse rather than fracture.

Table 7.1. A selection of pipeline girth weld workmanship-based defect acceptance levels (1986)

		Limit on length (mm)	
		Hollow bead	Lack of sidewall fusion
France	(GdF)	4	0
Norway	(DnV)	6	24
Japan	(JIS)	10	20
Holland	(Gasunie)	12	25
Italy	(SNAM)	13	25
Australia	(AS)	12	50
Canada	(CSA)	12	50
America	(API)	13	51
UK	(BSI)	50	50
UK	(BG)	100	100

Source: Jones et al. [3].

A number of different approaches to defect assessment in ductile structures are available, and two alternatives will be examined here, applied to girth and seam defects respectively, although this does not imply any superiority of either technique for the application illustrated. The girth defects will be assumed to be subject to bending only (the predominant load during laying), and the seam defects will be assumed to be subject to internal pressure loading only (the predominant load in service). In fact, both of these cases are likely to have a superimposed tension in practice, but this will be neglected for clarity.

The so-called 'C(T)OD Design Curve' (Harrison) [4] has been widely applied to defect assessment for pipeline girth welds. The approach is based upon the large amount of data which has been collected on wide cracked plates and the curve itself relates a 'normalised' COD to a normalised strain [5] (Fig. 7.1.).

The approach has been adopted in British Standards (PD6493 [6]) for weld defect assessment. The PD6493 method begins with a recategorisation procedure for defects which do not penetrate the thickness of the section, although plastic collapse due to through-thickness defects needs also to be checked. The process is illustrated in Fig. 7.2 for a pipeline, and commences with an idealisation of the defect shape, including if necessary any groups of defects. Using the worst case, where the defect is assumed to penetrate the full wall thickness, e, the first check is to ensure that the remaining ligament would be able to with-

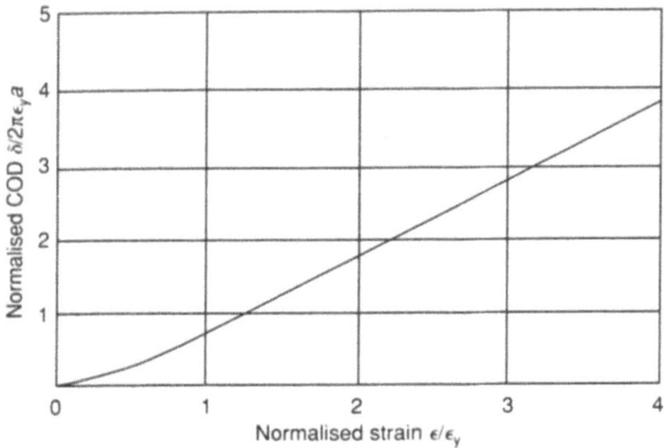

Fig. 7.1.   The C(T)OD design curve for specification of minimum acceptable values of toughness, δ. (From Harrison [4].)

stand yielding or plastic collapse, and this condition is given for cylindrical pressure vessels as being when the defect length,

$$l > 0.1\sqrt{re}$$

where r is the vessel radius.

If the defect passes this test, it will leak before break and does not need to be recategorised. If it fails, then the effect-

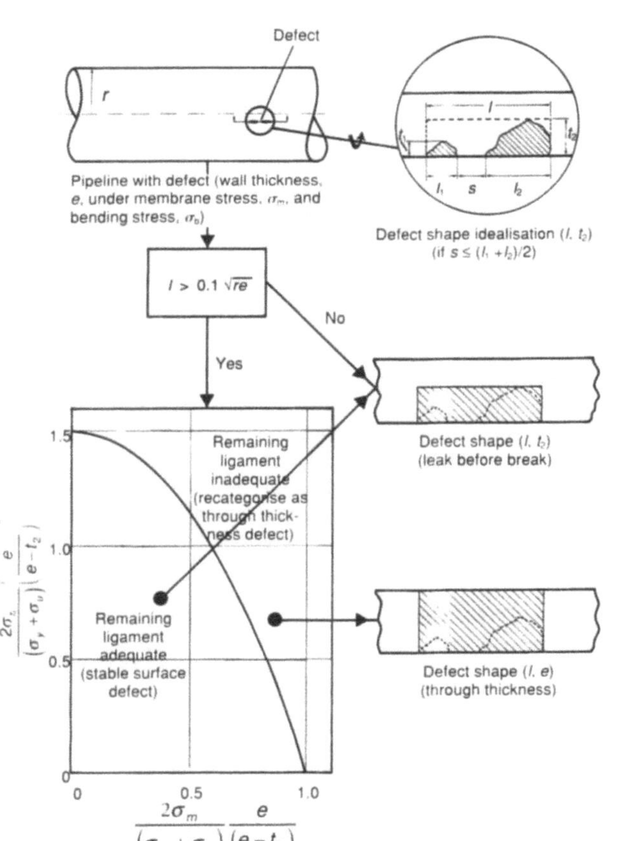

Fig. 7.2.   Schematic summary of the BS PD6493 method of defect assessment.

iveness of the ligament(s) between the defect and the walls of the cylinder needs to be checked. This is done by a graphical procedure in PD6493, illustrated in Fig. 7.2, and finally leads either to a recategorised through-thickness defect (if remaining ligament fails the check) or the adoption of the original idealised defect position and dimensions. Plastic collapse criteria for cylinders containing defects are fairly well established (e.g. Miller [7]). For cylinders containing circumferential defects under bending, the plain pipe collapse moment, M_{pp}, is useful as a baseline for normalisation of the results for defective cylinders:

$$M_{pp} = \sigma_f \pi R^2 t \quad \text{(for thin cylinders)}$$

where σ_f = the flow stress (usually taken to be the mean of yield stress and ultimate tensile strength)
R = the (thin) cylindrical radius
t = the wall thickness

Miller has reviewed limit load criteria for the circumferential defect under bending, and concludes that the results due to Kastner et al. [8] gave best results when compared with experimental observations:

$$\frac{M}{M_{pp}} = \frac{4}{\pi}\left[\cos\frac{\beta}{2} - \frac{1}{2}\sin\beta\right]$$

(for through-thickness defects)

also

$$\frac{M}{M_{pp}} = \frac{4}{\pi}\left[\cos\left(\frac{(1-\gamma)\beta}{2}\right) - \frac{(1-\gamma)\sin\beta}{2}\right]$$

$$\left(\text{for } \gamma < (\pi - \beta)/\beta \text{ or } \beta < \pi/(1+\gamma)\right)$$

and

$$\frac{M}{M_{pp}} = \frac{4}{\pi}\left[\gamma\sin\left(\frac{\pi - \beta(1-\gamma)}{2\gamma}\right) + \frac{(1-\gamma)\sin\beta}{2}\right]$$

$$\left(\text{for } \gamma > (\pi - \beta)/\beta \text{ or } \beta > \pi/(1+\gamma)\right)$$

(for surface defects)

where the geometric parameters β and γ are as shown in Fig. 7.3.

The ductile fracture assessment can now be made according to BS PD6493 by calculating a maximum tolera-

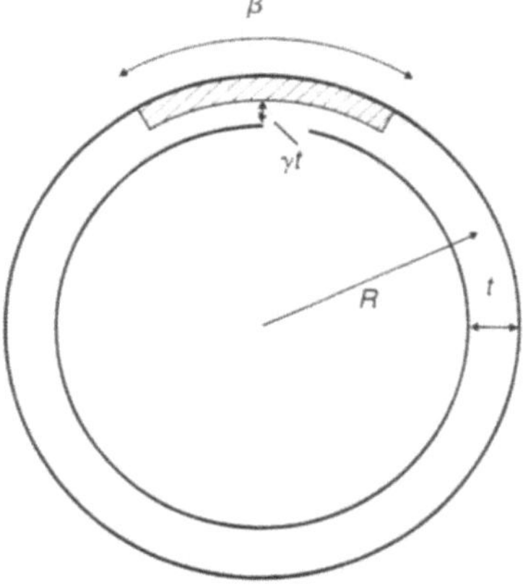

Fig. 7.3.   The geometric parameters required for assessment of collapse moment for a cylinder containing a circumferential defect under bending. (From Miller [7].)

ble defect parameter from

$$\bar{a}_m = \delta_c/[2\pi \, (\epsilon - 0.25\epsilon_y)]$$

for values of applied-strain-to-yield-strain ratio $(\epsilon/\epsilon_y) > 0.5$.

The defect parameter can be related to actual defect dimensions through rules given in BS PD6493.

As an example of a slightly different approach, it is possible to calculate the critical axial defect size in a pipe seam weld using the so-called CEGB R6 'two-criteria' approach. The basis of the R6 approach (e.g. Milne et al. [9]) is to interpolate between the failure conditions of linear-elastic fracture (characterised by K_{1c}) and plastic collapse (characterised by a limit load) using the interpolation formula,

$$K_r = S_r \, \{8/\pi \, [\ln \sec \, (\pi S_r/2)]\}$$

where K_r = a 'relative stress intensity factor' which describes the proportion of the way to linear elastic fracture and this can be expressed as the ratio of stress intensity factor to fracture toughness, K_1/K_{1c} – for defects of complex geometry, the linear elastic methods from BS PD6493 can be used;

and S_r = a 'relative plastic load' which describes the proportion of the limit load for the flawed structure which is taken up by the applied load. This can be expressed as the ratio of applied load to limit load, which, for a cylinder under pressure containing an axial defect, can be given by the ratio of internal pressure to collapse pressure for the defective pipe:

$$S_r = p/p_c$$

Miller [7] suggests that the equation of Kiefner et al. [10] is most suitable for the collapse pressure, p_c, of tubes containing axial defects under internal pressure:

$$p_c = \sigma_f t/RM(\mu) \quad \text{(for penetrating defects)}$$

and

$$p_c = \sigma_f t\gamma/R[1 - (1 - \gamma)/M(\mu)] \quad \text{(for surface defects)}$$

where $M(\mu) = (1 + 1.61\mu^2)^{1/2}$

and $\mu = c/\sqrt{(Rt)}$

and the geometric parameters are as shown in Fig. 7.4.

Miller [7] has pointed out that, since the second equation

is a local collapse estimate and is not continuous with the through-cracked estimate, the comparison of the two criteria allows the distinction between leak-before-break and catastrophic failures to be made.

The CEGB R6 equation can, of course, be programmed, but often a diagram (failure assessment diagram) is used and K_r and S_r calculated for a range of values of defect size. The point at which the curve of (S_r, K_r) pairs crosses the failure envelope then gives the critical defect size. This process is illustrated in Fig. 7.5.

The relevance of these methods to subsea pipelines is limited for circumferential defects to the cases where significant bending or tension is expected, for example during lay or under exceptional hydrodynamic loading. Whereas it is unlikely that either of these cases would lead to fracture except in grossly defective pipes, the methodology has been used to evaluate fatigue performance under laying conditions (Tenge and Karlsen [11]) and fatigue due to vortex shedding oscillation resulting from scour-induced freespans (Celant et al. [12]).

For the case of failure under internal pressure, Jones [3] has recommended the following fitness-for-purpose method for underwater pipelines:

Define the largest axial defect, a_i, which can conceivably survive the pre-service pressure test.

Define the smallest defect, a_f, which can cause failure at the maximum operating pressure.

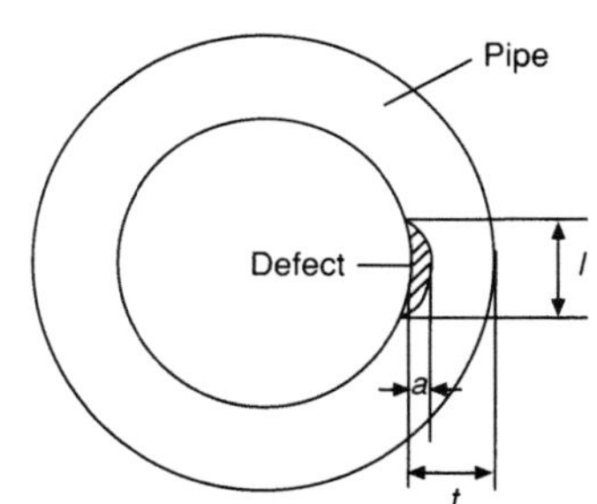

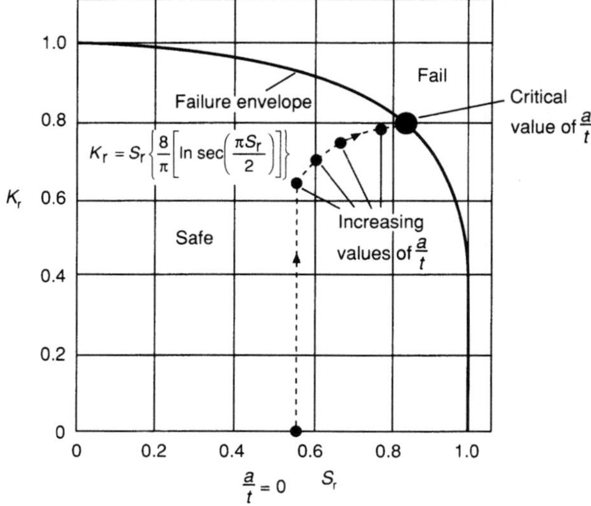

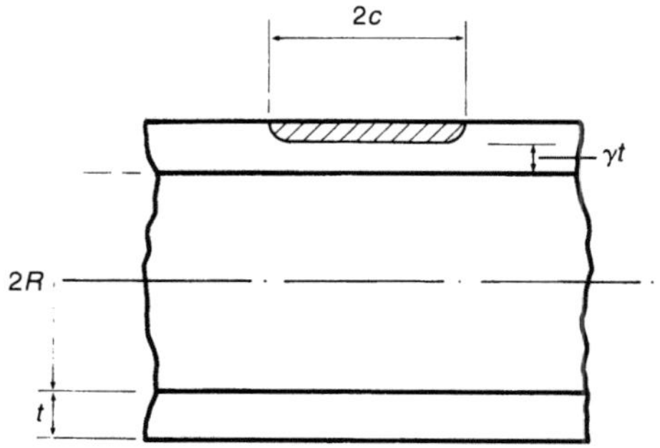

Fig. 7.4. The geometric parameters required for assessment of collapse pressure for a cylinder containing an axial defect under internal pressure. (From Miller [7].)

Fig. 7.5. Schematic illustration of the 'two-criteria' method for defect assessment for the simple case of a plain pipe containing an internal defect.

Calculate the time taken for a defect to extend in size from a_i to a_f under the design fatigue loading (due to pressure cycles).

As seen above, and as indicated by Jones, defect failure in modern linepipe steels is most likely under plastic collapse, and Jones further indicates that the fatigue life of largest possible surviving defects under typical pipeline operating conditions usually exceeds the normal design life of the pipeline by a large margin.

7.2 Fatigue of Tubular Joints for Offshore Structures

Many offshore structures for the exploitation of sub-seabed hydrocarbon resources are fabricated from tubular structural steel, often of very large diameter and thickness. Before these platforms became commonplace, there was very little experience in the world of such section sizes in such global geometries, particularly in such a severe environment. The main structural concern is that fatigue, under the cyclic hydrodynamic loading produced by wave action on the structure, may lead to failure by fatigue of the most heavily loaded (and hence most critical) nodal joints of the structure. As a result, offshore engineers have striven to develop design rules for tubular joints in offshore structures, and these are now quite sophisticated. The purpose of this study is to illustrate, using the example of such tubular joints, some of the most modern methods of fatigue design, which can of course be used in other applications.

The basis of the fatigue analysis of offshore structures is set out by Williams and Rinne [13], who indicate that such analysis revolves around three different relationships:

The S–N relationship between loading history and fatigue life (see Chapter 2).

The H–n relationship between wave height, H, and number of waves, n, of that height (equivalent to the number of cycles experienced in the Miner sum, Chapter 2).

The S–H relationship between wave height and stress range, essentially a measure of the structural response to wave loading.

Taking the last of these, it is clear that the S–H relationship will not be a simple one, owing to the complex nature of both the dynamic and the static features of the structure. In particular, the dynamic load factor will be a function of loading frequency, so that, strictly, the S–H relationship should be frequency-dependent. According to Williams and Rinne, the fundamental period of most offshore structures is well below the minimum period of waves large enough to contribute significantly to fatigue damage (Fig. 7.6); consequently, a general polynomial relationship,

$$S = aH^b$$

can be assumed where reasonable results [13] can be obtained taking $b = 1$, although the following development allows for superposition by taking the more general form.

The relationship between wave height and number of waves can be most easily obtained from a wave exceedance curve, which, as can be seen from Fig. 7.7, can be approximated by a logarithmic relationship of the form

$$H = H_m + m_2 \lg(n)$$

where $m_2 = -H_m/\lg(n_m)$

The exceedance curve shown in Fig. 7.7 represents the 100-year wave height distribution for waves from the

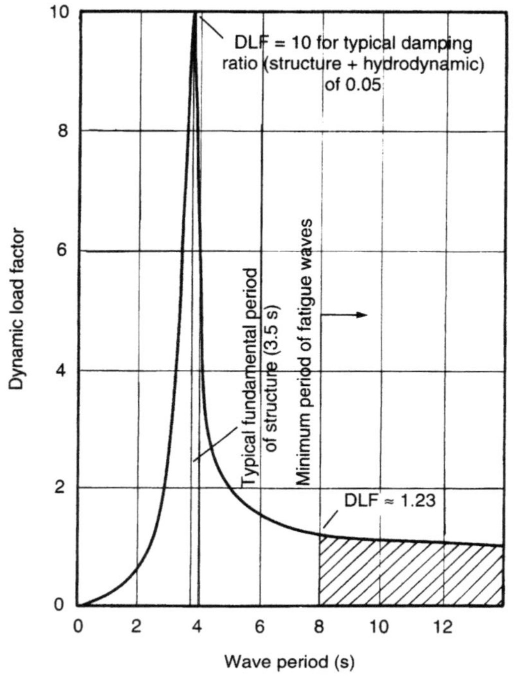

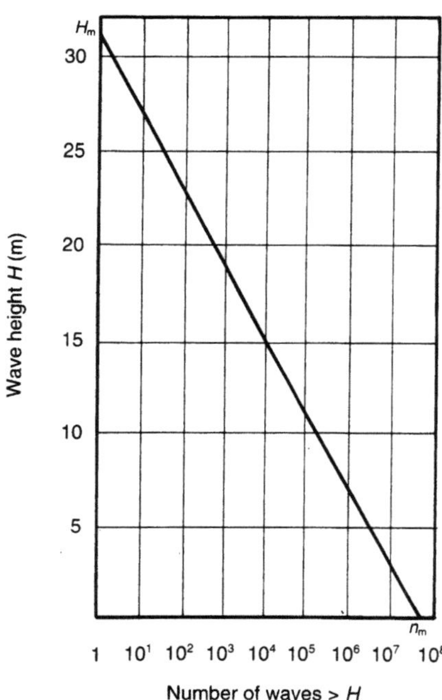

Fig. 7.6. Typical dynamic load factor vs. frequency for a steel jacket offshore structure. (From Williams and Rinne [13].)

Fig. 7.7. 100-year wave exceedance data for waves from a north-westerly direction in the North Sea. (From Williams and Rinne [13].)

north-west in the North Sea, so that n_m represents the nominal total number of waves in 100 years (number of exceedances of zero wave height) and H_m represents the height of the 'hundred-year wave'.

Recasting the wave exceedance curve into explicit form for number of waves,

$$n = n_m \exp (2.303H/m_2)$$

and differentiating gives the number of waves, dn, in a wave height range, dH:

$$\mathrm{d}n = - (2.303n_m/m_2) \exp (2.303H/m_2)\, \mathrm{d}H$$

Assuming a simple S–N curve, the final relationship required to assess the fatigue life can be expressed:

$$S = S_1 N^m$$

where S_1 and m are parameters of the S–N curve (Fig. 7.8).

The fatigue damage accumulated in the 100 years (or other period) of the wave statistics can now be obtained from a continuous version of the Miner sum:

$$D = \int_{H=0}^{\infty} \mathrm{d}n/N \quad \left(\text{formerly } \sum n/N\right)$$

Inserting the S–N, S–H and n–H relationships and integrating gives the damage accumulated in 100 years as

$$D_{100} = n_m(a/S_1)^{-1/m} (-2.303/m_2)^{b/m}\, \Gamma[1 - (b/m)]$$

where the gamma function, $\Gamma\{z\} = \int_0^{\infty} e^{-t} t^{z-1} dt$, is obtainable from mathematical tables.

Of course, unless the fatigue design is very conservative, D_{100} will be greater than unity and the fatigue life can be found from

$$L = 100/D_{100}$$

Since Williams and Rinne's original work, a number of developments have taken place in the knowledge of fatigue behaviour of large tubular joints.

First, extensive testing has lead to the adoption of a new fatigue curve in the UK (see, for example, Gurney [14]). This curve is bilinear (Fig. 7.9) and requires a reduction in fatigue strength with increasing thickness, for a given stress range. The 'basic' curve was obtained using material of 32-mm chord wall thickness (as most data were available for this thickness) and the S–N relationship for the basic curve

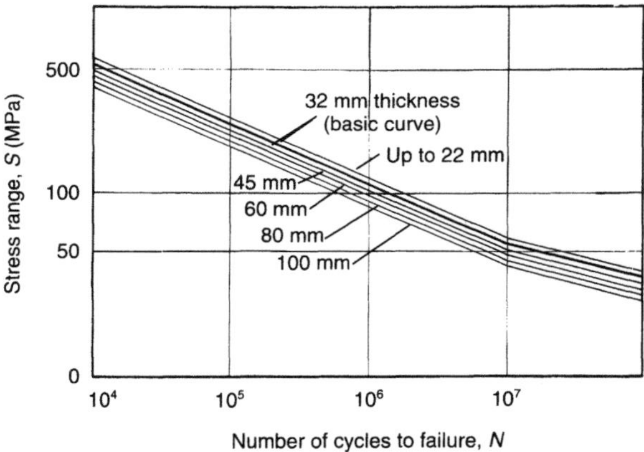

Fig. 7.9. The bilinear 'T' curve adopted for UK offshore structural design [14].

is:

$$\lg (N) = 12.164 - 3 \lg (S)$$

The chord wall thickness correction (for values of t greater than 16 mm only) is embodied in the modified equation:

$$\lg(N') = 12.164 - 3 \lg [S (32/t)^{-1/4}]$$

Improvements have also been made in the characterisation of the stress distribution around tubular welded intersections, and it is the 'hotspot', or maximum stress range in the joint, which is required as input to the S–N relationships.

A set of parametric equations has been developed for stress concentration factors in these joints. The equations are summarised in [15] for axial, in-plane and out-of-plane moment loadings. In general, the hotspot occurs somewhere along the weld toe, but may be in either the chord (main member) or the brace (subsidiary member), depending upon design. As an example, Fig. 7.10 shows the SCFs around a simple T-joint for each of the three loading conditions in both the chord and brace.

Because of the added complexity of these two developments, it is uncommon for analytical solutions to the damage integral to be sought, and most fatigue analyses are now carried out numerically.

For cases where it is necessary to predict fatigue crack growth from a known initial condition, the Paris law and its variants may be used, and some developments have also been made in this methodology.

Firstly, the law itself may be modified to correct for such effects as stress ratio, residual stress, threshold stress intensity factor and environment. Also, crack aspect ratios can alter significantly as the crack grows, and this makes treatment quite difficult because the function for stress intensity factor is already quite complex.

The approach is generally as outlined in Chapter 2 and consists of first selecting an appropriate crack-growth law of the Paris type and integrating this to give, for the unmodified Paris Law, for example:

$$N = C^{-1}\pi^{-m/2} S^{-m} \int_{a_i}^{a_f} Y^{-m} a^{-m/2}\, \mathrm{d}a$$

although modifications to the law itself and to the parameters C and m may be necessary in the light of the factors mentioned above. In particular, if there are many cycles

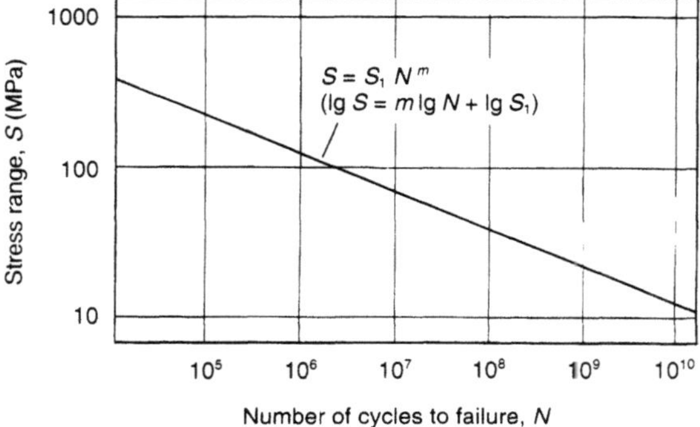

Fig. 7.8. Typical S–N curve for steel simplified for use in fatigue analysis model (see text).

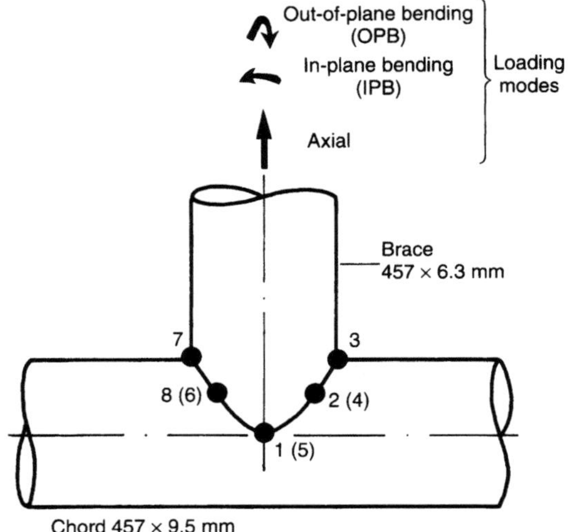

Stress concentration factors

Position	Axial		IPB		OPB	
	Chord	Brace	Chord	Brace	Chord	Brace
1	7.64	6.81	0.00	0.00	9.55	7.02
2	8.62	7.43	2.30	2.16	6.75	4.96
3	9.61	7.05	3.26	3.05	0.00	0.00
4	8.62	6.43	2.30	2.16	6.75	4.96
5	7.64	5.81	0.00	0.00	9.55	7.02
6	8.62	6.43	2.30	2.16	6.75	4.96
7	9.61	7.05	3.26	3.05	0.00	0.00
8	8.62	6.43	2.30	2.16	6.75	4.96

Fig. 7.10. Stress concentration factors around a simple intersecting tubular T-joint [15].

where the maximum value of SIF begins to approach K_{Ic} then the method is not applicable. Also, the method may be needlessly conservative if there are many cycles around the threshold stress intensity factor, ΔK_{th}. Variable-amplitude loading can be dealt with using an effective value of the SIF range:

$$\Delta K_{eff} = [(\Delta K_1^m + \Delta K_2^m + \ldots + \Delta K_n^m)/n]$$

where n is the number of cycles and m is the Paris exponent.

Recommended [15] values of the crack growth parameters for structural steel are given in Table 7.2.

Stress intensity factors can be calculated for tubular joints using a variety of semi-empirical methods (since numerical methods can be time-consuming and of variable accuracy). One such analysis might proceed by dividing the total SIF into contributions from the various geometric aspects of the joint [15]:

$$K = Y\sigma\sqrt{(\pi a)} \quad \text{(Chapter 2)}$$

where σ = the hotspot stress
a = the crack depth in the wall thickness direction
$Y = Y_S Y_W Y_G Y_C$

The various superimposed contributions to the geometric factor are summarised in Fig. 7.11 and are discussed below.

Table 7.2. Fatigue crack growth parameters for BS 4360 Grade 50D structural steel

In air: $C = 2.0 \times 10^{-11}$, $m = 3.0$
In seawater: $C = 6.0 \times 10^{-11}$, $m = 3.0$ } (in units of MN, m)

('In seawater' covers the case of cathodic protection up to a potential of −0.85 V with respect to the Ag/AgCl cell.)

$\Delta K_{th} = 6.01 - 4.55R$ MPa m$^{1/2}$ for unwelded steel.

$$\Delta K_{th} = \frac{1.45}{1 - 4.5/\sigma_y\pi\overline{a})^{1/2}} \text{ MPa m}^{1/2} \text{ for welded joints.}$$

Source: UEG [15].

Y_S is a well-known correction for a single-edge notch and the accepted value is 1.12 in the absence of any other modifications.

Y_W is used to allow for a finite width of wall through which the crack is propagating. Taking the crack depth as a and the wall thickness as t, the factors for wall-thickness and single edge notch can be combined to give [15]

$$Y_S Y_W = \left[(2t/\pi a) \tan (\pi a/2t)\right]^{1/2}$$
$$\times \left[\frac{0.752 + 2.02(a/t) + 0.37\left[1 - \sin (\pi a/2t)\right]^3}{\cos (\pi a/2t)}\right]$$

Y_G, for variable stress through the wall, is not conveniently expressed as a single multiplying factor. Rather, it is better to divide the hotspot stress between a membrane component, σ_m, which is constant across the wall, and a bending component, σ_b, which varies across the wall and is zero close to mid-wall (see for example BS PD6493 [6]). This leads to a modified equation for the SIF:

$$K = (Y_m\sigma_m + Y_b\sigma_b) \sqrt{(\pi a)}$$

where $Y_m = Y_S Y_W$ (from above)

and $Y_b = \left[(2t/\pi a) \tan (\pi a/2t)\right]^{1/2}$
$$\times \frac{0.923 + 0.199 \left[1 - \sin (\pi a/2t)\right]^4}{\cos (\pi a/2t)}$$

Y_C, the shell correction, is required to modify the excessive conservatism associated with flat-plate idealisations at large crack depths, and acts to allow for the restraint to move-

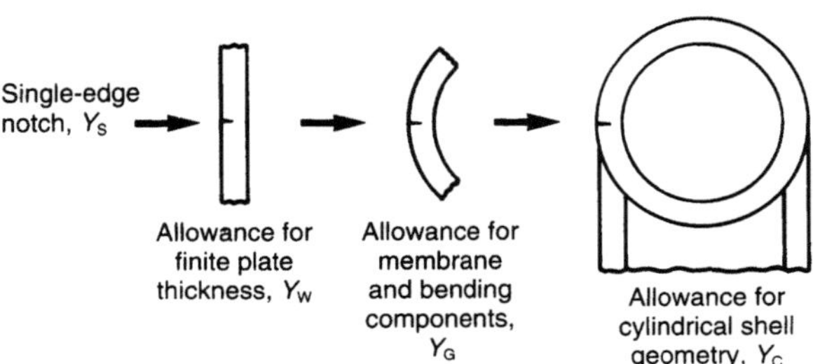

Fig. 7.11. Summary of the various geometric contributions to the stress concentration factor for a defect in tubular intersection. (Adapted from [5].)

ment of the cracked ligament offered by the cylindrical geometry. Both Y_m and Y_b need to be modified and the following has been suggested [15]:

$$Y_m' = \left[1.12 - 0.12(a/t)\right]f_1(a/t) + \frac{(6\pi a/t)f_2(a/t)Y_b}{(\pi a/t) + f_3(a/t)}$$

where

$$f_1(a/t) = [1 - 0.025(a/t)^2 + 0.06(a/t)^4][\sec (\pi a/2t)]^{1/2}$$

$$f_2(a/t) = (a/t) [0.236 (a/t) + 0.264]$$

$$f_3(a/t) = \{(a/t)/[1 - (a/t)]\}^2 \times [5.93 - 19.69(a/t) + 37.14(a/t)^2 - 35.84(a/t)^3 + 13.12(a/t)^4]$$

$$Y_b' = (\pi a/t)Y_b/[(\pi a/t) + f_3(a/t)]$$

Table 7.3 shows the variation in calculated values of the SIF if each of these factors is taken into account cumulatively and it is assumed that the bending stress is twice the membrane stress at the hotspot point. It can be seen that, in the final analysis, Y is in fact a relatively weak function of a/t for such cylindrical geometries, starting at the infinite plate single-edge-notch value of 1.12 and dipping slightly before increasing slowly.

Even this level of sophistication must be tempered with some engineering judgement for realism in the model. For

Table 7.3. Successive approximations to stress intensity factors in a tubular intersection containing a defect

a/t	Stress intensity factor geometric correction, Y			
	Single-edge notch	Finite wall thickness	Membrane and bending	Shell correction
0.0	1.12	1.12	1.12	1.12
0.1	1.12	1.20	1.09	1.02
0.2	1.12	1.37	1.15	0.995
0.3	1.12	1.66	1.28	1.02
0.4	1.12	2.11	1.53	1.08
0.5	1.12	2.83	1.93	1.16
0.6	1.12	4.04	2.61	1.26
0.7	1.12	6.38	3.94	1.38
0.8	1.12	11.99	7.11	1.52
0.9	1.12	34.72	19.39	1.69

Source: UEG [15].

example, if the cracked member forms part of an indeterminate structure, it will attract less load as it becomes more compliant and crack arrest may eventually occur.

Also, the model has assumed that the crack aspect ratio will remain constant as the crack grows. In fact, it is more likely that crack length will increase more rapidly than crack depth (Fig. 7.12), and the question has to be asked as to what constitutes the end of fatigue life.

Finally, as seen above, the stress will vary around the intersection, and it may also be necessary to include the effect of residual stress.

In summary, the fracture mechanics method has a number of advantages over the S–N approach for tubular joint design. However, the methodology has yet to mature fully [15], and designers and maintenance engineers should use calculated results with caution and in conjunction with observation.

7.3 Composite Structures for Marine Applications

A survey carried out in 1987 [16] reported very limited use of composite material in offshore engineering in the UK, although composites have been widely used in marine environments for many years. The reason for this is probably the conservatism in marine design occasioned by the need to ensure reliability. The acceptance of composite materials will therefore be slow, and will evolve through experience in some of the less critical marine applications.

This section illustrates some of the technology reported in the literature for the production and analysis of ship hulls, submersibles and piping to indicate the range and versatility of use of resin matrix composites in marine applications.

The survey referred to above has identified the major engineering applications of fibre-reinforced plastics, and some of these are summarised in Table 7.4. As can be seen, the major marine applications have been in craft of various sizes and in piping. However, examination of non-marine engineering applications points to a number of potential marine uses such as in ships' superstructures, walkways, cladding, tanks and vessels, accommodation and office modules on offshore units and even low-density drillstrings [16, 17].

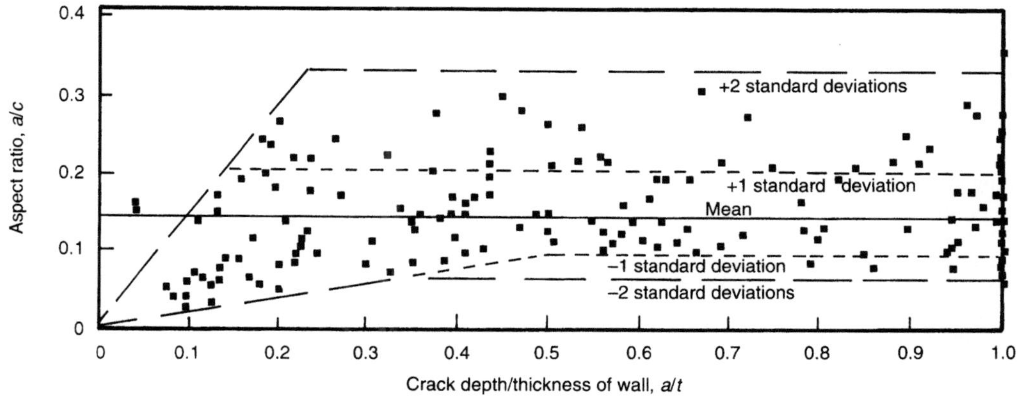

Fig. 7.12. Measured data for crack aspect ratio as a function of crack depth. (Simplified from [5].)

Table 7.4. Major marine engineering uses of fibre-reinforced plastics materials at 1987

Item	Materials used	Fabrication method/ product forms	Country of origin	Brief fabrication/ utilisation reasons
Military ships Mine countermeasure vessels (MCMV) and single-role mine hunters (SRMH) – Royal Navy	E-glass fibres Isophthalic polyester resin	Wet lay-up of glass fibre woven rovings (~ 270 tonnes)	UK (Italy)	Entire hull and superstructure in GFRP GFRP is non-magnetic Corrosion resistance (low maintenance) Good fatigue properties Characterised fire performance
Pleasure boats Motor launches Yachts Dinghies	Predominantly E-glass fibre/ polyester resin Some carbon and Kevlar fibres	Various wet lay-up and spray-up techniques	Numerous	Corrosion resistance Self-coloured Good fatigue properties. Lightweight, allowing higher speed, better planing characteristics Durable for water–shore transfer Manoeuvrable
Lifeboats	E-glass fibre/ polyester resin	Various wet lay-up and spray-up techniques	Numerous	Corrosion resistance Durable free-fall launching from vessels Thermally insulative Proven fire performance
Piping	Predominantly E-glass fibre with polyester or vinyl ester resins	Various winding, spray-up and casting techniques	Numerous	Corrosion resistance Pressurised piping to handle water, seawater sewerage some aqueous fluids some hydrocarbons Lightweight

Source: Fulmer Research Institute [16].

The list of types of composite materials in marine use is also quite modest, being mainly centred around polyester or vinyl ester resins reinforced almost invariably with glass.

The use of GRPs for piping is aided considerably by the ease of construction of the pipes (mainly by filament winding) and the relatively simple joining technology. Shipboard applications are widespread and, for example, Murtagh and Lemley [18] suggest the applications shown in Table 7.5. In general terms, the advantage of GRP for pressure vessel applications can be illustrated in terms of their 'storage efficiency' [19], defined as the ratio of energy stored to weight of the vessel. Grim [20] has reviewed the experience of one company in shipboard use of GRP piping. In particular, uses in very large crude carriers (VLCCs) were discussed and Fig. 7.13 shows a typical layout of the ballast and cargo piping in such a vessel. The piping material reported by this author was usually epoxide or vinyl ester reinforced with E-glass, and applications were mostly as cargo and ballast lines or as stripping lines.

The loading system in composite pipes is relatively simple, so that optimal design can be more easily achieved than it is for more intricate structural geometries. For a pipe subjected to internal pressure only, Fig. 7.14 shows the variation in stress with winding angle [19]. Given that it is important to arrange the fibres such that the stress normal to the fibres is minimised, it can be seen that the optimum winding angle is about 55 °. Of course, for other conditions of loading such as compression and bending this does not necessarily hold. For example, it has been found that

graphite-reinforced epoxy laminated tubes subject to compression could be optimised (with respect to weight) by using a 60% 0°, 40% ±45° laminate [21] (i.e one in which 60 per cent of the layers run in the direction of maximum stress and the rest are at ±45° to this direction).

Muscati and Bradford [22] have analysed a GRP pipe with integral flanges, the structure of which is illustrated in Fig. 7.15. They present a stress analysis for the plain pipe made up of three layers: a UPVC liner, a layer of resin impregnated chopped strand mat and an outer layer of resin impregnated rovings with unidirectional fibres along the hoop direction. They also present a finite-element stress analysis which includes the integral flange. They concluded that the burst pressure of the pipe could be severely compromised by lack of sufficient rovings adjacent to the flange, but that the local bending stress resulting from the flange geometry had little effect on the burst pressure. Both stress analyses tended to overestimate the failure pressure, highlighting the need to use appropriate factors of safety in such complex constructions.

The use of GRPs in ship and boat hulls has a long history, and development is driven principally by the requirement for larger structures, e.g. 'Brecon' class mine countermeasure vessels at 55 m [17], or for high performance, as in racing vessels {Norwood and Marchant [23]). For the former of these applications at least (as can be appreciated from Chapters 2, 3 and 4) the main considerations are to have a fabricable composite with good strength, toughness and water resistance. Furthermore, as with all

Table 7.5. Potential shipboard applications of glass-reinforced plastics

	Present approval status	Regulation change required	Serves vital equipment	Serves non-vital equipment	Physical strength critical	Flooding-failure/watertight integrity	Penetrates fire boundaries	Valve requirement	Electrostatic capabilities
Accommodation spaces									
Potable water (hot and cold)	No	Yes	No	Yes	No	No	Yes	Yes	No
Sanitary water	No	Yes	No	Yes	No	No	Yes	Yes	No
Chilled water	No	Yes	No	Yes	No	No	Yes	Yes	No
Sewage	No	Yes	No	Yes	No	No	Yes	Yes	No
Ballast									
In ballast and cargo tanks	Yes	No	No	No	Yes	No	Yes	Yes	Yes
In machinery spaces	No	No	No	No	Yes	Yes	Yes	Yes	Yes
In duct keels/ pipe tunnels	No	No	No	No	Yes	Yes	No	Yes	Yes
In pump rooms for ballast only	No	No	No	No	Yes	Yes	Yes	Yes	Yes
Bilge									
In double-bottom tanks	No	No	No	No	Yes	Yes	No	Yes	No
In duct keels/pipe tunnels	No	No	No	No	Yes	Yes	No	Yes	No
In machinery spaces	No	No	No	No	Yes	Yes	Yes	Yes	No
Crude oil washing									
In tanks/cargo	Yes	No	Yes	Yes	Yes	No	No	Yes	Yes
Fuel oil transfer									
In tanks	Yes	No	No	No	Yes	No	No	Yes	Yes
Inert gas									
On deck	No	No	Yes	No	Yes	No	Yes	Yes	Yes
In machinery spaces	No	No	Yes	No	Yes	No	Yes	Yes	Yes
Scrubber drain in tank	Yes	No	Yes	No	Yes	No	Yes	Yes	No
Low-pressure air									
On deck	Yes	No	No	Yes	Yes	No	Yes	Yes	Yes
Oily water separators									
Supply piping	No	No	No	Yes	Yes	No	No	No	No
Discharge piping	Yes	No	No	Yes	Yes	Yes	No	Yes	No
Salt water service									
To non-vital machinery	Yes	No	No	Yes	Yes	No	No	Yes	No
Tank cleaning									
Water on deck	Yes	No	No	No	Yes	No	Yes	Yes	Yes
Water in tanks	Yes	No	No	No	Yes	No	No	No	Yes
Tank vents									
From ballast tanks – wing tanks	Yes	No	No	No	Yes	No	Yes	No	No
From ballast tanks – voids or cargo	No	No	No	No	Yes	No	Yes	No	No

Source: Murtagh and Lemley [18].

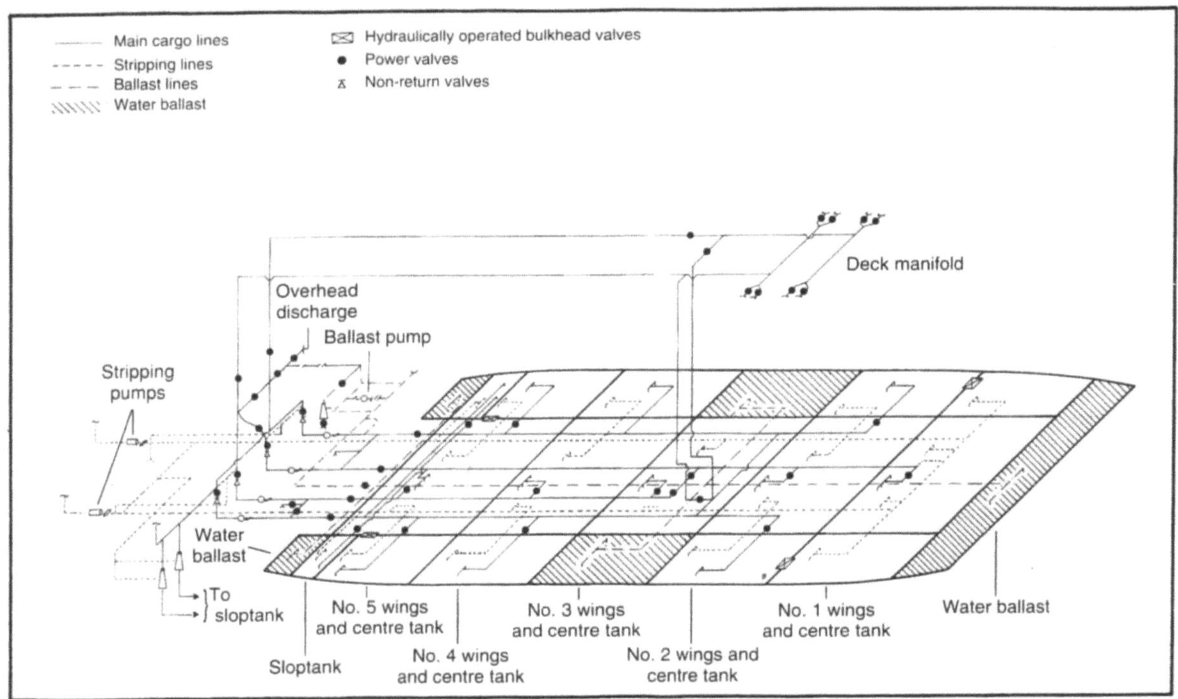

Fig. 7.13. Schematic diagram of the cargo and ballast piping systems aboard a typical large crude-oil carrier. (From Grim [20].)

transportation applications, 'structural efficiency', measured as some combination of Young's modulus or strength with density is of some importance. The cost of material is significant, being from three to five times that of steel, and fabrication costs are also high due to the labour-intensive nature of the construction process [24].

The mechanical design of such structures is complex, although some useful design solutions exist for some component geometries (e.g. Rhodes and Marshall [25]). Chalmers [24] has summarised some of the mechanical

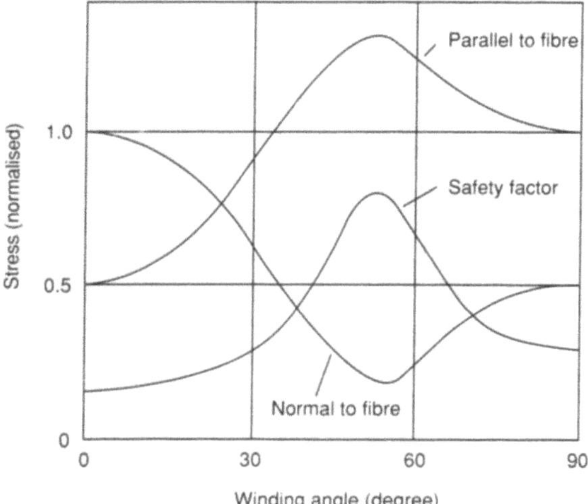

Fig. 7.14. Effect of winding angle on GRP pipe stresses. (From Cliffe [19].)

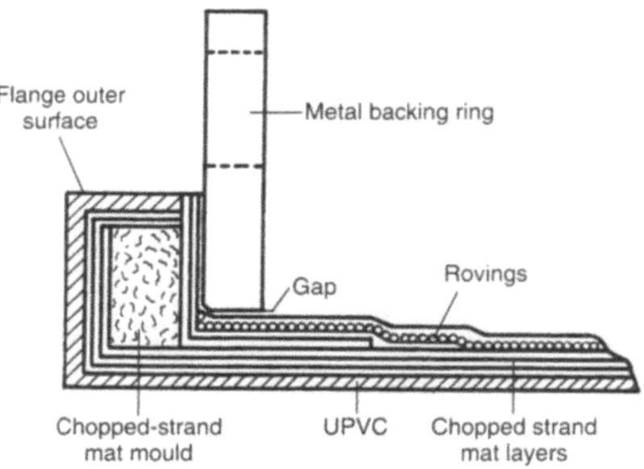

Fig. 7.15. Schematic diagram of composite pipe flange construction. (From Muscati and Bradford [22].)

aspects of the construction of ship hulls, while noting that warship design in the UK currently uses mostly woven roving for this application; Fig. 7.16 shows an example of the variation in the modulus and strength of glass-reinforced polyester resin laminates with direction for a given glass loading. Chalmers indicates that a useful failure criterion might be the orthotropic version of the von Mises yield criterion (see Chapter 2):

$$\frac{\sigma_x^2}{X^2} - \left[\frac{1}{X^2} + \frac{1}{Y^2} + \frac{1}{Z^2}\right]\sigma_x\,\sigma_y + \frac{\sigma_y^2}{Y^2} + \frac{\sigma_{xy}^2}{T^2} = 1$$

where the σ terms are the in-plane stresses and X, Y and T are the corresponding strengths. Z is the through-thickness

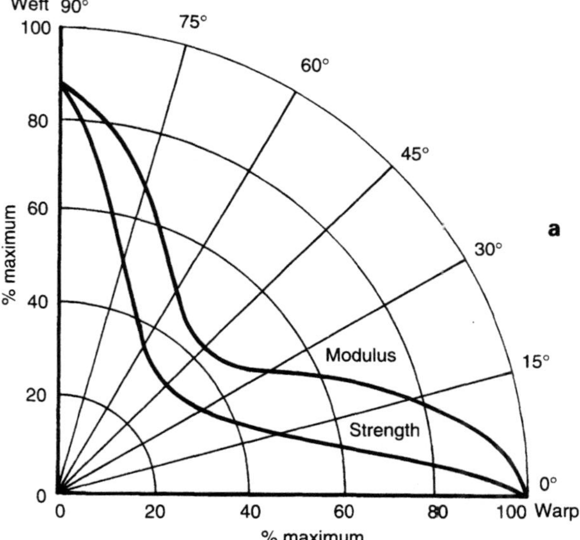

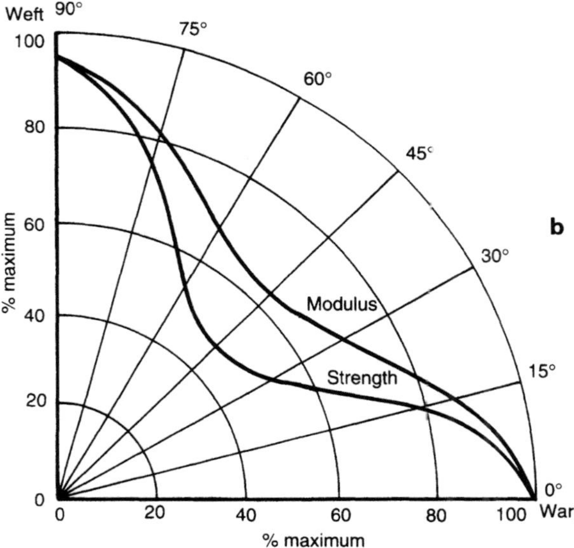

Fig. 7.16. Variation in modulus and strength with direction for glass-reinforced polyesters with 800 g/m² woven roving. **a** Tension. **b** Compression. (From Chalmers [24].)

Table 7.6. Factors of safety for glass-reinforced plastics in marine environments

	Ultimate strength/working strength
Static short-term loads	2
Static long-term loads	4
Variable or changing loads	4
Repetitive loads	6
Fatigue or load reversal	6
Impact loads	10

Source: Chalmers [24].

[23] have considered the relative advantages of the two and find that, with appropriate curing, polyester resins based on isophthalic acid show good water-resistant properties and that structural and economic optimisation is best achieved using reinforcements of mixed glass and polyaramid fibres. These authors illustrate this with examples of weight savings for the same structural performance. In one example of a 13-m, 30-knot patrol vessel, a 30 per cent weight saving for identical loading conditions and structural analysis was achieved by using a glass and polyaramid-reinforced polyester instead of the conventional glass-reinforced polyester.

Composite ship hulls were originally investigated in the early 1960s by the UK Ministry of Defence to replace the existing aluminium-framed wooden MCMVs. The eventual construction of HMS *Wilton* between 1970 and 1973 represented the culmination of this investigation, and the vessel, with a predicted hull life of 60 years, is still in service [17]. The design and construction of HMS *Wilton* are well documented (for example, Dixon et al. [26]). The hull was laid into a complete female mould using specially designed dispensers which pre-impregnated the 1.2-m (48-inch) wide 800 g/m² woven roving. The wetted cloth was hand-positioned and consolidated by hand rolling. For most of the ship's length there were 34 laminations at the keel, corresponding to a hull thickness of about 33 mm (1.29 inches). Weather and outerbottom surfaces were given a resin-rich outer layer.

As indicated by Dixon et al. [26], HMS *Wilton* was built ostensibly to prove the use of GRP as a hull material, and the experience gained was used in the development of the 'Hunt' class MCMVs, of which 13 have been built between 1975 and 1987. These vessels are 60 m long and are the largest GRP vessels afloat. A third generation of GRP minehunters are currently under development.

Mableson et al. [17] have summarised the British experience in large GRP marine constructions culminating in the third-generation MCMVs mentioned above. Nixon [27] has described the advances incorporated in the first of these third-generation vessels, HMS *Sandown*. As far as the hull is concerned these can be summarised as follows:

The experience of earlier vessels has engendered sufficient confidence to use a reduced and constant hull thickness.

The design of the stiffener has undergone modification to obviate the use of bolting.

Improvements in manufacturing technology include a modular building method and semi-automation of the laminating procedure.

These, along with a reduction in the complexity of the internal stiffening system, allowed the third-generation vessels to be built with a weight saving of 40 per cent, a cost

direct strength. However, Chalmers indicates that a slightly more conservative criterion might be safer for plane stress situations:

$$|\sigma_x|/X + |\sigma_y|/Y + |\sigma_{xy}|/T = \alpha$$

where α lies between 0.15 and 0.2.

Finally, Chalmers suggests that some design manuals recommend the factors of safety (ultimate strength/working strength ratio) shown in Table 7.6 for GRP in marine environments.

Materials selection for the hulls of large ships is still evolving, with the more traditional glass-fibre-reinforced polyesters having some competition from, for example, more durable polyesters reinforced with polyaramid or mixtures of polyaramid and glass. Norwood and Marchant

saving of 50 per cent with a reduced build time and superior shock resistance. Nixon suggests that the main improvent for future hulls of this type will be to corrugate the structure for improved stiffness.

A number of designers have found the properties of composites attractive for underwater pressure hull applications. For example, Liddle and George [28] have reviewed GRP applications for small (2 to 5 men) monobaric submersibles and for medium-to-large ROVs (see Chapter 6). One of the principal advantages of GRP is that it reduces the amount of buoyancy material which has to be used for a given payload. Table 7.7 summarises some of the applications of composites outlined by these authors along with their capabilities. Harruff and Sandman [29] describe a hybrid shell structure incorporating a polyimide syntactic foam between layers of carbon-fibre-reinforced epoxy for a pressure hull for an underwater ordnance application. The stated advantage of this construction was that it offered a considerable weight reduction over the operationally equivalent aluminium shell. Reduced self-noise was an additional advantage specific to this (military) application.

Table 7.7.   Some applications of GRPs in underwater intervention

Type/performance	GRP application
Five-man submersible with diver lock-out operating to 457 m	GRP used for fore-and-aft pressure hulls. (Centre DLO section is made of steel)
One-man ADS with four articulated limbs operating to 610 m	Toroidal section GRP hulls currently used. CFRP being evaluated for greater depth
Submersible remotely operated trenching vehicle	Chassis and hard ballast tanks
Inspection, construction and survey ROV for use to 1500 m	Large pressure vessel and frame in GRP providing buoyancy and large payload without the use of syntactic foam

Source: Liddle and George [28].

7.4   The Design of Cathodic Protection Systems for Subsea pipelines

Cathodic protection forms an important part of the external corrosion protection system for subsea pipelines. The extreme importance of the mechanical integrity of pipelines for the transport of hydrocarbons means that protection against all forms of degradation has to be assured with a high degree of reliability.

As mentioned in Chapter 4, most pipelines are manufactured from what, for the purposes of corrosion protection, can be considered as carbon–manganese steels, and therefore the best method for protection in the immersed condition is cathodic protection. In the design of subsea pipelines it is also usual to include an anti-corrosion coating, normally consisting of a coal-tar-based product reinforced with

a wrapping, usually of glass fibre, to an overall thickness of several millimetres. Thinner (usually fusion-bonded epoxy) coatings are used for some pipelines. On top of this, depending upon considerations of stability, a concrete weight coating is usually applied and this can influence the rate of degradation (at least physically) of the anti-corrosion coat. Because of the way in which subsea pipelines are laid, it is inevitable that discontinuities in the coatings are present at the ends of individual pipe sections to allow welding on the laybarge (Chapter 5). Such field joints are normally protected by priming and wrapping, followed by the pouring of mastic into a former to give roughly the same outer diameter as the concrete weight coating which stops typically around 0.25 m on either side of the girth weld. The field joints also contain the sacrificial anodes where necessary, and these anodes are commonly of the bracelet type illustrated schematically in Fig. 7.17.

Owing to their geometric simplicity, the design premise of cathodic protection systems for pipelines is normally a matter of using the electrochemical considerations outlined in Chapter 3 to decide upon the spacing of anodes on the line, usually expressed as the proportion of field joints which require to contain anodes. The normal design calculation consists of finding the required protective current for the entire line and dividing this by the maximum likely current output from each anode. This output current depends on the potential difference between anode and cathode, and also upon the anode-to-electrolyte resistance. Estimates of these quantities can be obtained from design rules such as those issued by DnV [30], and such rules normally also indicate protective current density requirements for various environments (seawater by geographical location, saline mud). For steel which is coated, the required protective current density can be multiplied by a quantity often referred to as the 'coating breakdown factor', a 0-per-cent breakdown factor being indicative of perfect coating and a 100-per-cents breakdown factor indicating a non-existent or totally degraded coating. Such coating breakdown factors are normally considered to increase linearly with time, usually with a high degree of conservatism. As well as having the required current-carrying capacity, it is also necessary for anodes to be able to supply the envisaged current profile (which, in general, will increase with time) for the lifetime of the pipeline and kilogram-per-ampere-year equivalents are well documented for all common anode material specifications.

In some cathodic protection cases attenuation (loss of protection with distance from anode) is a consideration but

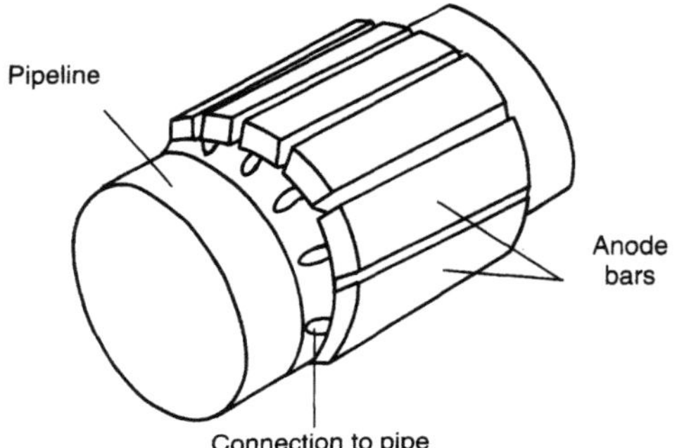

Fig. 7.17 Typical design of bracelet anode for subsea pipelines [30].

this is normally only so for high environmental resistivities, complex geometries or large anode spacings. It is, however, possible that loss of protection through attenuation may begin to take place in galvanic anode systems towards the end of their lives. As an example of the scale of the problem, Thomason et al. [31] have estimated the throwing power of galvanic anodes on coated structures to be at least 900 m. On the other hand, Rippon and Finnegan [32] have reported local exhaustion of galvanic anodes in a polyethylene-coated methanol pipeline where coating breakdown was much higher than was originally expected. This resulted in under-protection of nearly half of the length of the line, a situation which was remedied by retrofit with sled anodes. Apart from the necessity to assess conservatively coating breakdown, care must also be taken to ensure that low initial activity does not subsequently compromise the ability of the anode to supply current as the demand increases with coating degradation. Furthermore, there is the possibility that the CP system itself may accelerate the breakdown of the coating. Peterson et al. [33] have examined the former effect and conclude that resistive corrosion products formed on anodes operating at low current do result in a reduction in current capacity, but that this is so small that it is unlikely to be of any engineering significance. Thomason et al. [31] have suggested that the use of large-cored anodes may be more desirable for protection of coated structures, and that anode shapes could be configured to expose a greater surface area as they are consumed rather than a smaller one as is normally the case. One way of doing this would be to use a triangular cross-section with all but the tip coated at installation.

The following example of CP design for a typical concrete-coated and anti-corrosion-wrapped North Sea pipeline is given to illustrate some of the considerations involved.

The pipeline is of diameter 760 mm, with a corrosion-protective coating but is not buried. It is to be protected using zinc alloy bracelet anodes, of mass, W, 225 kg and exposed surface area, A, of 0.7 m^2, over its entire length of 120 km. Tables 7.8 to 7.11 provide the necessary data for the following design.

The pipeline will show an external surface area, A', of 2.39 m^2 per metre length and, from the data, should require 4.78 mA per metre of length at installation and 35.85 mA per metre of length towards the end of its life. From the

Table 7.9. Anode resistance formulae

Slender anodes	$R_a = \dfrac{\rho}{2\pi r}\left(\ln\dfrac{4l}{r} - l\right)$

ρ = resistivity (ohm cm)
l = length of anode (cm)
r = equivalent radius of anode (cm)
  = $\sqrt{\dfrac{a}{\pi}}$

a = cross-section of anode (cm^2)

Plate anodes	$R_a = \dfrac{\rho}{2S}$

S = mean length of anode sides (cm)
  = $\dfrac{b + c}{2}$, $b \geq 2c$

Other shapes	$R_a = \dfrac{0.315\rho}{\sqrt{A}}$

A = exposed area of anode (cm^2)

Source: Det norske Veritas [30]

electrochemical data and from the recommended resistivity for temperate waters (given by DnV [30]) of 33 Ω cm the anode resistance can be given by:

$$R_a = 0.315\ \rho/\sqrt{A} = 0.1242\ \Omega$$

and hence

$$I_a = E_d/R_a = 2\ A$$

(see Section 3.4.1.2), using a (conservative) driving potential of 250 mV as suggested by DnV.

Thus, using the coating breakdowns suggested by Mollan and Andersen [34], each anode should be capable of protecting a length, S of about 56 m of pipeline even towards the end of the design life of the pipeline. At the beginning of life, the current-carrying capacity would be sufficient to protect about nine times this length. The lifetime of the anodes working at the mean current given by

Table 7.8. Electrochemical parameters for some sacrificial anode alloys

Alloy	Environment	Driving potential (mV)	Current capacity (A h/kg)	Consumption rate (kg/A year)
Al–Zn–Hg	Seawater (5 to 30 °C)	200 to 250	2600 to 2800	3.1 to 3.4
Al–Zn–In	Seawater (5 to 30 °C)	250 to 300	2500 to 2700	3.2 to 3.5
Al–Zn–In	Saline mud (5 to 30 °C)	150 to 250	1300 to 2300	3.85 to 6.7
Al–Zn–In	Saline mud (30 to 90 °C)	100 to 200	400 to 1300	6.7 to 22
Zn	Seawater	200 to 250	760 to 780	11.2 to 11.5
Zn	Saline mud (0 to 60 °C)	150 to 200	760 to 780	11.2 to 11.5

Source: Det norske Veritas [30].

Table 7.10. Suggested minimum design current densities for the cathodic protection of bare steel (mA/m^2)

Location	Initial value	Mean value	Final value
North Sea (northern)	160	120	100
North Sea (southern)	130	100	90
Persian Gulf	120	90	80
India	120	90	80
Australia	120	90	80
Brazil	120	90	80
Gulf of Mexico	100	80	70
West Africa	120	90	80
Indonesia	100	80	70
Pipelines (burial specified)	50	40	30
Risers in shafts with flowing seawater	180	140	120
Risers in shafts with stagnant seawater	120	90	80
Saline mud (ambient temperature)	25	20	15

Source: Det norske Veritas [30].

$$I_m = 9 \times 10^{-3} A'S = 1.2 \text{ A}$$

is thus:

$$T = WU/EI_m = 12.2 \text{ years}$$

using a conservative consumption rate, E, from DnV of 11.5 kg/A year and a conservative utilisation factor (also from DnV) of 0.75. It should be pointed out that, if this pipeline was buried, a considerable reduction in the degree of coating breakdown would be expected along with a change in anode driving potential and electrolyte resistivity.

Some actual anode surface area, weight and distribution data, as used for North Sea subsea pipelines, have been published. Also, Chandler [35] has indicated that anode spacings in submarine pipelines are usually in the range 70 to 170 m, and anode sizes are normally constrained by manufacturing considerations and the need to keep thickness close to that of the weight coat. A rough idea of the variations in current practice can be obtained by examining published data for pipelines as indicated above and comparing these with a

Table 7.11. Suggested minimum design current densities for the cathodic protection of coated pipeline systems

Coating	Current density (mA/m^2)		
	Initial	Mean	Final
Corrosion protective coating and concrete weight coating, seawater	2	9	15
Corrosion protective coating and concrete weight coating, burial specified	1	3	3
Thick film coating, seawater	5	18	30
Thick film coating, burial specified	3	6	10

Source: Mollan and Andersen [34].

standard approach. Using the DnV mean values for design current densities (40 mA/m^2 and 110 mA/m^2 for buried and unburied lines respectively) and assuming a coating breakdown which increases linearly with time, i.e.

$$B = C_0 + C_t t$$

the mean anode current, I_m, over the lifetime, T, of the pipeline can be expressed as

$$I_m = I[C_0 + (C_t T)/2]AL/N_a$$

where I is the mean protective current density for bare steel, A is the surface area of pipeline per unit length, L is the total pipeline length and N_a is the total number of anodes. This gives a calculated anode lifetime of

$$T = -\frac{C_o}{C_t} + \frac{C_o^2}{C_t} + \frac{2WK}{C_t AS}$$

where K is a constant dependent upon the required protective current density (geography and whether or not the pipe is buried) and the current capacity of the anode (whether buried or not and anode material). Although detailed CP design criteria are rarely published for individual operators, the published data can be used to give an idea of the degree of conservatism employed in practice. Table 7.12 shows calculated values for lifetimes of cathodic protection systems for some North Sea submarine pipelines, assuming that the initial coating breakdown (C_o) can be taken as 1 per cent and using an annual accumulation of breakdown (C_t) of 0.25 per cent per annum. The range of figures given is for the buried and unburied cases respectively. Since all of the data relate to lines protected by zinc anodes, it can be seen that there is quite a wide range in the amount of anode material chosen for protection of subsea pipelines. The main factors affecting this decision are likely to be the original envisaged design life, whether or not the pipe is considered to be buried, and the coating breakdown pattern expected. It would appear that, for a nominal design life of 20 years, operators are choosing a breakdown factor slightly more conservative than that suggested above. Data on actual breakdown are difficult to obtain since they require substantial field experience (probably to anode exhaustion), but the conservatism is required to allow for the fact that the breakdown is unlikely to be spread uniformly along the line.

Table 7.12. Calculated anode lifetimes from published North Sea cathodic protection data

Anode mass per unit steel external area (kg/m²)	Calculated anode lifetime with C_o and C_t as given in text (years)
0.59	24 to 12
0.52	22 to 11
1.4	39 to 21
2.6	55 to 30
0.97	32 to 17
0.72	27 to 14
0.52	22 to 11
1.4	39 to 21

A number of methods are available for monitoring cathodic protection performance of submarine pipelines. Because it is rare to be able to have access to bare steel (as might be the case for an offshore platform), subsea pipeline surveys have to be made without direct contact. Such measurements usually indicate the pipeline potential along its length, and field gradient methods can also indicate anode current output [34]. A typical potential survey will show values in the range of around −870 mV vs. saturated calomel electrode (SCE) to around −1020 mV vs. SCE (e.g. Rippon and Finnegan [32]) and, aside from local coating damage, a number of other factors may affect the potential. For example, Backhouse [36] gives a case of a subsea pipeline where the onshore impressed current system was having an effect 20 to 30 miles out to sea, so that anodes were drawing current from the line rather than vice versa. Furthermore, anode currents and line potential are often seen to increase on the approach to an offshore platform. Figure 7.18 illustrates both of these effects.

Few cases of protection of subsea pipelines by impressed current are reported in the literature, but one interesting example is given by Wagner and Cremers [37], where an impressed-current system was used to protect a 25-year-

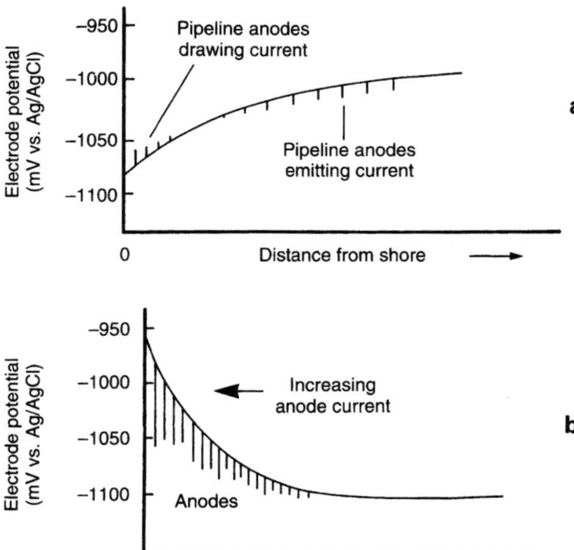

Fig. 7.18. Cathodic protection interference phenomena. **a** Effect of a shore-based impressed current system on pipe protection and anode activity. Last spike is at 30 km from shore. **b** Effect of a platform on the level of protection of a pipeline. (From Backhouse [36]).

old, 2-mile section of pipeline crossing Newark Bay near New York City after the sacrificial system had become exhausted. The reason for choosing an impressed-current system was that the coating had deteriorated to the point where the required protection was estimated to be about 50 to 100 amperes, which would need 120 kg of anode material spaced every 15 metres or so and giving a nominal lifetime of 8 to 16 years. Using the generalised attenuation equations (e.g. Parker and Peattie [38]), the authors were

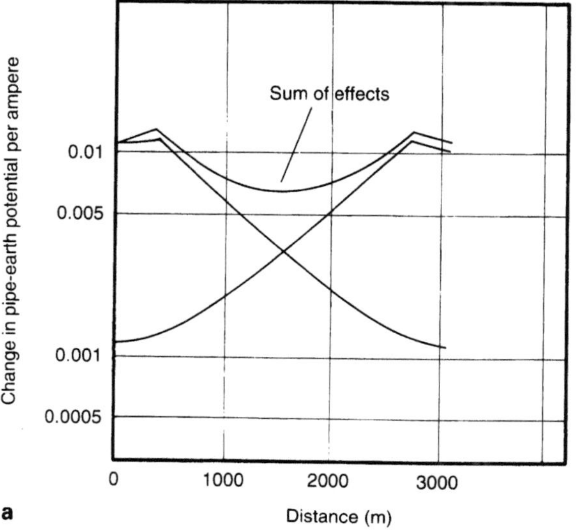

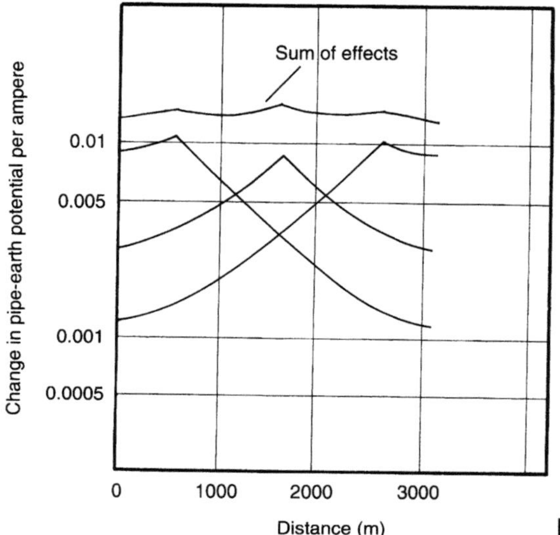

Fig. 7.19. Attenuation profiles for protection of a short sea-crossing pipeline using impressed current. **a** With current applied equally from two rectifiers with two negative returns at 350 m and 2650 m. **b** With current applied from a single remote anode using three balanced negative returns at 550 m, 1550 m and 2500 m. (From Wagner and Cremers [37].)

able to examine analytically the effectiveness of various possible anode placements. It was first shown that the coating conductance was such that a single anode located at one end of the line would be almost ineffective at the other end. The eventual choice was to place a rectifier at either end of the pipe, with one negative tap for each rectifier, although it was recognised that multiple negative taps even with one rectifier using resistors to balance the current drawn at each tap may be a more effective method. Figure 7.19 compares the attenuation profiles given by these authors for the two- and three-tap cases.

7.5 Flexible Pipes

Flexible pipes have found a large number of applications in offshore hydrocarbon production technology for functions such as reservoir water injection, risers, offshore and jetty loading systems and subsea applications. It is also probable that other marine activities such as seabed mining will benefit from the use of such technology because of the ability of flexible pipes to accommodate large excursions, especially useful at larger water depths. Flexible pipes are composite constructions, in that they are made of several layers of different materials and it is the selection of these materials that is the subject of this case study.

The key property of a flexible pipe is the provision of a low bending stiffness, but this has to be associated with the other functional requirements of the pipe such as its ability to withstand external and internal pressures, an acceptable axial stiffness and strength, and resistance to the internal and external environments. Durability is also a key factor, especially in dynamic applications, and the following discussion focuses on the application of flexible pipes for marine risers (the current major dynamic application), on the grounds that the varying loads, the internal pressure and the need for reliability due to the hazardous nature of the transported fluid combine to make this one of the most demanding applications which can be envisaged for flexible pipes in marine environments.

The properties of flexible pipes are generally obtained by the application of the philosophy used in wire rope design, suitably modified to provide the required conduit. The strength is provided principally by wire wound in a helical geometry, thus providing low bending stiffness but high tensile strength. Additional features are, of course, required for the pressure containment function. For high internal pressure applications, there are two main types of flexible pipes, generally referred to as of 'bonded' and 'unbonded' construction, examples being shown in simplified form in Fig. 7.20.

As can be seen, the principal difference between bonded and unbonded construction is that only the former uses an elastomeric material between the various reinforcements; this elastomer is cured by vulcanisation and thus bonds all the layers together. In the unbonded construction it is normal to use a thermoplastic as the interlayer material and each layer is free to slip over the others. This difference between the constructions of the two types of pipe means that the materials requirements are slightly different for each.

It is first necessary to examine the function and loading (mechanical and environmental) for each layer before considering the constraints on selection.

A typical unbonded pipe construction (Moore [39]) will have an external sheath (of a thermoplastic), which is designed to protect the inner parts of the pipe wall from abrasion or corrosion, and also serves to hold the underlying armour wires in place. The armouring is of double-crosswound construction, the two armour layers normally being in metal–metal contact. The long pitch of these armour wires provides good tensile and torsional resistance, and internal pressure is normally resisted by an additional steel spiral, this time close-wound and often of an interlocking nature. Another thermoplastic sheath is present between the armour and pressure-resisting layers and this reduces friction between these parts of the pipe. Under the pressure-resisting layer is the leakproof part of the pipe, consisting of another thermoplastic tube. Because this thermoplastic can allow the diffusion of gas, applications where this may occur require a metallic carcass to prevent

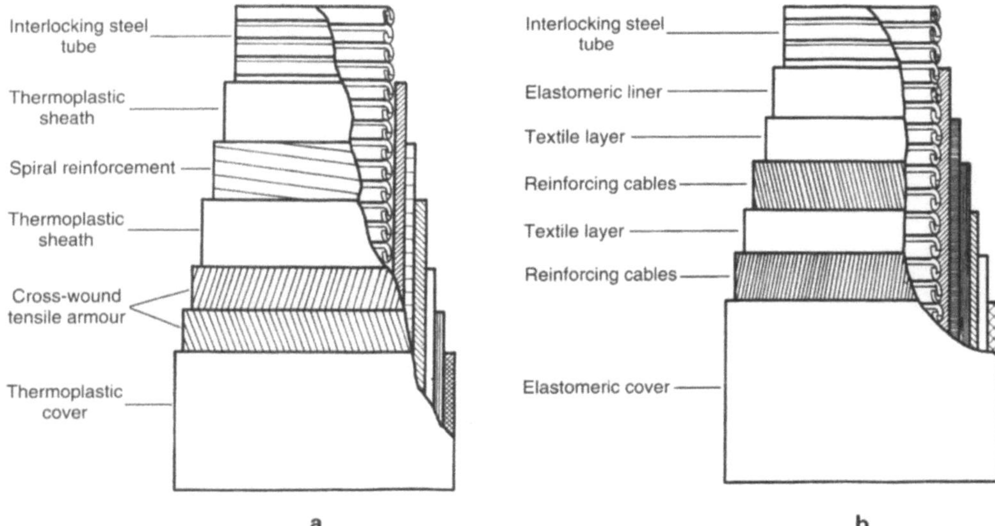

Fig. 7.20. Schematic diagram of bonded and unbonded flexible pipes. **a** Unbonded. **b** Bonded.

the inner sheath from collapsing into the bore under conditions of rapid decompression. In many designs this carcass is made from interlocking elements of stainless steel.

A typical bonded pipe (Griffiths [40]) will have an elastomeric cover, again with the function of protecting the underlying layers. The main internal pressure and strength elements are two sets of spirally applied balanced layers of cables, and these are encapsulated by elastomer which aids load-sharing between the layers, eliminates internal sliding and protects the cable wires. Underneath the cable layers are a set of spirally wound textile reinforced elastomeric layers which aid bonding of the entire pipe and act as hydraulic transfer layers. The next lowest layer is of a chemical-resistant elastomer, which provides a barrier against the permeation of gases through the pipe wall. The innermost layer is again an interlocking metallic carcass similar to that described above for unbonded pipes. The carcass provides hoop strength to resist point and distributed loads, external pressure, and radial forces resulting from tension in the reinforcement cables. The whole pipe is vulcanised, and this provides bonding between all of the structural layers except the interlock tube; this latter allows fluid to pass over it to the elastomeric liner, and it is this liner which provides the first leakproof element of the pipe. An additional function of the carcass then becomes its ability to resist the rupture of the elastomeric liner under conditions of rapid decompression.

Figure 7.21 shows functional diagrams of the two types of pipe discussed above.

Materials selection for such pipes is a complex process and obviously depends on the function of the particular layer and the overall design of the pipe. Some detailed aspects of the service conditions are also relevant to some of the materials selection, so that alternative materials, and even extra layers, may be specified in some cases. It is interesting to note that the basic types of materials in use for flexible pipe construction are steel, nylon and rubber, whose properties are radically different, and it is a tribute to the designers that such a disparity in properties can be combined for a number of useful purposes. Table 7.13 gathers some of the basic properties of flexible riser component materials.

Table 7.13. Some basic properties of materials used in flexible pipe production

	Steel wire	Nylon	Rubber
Elasticity, E (GPa)	200	2.5	0.006
Shear modulus, G (GPa)	80	0.06	0.002
Poisson's ratio	0.3	0.4	~ 0.5
Ultimate tensile stress (MPa)	~ 1200	~ 40	15 to 20
Thermal expansion coefficient ($10^{-6}/°C$)	~ 17	~ 90	~ 150

Source: MacFarlane [41].

The following discussion addresses each of the main functions of the pipe and considers the materials selection solutions which have been found by a number of manufacturers to solve each of these problems. The principal functions are seen from the above to be: reinforcement, internal and external environmental resistance, and the provision of a pressure barrier.

7.5.1 Reinforcing Materials

The over-riding consideration for reinforcing materials is strength, and most manufacturers have chosen materials used in wire rope technology for this purpose. This would seem sensible in view of the functional similarity in the two

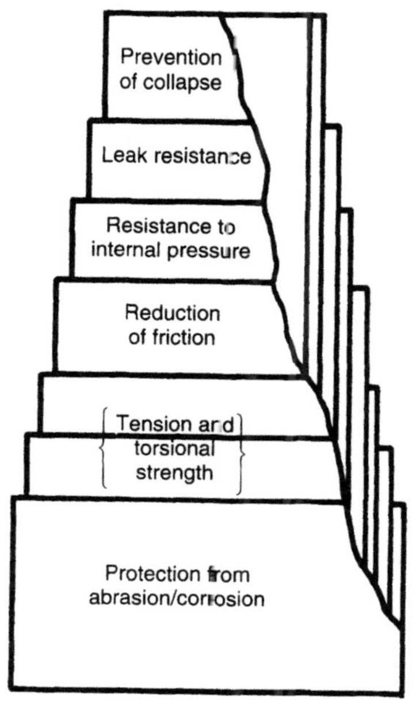

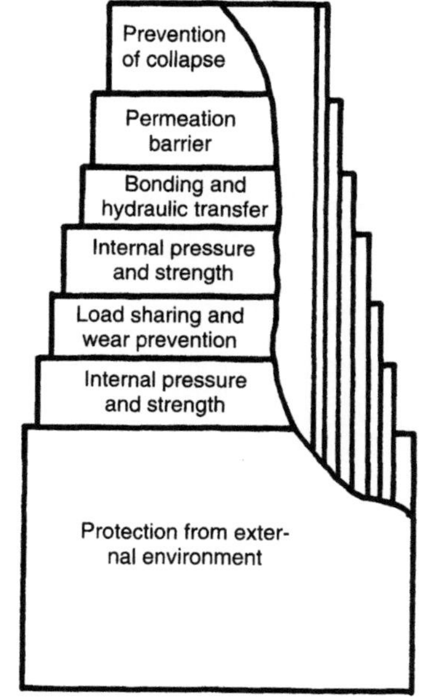

Fig. 7.21. Functional diagrams of unbonded and bonded flexible pipes. **a** Unbonded. **b** Bonded.

applications (e.g. Stacey [42]). Increased strength can be provided by increasing the amount of steel used or, to a certain extent, by improving the grade of the steel. One manufacturer [39] reports the use of three grades of wire material, depending on the local conditions, and some properties of these and some other rope wire specifications are shown in Table 7.14.

Table 7.14. High-strength carbon steel properties and applications in wire form
(These are not specifications – the original standards should be consulted)

Specification and Grade	UTS (MPa)	Applications
AFNOR FM15	784	Flexible riser reinforcement where sour conditions are possible
AFNOR FM35	850	Flexible riser reinforcement where sour conditions are possible
AFNOR FM72	1500	Flexible riser reinforcement for sweet production
ASTM A648/1	1300	Wire for reinforcing
ASTM A648/11	1500	Wire for reinforcing
BS 2763	Varies with diameter	Wire for rope

Although designed for strength, the steel elements of flexible pipes have a number of secondary selection considerations, principal among these being resistance to wear and/or fatigue and resistance to corrosion. Of these, the first is generally accepted to be the most likely failure mode for unbonded flexible pipes under dynamic conditions (Feret and Bournazel [43]). Under alternating wave loading, inter-wire contact between layers can reduce the wire thickness and, combined with the alternating tensile and bending loads, wire fatigue failures are possible. This problem is largely obviated by design to ensure that the endurance of the wires exceeds the service life of the pipe.

The presence of the surrounding low-viscosity elastomer in the bonded construction, and the incorporation by some manufacturers of anti-chaffing layers should reduce the wear element of this mechanism of wire failure (Cocks [44]).

Provided that the external and internal integrity of the pipe is maintained, there should be no corrosion problem at the armour layers. The construction of the bonded pipe is such that any gas which does penetrate the barrier elastomeric lining continues to pass through the pipe and penetration of seawater to the cables can only be local because of the low-viscosity elastomeric matrix. A brass coating on the wires presents further corrosion resistance. In the case of unbonded pipes the possibility of damage to the external sheath is recognised and here seawater could penetrate to a substantial area of armour. This might be aggravated by gas diffusion through the inner sheaths, giving rise to dissolved CO_2 or H_2S in the seawater, depending on the pipe service conditions. Thus for sour service, it is normal to use steel which is not susceptible to SSC (see Chapter 3) for armouring.

Recently, Lotveit and Ward [45] have questioned the use of steel for flexible pipe reinforcement. In particular, they suggest aramid as a competitive material for this application, and Table 7.15 shows a comparison of the strength properties of the two materials.

The perceived advantage of aramid fibre is primarily its superior strength-to-weight ratio, expressed either in terms of ultimate tensile strength or modulus of elasticity. However, its corrosion and chemical resistance, thermal and ultra-violet resistance and its resistance to creep and fatigue all need to be considered, and some of these are summarised in Table 7.16.

The other area which needs to be considered for aramid is the low ductility (short elongation to failure), which requires a larger safety factor than steel and may give problems in the termination area which is inevitably required in all flexible pipes.

Table 7.15. Comparison of aramid, steel and other potential reinforcement materials for flexible pipe applications

	Density (g/cm^3)	Tensile strength (MPa)	Elongation at break (%)	Elastic modulus (GPa)	Comment (example of trade name)
Aramid	1.44	2800	3.3	80	Typical values standard (Twaron or Kevlar)
Aramid HM	1.45	2800	2.0	125	High-modulus aramid
Steel cord	7.85	2550	1.9	160	High-strength steel wire rope (Ferenka)
Steel wire	7.85	<850	10	210	For flexible pipes transporting gaseous fluids with fractions of CO_2, H_2S or other corrosive fluids
Steel wire (non-corrosive environment)	7.85	1400	4	210	High-strength steel wires for water and stabilised crude pipes
Carbon fibre	1.78	3400	1.9	238	Standard carbon fibre (Tenax HTA)
Carbon fibre HM	1.83	2250	0.5	392	High-modulus carbon fibre (Tenax HM40)
E-glass	2.58	2000	2.0	73	Glass fibre

Source: Lotveit and Ward [45].

Table 7.16. Summary of some properties of aramid fibre for application as flexible riser reinforcement

Property	Comments
Creep	Resistance superior to other organics and close to steels
Fatigue	Resistance good where relatively large deformations take place
Thermal resistance	Good compared with other organics. Minor losses in property at 250 °C, short-term exposure may be possible at 400 °C
Light resistance	Protection against UV is necessary
Chemical resistance	Generally good. No loss of strength expected after 120 days at room temperature in gasoline, methanol or diglycol

Source: Lotveit and Ward [45].

7.5.2 Internal Surface

The principal function of the internal surface is that it resists the corrosive and/or erosive nature of the fluid carried, but under certain circumstances it may also be required to resist the collapse or blistering of overlying polymeric layers. The two circumstances under which this second requirement is not needed are where the inner layer provides a gas permeation barrier or where gas is not present in the transported fluid at a sufficiently high chemical potential to result in significant pressure differential should rapid decompression occur, for example in water or chemical injection or dead oil export. In such cases thermoplastic materials may be suitable as unsupported liners, and one manufacturer [39] uses HDPE for water injection flexibles where its high permeability to methane, relatively poor blistering resistance and low temperature resistance are not a disadvantage.

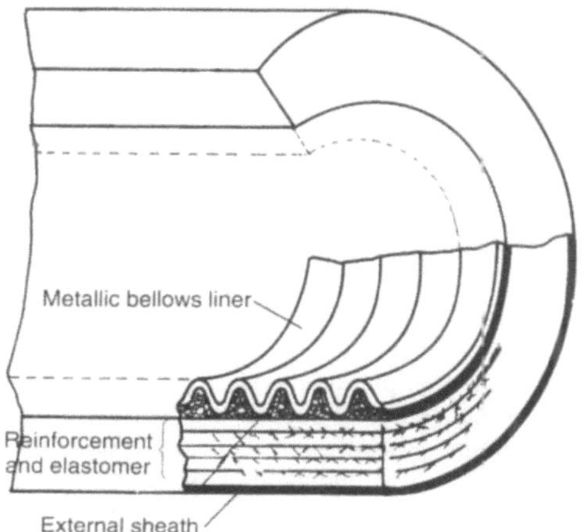

Fig. 7.22. Wall section of bonded flexible pipe with bellows liner. (From Neffgen [46].)

Another manufacturer [46] uses a gas-tight inner metallic layer in the form of a continuous metallic bellows (Fig. 7.22). The selection of the material for this bellows is the same as that for the interlocking carcass materials identified above. All manufacturers have chosen austenitic stainless steel for this application, owing primarily to its corrosion resistance but also probably in view of its cold formability for the relatively complex shapes required. This is particularly important for the interlocking tubes, which are formed from a continuous piece of strip material which is folded and spirally wound as part of the pipe manufacturing process. The corrosion resistance aspect depends upon pipe service variables such as temperature, presence of water, and dissolved oxygen, hydrogen sulphide, carbon dioxide and chloride contents. There appears to be general use of 316L stainless steel for this application and the presence of crevices in the construction and possibility of pitting conditions would certainly make this a sensible choice. Other selections are, however, used for very aggressive conditions. Depending upon the service conditions and the construction of the pipe (and hence the load transfer to the stainless steel liner), fatigue or corrosion fatigue of this component may become a life-limiting factor, especially where this liner is relied upon to provide a gas-tight layer.

7.5.3 External Surface

The external surface requires primarily to be resistant to marine corrosion and also to possible abrasion. In this context its functional requirement is more like that of an organic coating than anything else, and additional requirements such as fouling resistance, fire resistance and UV and ozone resistance are important. The issue of temperature resistance is more important in the inner layers of the pipe.

In bonded pipes, the exact formulation of the cover material does not seem to be a matter for public knowledge, although it is basically a textile-reinforced elastomer with anti-fouling additives. External steel strip armouring and fire-protective glass fibre fabric can also be added. The work of Seregely et al. [47] would, however, tend to suggest that at 20 °C natural rubber should only suffer a 50 per cent reduction in its most sensitive properties, even after a marine exposure of about 25 years.

In unbonded pipes HDPE is used as the external sheath for static flowlines and wellhead jumpers (Moore [39]). For dynamic applications, the increased flexibility displayed by the polyamides is beneficial (Lotveit and Ward [45]).

For very severe external mechanical service an external armour may be applied, and corrosion protection of this becomes a consideration.

7.5.4 Pressure Barrier Polymers

The subject about which there appears to be most debate in the market for flexible pipes for petroleum use is in the materials used for the pressure barrier. Leaving aside the pipes where this barrier is provided by a gas-tight metallic layer, the key considerations for this component are: permeabilities to the gases hydrogen sulphide, carbon dioxide and methane, temperature resistance, resistance to explosive decompression, and resistance to the various environments (hydrocarbons, chemicals and even water) which might be encountered during the service life. There seems to be a high degree of commercial sensitivity to the materials selection for this application, so that only a general discussion is possible.

The main materials in use appear to be polyamides (mostly PA11), and two only weakly defined materials, one referred to as 'a specially designed grade of PVdF' (Moore [39]) and the other as 'a modified polyolefin elastomer' (Griffiths [40]).

The suitability of polyamides and, in particular a plasticised PA11, has been discussed by Raveau and Simon [48] for application in unbonded pipes. Resistance tests in crude oil, gas, seawater and other oilfield fluids, along with data on the variation of properties with temperature, have led to recommendations on the useful service life of this form of PA11 when exposed to crude oil at a variety of temperatures. For continuous exposure at temperatures of up to 130 °C, the PVdF (polyvinylidene fluoride) polymer referred to above is recommended (Moore [39]), and temperature resistances of the four candidate polymers for use in unbonded pipes are shown in Table 7.17.

Table 7.17. Temperature effect on the properties of candidate polymers for unbonded flexible pipes

Polymer	Effects of increasing temperature
Nylon-11 (PA11)	Minor drop in shear modulus up to 90°C. Permeability to gases increases with temperature, but lower than PA12 and HDPE
Nylon-12 (PA11)	Minor drop in shear modulus up to 90 °C, but slightly more than PA11. Permeability to gases increases with temperature but lower than HDPE
High-density polyethylene (HDPE)	Melts at 120 °C. Resistance to blistering becomes poor at 60 °C. Very high permeability to methane
Polyvinylidene fluoride (PVdF)	Very high thermal stability allowing sustained use to 130 °C. Very low permeability to methane

Source: Moore [39].

Seregely et al. [47] have discussed the effect of a variety of possible offshore environments, ranging from seawater to hydrocarbon fluids on the filled vulcanisates of natural rubber, two polychloroprenes, ethylene–propylene copolymer, nitrile rubber, fully saturated hydrogenated nitrile rubber and a fluorelastomer. Testing in a highly aromatic oil (recognised as being particularly aggressive), these authors found best resistance to be offered by the fluorelastomer. Weston [49] goes further than this, stating that perfluorelastomers are the only possible choice for oilfield applications over 200 °C, although, of course, fluorelastomers are not excluded for service below such temperature. Figure 7.23 summarises some of these considerations. For bonded pipe application where there is no gas-tight stainless steel, the chosen elastomer (Griffiths

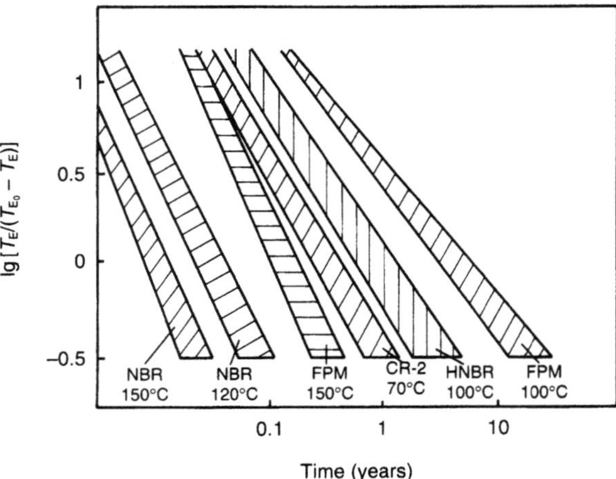

Fig. 7.23. Effect of ASTM no. 3 oil on a range of elastomers at a range of temperatures. T_E is breaking energy defined as tensile strength multiplied by elongation at break and is referred to its value at the commencement of the exposure test, T_{E_0}. NBR, nitrile rubber; FPM, a fluorelastomer; HNBR, fully saturated, hydrogenated nitrile rubber; CR-2, a polychloroprene rubber. (From Seregely et al. [47].)

[40]) does not appear to be a fluorelastomer, the stated properties being that it is chemically stable between –40 °C and +130 °C and that it would swell between +20% and +100% when freely immersed in crude oil, depending upon the aromatic content. It is claimed that such swelling is not necessarily a disadvantage in the bonded construction, since the carcass will constrain the rubber and hence build up hydrostatic pressure, so reducing permeability.

Permeability of the barrier material has a number of sometimes conflicting implications for flexible pipe design. It is helpful to start by defining the quantity, and this is most easily done for the simple case of steady-state permeation through a plane slab where the diffusion flux, J (measured, for example, in cm³/s), is given by:

$$J = PA\,(p_2 - p_1)/l$$

where A = the cross-sectional area over which diffusion is taking place

l = the thickness of the slab

p_2 and p_1 = the higher and lower pressures on either side of the slab (these may be partial pressures for dissolved gases or gas mixtures)

P = the permeability and is defined as the product of the solubility and the diffusion coefficient of the substance

Since both of the latter quantities are activated ones, the permeability shows Arrhenius behaviour with temperature:

$$P = P_0 \exp\,(Q/RT).$$

Figure 7.24 illustrates schematically the importance of permeability in high-pressure flexible pipes.

In the normal course of events, the pressure gradient of any contaminant gases dissolved in the production fluid across the pipe wall will be fixed by the internal and external conditions, so that the effect of increased average wall permeability will simply mean that gas will pass more rapidly through the wall (increase in J). However, if water penetrates to internal parts of the pipe wall, this water may

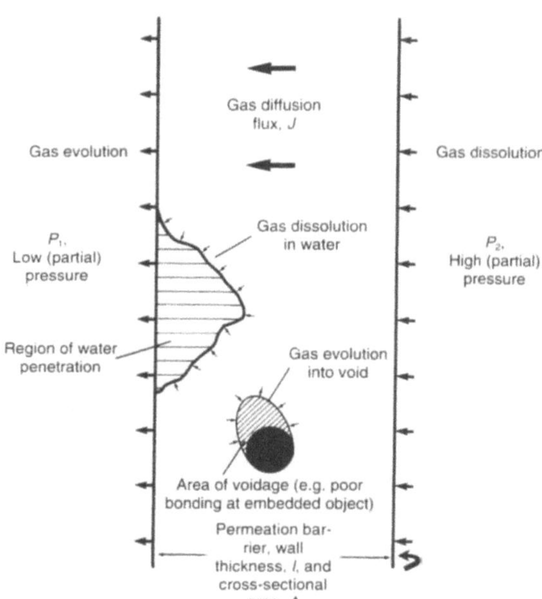

Fig. 7.24. Schematic diagram of the importance of permeability in high-pressure polymer-based pipes where dissolved gas is present in the fluid.

equilibrate with the local partial pressure (which will depend upon where in the section this occurs and on the remaining integrity of the wall), which, in the case of aggressive gases, may result in internal damage by corrosion. Depending upon whether or not the corrosion reaction is rate-limited by the supply of aggressive agent (see Chapter 3), the permeability of the pipe wall to aggressive gases may or may not be important. The thermal conductivity of the pipe will also be an issue here, because the local temperature will affect the corrosion rate and the temperature gradient will cause a variation in permeability. Campion [50] and Moore [39] provide some data on permeation of aggressive oilfield gases through polymers and elastomers.

Whereas the aggressive gases are likely to be present at a relatively low partial pressure, it is possible that hydrocarbon gases such as methane might be present at a partial pressure which is a substantial proportion of the internal pressure. The question then arises as to what may happen if there is a sudden drop in internal pressure, normally referred to as 'explosive decompression'. The term is an appropriate one, since gas will be trapped in solution in the polymer materials (such gases do not dissolve in metals) and there is a danger that they may produce internal rupture as they undissolve in a process not unlike the 'bends' experienced by divers who are inappropriately decompressed (see Chapter 6). The phenomenon also has some elements in common with hydrogen blistering and hydrogen induced cracking resulting from the de-solution of hydrogen in metals (see Chapter 3). Campion [50] makes the distinction between a bubble which may form inside an elastomeric material from a pre-existing flaw and an internal failure by fracture. In the former case the bubble will eventually dissipate, allowing the material to return to its former condition, but in the latter permanent internal damage is done. This

fact has led to the development of routine test procedures for candidate polymeric materials for explosive decompression, the tests normally consisting of the measurement of property loss resulting from a given number of decompressions from a given pressure in either oil or gas (e.g. Moore [39] and Rispin and Phelan [51]). It is questionable, however, whether such tests should be used to compare different pipe constructions, since many features of the pipe, such as the presence or absence of bonding, swelling and the provision of mechanical collapse support will all affect the likely degree of damage owing to explosive decompression. On the other hand, free explosive decompression tests are very helpful in selecting the most suitable material for a given type of pipe.

7.5.5 Other Materials

Other materials used within bonded flexible pipes include the various textile reinforcements and the elastomer used to bond the various layers together. Since the difficult-to-obtain properties are provided by the outer cover and the inner lining, this elastomer need only provide good transfer and bonding properties, and this can be achieved by one of the natural rubbers with appropriate additives. The good mastication properties of NR are helpful in the dynamic application where cycling is expected.

The intermediate sheaths in unbonded pipes need, in principle, only provide anti-friction properties, but they may be designed as a second level of defence should damage occur to one of the outer or inner layers. Even if this is the case, it is probable that the temperature requirement would be relaxed, so that PA11 is the most likely material to be used here.

7.6 Elastomers in Dynamic Marine Applications

Quite a different application of elastomers in marine environments exploits their energy-absorbing properties (referred to briefly in Chapter 2). These properties have been exploited for almost as long as rubber has been available, for example as fenders on boats and jetties, but a number of far more sophisticated applications exist, from 'engineered fending' to vibration isolation for very heavy machinery.

It is, of course, the ability of rubbers to undergo large elastic deformations which makes them suitable for such applications, and the additional degree of design freedom introduced by laminating the rubber with steel plates (e.g. Fig. 7.25) allows for a certain amount of tuning of properties. In the bonded laminate illustrated in Fig. 7.25, the effective shear, K_s, and compression K_c, stiffnesses can be given by [52, 53]:

$$K_s = GA/nt$$

$$K_c = E_c A/nt$$

where n is the number of laminations each of thickness t, and A is the loaded area as shown in Fig. 7.25.

The behaviour of rubber–steel laminates is thus seen to be almost the same as the unlaminated material of thickness nt, and is controlled by the shear modulus, G, of the material.

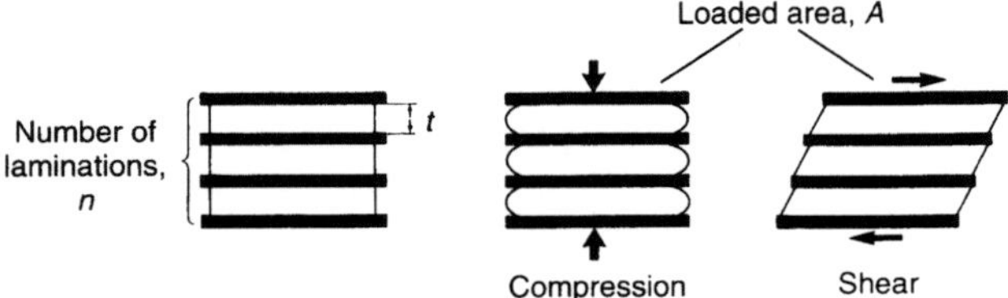

Fig. 7.25. Effective compression and shear stiffnesses in bonded steel–rubber laminates. (From McCrum [52].)

On the other hand, the compressive behaviour of the laminate is influenced by the facts that bonding to the steel plates inhibits lateral expansion and that the bulk modulus of elastomeric materials is much higher than their shear modulus. The effective compression modulus, E_c, can be given by [52]:

$$E_c = E (1 + 2kS^2)$$

where E is Young's modulus for the rubber and k is a factor which decreases from 0.93 to 0.53 as rubber hardness increases from 30 to 70 IRHD [52]. The shape factor, S, is defined as the loaded area divided by the free area, as illustrated in Fig. 7.26. A number of other expressions for effective modulus have been published [53] for shape factors outwith the range 1 to 10 and for other geometries. This ability to control elastic behaviour means that laminate design is a powerful technique for controlling the behaviour of rubber engineering elements. Some care, however, has to be exercised over the non-linear elastic behaviour of rubbers, especially if large strains are expected.

For dynamic applications, the effects of temperature, frequency and amplitude on the dynamic modulus need to be considered and creep stress relaxation may be important for some long-term, highly loaded applications.

Table 7.18 collects some data on important properties of rubbers for engineering applications.

The following examples have been drawn from the literature to illustrate some of the materials engineering considerations for structural elastomers in marine technology.

Table 7.18. Viscoelastic data for some natural rubber vulcanisates (G is the dynamic shear modulus (MN m^{-2}) and A is the loss angle (degrees))

Carbon black loading, %	Hardness IRHD*	Static shear modulus (MN m^{-2}) at			
		2%	10%	50%	100%
0	40	0.48	0.48	0.45	0.44
20	50	0.70	0.69	0.63	0.60
40	60	0.97	0.94	0.86	0.84
65	65	1.27	1.19	0.92	0.84
60[†]	75	1.82	1.68	1.25	1.45

Hardness IRHD	Resilience (%)	Temperature (°C)	Dynamic data at 10% shear strain			
			1 Hz		15 Hz	
			G	A	G	A
40	87	100	0.55	1.0	0.56	0.3
		23	0.50	1.2	0.51	1.2
		−25	0.53	7.7	0.81	27
50	78	100	0.66	1.5	0.68	2.3
		23	0.75	2.1	0.79	2.5
		−25	0.88	11	1.35	29
60	73	100	1.16	2.3	1.20	2.3
		23	1.32	4.2	1.40	5.0
		−25	1.70	12	2.43	27
65	71	100	1.19	3.7	1.26	3.7
		23	1.38	6.2	1.48	6.8
		−25	1.92	14	2.75	29
75	67	100	1.55	2.5	1.62	2.8
		23	1.70	5.7	1.87	7.3
		−25	2.52	16	3.26	27
		Dynamic data at 2% strain				
40	87	23	0.52	1.2	0.53	1.2
50	78	23	0.80	2.0	0.84	2.3
60	73	23	1.54	5.5	1.64	6.0
65	71	23	1.71	6.3	1.83	7.0
75	67	23	2.34	8.0	2.68	8.5

Source: Teo and Coveney [54].
* International rubber hardness degrees.
[†] Level of sulphur increased from 2.5 to 3.25%.

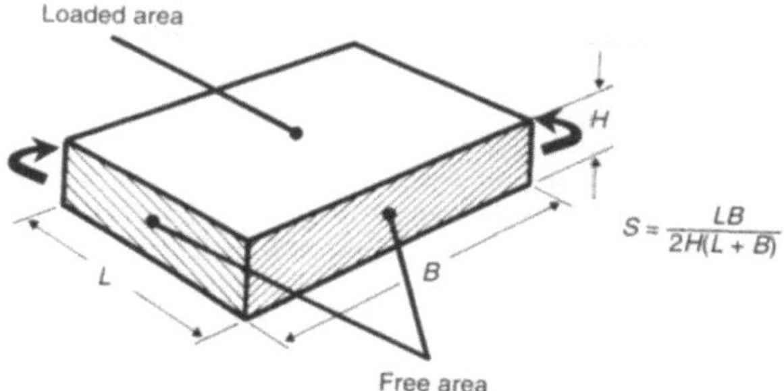

Fig. 7.26. Loaded and free areas in rubber laminae.

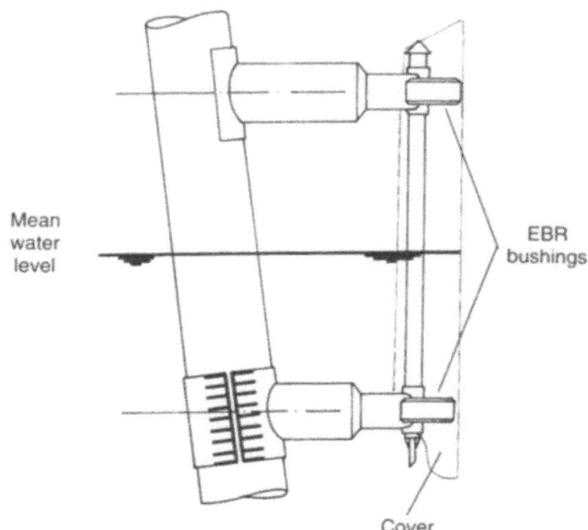

Fig. 7.27. Design of an elastomer-based barge fender. (From Arnott [56].)

7.6.1 Elastomers for Energy Absorption and Vibration Isolation

The high elastic deformations permitted by elastomers allow them to store significant amounts of elastic energy. The introduction of a pneumatic element or the exploitation of buckling and lamination mean that elastomeric-based systems can be used to absorb kinetic energy over a wide range of frequencies from ship–platform collision to machine vibrations.

Barratt and Brendling [55] have analysed ship–platform collision data for the North Sea, and their findings for mean and 10 per cent exceedance collision velocities for a number of different scenarios are shown in Table 7.19. The degree of potential damage is also dependent upon the effective mass of the colliding vessel(s), and these authors indicate that about one collision per ten platform years could be described as requiring structural repair. They also note that the distribution of collision velocities was insensitive to vessel size and that their data were consistent with current guidance, which requires fendering to be adequate to withstand a ship of 2500 tonnes' displacement coming into contact with a platform at 0.5 m s^{-1}.

Arnott [56] describes a barge fender which incorporates two distinct elastomeric components and is designed to respond to both axial and lateral loadings. The design of this fender is shown in Fig. 7.27. The elastomer is reported to be a 'skilfully compounded' natural rubber which is vul-

canised and bonded to the steel components and the combined fender has a tested performance of 58 metre-tonnes energy absorption with a reaction of 260 tonnes after a deflection of 660 mm.

A slightly more prosaic fendering application is described by Hinds and Turner [57] for buoys, which according to current regulations should be tolerant to impact with a trawler travelling at 10 knots, corresponding to about 1.5 metre-tonnes of impact energy for a 1.5-tonne buoy. The authors describe a pneumatic toroidal fender built like a large crossply tyre but with a more rounded crown shape. The pneumatic design of the fender means that, whatever the applied load, it is very unlikely that the nylon reinforcing cords will be overloaded, and buoyancy-to-mass ratio was superior to competing fendering systems.

Wreglesworth [58] has described the key factors in the design of isolation mountings for offshore and marine machinery such as propulsion engines, generator sets, ventilation fans, compressors and other service machinery. The main considerations were seen to be limitation to

Table 7.19. Some North Sea ship–platform collision data

Scenario	Mean collision velocity (m/s)	10% exceedance velocity (m/s)
1. Heave collision at the stern	0.83	1.53
2. Collision when alongside (moored or joystick positioned)		
(a) Stern surge collision	0.39	0.73
(b) Stern sway collision	0.37	0.70
(c) Side sway collision	0.28	0.54
3. Collision when manoeuvring	0.74	1.29
4. Collision of drifting vessel		
(a) Sideways drifting: impact amidships	0.76	0.98
(b) Sideways drifting: bow or stern impact	0.83	1.44
(c) Forward drifting: box impact	1.18	1.82

Source: Barratt and Brendling [55].

movement and the dynamic behaviour of the mounting equipment itself, and it is suggested that mounting systems which give good isolation properties generally allow unacceptable levels of movement under extreme seaway conditions requiring a secondary system of limit stops (buffers). Harris [59] has provided a more quantitative description of the performance of single-degree-of-freedom mass–spring systems incorporating a complex stiffness:

$$K^* = K_1 + jK_2$$

The transmissibility, T, is defined as the ratio of the magnitude of the response to that of the input excitation and, for this case, is given by:

$$T = [\{1 + \tan^2 \delta\}/\{(1 - r^2 K_{1n}/K_1)^2 + \tan^2 \delta\}]^{1/2}$$

where δ is the loss angle, $\tan^{-1} (K_2/K_1)$, and K_{1n} is the real part of the stiffness at the natural frequency ($r = 1$). More usefully, the transmissibility can be expressed in terms of the vibration frequency, ω, and the vibrating mass, m:

$$T = [(K_1^2 + K_2^2 - \omega^2 mK_1)^2 + \omega^4 m^2 K_2^2]^{1/2}/[(K_1 - \omega^2 m)^2 + K_2^2]$$

and the natural frequency, f_n, is given iteratively by

$$f_n = (1/2\pi)\sqrt{(K_d/m)}$$

first estimating the dynamic modulus, K_d, at f_n.

Of course, the design of the isolation system also requires a consideration of static deflection.

7.6.2 Elastomers for Articulations and Bearings

Aside from the element of energy absorption, the large deflections of elastomeric components can be useful in other compliant applications. Numerous examples of this can be found in the literature, and the following three have been chosen to illustrate the range of possible applications.

Moore [60] has described the design process for tether flexible joints for use in tension leg platforms. Such platforms are of compliant design, incorporating a buoyant platform which is held below its equilibrium position by tensioned flexible legs. The platform is intended to comply with wave forces rather than resist them, and this obviously requires that there be articulations at the ends of the tethers. In the design of the Hutton TLP (e.g. Salama and Ellis [61]) the lower articulation was provided by an elastomeric joint (Fig. 7.28) made up of a spherical laminate of 27 nitrile rubber laminations each 4 mm thick and reinforced with part-spherical martensitic precipitation hardening stainless steel plates. The joint gave a maximum angular offset of 16.6 °, a maximum axial load capability of 2400 tonnes, a maximum angular stiffness of 35 Nm/degree and a minimum axial stiffness of 1.75 GN/m. The methodology outlined by Moore [60] includes laminate and stress analysis to allow control of static characteristics and fatigue life.

Van Schaik [62] has outlined the design of a jack-type bearing for a seawater dam project in the Netherlands. The bearing was intended to support the sill beams for the dam, and Fig. 7.29 shows the deployment and detail of the bearings. As can be seen, the bearings consist of rubber blocks supported on a flat steel jack which is inflated at installation with high pressure grout. One extraordinary feature of this bearing is its design life of 200 years. Two rubber materials were used in the design, one a cover for the jack and bearing assembly and the other for the bearing block itself. The basic functional requirements of the cover rubber were to

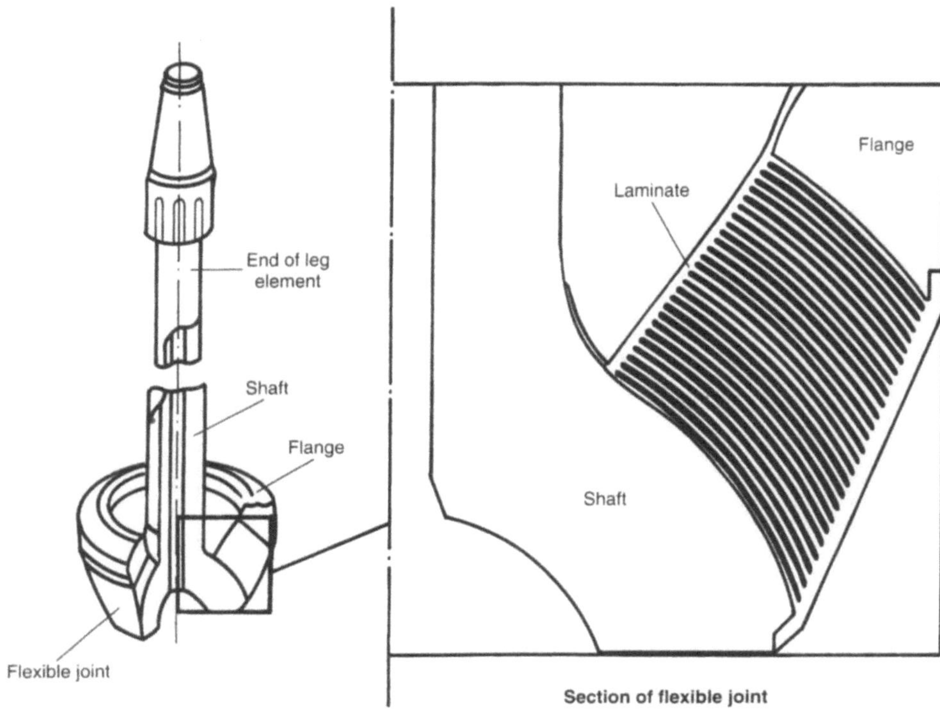

Fig. 7.28. Schematic of the design of a tension leg platform lower articulation flexible joint. (Adapted from Salama and Ellis [61].)

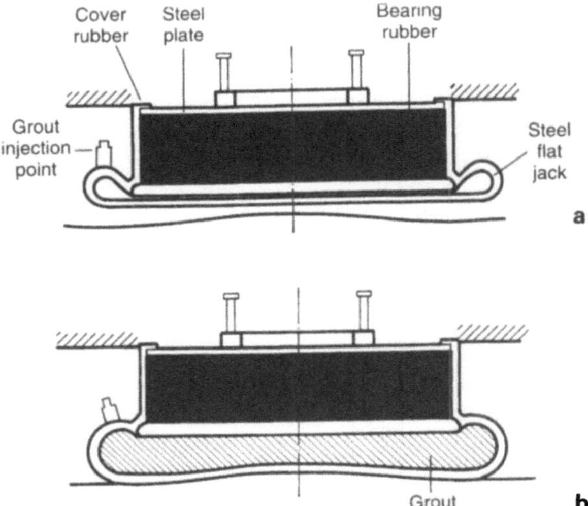

Fig. 7.29. Jack-type bearing for the sill beams of a sea-water dam. **a** Prior to filling. **b** Filled. (From Van Schaik [62].)

exclude water from the steel parts and also from the bearing rubber to prevent it from absorbing water. Durability against mechanical influences, and the ability to sustain permanent strains of up to 75 per cent were also essential. The designers considered the use of EPDM, butyl rubber, chloroprene and partly deproteinised natural rubber. After extensive testing, it was decided that the natural rubber was the most suitable choice. For the bearings themselves, a high-quality natural rubber was also chosen, primarily for its low rate of stress relaxation and high resistance to crack growth under dynamic loading.

The final example of elastomer use in marine environments is provided by Duckers et al. [63], and concerns a wave energy device called 'Clam'. This consists of a toroidal spine which carries a series (twelve in the design reported) of air-filled cells which are compressed by the wave action and drive an air turbine in the cell. The walls of the cells are made of reinforced rubber, and the requirement of reasonable operational cost dictates at least a five-year lifetime for the membrane. The material choice for the membrane was rubber reinforced with polyaramid fibres. The direction of the reinforcement was seen as being important, with the membrane requiring to be compliant in the horizontal direction (circumferential direction of the torus) and rather stiffer in the vertical direction to avoid volumetric losses during operation and also to support the vertical buoyancy forces without excessive distortion.

7.7 Economic Considerations for Materials in Marine Technology

The sheer scale of some projects in marine technology often makes the economics of some details of the manufacturing process or the service life an interesting point for analysis. This case study gives two such examples, one related to a strategy for corrosion control and the other related to the manufacture of large tubular intersections by welding.

7.7.1 Economics of Corrosion Control for Downhole Tubulars

The economic importance of corrosion control has never been in doubt, and many figures about the world annual cost of corrosion to industry have been published. This example focuses rather on what can be done at the detail design level to avoid some of the waste associated with lack of corrosion control.

The American National Association of Corrosion Engineers (NACE) has published a recommended practice for carrying out economic appraisals of corrosion control measures [64]. The approach recommended by NACE is discounted cash flow, and this example, taken from the work of Tischuk and Huber [65], follows this methodology.

Tischuk and Huber's example concerns the completion strings for a North Sea development, consisting of seven wells for a projected field life of ten years. Apart from a few details of the completion assemblies, the analysis consisted of considering the benefits, or otherwise, of using a 13 per cent chromium stainless steel in place of the conventional carbon–manganese steel tubing (N-80) which might normally be used in this application. The analysis was made necessary because of the fact that the well fluid (gas) was known to have a particularly high carbon dioxide content, with the consequent dangers of corrosion in areas where an aqueous phase is present (see Chapter 3).

In total, six possible options were considered for the corrosion control strategy:

1. Use carbon–manganese (N-80) steel and take action only when an annual wireline survey indicates that an unacceptable degree of corrosion has taken place.

2. Use stainless (13%Cr) steel and carry out no surveys.

3. Assess the likelihood of condensation on the tubing walls and use 13%Cr only in those areas which are expected to be permanently wet.

4. Use N-80 tubing with a monthly batch inhibitor treatment.

5. Use N-80 tubing with a three-monthly inhibitor squeeze treatment.

6. Use N-80 tubing with continuous inhibitor injection.

All of the options except no. 2 require the provision of an annual caliper survey for internal corrosion using wireline techniques. In addition, Options 4 to 6 require daily monitoring of inhibitor effectiveness by produced water analysis.

The appraisal commenced with a simple calculation of the net present value (*NPV*) of the annual costs (*AC*) of the various options from:

$$NPV = \sum_{N=1}^{L} AC/(1 + D)^N$$

where L is the field life and D is the discount rate.

The addition of the *NPV* of the annual costs to the capital cost allows a calculation of the net project cost. One unknown in this analysis is the discount rate D, and Fig. 7.30 shows the effect of discount rate on the net project cost for each of the options. As can be seen, all of the inhibitor treatments except for Option 4 were so expensive (owing to the high annual costs) that they were precluded from further analysis.

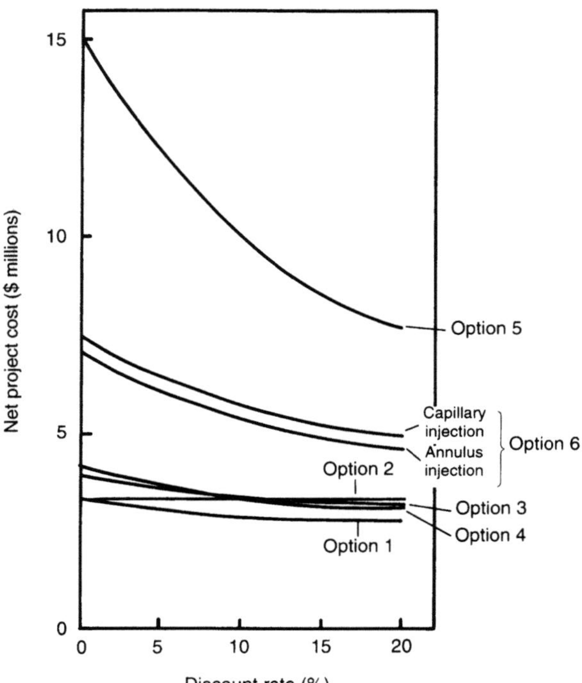

Fig. 7.30. Effect of discount rate on net project cost for some alternative schemes for production tubulation in an offshore hydrocarbon reservoir. (From Tischuk and Huber [65].)

Having narrowed the number of possible options to four, it is now necessary to refine the analysis. The reason for this is that the foregoing has assumed that all options are of equal risk. This is of course not true, since the choice of a corrodable material (as in Option 1) with no corrosion control measures carries with it the risk that expensive remedial treatment may become necessary at some time in the field life, carrying with it a cost consequence. Such 'risk costs' can be calculated, provided that some reasonable assessment of the risks and their associated cost consequences can be made.

The total risk cost, C_r, for each of the options can be estimated from:

$$C_r = \Sigma_i(P_iC_i)$$

where the summation is taken over all conceivable failure events and the P_i and C_i are the probability and cost consequence (discounted as appropriate) respectively of the event.

A constant risk associated with mechanical damage was envisaged for all options requiring a wireline survey and the corrosion risks were assessed by the following representative risks (based on past operating experience):

Option 1.

$P = 0.1$. That the performance of the string is acceptable over the entire field life.

$P = 0.3$. That monitoring shows that inhibitor batch treatment is required from Year 2 and that no further action is required.

$P = 0.3$. That, as well as the batch treatment indicated above, a workover is required in one of the wells in Year 2

but that no further action is required.

$P = 0.3$. That the corrosion rate is such that all seven wells can be left for four years without treatment and that a complete workover is carried out then, the replacement strings lasting for the remaining field life.

Option 2.

$P = 0.999$. That performance is acceptable over the entire field life.

$P = 0.001$. That a complete workover is required in Year 5 owing to localised corrosion.

Option 3.

$P = 0.73$. That performance is acceptable.

$P = 0.07$. That accidental mixing of tubing results in the requirement for a workover.

$P = 0.2$. That an error in the hydraulic analysis has resulted in the incorporation of N-80 tubing in a permanently wet area requiring batch treatment with inhibitor from Year 2.

Option 4.

$P = 0.9$. That performance is acceptable.

$P = 0.1$. That pitting failure occurs in one well requiring a workover in Year 5.

These scenarios were not intended to be exhaustive and were simply chosen to give a representation of the relative risk between the options. The final total calculated costs are shown in Fig. 7.31, and these were taken as evidence that the reduction in corrosion risk justified the initial capital outlay on the stainless steel tubing.

7.7.2 Economics of Production Welding

Although automatic welding can be used for a number of the steps in the welding of steel jacket platforms for offshore hydrocarbon production, there are inevitably a substantial number of connections which require to be made manually. The size of these constructions and the total amount of deposition required mean that an economic appraisal of any proposed process is often beneficial.

It is not uncommon to use economic analysis for production welding, and the methodology is generally to express

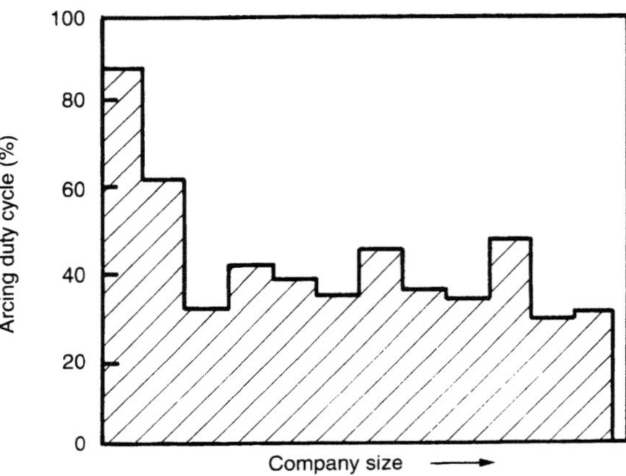

Fig. 7.32. Measured manual welder duty cycles as a function of company size. (From Salter [66].)

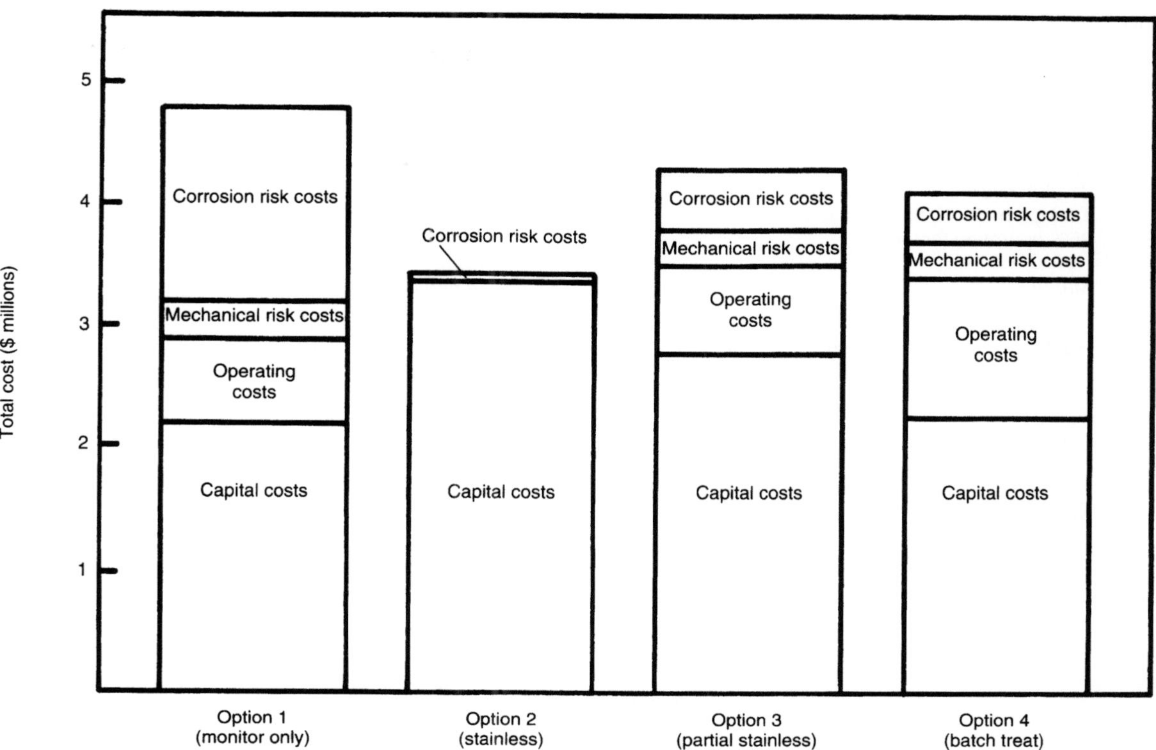

Fig. 7.31. Total capital plus operating plus risk costs for some alternative schemes for production tubulation in an offshore hydrocarbon reservoir using 10-per-cent discount rate and 10-year project life. (From Tischuk and Huber [65].)

all costs in terms of the product of the welding operation, i.e. deposited weld metal. This usually means that costs are expressed per unit length or per unit mass of finished weld.

Generally, when comparing welding with other possible manufacturing processes, it would be necessary to consider the costs associated with joint preparation, pre- and post-weld cleaning, pre-heating, post-weld heat treatment, straightening, stress relieving and non-destructive testing. When comparing welding processes with each other, it is often practice to use an operating factor which is an expression of the percentage of the welder's time which is spent in depositing metal. For example, Salter [66] has analysed welder duty cycles in a number of industries and has found that usually only about 25 to 40 per cent of paid time is spent depositing metal (Fig. 7.32). Other authors (e.g. Gitlevich [67]) divide this effect between a coefficient of utilisation of the equipment and a coefficient of utilisation of welder time.

Apart from the labour costs, welding is also costed in terms of its consumables, which may include power, although this is usually small compared with the costs of electrodes, flux and shielding gas.

Thus, the total cost per metre of deposited metal, C, can be expressed as the sum of the materials cost, C_m, and the labour cost, C_l, both again expressed per metre of weld [68]:

$$C = C_m + C_l$$

The material costs can be divided into the usage rates of the main consumables, W_e (mass of electrode required per unit length of weld), W_f (mass of flux required per unit length of weld) and V_g (volume of shielding gas required

per unit length of weld), and the costs of these materials per unit mass (or volume in the case of gas), C_e, C_f and C_g:

$$C_m = W_e C_e + W_f C_f + V_g C_g$$

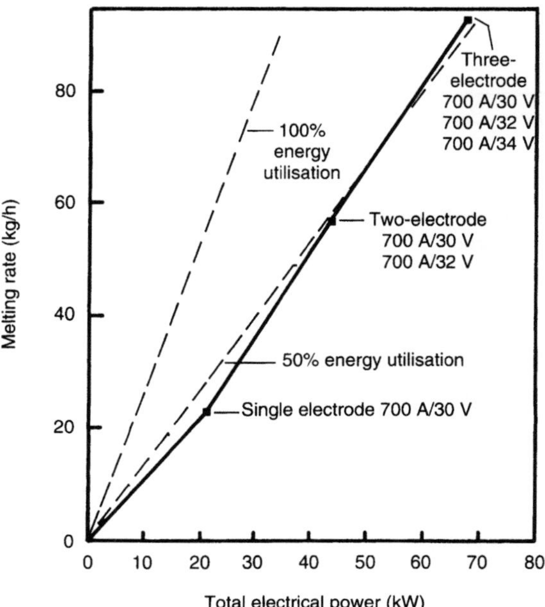

Fig. 7.33. Melting rates in single-, double- and treble-electrode submerged-arc welds. (From Duren et al. [70].)

Table 7.20.   Cost comparison of the Innershield and MMA welding processes for the production of large tubular intersection joints

---

1. MMA 8018-C1 5/32
    (a) *Assumptions*:
         £0.542/kg of rod
         60% deposit efficiency
         £2.50/h direct labour
         £1.25/h overhead
         1.8 kg/h melt-off, no waste factor
         25% operating factor

    (b) *Material cost*:

$$\frac{£}{kg\ deposit} = \frac{£}{kg\ electrode} \times \frac{kg\ electrode}{kg\ deposit}$$

$$= £0.542 \times \frac{1}{0.60} = £0.903$$

    (c) *Labour cost*:

$$\frac{£}{kg\ deposit} = \frac{£}{h} \times \frac{h}{h\ worked} \times \frac{h\ worked}{kg\ deposit}$$

$$= £3.75 \times \frac{1}{0.25} \times \frac{1}{1.8 \times 0.60}$$

$$= £13.88/kg\ deposit$$

2. Innershield NR203Ni–C 5/64
    (a) *Assumptions*:
         £1.334/kg of wire
         90% deposit efficiency
         £2.50/h direct labour
         £1.25/h overhead
         2.35 kg/h melt-off
         45% operating factor

    (b) *Material cost*:

$$\frac{£}{kg\ deposit} = \frac{£}{kg\ wire} \times \frac{kg\ wire}{kg\ deposit}$$

$$= £1.334 \times \frac{1}{0.90} = £1.482$$

    (c) *Labour cost*:

$$\frac{£}{kg\ deposit} = \frac{£}{h} \times \frac{h}{h\ worked} \times \frac{h\ worked}{kg\ deposit}$$

$$= £3.75 \times \frac{1}{0.45} \times \frac{1}{2.25 \times 0.90}$$

$$= £4.115/kg\ deposit$$

Hence:
Total cost per kilogram of weld metal deposited with 8018-C1 electrode operating at 175 A will be £0.903 (material) + £13.88 (labour)   = £14.791

Total cost per kilogram of weld metal deposited with NR203Ni–C operating at 190 A will be £1.482 (material) + £4.115 (labour)   = £5.597

Conclusion:
There is a saving of 62% in the cost of weld metal deposited using the Innershield process. For example, a butt weld in 25-mm plate with a 60° included angle and a 3.2-mm gap would require 4 kg of weld metal per metre of weld. The comparative cost would be:
8018-C1 £14.791 × 4 = £59.16
NR203 £5.597 × 4 = £22.39
Net saving = £36.77 per metre of weld

Source: Adapted from Keeler [69].

The labour costs can simply be calculated from the time taken to carry out the deposition of one metre of weld, T (the reciprocal of welding speed times number of passes), the hourly labour costs including any overhead, C_r, and the operating factor, OF, defined above (p. 243):

$$C_l = (T \times C_r)/OF$$

Keeler [69] has carried out such an economic analysis to compare welding costs between the 'Innershield' process (essentially flux-cored arc welding) and the manual metal-arc (MMA) process for the fabrication of offshore structures. His findings (at 1981 prices and costs) are summarised in Table 7.20, and it can be seen that a substantial advantage accrues from the use of the Innershield process. As can be seen from the details in Table 7.20, this is despite higher consumable cost and is almost entirely associated with the increased operating factor (due to semi-automatic operation), but contributions also come from the increased deposition rate and deposition efficiency associated with the Innershield process. In this last context, it is useful to note the deposition rates achievable with submerged-arc welding (which might be used for the shop fabrications of the same platforms), and Fig. 7.33 shows some typical data for this process [70]. A similar, or better, operating factor might be expected for the submerged-arc process, depending upon the organisation of supporting tasks such as preparation, cutting and plate rolling.

References

1. API 1104. Welding of steel pipelines and related facilities. American Petroleum Institute, Dallas, 1973
2. BS 4515. Welding of steel pipelines on land and offshore. BSI, London, 1984
3. Jones DG, Hopkins P, Clyne AJ. Assessment of weld defects in offshore pipelines. Proceedings. Conference Offshore

Pipeline Technology, Stavanger, Norway, January 1988. IBC/NPD, 1988

4. Harrison JD. The COD approach and its applications to welded structures. Welding Institute Research Report SS/1978/L. Welding Institute, Cambridge, 1978

5. Knott JF, Elliott D. Worked examples in fracture mechanics, Institution of Metallurgists Monograph no. 4. IOM, 1979

6. BS PD6493. Guidance on some methods for the derivation of acceptance levels for defects in fusion welds. BSI, London, 1980

7. Miller AG. Review of limit loads of structures containing defects, CEGB Report TPRD/B/0093/N82. Central Electricity Generating Board, Berkeley, 1987

8. Kastner W, Roehrich E, Schmitt W, Steinbuch R. Critical crack sizes in ductile piping. International Journal of Pressure Vessels and Piping 1981; 9: 197–219

9. Milne I, Loosemore K, Harrison RP. A procedure for assessing the significance of flaws in pressurised components. In: Tolerance of flaws in pressurised components. Mechanical Engineering Publications, London, 1978, pp 271–276

10. Kiefner JF, Maxey WA, Eiber RJ, Duffy AR. Failure stress levels of flaws in pressurised cylinders. In: Progress in flaw growth and fracture toughness testing, ASTM STP 536. ASTM, Philadelphia, 1973, pp 461–481

11. Tenge P, Karlsen A. Influence of weld defects on low-cycle, high-strain fatigue properties of welds in offshore pipelines. Norwegian Maritime Research, no. 3, 2–15, 1977

12. Celant M, Re G, Venzi S. Fatigue analysis for submarine pipelines. Proceedings, 14th Offshore Technology Conference. Houston, paper no. OTC 4233, 1982

13. Williams AK, Rinne JE. Fatigue analysis of steel offshore structures. Proceedings, Institute of Civil Engineers 1976; 60(1), 635–654

14. Gurney TR. The basis of the revised fatigue design rules in the Department of Energy Guidance Notes. Proceedings, Conference on Welded Offshore Structures. Welding Institute, Cambridge, 1983, pp 55-1–55-11

15. UEG. Design of tubular joints for offshore structures, UEG Publication UR33. UEG/CIRIA, London, 1985

16. Fulmer Research Institute and Wimpey Offshore. Study on lightweight materials for offshore structures, report no. WOL 161/87 prepared for Department of Energy Offshore Supplies Office. 1987

17. Mableson AR, Osborn RJ, Nixon JA. Structural use of polymeric composites in ships and offshore. Proceedings, Second International Conference Polymers in a Marine Environment. Marine Management (Holdings), London, 1989, pp 79–86

18. Murtagh MM, Lemley NW. Guidelines governing the use of fibreglass pipe on United States Coast Guard inspected vessels. Proceedings, Conference on Polymers in a Marine Environment. Marine Management (Holdings), London, 1989, pp 7–16

19. Cliffe DA. Designing composites for engineering applications. Proceedings, Symposium on Composites at Sea. European Pultrusion Association, Rotterdam, 1990, paper no. 2

20. Grim GC. Shipboard experience with GRP pipes in Shell fleet vessels. Proceedings, Second International Conference Polymers in a Marine Environment. Marine Management (Holdings), London, 1989, pp 47–48

21. Weeton JW, Peters DM, Thomas KL (eds). Engineers' guide to composite materials. ASM, Metals Park, Ohio, 1987

22. Muscati A, Bradford R. A comparison of the failure pressure predicted by FE stress analysis with the results of full scale burst tests on GRP flanges. In: Marshall IH (ed.) Composite structures. Applied Science Publishers, London, 1981, pp 690–703

23. Norwood LS, Marchant A. Recent developments in polyester matrices and reinforcements for marine applications. In: Marshall IH (ed.) Composite structures. Applied Science Publishers, London, 1981, pp 158–181

24. Chalmers DW. The properties and uses of marine structural materials. Marine Structures 1988; 1: 47–70

25. Rhodes J, Marshall IH. On the use of effective width concept for composite plates. In: Marshall IH (ed.) Composite structures. Applied Science Publishers, London, 1981, pp 335–351

26. Dixon RH, Ramsay BW, Usher PJ. Design and build of the GRP hull of HMS Wilton. Proceedings, Symposium on GRP Ship Construction. RINA, London, 1973, pp 1–32

27. Nixon JA. Composite ship design and construction. Proceedings, Symposium on Composites at Sea. European Pultrusion Association, Rotterdam, 1990, paper no. 4

28. Liddle D, George AP. The application of composite materials for underwater vehicles and associated marine equipment. Proceedings, Second International Conference Polymers in a Marine Environment. Marine Management (Holdings), London, 1989, pp 87–92

29. Harruff PW, Sandman BE. Carbon–epoxy composite structures for underwater pressure hull applications. Proceedings, 28th National SAMPE Symposium, vol. 28: Materials and processes – continuing innovations, 1983, pp 40–49

30. Det norske Veritas. Cathodic protection evaluation, Technical Note TNA 703. DnV, Hovik, 1981

31. Thomason WH, Pape SE, Evans S. The use of coatings to supplement cathodic protection of offshore structures. Materials Performance 1987; 26 (11), 22–27

32. Rippon IJ, Finnegan JE. Polyethylene coating damage on an underwater pipeline in the Southern North Sea. Proceedings, 16th International Offshore Technology Conference, Houston, paper no. OTC 4668, 1984, pp 179–188

33. Peterson MH, Lennox TJ, Groover RE. The effect of initial low anode current densities on the subsequent performance of galvanic anodes. Proceedings, Conference Corrosion '75, paper no. 17. NACE, Houston, 1975, pp 48–55

34. Mollan R, Andersen TR. Design of cathodic protection systems. Proceedings, Conference Corrosion '86, paper no. 286. NACE, Houston, 1986, pp 286-1–286-10

35. Chandler KA. Marine and offshore corrosion. Butterworths, London, 1985

36. Backhouse GH. Equipment for offshore measurements. In: Ashworth V, Booker CJL (eds) Cathodic protection. Ellis Horwood, Chichester, 1986, pp 235–248

37. Wagner J, Cremers JF. Cathodic protection for older bay line was design challenge. Pipe Line Industry 1980; July: 55–60

38. Parker ME, Peattie EG. Pipeline corrosion and cathodic protection. Gulf Publishing Company, Houston, 1984

39. Moore F. Materials for flexible riser systems: problems and solutions. Engineering Structures 1989; 11(4): 208–216

40. Griffiths AD. Elastomeric high pressure flexible pipes for high temperature applications. Proceedings, Conference on Polymers in Offshore Engineering. Plastics and Rubber Institute, London, 1988, pp 16/1–16/18

41. MacFarlane CJ. Flexible riser pipes: problems and unknowns. Engineering Structures 1989; 11(4): 281–289

42. Stacey AG. The technology of steel wire ropes. Metals and Materials 1987; 3(12): 706–711

43. Feret JJ, Bournazel CL. Evaluation of flexible pipes' life expectancy under dynamic conditions. Proceedings, 18th International Offshore Technology Conference, Houston, paper no. OTC 5230, 1986, pp 83–90

44. Cocks PJ. Testing and structural integrity of flexible pipes. Engineering Structures 1989; 11(4): 217–222

45. Lotveit SA, Ward H. Fibre reinforced flexible pipe materials selection and design methods. Proceedings, Conference Offshore and Polar Engineering, Edinburgh, 1991

46. Neffgen JM. Integrity monitoring for flexible pipes. Pipes and Pipelines International 1988; 33(3): 7–14

47. Seregely ZI, Nagy TT, Pfisztner N. Ageing of elastomers under simulated offshore conditions. Proceedings, Conference on Polymers in Offshore Engineering. Plastics and Rubber Institute, London, 1988, pp 15/1–15/11

48. Raveau G, Simon JP. Polyamide 11 for use in collecting pipes and submarine umbilicals. Proceedings, Conference on Polymers in Offshore Engineering. Plastics and Rubber Institute, London, 1988, pp 14/1–14/25

49. Weston RJ. A comparison of perfluoroelastomers and other elastomers tested in oilfield media. Proceedings, Conference on Polymers in Offshore Engineering. Plastics and Rubber Institute, London, 1988, pp 6/1–6/19

50. Campion RP. Failure mechanisms in elastomers in high pressure oil and gas conditions. Proceedings, Conference on Polymers in Offshore Engineering. Plastics and Rubber Institute, London, 1988, pp 5/1–5/14

51. Rispin A, Phelan P. The introduction of an improved elastomer for high pressure pipework fittings. Proceedings, Conference on Polymers in Offshore Engineering. Plastics and Rubber Institute, London, 1988, pp 9/1–9/10

52. McCrum NG, Buckley CP, Bucknall CB. Principles of polymer engineering, Oxford University Press, Oxford, 1988, pp 14–37

53. Stevenson A. Design principles and fatigue life determinations for structural rubber bearings. In: Stevenson A (ed.) Rubber in offshore engineering. Adam Hilger, Bristol, 1984, pp 14–37

54. Teo SC, Coveney VA. Engineering uses of rubber. Metals and Materials 1987; 3(11): 668–671

55. Barratt MJ, Brendling WJ. Impact scenarios for offshore structures. Proceedings, Conference on Polymers in Offshore Engineering. Plastics and Rubber Institute, London, 1988, pp 33/1–33/10

56. Arnott T. The development of energy absorbing barge fenders. In: Stevenson A (ed.) Rubber in offshore engineering. Adam Hilger, Bristol, 1984, pp 174–185

57. Hinds M, Turner DM. Rubber and rubber/cord composites for energy absorption offshore. In: Stevenson A (ed.) Rubber in offshore engineering. Adam Hilger, Bristol, 1984, pp 166–173

58. Wreglesworth J. Vibration isolation of machinery in offshore and marine applications. Proceedings, Conference on Polymers in Offshore Engineering. Plastics and Rubber Institute, London, 1988, pp 37/1–37/10

59. Harris JA. Design principles for vibration isolation and damping with elastomers including nonlinearity. Proceedings, Conference on Polymers in Offshore Engineering. Plastics and Rubber Institute, 1988, pp 36/1–36/15

60. Moore AF. Development of UK flexible tether joint. Proceedings, Conference on Polymers in Offshore Engineering. Plastics and Rubber Institute, London, 1988, pp 30/1–30/15

61. Salama MM, Ellis N. Hutton TLP – a materials challenge. Proceedings, 4th International Offshore Mechanics and Arctic Engineering Symposium, Dallas, 1985, pp 1–13

62. Van Schaik HH. Structural rubber bearings for the Dutch delta project. Proceedings, Conference on Polymers in Offshore Engineering. Plastics and Rubber Institute, London, 1988, pp 29/1–29/10

63. Duckers LJ, Lockett FP, Loughridge BW, Peatfield AM, West MJ, White PRS. Reinforced rubber membranes for the 'Clam' wave energy device. Proceedings, 2nd International Conference Polymers in a Marine Environment. Marine Management (Holdings), London, 1989, pp 219–224

64. NACE-RP-02-72. Recommended practice for direct calculation of economic appraisals of corrosion control measures. NACE, Houston, 1972

65. Tischuk JL, Huber DS. Use of economic analysis to select the most cost effective method of downhole corrosion control. Proceedings, Conference UK Corrosion 1984. Institution of Corrosion Science and Technology, London, 1984, pp 13–18

66. Salter GR. An economic view of arc welding. Metal Construction 1984; 16(2): 74–78

67. Gitlevitch AD. Fields in which methods of arc welding are an economic proposition. Avtomatika Svarka 1982; no. 12: 54–59

68. Lincoln Electric Company. Procedure handbook of arc welding. Lincoln Electric, Cleveland, 1973

69. Keeler T. Innershield welding – development and applications. Metal Construction 1981; 13: 667–673

70. Duren CF, Hieber G, Wiederhoff WW. Development of welding technique for longitudinal welded large diameter pipe production. Proceedings, Welding Research Symposium at AWS 57th Annual Meeting, St Louis, 1976, pp 3–55

Subject Index